Dzyadyk/Shevchuk

Theory of Uniform Approximation of Functions by Polynomials

Vladislav K. Dzyadyk
Igor A. Shevchuk

Theory of Uniform Approximation of Functions by Polynomials

Walter de Gruyter · Berlin · New York

Authors
Vladislav K. Dzyadyk (†)
Igor A. Shevchuk, Shevchenko Kiev National University, Kiev, Ukraine

Translated and typeset by
Dmitry V. Malyshev and Peter V. Malyshev,
Institute of Mathematics, Ukrainian Academy of Sciences, Kiev, Ukraine
Vladimir V. Gorunovich (†)

Mathematics Subject Classification 2000
00-01, 00-02, 41-01, 41-02, 41A10, 41A17, 41A20, 41A25, 41A27, 41A50, 41A52, 42A05, 42A10

Keywords
Best uniform approximation, Chebyshev systems, alternation theorem, de la Vallée Poussin theorem, Remez algorithm, uniform approximation, Weierstrass theorem, Stone's theorem, smoothness of functions, Hölder continuous functions, Langrange interpolation, moduli of continuity of higher order, smooth continuation of functions, periodic functions, Fourier series, inverse estimates, Bernstein inequality, shape-preserving approximation

♾ Printed on acid-free paper which falls within the guidelines of the ANSI
to ensure permanence and durability.

ISBN 978-3-11-020147-5

Bibliographic information published by the Deutsche Nationalbibliothek

The Deutsche Nationalbibliothek lists this publication in the Deutsche Nationalbibliografie;
detailed bibliographic data are available in the Internet at http://dnb.d-nb.de.

Printed in Germany.
Cover design: Martin Zech, Bremen.
Printing and binding: Hubert & Co. GmbH & Co. KG, Göttingen.

Preface

The classical monograph *Introduction into the Theory of Uniform Approximation of Functions by Polynomials* [Nauka, Moscow (1977)] by V. K. Dzyadyk has never been translated into English. The book offered to your attention is a translation of selected chapters of Dzyadyk's monograph complemented by several chapters from my monograph *Approximation by Polynomials and Traces of Functions Continuous on a Segment* [Naukova Dumka, Kyiv (1992)]. Many facts presented in our book were included in the excellent encyclopedic monographs *Constructive Approximation* [Springer (1993)] by R. A. DeVore and G. G. Lorentz and *Constructive Approximation: Advanced Problems* [Springer (1996)] by G. G. Lorentz, M. von Golitschek, Y. Makovoz. The main attention in our book is given to approximation in the uniform metric, which, augmented by Dzyadyk's masterful style of presentation, makes it still as interesting today as when the Russian edition was published. This book does not include the last chapter of Dzyadyk's monograph and the last chapter of my monograph. Both these chapters are devoted to the problem of the constructive characteristic of functions on closed sets of a complex domain. Most results of these two chapters were included in the monograph *Conformal Invariants in Constructive Theory of Functions of Complex Variable* [World Federation Publishers (1995)] by V. V. Andrievskii, V. I. Belyi, and V. K. Dzyadyk.

V. K. Dzyadyk, a brilliant mathematician and my teacher, passed away in 1998. The final preparation of this monograph for the publication has been done by me, and I bear the entire responsibility for all shortcomings of the book.

A few words should be said about the enumeration of formulas, theorems, lemmas, etc. Within each section of the book, they are numbered consecutively by double numbers (section and item numbers). To refer to an item from a different chapter, we use triple numbers (chapter, section, and item numbers). Figures are numbered consecutively throughout the book by ordinary numbers.

In some chapters, additional information about the contribution of different mathematicians into one result or another is given in remarks at the end of the corresponding chapter. The reference marks to these remarks appear as Arabic superscripts.

I am deeply grateful to Dr. P. V. Malyshev and Dr. D. V. Malyshev, the translators and editors of this book, for their endless patience and firm insistence without which the book would have never come to life. I am also grateful to Dr. A. Prymak, who pointed out numerous misprints and mistakes.

I. A. Shevchuk
2008

Preface to
Introduction into the Theory of Uniform Approximation of Functions by Polynomials
by V. K. Dzyadyk

In almost all fields of mathematics, an important role is played by problems of approximation of more complex objects by less complex ones. In most cases of this sort, it is very helpful to know the main methods, results, and problems of the theory of approximation of functions. At present, approximation theory mainly deals with the approximation of individual functions and classes of functions with the use of given subspaces each of which consists of functions that are, in a certain sense, simpler than the functions being approximated. The role of these subspaces is most often played by the set of algebraic polynomials or (in the periodic case) by the set of trigonometric polynomials of a given degree n.

The present monograph is devoted to the problem of approximation of functions in the uniform metric. Investigations in other linear normed spaces are carried out, as a rule, only in the cases where the results obtained there enable one to solve some problems in the uniform metric or to consider these problems from a more general point of view.

For reader's convenience, the first several chapters contain mainly classical results; the share of new results here is less than one third. In the other chapters, on the contrary, the author gives full rein to his preference. Therefore, these chapters mainly contain results obtained with active participation of the author. For this reason, little or no attention is given here to such problems as approximation of functions of many variables, approximation of functions defined on unbounded sets, approximation in spaces different from C and L, interpolation theory, quadrature formulas, spline theory, widths of sets, etc. For detailed information on these and many other important problems, we refer the reader to the monographs (see the references at the end of the book) by de la Vallée Poussin (1919), Bernstein (1937), Jackson [(1930), (1941)], Walsh (1961), Goncharov (1954), Akhiezer (1965), Natanson (1949), S. Nikol'skii [(1969), (1974)], Krylov (1959), Szegő (1959), Korovkin (1959), Timan (1960), Smirnov and Lebedev (1964), Rice [(1964), (1969)], Cheney (1966), G. Lorentz (1966), Davis (1965), Remez (1969), Butzer and Nessel (1971), Ibragimov (1971), Ahlberg, Nilson, and Walsh (1967), Krein and Nudel'man (1973), Laurent (1972), Tikhomirov (1976), Korneichuk (1976), Stechkin and Subbotin (1976), Karlin and Studden (1966), etc.

Unfortunately, the limited size of the book did not allow the author to include topics related to the fusion of methods and results of the theory of approximation of functions on the one side and methods and results of the theory of ordinary differential equations and computational mathematics on the other (see author's works [(1970), (1973a), (1974), (1976)] and the papers of Denisenko, Krochuk, Podlipenko, Stolyarchuk, etc.). For the same reason, the author did not include in this monograph results related to the best approximation of absolutely monotone functions and to the least upper bounds of the best approximations on classes of r times differentiable functions (see the papers of Weyl (1917), Bohr (1935), Hardy and Littlewood [(1928), (1932)], Favard [(1936), (1937)], Akhiezer and Krein (1937), Szőkefalvi-Nagy (1938), Dzyadyk [(1953), (1955), (1959a), (1961), (1974b), (1975a)], S. Nikol'skii (1946), Stechkin (1956), Sun Yun-shen [(1958), (1959), (1961)], Korneichuk [(1961), (1971)], Babenko (1958), Taikov (1963), Dzhrbashyan (1966), Bushanskii [(1974), (1974a)], etc.).

Most results of the monograph have been delivered many times in special courses regularly given by the author at Shevchenko Kiev State University since 1961. For this reason, much attention in formulations and proofs has always been given to the clarity and consistency of presentation.

Main attention in this monograph is given to the following questions:

(i) the Chebyshev theory of uniform approximation of functions and its development;

(ii) the constructive characteristic of functions of real and complex variables;

(iii) linear methods for summation of Fourier series and their generalizations.

Since 1961, all new results presented in this monograph have been regularly delivered at seminars on approximation theory held at the Department of the Theory of Functions at the Institute of Mathematics of the Ukrainian Academy of Sciences under the guidance of the author. For many years, an active part in these seminars has been taken by the members of the department, namely V. T. Gavrilyuk, V. N. Konovalov, V. D. Koromyslichenko, Yu. I. Mel'nik, I. P. Mityuk, R. V. Polyakov, A. I. Stepanets, P. M. Tamrazov, and I. A. Shevchuk, as well as by mathematicians from other institutes and disciples of the author: G. A. Alibekov, M. I. Andrashko, P. E. Antonyuk, V. I. Belyi, V. A. Borodin, V. P. Burlachenko, A. V. Bushanskii, Yu. I. Volkov, N. N. Vorob'ev, D. M. Galan, V. B. Grishin, P. N. Denisenko, R. N. Koval'chuk, L. I. Kolesnik, V. V. Krochuk, E. K. Krutigolova, P. D. Litvinets, V. A. Panasovich, Yu. K. Podlipenko, A. S. Prypik, G. S. Smirnov, V. K. Stolyarchuk, T. Tugushi, L. I. Filozof, M. I. Hussein, A. I. Shvai, L. B. Shevchuk, etc. Most of them contributed in one way or another to the improvement of this book, and I express my gratitude to all of them. I am especially grateful to I. A. Shevchuk, Yu. I. Mel'nik, and G. A. Alibekov, who thoroughly read selected chapters of this book.

I am deeply grateful to S. A. Telyakovskii and L. V. Taikov, who read the entire manuscript and made valuable remarks, which were taken into account in the final version.

I also want to thank I. F. Grigorchuk and É. A. Storozhenko, who made a number of remarks concerning the first chapters, and V. É. Gontkovskaya and S. F. Karpenko, who helped much in the preparation of the manuscript.

In conclusion, I express my sincere gratitude to S. M. Nikol'skii, my teacher and old friend. More than 20 years ago, he set me a series of important and interesting problems, and his valuable advices, remarks, and suggestions made in our numerous subsequent discussions of my results were exceptionally helpful and cannot be overestimated.

V. K. Dzyadyk
1977

Contents

Chapter 1
Chebyshev theory and its development

For any two elements x and y of a linear normed space L, the norm of the difference $x - y$ is called the *distance* $\rho(x; y)$ between these elements or the *deviation* of one of these elements from the other. If $\mathfrak{M}$ is a subset of L, then the distance of an element $y \in L$ from $\mathfrak{M}$ is defined in the standard way (for metric spaces) and is denoted by $\rho(x; \mathfrak{M})$, i.e.,

$$\rho(y; \mathfrak{M}) := \inf_{x \in \mathfrak{M}} \rho(y; x).$$

As a rule, the role of $\mathfrak{M}$ is played by a finite-dimensional linear subspace $\mathfrak{M}_n$ spanned over a field P of real or complex numbers by a properly chosen system of finitely many linearly independent elements $x_i \in L$ whose number depends on n:

$$\mathfrak{M}_n = \left\{x\colon x = \sum c_i x_i,\ c_i \in P\right\}.$$

Most often, the dimensionality of the subspace $\mathfrak{M}_n$ spanned by a certain system $\{x_i\}_{i=0}^{n}$ is equal to $n+1$. However, if L consists solely of periodic functions, then, for convenience, we take a system $\{x_i\}$ of $2n+1$ linearly independent elements and denote the corresponding subspace spanned by this system by $\tilde{\mathfrak{M}}_n$. In both cases, the quantity $\rho(y; \mathfrak{M}_n)$ [or $\rho(y; \tilde{\mathfrak{M}}_n)$] is denoted by $E_n(y)$ and called the *value of the best approximation* of an element y by the subspace $\mathfrak{M}_n$ (or $\tilde{\mathfrak{M}}_n$, respectively). If there exists an element $x^* \in \mathfrak{M}_n$ (or $x^* \in \tilde{\mathfrak{M}}_n$) for which the equality $\rho(y; x^*) := \rho(y; \mathfrak{M}_n)$ [or $\rho(y; x^*) := \rho(y; \tilde{\mathfrak{M}}_n)$] is true, then this element is called either the *best approximating element* for the element y in $\mathfrak{M}_n$ (or in $\tilde{\mathfrak{M}}_n$) or the element with the least deviation from y. In what follows, the role of $\mathfrak{M}_n$ is almost always played by the set $\mathcal{P}_n$ of algebraic polynomials of degree $\leq n$ and the role of $\tilde{\mathfrak{M}}_n$ by the set $\mathcal{T}_n$ of trigonometric polynomials of degree $\leq n$.

By C_D we denote the class of all possible real or complex functions f continuous on a set D. The uniform norm of each of these functions is denoted by $\|f\|_D$ or, simply, by $\|f\|$, i.e.,

$$\|f\| = \|f\|_D := \sup_{x \in D} |f(x)|.$$

The value of the best uniform approximation for a function $f \in C_D$ by the set $\mathcal{P}_n$ of algebraic polynomials of degree $\leq n$ is denoted by $E_n(f)_D$ or, simply, by $E_n(f)$, i.e.,

$$E_n(f) = E_n(f)_D := \rho(y; \mathcal{P}_n) = \inf_{p \in \mathcal{P}_n} \|f - p\|_D.$$

In the first chapter, we study the problems of existence, uniqueness, and characterization of the best approximating elements. Note that these problems were posed, explicitly or implicitly, by P. Chebyshev who solved them in numerous important cases.

1. Chebyshev theorems

In this section, we present the most important, in our opinion, Chebyshev approximation theorems and some related results of the other mathematicians.

1.1. Existence of the element of the best approximation

Theorem 1.1 (on existence [Borel (1905)]). *If f is a function continuous on a segment $[a, b]$, then, for any $n = 0, 1, 2, \ldots$, the subspace $\mathcal{P}_n$ of polynomials p_n of degree $\leq n$ of the form*

$$p_n(x) = \sum_{j=0}^{n} a_j x^{n-j} \tag{1.1}$$

contains a polynomial (i.e., an element of $\mathcal{P}_n$) of best approximation to the function f.

Proof. By the definition of infimum, for any positive integer N, there exists an element $p_N \in \mathcal{P}_n$ such that

$$\|f - p_N\|_{[a,b]} \leq E_n(f)_{[a,b]} + \frac{1}{N}. \tag{1.2}$$

Since

$$E_n(f) \leq \|f - 0\| = \|f\|,$$

we have

$$\|p_N\| \leq \|p_N - f\| + \|f\| \leq E_n(f) + \frac{1}{N} + \|f\| \leq 2\|f\| + 1 = \text{const}, \quad (1.3)$$

i.e., the sequence $\{p_N\}_{N=1}^{\infty}$ is bounded and, consequently, belongs to a closed ball in the space $\mathcal{P}_n$.

Since, by virtue of the compactness criterion for a finite-dimensional space $\mathcal{P}_n$, each closed bounded set is compact, one can select a subsequence $\{p_{N_k}\}_{k=1}^{\infty}$ of $\{p_N\}_{N=1}^{\infty}$ convergent to a certain element $p^* \in \mathcal{P}_n$. Therefore, in view of the fact that

$$\left\|f - p^*\right\| \leq \left\|f - p_{N_k}\right\| + \left\|p_{N_k} - p^*\right\| \leq E_n(f) + \frac{1}{N_k} + \left\|p_{N_k} - p^*\right\|,$$

we can pass to the limit in this inequality and get

$$\left\|f - p^*\right\| \leq E_n(f). \quad (1.4)$$

This means that the element (polynomial) $p^* \in \mathcal{P}_n$ is an element of the best approximation for the function f.

■

Theorem 1.1′. *For any* 2π*-periodic continuous function* f *and any* $n = 0, 1, 2, \ldots,$ *there exists a trigonometric polynomial* T_n^* *of the best approximation to* f.

The proof of this assertion coincides with the proof of Theorem 1. Note that one must take into account the fact that the functions $1, \cos t, \sin t, \ldots, \cos nt,$ and $\sin nt$ are linearly independent (see Corollary 1.2 in what follows).

Similarly, one can prove the following general result for an arbitrary linear normed space:

Theorem 1.1″. *Assume that an arbitrary linear normed space* E *contains* $n + 1$ *linearly independent elements* $g_0, g_1, \ldots, g_n$. *Then, for any* $x \in E$, *the set of all polynomials* $P_n(c; g)$ *of the form*

$$P_n(c; g) = \sum_{k=0}^{n} c_k g_k, \quad (1.5)$$

where c_k *are arbitrary real (or complex) numbers, contains at least one polynomial*

$$P_n^*(c^*;g) = \sum_{k=0}^{n} c_k^* g_k$$

of the best approximation to the element x.

1.2. Chebyshev alternation theorem

We now proceed to the main theorem of the present section. This theorem establishes the necessary and sufficient conditions for a polynomial P_n^* to be a polynomial of best approximation of degree n to a continuous real function f defined on $[a, b]$. This theorem was proved by Chebyshev in 1854 and marked the onset of development of the theory of approximation of functions.

Theorem 1.2 [Chebyshev (1854)]. *Assume that a continuous real function* f *is defined on a segment* $[a, b]$. *In order that a polynomial* P_n^* *of degree* $\leq n$ *be a polynomial of the best approximation to* f, *it is necessary and sufficient that there exist at least one system of* $n+2$ *points* x_i, $a \leq x_1 < x_2 < \ldots < x_{n+2} \leq b$, *such that the difference* $f(x) - P_n^*(x) =: r_n(x)$

(i) consecutively takes alternating values at the points x_i,

(ii) attains its maximum absolute value on $[a, b]$ *at the points* x_i, *i.e.*,

$$r_n(x_1) = -r_n(x_2) = r_n(x_3) = \ldots = (-1)^{n+1} r_n(x_{n+2}) = \pm \| r_n \|_{[a,b]}. \tag{1.6}$$

A system of points $\{x_j\}_{j=1}^{n+2}$ satisfying equalities (1.6) is called *alternation* or *Chebyshev alternation.*

Example 1.1. Let $f(x) = \sin x$. We show that the polynomial $P_6^*(x) \equiv 0$ is a polynomial of best approximation of degree six to this function on $[-4\pi, 4\pi]$.

This follows from the fact that the difference

$$r_6(x) = \sin x - P_6^*(x)$$

satisfies the conditions of Theorem 1.2 at $8 = 6 + 2$ points $x_k = -\frac{7}{2}\pi + k\pi$ $(k = 0, 1,$

$2, \ldots, 7)$ from $[-4\pi, 4\pi]$, namely,

$$r_6(x_k) = \sin\left(k\pi - \frac{7}{2}\pi\right) = (-1)^k = (-1)^k \|r_6\|.$$

Proof. *Necessity.* Since the case where the function f is itself a polynomial of degree at most n is trivial, we omit its analysis for the sake of convenience. Prior to proving the theorem, we introduce the following definitions:

A point x_0 is called the *point of maximum deviation* for the difference $f(x) - P_n^*(x) = r_n(x)$ or the *e-point provided that*

$$|r_n(x_0)| = \max_{a \le x \le b} |r_n(x)| = \|r_n\|. \tag{1.7}$$

If, in addition,

$$r_n(x_0) = \|r_n\|, \tag{1.8}$$

then the point x_0 is called a *point of positive deviation* or a (+)-*point.* Further, if

$$r_n(x_0) = -\|r_n\|, \tag{1.9}$$

then x_0 is called a *point of negative deviation* or a (–)-*point.*

Consider a polynomial P_n^* of the best approximation for a function f. Since the function r_n is continuous, there exists at least one e-point for r_n on the segment $[a, b]$.

Let us show that the segment $[a, b]$ contains both (+)- and (–)-points. Indeed, if the analyzed segment contains no (–)-points, then the least value of the continuous function r_n on $[a, b]$ is greater than $-\|r_n\|$ and, hence, there exists h, $0 < h < \|r_n\|$, such that

$$-\|r_n\| + h \le r_n(x) = f(x) - P_n^*(x) \le \|r_n\| \tag{1.10}$$

for all $x \in [a, b]$. We set $Q_n(x) = P_n^*(x) + h/2$. Inequalities (1.10) now imply that

$$-\|r_n\| + h/2 \le f(x) - Q_n(x) = \|r_n\| - h/2, \qquad x \in [a, b],$$

i.e.,

$$\|f - Q_n\| = \|r_n\| - h/2 = \|f - P_n^*\| - h/2 < \|f - P_n^*\|,$$

which is impossible since P_n^* is, by assumption, the polynomial of the best approximation for f.

Let us show that there exists a system of $n+2$ e-points on $[a, b]$ satisfying relations (1.6). For this purpose, we first demonstrate that the segment $[a, b]$ can be split into $m+1$ subsegments

$$[a, z_1],\ [z_1, z_2],\ [z_2, z_3],\ \ldots\ ,\ [z_{m-1}, z_m],\ [z_m, b] \tag{1.11}$$

alternatively containing either solely (+)-points or solely (–)-points. Note that, in view of the continuity of the function r_n on the segment $[a, b]$, the number $m+1$ of these subsegments is finite.

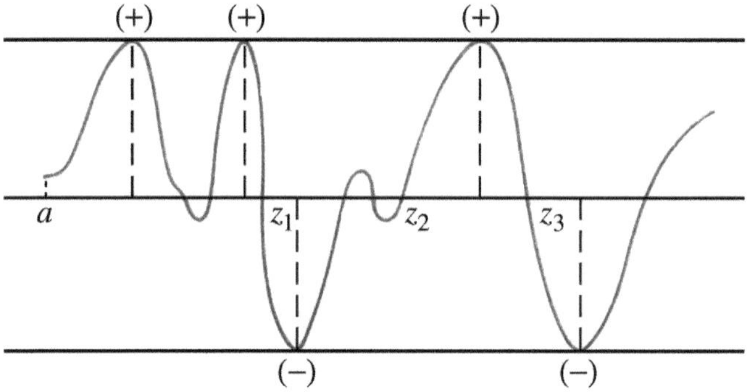

Fig. 1

The desired partition is realized as follows (see Fig. 1):

For the sake of definiteness, we assume that the first e-point encountered in moving along the segment $[a, b]$ from a to b is a (+)-point. Let z_1 be the rightmost zero of the difference $f - P_n^* = r_n$ located between the point a and the first (–)-point encountered after this point in the indicated direction.

Let z_2 be the rightmost zero of the difference $f - P_n^* = r_n$ located between the point z_1 and the first (+)-point after this point. If there are no (+)-points on the segment $[z_1, b]$, then we set $z_2 = b$.

Let z_3 be the rightmost zero of the difference $f - P_n^* = r_n$ located between the point z_1 and the first (–)-point after this point. We set $z_3 = b$ if there are no (–)-points on the segment $[z_2, b]$, etc.

Let z_m be the last point in the constructed sequence of points z_j other than b. Note that each segment in the proposed sequence alternatively contains (+)- or (–)-points and, hence, in order to prove the necessity of the condition of Theorem 1.2, it suffices to show that $m+1 \geq n+2$.

By contradiction, we assume that $m+1 < n+2$, i.e., that $m \leq n$. Let us show that, under this assumption, it is possible to construct a polynomial Q_n deviating from f less than the polynomial P_n^*. Indeed, since the segment $[a, z_1]$ does not contain (–)-points, the segment $[z_1, z_2]$ does not contain (+)-points, etc., we conclude that, in view of the continuity of r_n and the fact that the number of segments (1.11) is finite, there exists a number h, $0 < h < \|r_n(x)\|$, such that

$$\begin{aligned} -\|r_n\| + h \leq r_n(x) \leq \|r_n\|, \quad x \in [a, z_1], \\ -\|r_n\| \leq r_n(x) \leq \|r_n\| - h, \quad x \in [z_1, z_2], \end{aligned} \tag{1.12}$$

etc. Thus, we set

$$p_m(x) = \delta(x - z_1)(x - z_2)\ldots(x - z_m), \tag{1.13}$$

where δ is chosen so that

$$\begin{aligned} &\text{(i)} \ \|p_m\| \leq h/2, \\ &\text{(ii)} \ \operatorname{sgn} p_m(x) = 1 \quad \text{for} \quad x \in [a, z_1). \end{aligned} \tag{1.14}$$

This means that the polynomial p_m changes its sign on passing through the points z_j and has the same sign as the difference r_n at all (+)- and (–)-points. More precisely, $p_m(x) > 0$ for $x \in [a, z_1)$, $p_m(x) < 0$ for $x \in (z_1, z_2)$, etc. Therefore, by setting

$$Q_n(x) = P_n^*(x) + p_m(x),$$

we get a polynomial of degree at most n (since $m \leq n$) such that

$$f(x) - Q_n(x) < \|r_n\|, \quad x \in [a, z_1].$$

At the same time, in view of inequalities (1.12) with $x \in [a, z_1]$, we find

$$f(x) - Q_n(x) \geq r_n(x) - p_m(x) \geq -\|r_n\| + h - \|p_m\| \geq -\|r_n\| + h/2,$$

i.e., for all $x \in [a, z_1]$,

$$|f(x) - Q_n(x)| < \|r_n\|. \tag{1.15}$$

In exactly the same way, we conclude that, for $x \in [z_1, z_2]$,

$$f(x) - Q_n(x) = r_n(x) - p_m(x) - \|r_n\|,$$

and, at the same time,

$$f(x) - Q_n(x) \leq \|r_n\| - h - p_m(x) \leq \|r_n\| - h/2.$$

Therefore,

$$|f(x) - Q_n(x)| < \|r_n\|$$

for all $x \in [z_1, z_2]$.

Similarly, we can show that inequality (1.15) holds on all other segments of system (1.11), i.e., for all $x \in [a, b]$. This means that the polynomial Q_n approximates the function f better than P_n^*, i.e., we arrive at a contradiction. The necessity of the condition of Theorem 1.2 is proved.

Sufficiency. Assume that the polynomial P_n^* satisfies, for a certain system of points $a \leq x_1 < x_2 < \ldots < x_{n+2} \leq b$, the condition

$$f(x_1) - P_n^*(x_1) = -[f(x_2) - P_n^*(x_2)] = \ldots$$

$$= (-1)^{n+1}[f(x_{n+2}) - P_n^*(x_{n+2})] = \|f - P_n^*\|.$$

Let us show that P_n^* has the least deviation from f. By contradiction, assume that there exists a polynomial Q_n approximating the function f better than P_n^*, i.e.,

$$\|f - Q_n\| < \|f - P_n^*\|$$

and, specifically,

$$|f(x_j) - Q_n(x_j)| < \|f(x_j) - P_n^*(x_j)\| = \|f - P_n^*\|. \tag{1.16}$$

It follows from inequality (1.16) that, at all points x_j $(j = 1, 2, \dots, n+2)$, the difference $Q_n(x_j) - P_n^*(x_j) = [f(x_j) - P_n^*(x_j)] - [f(x_j) - Q_n(x_j)]$ has the same sign as $f(x_j) - P_n^*(x_j)$, i.e., the difference $Q_n - P_n^*$ changes its sign on $[a, b]$ at least $n+1$ times. This means that the polynomial $Q_n - P_n^*$ has at least $n+1$ zeros on the segment $[a, b]$ but this is impossible because $Q_n - P_n^*$ is a polynomial of degree n by virtue of the assumption that $Q_n \not\equiv P_n^*$.

■

Remark 1.1. The uniqueness of the best approximating polynomial P_n^* for the function f can, in general, be proved by analogy with the proof of sufficiency in the Chebyshev theorem. However, we do not present the proof here because somewhat later, in Theorem 2.6, the problem of uniqueness is analyzed more completely.

A similar theorem and remark also hold for the approximation of continuous 2π-periodic functions by trigonometric polynomials. The only difference is that the above-mentioned points x_j must be located in a certain half interval $(a, a+2\pi]$ of length 2π and the number of these points must be equal to $2n+2$, i.e., the following theorem is true:

Theorem 1.2′. *Assume that a* 2π*-periodic continuous real function* f *is defined on the real axis. Then, in order that a trigonometric polynomial* T_n^* *of degree* $\le n$ *be the polynomial of the best approximation for* f, *it is necessary and sufficient that some (and, hence each) period* $[a, a+2\pi)$ *contain* $2n+2$ *points* t_j,

$$a \le t_1 < t_2 < \dots < t_{2n+2} < a + 2\pi,$$

such that the difference $f(t) - T_n^*(t) =: r_n(t)$

(i) alternatively changes its sign at the points t_j,

(ii) attains its maximum absolute value on $[a, a+2\pi)$ *at the points* t_j, *i.e., the following conditions are satisfied:*

$$r_n(t_1) = -r_n(t_2) = \dots = -r_n(t_{2n+2}) = \pm\|r_n\|.$$

The proof of this theorem is absolutely similar to the proof of Theorem 1.2.

However, it is necessary to take into account the following specific features of the analyzed case:

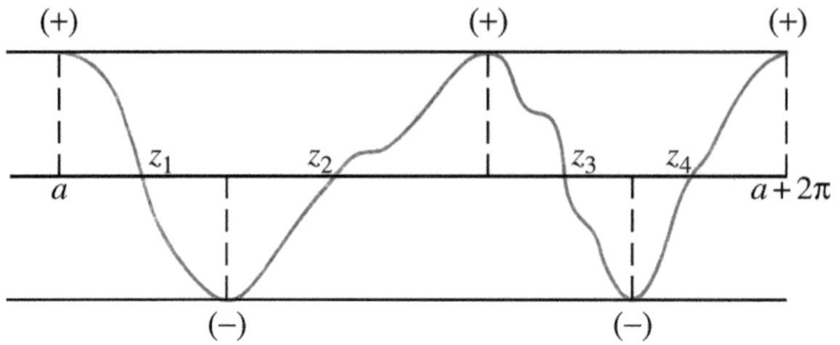

Fig. 2

(i) the number n should be replaced by $2n$;

(ii) it is reasonable to choose the left end of the half segment $[a, a+2\pi)$, i.e., the point a, at a (+)-point; in this case, in view of the fact that the point $a+2\pi$ must also be a (+)-point, the number m of the points z_k cannot be odd (Fig. 2);

(iii) the polynomial p_m (1.13) should be replaced by a trigonometric polynomial t_m of degree $m/2 \leq n$ of the form

$$t_m = \delta \sin\frac{t-z_1}{2} \sin\frac{t-z_2}{2} \ldots \sin\frac{t-z_m}{2}$$

$$= \frac{\delta}{2^{m/2}} \left[\cos\frac{z_2-z_1}{2} - \cos\left(t - \frac{z_2-z_1}{2}\right)\right]$$

$$\times \ldots \times \left[\cos\frac{z_m - z_{m-1}}{2} - \cos\left(t - \frac{z_m - z_{m-1}}{2}\right)\right],$$

(iv) by virtue of Theorem 1.3 presented in what follows, if a trigonometric polynomial T_n has more than $2n$ zeros in the half interval $[a, a+2\pi)$, then $T_n(t) \equiv 0$.

1.3. On the number of zeros of trigonometric polynomials

Unlike algebraic polynomials, trigonometric polynomials of any degree may have no zeros in $\mathbb{C}$. Thus, e.g., this is true for the polynomials $\cos n\xi + i \sin n\xi = e^{in\xi}$.

Nevertheless, the numbers of zeros of algebraic and trigonometric polynomials are closely correlated. The following theorem establishes the indicated relationship:

Theorem 1.3 (on the number of zeros of trigonometric polynomials). *In any vertical strip R_α of width 2π:*

$$R_\alpha := \{z \in \mathbb{C} : \alpha \le \operatorname{Re} z < \alpha + 2\pi\}, \quad \text{where } \alpha \text{ is a real number,}$$

the number of zeros z_j of any trigonometric polynomial T_n of the form

$$T_n(z) = \sum_{k=-n}^{n} c_k e^{ikz} = e^{-inz} \sum_{l=0}^{2n} c_{l-n} e^{ilz} \tag{1.17}$$

(counting multiplicities) is equal to the number of nonzero zeros ξ_j of the algebraic polynomial

$$P_{2n}(\xi) = \sum_{l=0}^{2n} c_{l-n} \xi^l \tag{1.18}$$

in $\mathbb{C}$. Furthermore, the following equality is true:

$$z_j = -i \ln_\alpha \xi_j, \tag{1.19}$$

where $-i \ln_\alpha \xi_j$ is the only number from the sequence

$$-i\left(\ln|\xi_j| + i \arg \xi_j + 2k\pi i\right) = -i \ln|\xi_j| + \arg \xi_j + 2k\pi,$$

$$k = 0, 1, -1, 2, -2, \ldots,$$

contained in the strip R_α.

Proof. The function

$$\xi = \xi(z) = e^{iz} \tag{1.20}$$

realizes a one-to-one mapping of any line $z = z(y) = x_0 + iy$, $x_0 \in [\alpha, \alpha + 2\pi)$, $y \in (-\infty, \infty)$, (i.e., of the line $x = x_0$) from the strip R_α onto a half line $\xi = \xi_{x_0}(r) = e^{ix_0} r$,

$0 < r = e^{-y} < \infty$, and hence, bijectively maps the entire strip R_α from the plane z into the plane ξ with deleted point $\xi = 0$. Therefore, by virtue of the identity

$$P_{2n}(\xi) = e^{inz}\,T_n(z), \quad \xi = e^{iz}, \tag{1.21}$$

which follows from (1.17) and (1.18), there exists a one-to-one correspondence between the different zeros $z_j = -i\ln_\alpha \xi_j \in R_\alpha$ of the polynomial $T_n(z)$ and different zeros $\xi_j = e^{iz_j} \neq 0$ of the polynomial P_{2n}, and vice versa.

Thus, to complete the proof of theorem, it remains to show that each zero ξ_j of the polynomial P_{2n} of multiplicity $r > 1$ corresponds to the zero $z_j = -i\ln_\alpha \xi_j$ of the polynomial T_n of the same multiplicity, and vice versa. To this end, we differentiate identity (1.21) and arrive at the following identities:

$$i P'_{2n}(\xi)\, e^{iz} \equiv e^{inz}\left[inT_n(z) + T'_n(z)\right],$$

or

$$i P'_{2n}(\xi) \equiv e^{i(n-1)z}\left[\alpha_0^{(1)} T_n(z) + T'_n(z)\right] \tag{1.21′}$$

and, similarly,

$$i^2 P''_{2n}(\xi) \equiv e^{i(n-2)z}\left[\alpha_0^{(2)} T_n(z) + \alpha_1^{(2)} T'_n(z) + T''_n(z)\right], \tag{1.21″}$$

....

$$i^{r-1} P_{2n}^{(r-1)}(\xi)$$

$$\equiv e^{i(n-r+1)z}\left[\alpha_0^{(r-1)} T_n(z) + \alpha_1^{(r-1)} T'_n(z) + \ldots + T_n^{(r-1)}(z)\right], \tag{1.21$^{(r-1)}$}$$

where $\xi_j = e^{iz}$ and $\alpha_j^{(k)}$ are certain numbers. Identities (1.21), … , (1.21$^{(r-1)}$) immediately imply that if

$$P_{2n}(\xi_j) = P'_{2n}(\xi_j) = \ldots = P_{2n}^{(r-1)}(\xi_j) = 0$$

at a point $\xi_j \neq 0$, then $T_n(z_j) = 0$, $T'_n(z_j) = 0, \ldots,$ $T_n(z_j) = 0$ at the corresponding point $z_j = -i\ln_\alpha \xi_j$, and vice versa. Hence, the multiplicities of the zeros ξ_j of the polynomial $P_{2n}(\xi)$ and the multiplicities of the corresponding zeros $z_j = -i\ln_\alpha \xi_j$ of the polynomial $T_n(z)$ are identical. This completes the proof of Theorem 1.3.

■

Corollary 1.1. *If a trigonometric polynomial T_n of degree n of the form*

$$T_n(z) = \frac{a_0}{2} + \sum_{k=1}^{n} \left(a_k \cos kz + b_k \sin kz\right) \tag{1.22}$$

possesses the property

$$a_n^2 + b_n^2 \neq 0, \tag{1.23}$$

then it has exactly $2n$ zeros (counting multiplicities) in any strip $R_\alpha = \{\alpha \le \operatorname{Re} z < \alpha + 2\pi\}$. If condition (1.23) is not true, i.e., $a_n^2 + b_n^2 = 0$, then $T_n(z)$ has less than $2n$ zeros in R_α.

Indeed, by applying Euler's formulas

$$\cos kz = \frac{e^{ikz} + e^{-ikz}}{2} \qquad \text{and} \qquad \sin kz = \frac{e^{ikz} - e^{-ikz}}{2i},$$

we can represent the polynomial T_n as

$$T_n(z) = \sum_{k=-n}^{n} c_k e^{ikz} = e^{-inz} \sum_{l=0}^{2n} c_{l-n} e^{ilz}, \tag{1.24}$$

where

$$c_0 = \frac{a_0}{2}, \quad c_k = \frac{a_k - ib_k}{2}, \quad c_{-k} = \frac{a_k + ib_k}{2}, \quad k = 1, 2, \ldots, n. \tag{1.25}$$

Further, since $a_n^2 + b_n^2 = 4 c_n c_{-n} \neq 0$ by virtue of condition (1.23), the coefficients of the polynomial

$$P_{2n}(\xi) = \sum_{l=0}^{2n} c_{l-n} \xi^l, \quad \xi = e^{iz},$$

corresponding to $l = 0$ and $l = 2n$ are nonzero. Hence, the indicated polynomial possesses $2n$ nonzero zeros (counting multiplicity).

By virtue of Theorem 1.3, the number of zeros of the polynomial T_n in the strip R_α and the polynomial P_{2n} is identical.

Corollary 1.2. *If two trigonometric polynomials T_1 and T_2 of degree $\leq n$ take identical values at more than $2n$ points z_k from the same strip R_α (e.g., for* $\operatorname{Re} z \in [0, 2\pi)$), *then*

$$T_2(z) \equiv T_1(z).$$

Corollary 1.3 (on the trigonometric polynomial factorization). *Under the conditions of Corollary 1.1, the polynomial T_n admits the following representation in terms of its zeros z_k, $k = 1, 2, \dots, 2n$, from a strip R_α:*

$$T_n(z) = A \prod_{k=1}^{2n} \sin \frac{z - z_k}{2}, \quad A = \text{const}. \tag{1.26}$$

Indeed, if we denote the (nonzero) zeros of the polynomial P_{2n} by ξ_k and the zeros of the polynomial T_n satisfying the equality $\xi_k = e^{iz_k}$ by z_k, then in view of (1.24), we get

$$P_{2n}(\xi) = a_0 \prod_{k=1}^{2n} (\xi - \xi_k), \quad a_0 = \text{const},$$

$$T_n(z) = a_0 e^{-inz} \prod_{k=1}^{2n} \left(e^{iz} - e^{iz_k}\right) = a_0 \prod_{k=1}^{2n} e^{iz_k/2} \prod_{k=1}^{2n} \left(e^{i(z-z_k)/2} - e^{-i(z-z_k)/2}\right).$$

This yields the required relation (1.26).

2. Chebyshev systems of functions

The Chebyshev theorem and numerous other results obtained by P. Chebyshev play an important role in the approximation of functions by the so-called generalized Chebyshev polynomials. In the present section, we study this class of functions.

2.1. Definitions, examples

Definition 2.1. *Consider a set $\mathfrak{M}$ (containing at least $n + 1$ points) in a metric space of continuous functions. A system of functions*

$$\varphi_0, \varphi_1, \dots, \varphi_n, \tag{2.1}$$

defined on this set is called a Chebyshev system or a T-system on the set $\mathfrak{M}$ *if any generalized polynomial* $P_n(x) = P_n(\varphi_k, c_k; x)$ *of the form*

$$P_n(x) = \sum_{k=0}^{n} c_k \varphi_k(x), \tag{2.2}$$

where c_k, $k = 0, \dots, n$, *is a collection of numbers at least one of which is nonzero, has at most* n *different zeros on* $\mathfrak{M}$.

In what follows, the generalized polynomial (2.2) with respect to system (2.1) is sometimes simply called a *polynomial* or a *T-polynomial.*

Clearly, if (2.1) is a Chebyshev system of functions on $\mathfrak{M}$, then it is also a Chebyshev system on any subset of $\mathfrak{M}$ containing at least $n+1$ points.

Examples of Chebyshev systems

1. The collection of functions

$$1,\ z,\ z^2, \dots,\ z^n \tag{2.3}$$

is a Chebyshev system in the entire complex plane.

2. By virtue of Theorem 1.3, the family of functions

$$1,\ e^{iz},\ e^{-iz}, \dots,\ e^{inz},\ e^{-inz} \tag{2.4}$$

and, hence, the family

$$1,\ \cos z,\ \sin z, \dots,\ \cos nz,\ \sin nz \tag{2.5}$$

are Chebyshev systems in any strip $\alpha \le \operatorname{Re} z < \alpha + 2\pi$, where α is a real number; moreover, each of these systems is a Chebyshev system in any half segment (period) $[\alpha, \alpha + 2\pi)$.

3. The system

$$1,\ \cos t,\ \sin t, \dots,\ \cos nt,\ \sin nt \tag{2.5$'$}$$

is not a Chebyshev system on the segment $[0, 2\pi]$ because the polynomial

$$P_{2n}(t) = a_0 + a_1 \cos t + a_2 \sin t + \ldots + a_{2n-1} \cos nt + a_{2n} \sin nt$$

with $a_0 = a_1 = \ldots = a_{2n-1} = 0$ and $a_{2n} = 1$ has $2n+1$ zeros $t_k = k\pi/n$, $k = 0, \ldots, 2n$, on $[0, 2\pi]$.

Nevertheless, system (2.5′) turns into a Chebyshev system on a closed set $[0, 2\pi]$ or even on the entire line if we "identify" the points lying at distances equal to multiples of 2π. It is known that, for this purpose, it suffices to identify the argument t of the functions in (2.5′) with a point of the unit circle centered at the origin. Note that this circle is a compact set and the angle made by its radius vector with the OX-axis is denoted by t.

In this connection, we introduce the following definition:

Definition 2.2. *A system of functions* $\{\varphi_k\}_{k=0}^n$ *continuous on the entire axis and periodic with period* θ *is called a Chebyshev system (or a T-system) if any generalized polynomial*

$$P_n(t) = \sum_{k=0}^{n} c_k \varphi_k(t),$$

where c_k, $k = 0, \ldots, n$, *is a collection of numbers at least one of which is nonzero, has at most* n *different zeros on a certain (and, hence, on each) half interval* $(a, a+\theta]$ *or a half segment* $[a, a+\theta)$.

We now present several more examples:

4. The fact that (2.5) is a Chebyshev system on $[-\pi, \pi)$ readily implies that the family of functions

$$1,\ \cos t,\ \cos 2t,\ \ldots,\ \cos nt \tag{2.6}$$

is a Chebyshev system on $[0, \pi)$ and the family

$$\sin t,\ \sin 2t,\ \ldots,\ \sin nt \tag{2.7}$$

is a Chebyshev system on $(0, \pi)$.

5. If a family of functions $\{\varphi_k\}_{k=0}^n$ is a Chebyshev system on $\mathfrak{M}$ and $p(x)$ is a positive function continuous on $\mathfrak{M}$, then $\{p\varphi_k\}_{k=0}^n$ is also a Chebyshev system on $\mathfrak{M}$.

6. The set of two functions $\varphi_0(x) = 1$ and $\varphi_1(x) = x^3$ is a Chebyshev system on the real axis.

7. The set formed by the functions $\varphi_0(x) = x^{-3}$ and $\varphi_1(x) \equiv 1$ is a Chebyshev system on the semiaxis $(0, \infty)$.

8. The set of functions $\varphi_0(x) = x^2 - x$, $\varphi_1(x) = x^2 + x$, and $\varphi_2(x) = x^2 + 1$ forms, clearly, a Chebyshev system on the axis; at the same time neither two functions from this set φ_j, $j = 0, 1, 2$, form a Chebyshev system on the axis.

Note that the real T-systems of functions $\{\varphi_k\}_{k=0}^n$ do not exist in the case of functions of many variables on sets with interior points. This becomes clear from the following result by Mairhuber (1956) presented without proof:

If it is possible to specify a T-system $\left\{\varphi_j(x)\right\}_{j=0}^n$, where $n > 0$, on an abstract compact set, then this set is homeomorphic either to a circle or to its part.

Definition 2.3. *A T-system $\{\varphi_k\}_{k=0}^n$ is called a Markov system if all its subsystems $\{\varphi_k\}_{k=0}^m$, $0 < m < n$, are also T-systems.*

It is possible to show (see, e.g., [M. Krein and Nudel'man (1973)]) that each T-system on the segment $[a, b]$ can be replaced by an equivalent system $\left\{\varphi_j^*\right\}_{j=0}^n$, $\varphi_j^*(x) = \sum_{k=0}^n c_{jk}\,\varphi_k(x)$, which is a Markov system on any segment $[a_1, b_1] \subset (a, b)$.

2.2. Basic properties of the Chebyshev systems

For numerous other properties and applications of the T-systems, the reader is referred to the monograph [Karlin and Studden (1976)].

Theorem 2.1. *In order that a collection of functions (2.1) defined on a set $\mathfrak{M}$ be a Chebyshev system on this set, it is necessary and sufficient that the determinant*

$$D(\varphi; x) = D\begin{pmatrix} \varphi_0 & \varphi_1 & \cdots & \varphi_n \\ x_0 & x_1 & \cdots & x_n \end{pmatrix}$$

$$:= \begin{vmatrix} \varphi_0(x_0) & \varphi_0(x_1) & \dots & \varphi_0(x_n) \\ \varphi_1(x_0) & \varphi_1(x_1) & \dots & \varphi_1(x_n) \\ \dots & \dots & \dots & \dots \\ \varphi_n(x_0) & \varphi_n(x_1) & \dots & \varphi_n(x_n) \end{vmatrix} \tag{2.8}$$

be nonzero for any family of $n+1$ *different points* $\{x_j\}_{j=0}^n$ *from the set* $\mathfrak{M}$.

The assertion of this theorem follows from the fact that (2.1) is a Chebyshev system if and only if the system of equations

$$A_0\varphi_0(x_j) + A_1\varphi_1(x_j) + \ldots + A_n\varphi_n(x_j) = 0, \quad j = 0,\ldots,n,$$

possesses only the trivial solution $A_0 = A_1 = \ldots = A_n = 0$ for any system of $n+1$ different points $x_j \in \mathfrak{M}$, $j = 0,\ldots,n$. It is known that this condition is equivalent to the requirement that the determinant $D\begin{pmatrix} \varphi_0 & \varphi_1 & \dots & \varphi_n \\ x_0 & x_1 & \dots & x_n \end{pmatrix}$ of system (2.8) must differ from zero.

Corollary 2.1. *If a collection of functions* $\{\varphi_k\}_{k=0}^n$ *is a Chebyshev system on* $[a, b]$ *and* $\{x_j\}_{j=0}^n$ *is a set of different points from* $[a, b]$ *such that the distance between any two points from this set is not smaller than* $c > 0$, *i.e.,*

$$\min_{\substack{j,k \\ j\neq k}} |x_j - x_k| \geq c,$$

then

$$\left| D\begin{pmatrix} \varphi_0 & \varphi_1 & \dots & \varphi_n \\ x_0 & x_1 & \dots & x_n \end{pmatrix} \right| \geq r > 0, \tag{2.9}$$

where $r = r(c)$ *does not depend on the chosen set of points.*

Proof. Indeed, if it is impossible to find a number $r > 0$ for which inequality (2.9) is satisfied, then there exists a sequence of systems of points $\{x_k^j\}_{k=0}^n$, $j = 1, 2, \ldots$, such that

(i) $\min_{\substack{k,l \\ k\neq l}} \left| x_k^j - x_l^j \right| \geq c$, and

(ii) $\lim\limits_{j\to\infty} D\begin{pmatrix} \varphi_0 & \varphi_1 & \dots & \varphi_n \\ x_0^j & x_1^j & \dots & x_n^j \end{pmatrix} = 0.$

In this case, one can find a subsequence $\{j_r\}$ such that each subsequence $\left\{x_k^{j_r}\right\}$, $k = 0,\dots,n$, has a limit $x_k^0 \in [a, b]$, $k = 0,\dots,n$, and, for the system of points $\left\{x_k^0\right\}_{k=0}^n$, we get

$$\min_{\substack{k,l \\ k\neq l}} \left| x_k^0 - x_l^0 \right| \geq c,$$

and

$$D\begin{pmatrix} \varphi_0 & \varphi_1 & \dots & \varphi_n \\ x_0^0 & x_1^0 & \dots & x_n^0 \end{pmatrix} = \lim_{r\to\infty} D\begin{pmatrix} \varphi_0 & \varphi_1 & \dots & \varphi_n \\ x_0^{j_r} & x_1^{j_r} & \dots & x_n^{j_r} \end{pmatrix} = 0.$$

This means that we arrive at a contradiction to Theorem 2.1.

■

Theorem 2.2 (interpolation). *Assume that a Chebyshev system of functions* $\{\varphi_k\}_{k=0}^n$ *is given on a set* $\mathfrak{M}$ *of a metric space. Then, for any collection of* $n+1$ *different points* $x_j \in \mathfrak{M}$, $j = 0,\dots,n$, *and any collection of* $n+1$ *numbers* y_j, $j = 0,\dots,n$, *there exists a unique generalized polynomial* P_n *of the form (2.2) satisfying the conditions*

$$P_n(x_j) = y_j, \quad j = 0,\dots,n. \tag{2.10}$$

This polynomial is called interpolation polynomial and admits the following representation:

$$P_n(x) = \frac{\begin{vmatrix} 0 & \varphi_0(x) & \varphi_1(x_0) & \dots & \varphi_n(x_0) \\ y_0 & \varphi_0(x_0) & \varphi_1(x_1) & \dots & \varphi_n(x_1) \\ \dots & \dots & \dots & \dots & \dots \\ y_n & \varphi_0(x_n) & \varphi_1(x_n) & \dots & \varphi_n(x_n) \end{vmatrix}}{\begin{vmatrix} \varphi_0(x_0) & \varphi_1(x_0) & \dots & \varphi_n(x_0) \\ \varphi_0(x_1) & \varphi_1(x_1) & \dots & \varphi_n(x_1) \\ \dots & \dots & \dots & \dots \\ \varphi_0(x_n) & \varphi_1(x_n) & \dots & \varphi_n(x_n) \end{vmatrix}}. \tag{2.11}$$

The problem of construction of a polynomial satisfying conditions (2.10) is called *interpolation problem.*

In fact, it follows from Theorem 2.1 that the denominator on the right-hand side of (2.11) differs from zero, therefore, the right-hand side of (2.11) always specifies a generalized polynomial of degree n.

Subtracting the $(j+1)$th row from the first row in the determinant appearing in the numerator on the right-hand side of (2.11) and setting $x = x_j$, we immediately conclude that the polynomial P_n satisfies all $n+1$ conditions (2.10). The uniqueness of this polynomial follows from the fact that $\{\varphi_k\}_{k=0}^n$ is a Chebyshev system.

It is easy to see that the converse assertion is also true: If, under the conditions of Theorem 2.2, the sole condition imposed on a system of functions $\{\varphi_k\}_{k=0}^n$ is that the functions φ_k are continuous and, at the same time, the interpolation problem (2.10) is always uniquely solvable, then $\{\varphi_k\}_{k=0}^n$ is a Chebyshev system.

2.3. Chebyshev systems of real functions on an interval

In what follows, for $a < b$, we agree to use the term "interval" and write $\langle a, b\rangle$ to denote any of the following four sets: the segment $[a, b]$, the half interval $(a, b]$, the half segment $[a, b)$, or the interval (a, b).

To prove Theorem 2.3 formulated somewhat later, we need the following auxiliary assertions:

Definition 2.4. *Assume that* $\{\varphi_k\}_{k=0}^n$ *is a Chebyshev system of real functions on an interval* $\langle a, b\rangle$ *and* P_n *is a polynomial in this system. A zero* $x_0 \in (a, b)$ *of this polynomial is called odd if the polynomial* P_n *changes its sign in passing through* x_0 *(i.e.,* $\operatorname{sgn} P_n(x_0 - 0) = -\operatorname{sgn} P_n(x_0 + 0)$*) and even if the polynomial* P_n *preserves its sign (i.e.,* $\operatorname{sgn} P_n(x_0 - 0) = \operatorname{sgn} P_n(x_0 + 0)$*). If a zero* $x_k \in \langle a, b\rangle$ *coincides with one of the ends of the interval* $\langle a, b\rangle$*, i.e.,* a *or* b*, then it is always regarded as odd.*

We now establish the following auxiliary lemmas, which are also of independent interest:

Lemma 2.1. *Assume that* $\{\varphi_k\}_{k=0}^n$ *is a Chebyshev system of real functions on an interval* $\langle a, b\rangle$*. Then, for any* n *different points* $x_k \in \langle a, b\rangle$*,*

$$a \le x_1 < x_2 < \ldots < x_n \le b,$$

the generalized polynomial

$$P_n(x) = D\begin{pmatrix} \varphi_0 & \varphi_1 & \ldots & \varphi_n \\ x & x_1 & \ldots & x_n \end{pmatrix}$$

$$= \begin{vmatrix} \varphi_0(x) & \varphi_0(x_1) & \ldots & \varphi_0(x_n) \\ \varphi_1(x) & \varphi_1(x_1) & \ldots & \varphi_1(x_n) \\ \cdots & \cdots & \cdots & \cdots \\ \varphi_n(x) & \varphi_n(x_1) & \ldots & \varphi_n(x_n) \end{vmatrix} \tag{2.12}$$

vanishes at all points x_k *and these points exhaust the set of its zeros. Moreover, these zeros are odd. Any other polynomial* $\tilde{P}_n$ *vanishing at all points* x_k *has the form* $\tilde{P}_n = CP_n$, $C = \text{const}$.

Proof. Indeed, the polynomial P_n vanishes at all points x_k, $k = 1, \ldots, n$, and, in view of the fact that (2.1) is a Chebyshev system, solely at these points. Therefore, the polynomial P_n preserves its sign on each interval (a, x_1), $(x_1, x_2), \ldots, (x_n, b)$. Since, for any fixed x, P_n can be regarded as a polynomial with respect to any point x_k, it preserves the sign under any variations of positions of the points x, x_1, $x_2, \ldots, x_n$ provided that their order in the sequence remains unchanged. Thus, if we now set, e.g., $x' \in (x_{k-1}, x_k)$ and $x'' \in (x_k, x_{k+1})$, then we get

$$\operatorname{sgn} P_n(x') = \operatorname{sgn} D\begin{pmatrix} \varphi_0 & \varphi_1 & \ldots & \varphi_n \\ x' & x_1 & \ldots & x_n \end{pmatrix}$$

$$= (-1)^{k-1} D\begin{pmatrix} \varphi_0 & \varphi_1 & \ldots & \varphi_{k-2} & \varphi_{k-1} & \varphi_k & \ldots & \varphi_n \\ x_1 & x_2 & \ldots & x_{k-1} & x' & x_k & \ldots & x_n \end{pmatrix}$$

$$= (-1)^{k-1} D\begin{pmatrix} \varphi_0 & \varphi_1 & \ldots & \varphi_{k-1} & \varphi_k & \varphi_{k+1} & \ldots & \varphi_n \\ x_1 & x_2 & \ldots & x_k & x'' & x_{k+1} & \ldots & x_n \end{pmatrix}$$

$$= -\operatorname{sgn} D\begin{pmatrix} \varphi_0 & \varphi_1 & \ldots & \varphi_n \\ x'' & x_1 & \ldots & x_n \end{pmatrix} = -\operatorname{sgn} P_n(x'')$$

and, hence,

$$\operatorname{sgn} P_n(x_k - 0) = -\operatorname{sgn} P_n(x_k + 0).$$

This means that the zero x_k is odd. Now assume that the conditions of Lemma 2.1 are satisfied not only by P_n but also by a polynomial $\tilde{P}_n(x) \not\equiv P_n(x)$. Let $x_0 \in (a, b)$ be a point at which $\tilde{P}_n(x_0) \neq P_n(x_0)$. Since, clearly, $x_0 \neq x_k$, $k = 1, \ldots, n$, we have $\tilde{P}_n(x_0) \neq 0$ and $P_n(x_0) \neq 0$. We also set

$$\check{P}_n(x) = \tilde{P}_n(x_0) P_n(x) - P_n(x_0) \tilde{P}_n(x), \quad x \in [a, b].$$

The polynomial $\check{P}_n$ vanishes at $n+1$ points $x, x_1, x_2, \ldots, x_n$. Therefore, $\check{P}_n(x) \equiv 0$ and $\tilde{P}_n(x) \equiv C P_n(x)$, where

$$C = \frac{\tilde{P}_n(x_0)}{P_n(x_0)} = \text{const}.$$

■

Lemma 2.2. *If a T-polynomial* P_n *has* m' *odd zeros in an interval* $\langle a, b\rangle$ *and* m'' *even zeros in the interval* (a, b), *then*

$$m' + 2m'' \leq n. \tag{2.13}$$

Proof. Denote by $x_1 < x_2 < \ldots < x_{m'}$ the odd zeros of the polynomial $P_n(x)$ and by $y_1 < y_2 < \ldots < y_{m''}$ its even zeros. Let (a, c) be an interval containing neither odd zeros x_j, $j = 1, \ldots, m'$, nor even zeros y_k, $k = 1, \ldots, m''$. In the interval (a, c), we choose $n - m'$ different auxiliary points z_i, $i = 1, \ldots, n - m'$, and construct an auxiliary generalized polynomial of the form

$$\check{P}_n(x) = \pm D\begin{pmatrix} \varphi_0 & \varphi_1 & \varphi_2 & \cdots & \varphi_{n-m'} & \varphi_{n-m'+1} & \varphi_{n-m'+2} & \cdots & \varphi_n \\ x & z_1 & z_2 & \cdots & z_{n-m'} & x_1 & x_2 & \cdots & x_{m'} \end{pmatrix}. \tag{2.14}$$

All zeros of this polynomial are odd according to Lemma 2.1. Since all zeros $x_j > a$, and y_k of the polynomial P_n lie outside the interval (a, c) containing the zeros z_i of $\check{P}_n$ and the polynomial P_n changes its sign only at the points $x_j \in (a, b)$, by choosing the proper sign ("+" or "–") of the right-hand side of relation (2.14), we can guarantee that the polynomial $\check{P}_n$ has the same sign as P_n in a sufficiently small neighborhood of each zero $x_j > a$ for $x \neq x_j$. However, in this case, we also have $\operatorname{sgn} \check{P}_n(x) = \operatorname{sgn} P_n(x)$ at all points $x \notin (a, c)$, $x \neq y_k$, $k = 1, \ldots, m''$, and $\check{P}_n(y_k) \neq 0$, $P_n(y_k) = 0$ at the points y_k. This means that, for sufficiently small $\lambda > 0$, the graph of the polynomial

$\lambda \check{P}_n$ crosses the graph of the polynomial P_n at least twice in a sufficiently small neighborhood of each point y_k. Hence, the polynomial $P_n - \lambda \check{P}_n \not\equiv 0$ vanishes at all points x_j, $j = 1, \ldots, m'$, and, in addition, twice in the neighborhood of each point y_k, $k = 1, \ldots, m''$, i.e., has at least $m' + 2m''$ zeros. This yields inequality (2.13).

■

Theorem 2.3. *Assume that* $\{\varphi_k\}_{k=0}^{n}$ *is a Chebyshev system of real functions on an interval* $\langle a, b\rangle$. *Then, for any two mutually disjoint collections of points from the interval* $\langle a, b\rangle$, *i.e.,* $\{x_i\}_{i=1}^{k} \subset \langle a, b\rangle$ *and* $\{y_j\}_{j=1}^{l} \subset (a, b)$, *whose number satisfies the equality*

$$n - (k + 2l) = 2r, \qquad r = 0, 1, 2, \ldots,$$

there exists a T-polynomial P_n *such that the sets of its odd and even zeros on the interval* $\langle a, b\rangle$ *are exhausted by the points* x_i, $i = 1, \ldots, k$, *and* y_j, $j = 1, \ldots, l$, *respectively. If* $\langle a, b\rangle = [a, b]$, *then this assertion remains true in the case where* $n - (k + 2l) = 2r + 1$, $r = 0, 1, \ldots$, *i.e., in all cases where* $k + 2l \leq n$.

Proof. 1. First, we consider the case $k + 2l = n$ and choose a positive number ε so small that the neighborhoods $(y_j - \varepsilon, y_j + \varepsilon)$ of different points y_j do not intersect and contain neither the ends a and b of the interval $\langle a, b\rangle$ nor the points x_i, $i = 1, \ldots, k$. For fixed j, by $y_j^{\nu -}$ and $y_j^{\nu +}$ we denote sequences of points convergent to y_j and contained in the intervals $(y_j - \varepsilon, y_j)$ and $(y_j, y_j + \varepsilon)$, respectively. Consider a sequence of generalized polynomials $\{P_n^{\nu}\}_{\nu=1}^{\infty}$ of the form

$$P_n^{\nu}(x) = \sum_{k=0}^{n} c_k^{\nu}\varphi_k(x) = \gamma_\nu D\begin{pmatrix} \varphi_0 & \varphi_1 & \varphi_2 & \cdots & \varphi_k & \varphi_{k+1} & \varphi_{k+2} & \cdots & \varphi_{n-1} & \varphi_n \\ x & x_1 & x_2 & \cdots & x_k & y_1^{\nu -} & y_1^{\nu +} & \cdots & y_l^{\nu -} & y_l^{\nu +} \end{pmatrix},$$

where the numbers $\gamma_\nu > 0$ are chosen to guarantee the validity of the equalities

$$\sum_{k=0}^{n} \left|c_k^{\nu}\right| = 1 \tag{2.15}$$

for all $\nu = 1, 2, \ldots$. By virtue of these equalities, one can find a subsequence $\{P_n^{\nu_i}\}_{i=1}^{\infty}$ of the sequence $\{P_n^{\nu}\}$ pointwise convergent to a nontrivial polynomial P_n satisfying the conditions of Theorem 2.3.

We denote this polynomial (or one of these polynomials) as follows:

$$P_n^*(x) := P_n^*(x; x_1, x_2, \ldots, x_k; y_1, y_2, \ldots, y_l) = \lim_{i \to \infty} P_n^{\nu_i}(x), \tag{2.16}$$

$$k + 2l = n, \quad x \in \langle a, b \rangle.$$

2. If $n - (k + 2l) = 2r$, where r is a positive integer, then we additionally fix any two sets of r different points $y_\nu^0 \in (a, \alpha)$, $\nu = 1, \ldots, r$, and $y_\nu^* \in (\beta, b)$, $\nu = 1, \ldots, r$, where $\alpha \in (a, b)$ and $\beta \in (a, b)$ are chosen so that the interval $(a, \alpha) \cup (\beta, b)$ does not contain the zeros x_i and y_j, $i = 1, \ldots, k$, $j = 1, \ldots, l$. Further, we construct polynomials

$$\begin{aligned} P_n^{(1)}(x) &= P_n^*(x; x_1, \ldots, x_k; y_1, \ldots, y_l; y_1^0, \ldots, y_r^0), \\ P_n^{(2)}(x) &= P_n^*(x; x_1, \ldots, x_k; y_1, \ldots, y_l; y_1^*, \ldots, y_r^*), \end{aligned} \tag{2.17}$$

and easily show that $\operatorname{sgn} P_n^{(1)}(x) = \operatorname{sgn} P_n^{(2)}(x)$ for almost all x and the polynomial

$$P_n = P_n^{(1)} + P_n^{(2)} \tag{2.18}$$

satisfies the conditions of the theorem.

3. If $\langle a, b \rangle = [a, b]$ and $n - (k + 2l) = 2r + 1$, where r is a nonnegative integer, then we again represent the polynomial P_n in the form (2.18). However, in constructing the polynomial $P_n^{(1)}$, we supplement the collection of zeros x_i, y_j, and y_ν^0 in relation (2.17) with the zero $x_0 = a$ and, in constructing the polynomial $P_n^{(2)}$, the collection of zeros is supplemented with the zero $x_{k+1} = b$.

■

Corollary 2.2. *For any Chebyshev system* $\{\varphi_k\}_{k=0}^n$ *on a segment* $[a, b]$, *one can indicate a polynomial (with respect to this system) positive for all* $x \in [a, b]$ *and, hence, a polynomial* π_n *such that*

$$1 \le \pi_n(x) \le M, \quad M = \text{const}, \tag{2.19}$$

for all $x \in [a, b]$.

Note that

(i) it is easy to see that the Chebyshev system of functions

$$\varphi_0(x) = x(x-2) \quad \text{and} \quad \varphi_1(x) = (x-2)(x-3), \quad x \in (0,2]$$

possesses the following property: any polynomial of the form $c_0\varphi_0 + c_1\varphi_1$ with $c_0^2 + c_1^2 > 0$ has exactly one zero on $(0,2]$; therefore, the statement of Theorem 2.3 for $n-(k+2l) = 2r+1$, $r = 0, 1, 2, \ldots$, formulated for the segment cannot be generalized to the interval $\langle a, b\rangle \neq [a, b]$;

(ii) even if $k + 2l = n$ with $l > 0$, the polynomial P_n satisfying the conditions of Theorem 2.3 is, in general, not unique to within a constant factor.

Indeed, consider the following system of functions on $[-1, 1]$:

$$\varphi_0(x) \equiv 1, \quad \varphi_1(x) = \begin{cases} 2|x| & \text{for } x \in [-1,0], \\ x^2 & \text{for } x \in [0,1], \end{cases} \quad \text{and} \quad \varphi_2(x) = \varphi_1(-x).$$

It is easy to see that this is a Chebyshev system of functions. At the same time, for $y_1 = 0$ (i.e., $k = 0$, $l = 1$, and $n = 2$), the zero $y_1 = 0$ is even for the polynomials φ_1 and φ_2 but $\varphi_2(x) \not\equiv c\,\varphi_1(x)$ for any $c = \text{const}$.

By using Theorem 2.3, one can easily show that Theorems 1.2 and 1.2′ admit the following generalization:

Theorem 2.4. *Assume that a Chebyshev system of functions* $\{\varphi_k\}_{k=0}^n$ *and a continuous function* f *are given on a segment* $[a, b]$. *Then, in order that* P_n^* *be a polynomial of the best approximation for the function* f *as compared with all other polynomials with respect to the system* $\{\varphi_k\}_{k=0}^n$, *it is necessary and sufficient that there exist at least one system of* $n+2$ *points* x_j *(alternation)*

$$a \le x_1 < x_2 < \ldots < x_{n+2} \le b$$

such that the difference

$$f(x) - P_n^*(x) =: r_n(x)$$

(i) alternatively changes its sign at the points x_i;

(ii) attains its maximum absolute value on $[a, b]$ *at the points* x_i, *i.e.,*

$$r_n(x_1) = -r_n(x_2) = \ldots = (-1)^{n+1} r_n(x_{n+2}) = \pm \|r_n\|_{[a,b]}.$$

The proof of this theorem, in fact, repeats the proof of the Chebyshev theorem with the following changes:

(i) after inequality (1.10), the auxiliary polynomial Q_n should be constructed in the form

$$Q_n = P_n^* + \frac{h}{2M}\pi_n(x)$$

(instead of $Q_n = P_n^* + h/2$; see (2.19)); this yields

$$-\|r_n\| + \frac{h}{2M} \le -\|r_n\| + \frac{h}{2} \le -\|r_n\| + h - \frac{h}{2M}\pi_n(x)$$

$$\le f(x) - Q_n(x) \le \|r_n\| - \frac{h}{2M}, \qquad x \in [a, b],$$

i.e.,

$$\|f - Q_n\| \le \|r_n\| - \frac{h}{2M} \le \left\|f - P_n^*\right\| - \frac{h}{2M};$$

(ii) the auxiliary polynomial P_m in (1.13) should be replaced by a polynomial

$$P_m(x) = \delta \check{P}_n(x),$$

where $\check{P}_n$ is a polynomial for which the points $z_1, z_2, \ldots, z_m$ are the odd zeros and exhaust the set of all its zeros on $[a, b]$ and the number δ is chosen in exactly the same way as above.

Note that the Chebyshev Theorem 1.2″ can be generalized in a similar way to the case of approximation of periodic continuous functions by polynomials with respect to periodic Chebyshev systems. We leave the proof of this assertion to the reader.

2.4. Uniqueness of the best approximation by generalized polynomials

Theorem 2.5. *For any Chebyshev system of functions* $\{\varphi_k\}_{k=0}^n$ *on a segment* $[a, b]$ *and any continuous function* f *on* $[a, b]$, *there exists a unique generalized polynomial* P_n^* *of the best approximation to this function.*

This theorem is a consequence of the following (very general) Haar theorem specifying the system of continuous functions $\{\varphi_k\}_{k=0}^n$ required for the existence, for any continuous function f, of a unique polynomial P_n^* of its best approximation of the form

$$P_n^*(x) = \sum_{k=0}^n c_k^* \varphi_k(x) .$$

Theorem 2.6 [Haar (1918); Kolmogorov (1948)]. *Assume that a system* $\{\varphi_k\}_{k=0}^n$ *of real or complex continuous functions is given on a closed bounded set* $\mathfrak{M} \subset R^m$ *containing more than* $n + 1$ *points. Then, in order that a unique polynomial* P_n^* *of the best uniform approximation of the form*

$$P_n^*(x) = \sum_{k=0}^n c_k^* \varphi_k(x) \tag{2.20}$$

exist for an arbitrary continuous function f *on* $\mathfrak{M}$, *it is necessary and sufficient that the system of functions* $\{\varphi_k\}_{k=0}^n$ *be a Chebyshev system on* $\mathfrak{M}$.

Proof.[1] *Necessity.* Without loss of generality, we can assume that the system $\{\varphi_k\}_{k=0}^n$ is linearly independent. Suppose that, for any continuous function f, there exists a unique polynomial of the best approximation (2.20) but, contrary to the assertion of the theorem, $\{\varphi_k\}_{k=0}^n$ is not a Chebyshev system. This means that there exists a polynomial

$$P_n^0(x) = \sum_{k=0}^n a_k^0 \varphi_k(x), \qquad x \in \mathfrak{M}, \tag{2.21}$$

not identically equal to zero with at least $n + 1$ different zeros $x_0, x_1, \ldots, x_n$ in the set $\mathfrak{M}$:

$$P_n^0(x_0) = a_0^0 \varphi_0(x_0) + a_1^0 \varphi_1(x_0) + \ldots + a_n^0 \varphi_n(x_0) = 0,$$

$$P_n^0(x_1) = a_0^0 \varphi_0(x_1) + a_1^0 \varphi_1(x_1) + \ldots + a_n^0 \varphi_n(x_1) = 0,$$

$$\ldots\ldots\ldots\ldots\ldots\ldots\ldots\ldots\ldots\ldots\ldots\ldots$$

$$P_n^0(x_n) = a_0^0 \varphi_0(x_n) + a_1^0 \varphi_1(x_n) + \ldots + a_n^0 \varphi_n(x_n) = 0.$$

In this case, we have

$$D\begin{pmatrix}\varphi_0 & \varphi_1 & \dots & \varphi_n \\ x_0 & x_1 & \dots & x_n\end{pmatrix} = \begin{vmatrix}\varphi_0(x_0) & \varphi_1(x_0) & \dots & \varphi_n(x_0) \\ \varphi_0(x_1) & \varphi_1(x_1) & \dots & \varphi_n(x_1) \\ \dots & \dots & \dots & \dots \\ \varphi_0(x_n) & \varphi_1(x_n) & \dots & \varphi_n(x_n)\end{vmatrix} = 0. \tag{2.22}$$

As follows from relation (2.22), the rows of the determinant are not linearly independent and, hence, there exists a system of nonzero, generally speaking, complex numbers $b_0, b_1, \dots, b_n$ such that

$$b_0\varphi_k(x_0) + b_1\varphi_k(x_1) + \dots + b_n\varphi_k(x_n) = 0, \quad k = 0, \dots, n.$$

These equalities imply that an arbitrary generalized polynomial of the form

$$P_n(x) = \sum_{k=0}^{n} c_k\varphi_k(x) \tag{2.20$'$}$$

satisfies the equality

$$\sum_{j=0}^{n} b_j P_n(x_j) = \sum_{k=0}^{n} c_k \sum_{j=0}^{n} b_j \varphi_k(x_j) = 0. \tag{2.23}$$

Let us show that, in this case, one can find a function f continuous on $\mathfrak{M}$ for which the polynomial of its best approximation is not unique.

Denote by g a function continuous on $\mathfrak{M}$ and satisfying the conditions

$$\begin{aligned} &|g(x)| \le 1, \quad x \in \mathfrak{M}, \\ &g(x_j) = \operatorname{sgn} \bar{b}_j, \quad j = 0, \dots, n, \end{aligned} \tag{2.24}$$

where we set $\operatorname{sgn} c = 0$ for $c = 0$ and

$$\operatorname{sgn} c = \frac{c}{|c|} = e^{i \arg c}$$

for $c \neq 0$.[†]

[†] To construct a function of this sort, we denote the minimum distance between the points x_j by 2ε and set

$$g(x) = \operatorname{sgn}\bar{b}_j\left(1 - \frac{|x - x_j|}{\varepsilon}\right), \quad x \in U(x_j, \varepsilon), \quad |x - x_j| = \rho(x, x_j),$$

in the ε-neighborhoods $U(x_j, \varepsilon)$ of the points x_j and $g(x) = 0$ outside these neighborhoods.

By using the indicated function g and the polynomial P_n^0, we construct the function

$$f(x) = g(x)\left[1 - \lambda\left|P_n^0(x)\right|\right], \qquad x \in \mathfrak{M}, \tag{2.25}$$

where λ is a positive number chosen to guarantee the validity of the equality

$$\max_{x \in \mathfrak{M}} \lambda\left|P_n^0(x)\right| = 1.$$

By virtue of relations (2.24) and (2.25), we obtain $f(x_j) = \operatorname{sgn} \bar{b}_j$ and $|f(x)| \le 1$, $x \in \mathfrak{M}$. Hence, in order that the inequality

$$\max_{x \in \mathfrak{M}} |f(x) - P_n(x)| < 1$$

(this inequality implies, in particular, that there is $j \in 0, \ldots, n$ such that $\bar{b}_j P_n(x_j) \neq 0$) be true for a polynomial $P_n(x)$, it is necessary that the conditions

$$\left|\arg P_n(x_j) - \arg f(x_j)\right| = \left|\arg P_n(x_j) - \arg \bar{b}_j\right| < \pi/2$$

be satisfied for all $j \in 0, \ldots, n$ such that $\bar{b}_j P_n(x_j) \neq 0$. (Indeed, otherwise, we get

$$\left|P_n(x_j) - f(x_j)\right| = \left|P_n(x_j) - \operatorname{sgn} \bar{b}_j\right| \ge 1.)$$

However, by virtue of (2.23), this is impossible because the inequality

$$\left|\arg P_n(x_j) - \arg f(x_j)\right| < \pi/2$$

implies that

$$\left|\arg P_n(x_j) b_j\right| = \left|\arg P_n(x_j) - \arg \bar{b}_j\right| < \pi/2$$

for any j and, consequently,

$$\operatorname{Re} \sum_{j=0}^{n} b_j P_n(x_j) > 0.$$

Therefore,

$$E_n := \inf_{P_n} \| f - P_n \|_{\mathfrak{M}} = 1.$$

On the other hand, for any $\varepsilon \in (0, 1]$ and $x \in \mathfrak{M}$, we find

$$\begin{aligned} \left| f(x) - \varepsilon\lambda P_n^0(x) \right| &\le |f(x)| + \varepsilon\lambda \left| P_n^0(x) \right| \\ &= |g(x)| \left[1 - \lambda \left| P_n^0(x) \right| \right] + \varepsilon\lambda \left| P_n^0(x) \right| \\ &\le 1 - \lambda \left| P_n^0(x) \right| + \varepsilon\lambda \left| P_n^0(x) \right| \\ &= 1 - \lambda \left| P_n^0(x) \right| (1 - \varepsilon) \le 1. \end{aligned}$$

Thus, contrary to the assumption of the theorem, $\varepsilon\lambda P_n^0$ is a polynomial of the least deviation from the function f for any $\varepsilon \in (0, 1]$.

Sufficiency. Assume that $\{\varphi_k\}_{k=0}^n$ is a Chebyshev system and P_n^* is the polynomial of the best approximation for the function f on $\mathfrak{M}$. Let us show that, in this case, one can find at least $n + 2$ different points $x_1, x_2, \ldots, x_{n+1} \in \mathfrak{M}$ of validity of the equality

$$\left| f(x_k) - P_n^*(x_k) \right| = \left\| f - P_n^* \right\|, \quad k = 1, \ldots, n + 2. \tag{2.26}$$

This follows from the fact that if the number of these points is $n_1 < n + 2$, then we can apply Theorem 2.2 and construct a polynomial π_n such that

$$\pi_n(x_k) = f(x_k) - P_n^*(x_k), \quad k = 1, \ldots, n_1.$$

Thus, one can easily show that, for sufficiently small $\delta > 0$, the polynomial $P_n^* + \delta\pi_n$ deviates from the function f less than P_n^*, i.e., we arrive at a contradiction.

Contrary to the assumption, we now assume that, parallel with P_n^*, there exists one more polynomial Q_n^* of the best approximation to the function f:

$$\left\| f - P_n^* \right\| = \left\| f - Q_n^* \right\| = \min_{P_n} \| f - P_n \| = E_n.$$

In this case, we get

$$E_n = \left\| f - \frac{P_n^* + Q_n^*}{2} \right\| \leq \frac{1}{2} \left\| f - P_n^* \right\| + \frac{1}{2} \left\| f - Q_n^* \right\| = E_n,$$

i.e., $\frac{1}{2}\left[P_n^* + Q_n^*\right]$ is also a polynomial of the best approximation for the function f and, moreover, as already proved, there exist at least $n+2$ points $x_j \in \mathfrak{M}$ of validity of the equality

$$E_n = \left| f(x_j) - \frac{P_n^*(x_j) + Q_n^*(x_j)}{2} \right| = \frac{1}{2}\left[\left| f(x_j) - P_n^*(x_j)\right| + \left| f(x_j) - Q_n^*(x_j)\right|\right].$$

However, the last equality is possible only in the case where, first,

$$\left| f(x_j) - P_n^*(x_j)\right| = \left| f(x_j) - Q_n^*(x_j)\right| = E_n$$

and, second, the arguments of the numbers $f(x_j) - P_n^*(x_j)$ and $f(x_j) - Q_n^*(x_j)$ are also equal, i.e., $f(x_j) - P_n^*(x_j) = f(x_j) - Q_n^*(x_j)$. Hence,

$$P_n^*(x_j) = Q_n^*(x_j), \quad j = 1, 2, \ldots, n+2.$$

In view of the fact that $\{\varphi_k\}_{k=0}^n$ is a Chebyshev system, this situation is possible only for $Q_n^*(x) \equiv P_n^*(x)$. The proof of the Haar theorem is completed.

■

Example 2.1. Let $\varphi_0(x, y) = 1$, $\varphi_1(x, y) = x$, $\varphi_2(x, y) = y$ and let $\mathfrak{M} = \{(x, y)\colon x, y \in [-1, 1]\}$. Then the set $\mathfrak{M}$ contains continuous functions for which a polynomial of the best approximation is not unique.

Remark 2.1. The uniqueness theorem and the necessity part of Theorem 2.4 (Chebyshev) are not true if the system $\{\varphi_k\}_{k=0}^n$ is defined not on a segment $[a, b]$ but on a half interval $(a, b]$ or $[a, b)$, or on an interval (a, b).

Thus, in particular, the family of functions 1, $\cos t$, $\sin t, \ldots, \cos nt$, $\sin nt$ is a Chebyshev system in the half interval $(-\pi, \pi]$. Nevertheless, one can easily see that any trigonometric polynomial whose plot on $(-\pi, \pi]$ lies inside the strip bounded by the lines $y = t - \pi$ and $y = t + \pi$ is the polynomial of the best approximation for a function $f(t) = t$ continuous on $(-\pi, \pi]$.

At the same time, if a continuous function f is 2π-periodic, i.e., the points $-\pi$ and π can be identified, then the entire proof of Theorem 2.6 remains valid, hence, the polynomial of the best approximation is unique.

We now present a theorem which shows that if a polynomial of the best approximation is unique, then it continuously depends on the approximated function.

Theorem 2.7 [S. Nikol'skii (1947)]. *If, for any continuous function f on a compact set $\mathfrak{M}$, there exists a unique polynomial of its best approximation $P_n^*(f;\cdot)$ with respect to a given system $\{\varphi_k\}_{k=0}^n$ of linearly independent continuous functions, then this polynomial continuously depends on the approximated function f in the sense that, for any $\varepsilon > 0$, one can find $\delta = \delta(f;\varepsilon) > 0$ such that*

$$\left\| P_n^*(f;\cdot) - P_n^*(f_1;\cdot) \right\|_{C(\mathfrak{M})} < \varepsilon$$

whenever $\| f - f_1 \|_{C(\mathfrak{M})} < \delta$.

Proof.[2] By $\{f_k\}_{k=1}^\infty$ we denote a sequence of functions continuous on $\mathfrak{M}$. Assume that this sequence uniformly converges to f. Note that, for any $k = 1, 2, \ldots,$ we can write

$$\left\| P_n^*(f_k;\cdot) \right\| - \| f_k \| \leq \left\| P_n^*(f_k;\cdot) - f_k \right\| \leq \| f_k \|.$$

Therefore,

$$\left\| P_n^*(f_k;\cdot) \right\| \leq 2\,\| f_k \| \leq A = \text{const}, \qquad k = 1, 2, \ldots .$$

As in the proof of the Borel theorem (Theorem 1.1), one can select a subsequence $\left\{ P_n^*\left(f_{k_j};\cdot\right) \right\}_{j=1}^\infty$ of the sequence $\left\{ P_n^*(f_k;\cdot) \right\}_{k=1}^\infty$ uniformly convergent to a polynomial

$$\check{P}_n(x) = \lim_{j \to \infty} P_n^*\left(f_{k_j};x\right). \tag{2.27}$$

The inequality

$$\left\| f_{k_j} - P_n^*\left(f_{k_j};\cdot\right) \right\| \leq \left\| f_{k_j} - P_n \right\|$$

holds for any polynomial

$$P_n(x) = \sum_{k=0}^{n} c_k \varphi_k(x)$$

by the definition of polynomials of the best approximation. As a result of the limit transition in this inequality as $j \to \infty$, we obtain

$$\| f - \check{P}_n \| \leq \| f - P_n \|,$$

i.e., we see that $\check{P}_n(x)$ is a polynomial of the best approximation for the function $f(x)$. By the assumption, the polynomial of the best approximation is unique. Hence, $\check{P}_n(x) = P_n^*(f; x)$, $x \in \mathfrak{M}$, and, by virtue of (2.27),

$$\lim_{j \to \infty} P_n^*\left(f_{k_j}; x\right) = P_n^*(f; x) \tag{2.28}$$

uniformly in $x \in \mathfrak{M}$. Since the sequence $\left\{P_n^*(f_k; \cdot)\right\}$ is a compact set and the right-hand side of (2.28) is independent of the subsequence $\{k_j\}$, we conclude that

$$\lim_{k \to \infty} P_n^*(f_k; x) = P_n^*(f; x)$$

uniformly in $x \in \mathfrak{M}$. The theorem is proved.

■

The reader can easily check the validity of Theorem 2.7 and its proof for the approximation of functions in Banach spaces (see [S. Nikol'skii (1947), p. 47]).

2.5. De la Vallée Poussin theorem

As a rule, the Chebyshev theorems enable one only to check whether a given polynomial is the polynomial of the best approximation for a given continuous function f. At the same time, the problem of finding these polynomials and estimation of the value of the best approximation $E_n(f)$ proves to be extremely difficult. In this connection, it seems reasonable to analyze the possibility of approximate construction of these polynomials and finding the upper and lower bounds of the quantity $E_n(f)$. The corresponding upper estimates are studied in Chapters 4 and 6. For the lower bounds, we now present the following two theorems:

Theorem 2.8 [de la Vallée Poussin (1919)]. *If the difference $f - P_n$ between a function f continuous on $[a, b]$ and a polynomial P_n with respect to a Chebyshev system $\{\varphi_k\}_{k=0}^n$ takes alternating values at points x_j of an ordered sequence of $n+2$ points $a \le x_1 < x_2 < \ldots < x_{n+2} \le b$, i.e.,*

$$\operatorname{sgn}\{f(x_j) - P_n(x_j)\} = -\operatorname{sgn}\{f(x_{j+1}) - P_n(x_{j+1})\},$$

then

$$E_n(f) \ge \min_{1 \le j \le n+2} \left| f(x_j) - P_n(x_j) \right|. \tag{2.29}$$

The proof is similar to the proof of sufficiency in the Chebyshev theorem (Theorem 1.2).

Indeed, by contradiction, we can assume that

$$E_n(f) < \min_{1 \le j \le n+2} \left| f(x_j) - P_n(x_j) \right|.$$

Let P_n^* be the polynomial of degree at most n of the best approximation to the function f. Then

$$\left| f(x_j) - P_n^*(x_j) \right| < \left| f(x_j) - P_n(x_j) \right| \tag{2.30}$$

at all points x_j. Hence, the difference

$$P_n^*(x_j) - P_n(x_j) = [f(x_j) - P_n(x_j)] - [f(x_j) - P_n^*(x_j)]$$

has the same sign as $f(x_j) - P_n(x_j)$ at all points x_j, $j = 1, \ldots, n+2$, i.e., its sign changes on the analyzed segment at least $n+1$ times.

Therefore, the polynomial $P_n^* - P_n$ has at least $n+1$ zeros on the segment $[a, b]$ and, consequently, $P_n^*(x) - P_n(x) \equiv 0$, contrary to inequality (2.30).

For the case of periodic functions approximated by periodic polynomials, the theorem is formulated and proved similarly.

Theorem 2.8′ [de la Vallée Poussin (1919)]. *If the difference $f - T_n$ between a 2π-periodic continuous function $f(t)$ and a trigonometric polynomial T_n of degree n takes alternating values at points t_j of an ordered sequence of $2n+2$ points ($t_1 < t_2 < \ldots < t_{2n+2} < t_1 + 2\pi$), then*

$$E_n(f) \geq \min_{1 \leq j \leq 2n+2} \left| f(t_j) - T_n(t_j) \right|. \qquad (2.29')$$

2.6. Snakes

We present some important results related to the Chebyshev Theorem 1.2.

Definition 2.5. *For any two continuous functions g_0 and g_1 defined on a segment $[a, b]$ and such that $g_1(x) < g_0(x)$, $x \in [a, b]$, and a Chebyshev system of functions $\{\varphi_j\}_{j=0}^n$ given on this segment, a polynomial $\overline{P}_n$ ($\underline{P}_n$) of the form*

$$\overline{P}_n(x) = \sum_{j=0}^{n} \overline{c}_j \varphi_j(x) \quad \left(\sum_{j=0}^{n} \underline{c}_j \varphi_j(x) \right),$$

is called an upper (lower) snake if it exists and satisfies the conditions:

(a) the inequalities $g_1(x) \leq \overline{P}_n(x) \leq g_0(x)$ ($g_1(x) \leq \underline{P}_n(x) \leq g_0(x)$) hold for all $x \in [a, b]$;

(b) there exists at least one sequence of $n+1$ points $\{\overline{x}_j\}_{j=1}^{n+1}$ $\left(\{\underline{x}_j\}_{j=1}^{n+1}\right)$:

$$a \leq \overline{x}_1 < \overline{x}_2 < \overline{x}_3 < \ldots < \overline{x}_{n+1} \leq b$$

such that the following conditions are satisfied:

$$\overline{P}_n(\overline{x}_{2i-1}) = g_0(\overline{x}_{2i-1}), \quad i = 1, 2, \ldots,$$

$$\overline{P}_n(\overline{x}_{2i}) = g_1(\overline{x}_{2i}), \quad i = 1, 2, \ldots,$$

$$\left(\underline{P}_n(\underline{x}_{2i-1}) = g_1(\underline{x}_{2i-1}) \quad \text{and} \quad \underline{P}_n(\underline{x}_{2i}) = g_0(\underline{x}_{2i}) \right).$$

Theorem 2.9 (on snakes [Karlin (1963); Karlin and Studden (1966)]).[3] *Let g_0 and g_1 be two continuous functions given on a segment $[a, b]$ and such that $g_1(x) < g_0(x)$, $x \in [a, b]$, and let $\{\varphi_j\}_{j=0}^n$ be a Chebyshev system of functions on this segment. If there exists at least one polynomial*

$$\check{P}_n(x) = \sum_{j=0}^{n} c_j \varphi_j(x), \quad x \in [a, b],$$

such that $g_1(x) < \check{P}_n(x) < g_0(x)$ *for all* $x \in [a, b]$, *then the functions* g_ν, $\nu = 0, 1$, *and the T-system* $\{\varphi_j\}_{j=0}^{n}$ *possess unique upper and lower snakes. The points of contact* $\{\bar{x}_j\}$ *and* $\{\underline{x}_j\}$ *of these snakes are alternating.*

Proof. I. First, we prove the theorem under assumption that $\check{P}_n(x) \equiv 0$ and, hence, $g_1(x) < 0$ and $g_0(x) > 0$, $x \in [a, b]$.

It suffices to prove the existence of a snake only for the upper snakes $\overline{P}_n$ because if $\overline{P}_n$ is an upper snake for the functions $-g_1$ and $-g_0$, then $-\overline{P}_n$ is a lower snake for the functions g_0 and g_1. For the sake of simplicity, we set $a = 0$ and $b = 1$. Consider the set $\mathcal{P}_n$ of all possible *T*-polynomials P_n satisfying the conditions:

(i) each polynomial $P_n \in \mathcal{P}_n$ has n zeros x_j in the interval $(a, b) = (0, 1)$;

(ii) each polynomial $P_n \in \mathcal{P}_n$ satisfies the inequalities

$$c \le \max_{x \in [x_j, x_{j+1}]} |P_n(x)| \le C, \quad j = 0, 1, 2, \dots, n, \tag{2.31}$$

where $x_0 := 0$, $x_{n+1} := 1$, and c and C are positive constants specified in what follows;

(iii) $P_n(0) > 0$ for all $P_n \in \mathcal{P}_n$.

By P_n^0 we denote a polynomial given by relation (2.12) for which $x_j = x_j(P_n^0) = \frac{j}{n+1}$, $j = 1, 2, \dots, n$. After this, we determine the constants c and C in inequality (2.31) as follows:

$$c_0 := \min_{0 \le j \le n+1} \max_{x \in [j/(n+1), (j+1)/(n+1)]} |P_n^0(x)|, \qquad C_0 = \| P_n^0 \|,$$

$$c_0' = \min\{ \min_{x \in [0,1]} g_0(x), \min_{x \in [0,1]} |g_1(x)| \},$$

$$C_0' = \max\{ \max_{x \in [0,1]} g_0(x), \max_{x \in [0,1]} |g_1(x)| \},$$

$$c = \min\left\{c_0, \frac{c_0'}{2}\right\}, \qquad C = \max\{c_0, 2c_0'\}.$$

In this case, the class $\mathcal{P}_n$ is nonempty. Moreover, if a snake exists, then, first, this snake and the polynomials P_n, which are almost snakes, belong to the class $\mathcal{P}_n$ and, second, in view of the linear independence of the system of functions $\{\varphi_j\}_{j=0}^n$ and the second inequality in (2.31), the coefficients of any polynomial $P_n \in \mathcal{P}_n$ are bounded in modulus by the same number M.

The *defect* $\Delta_j = \Delta_j(P_n)$ of a polynomial $P_n \in \mathcal{P}_n$ on a segment $[x_j, x_{j+1}]$, $j = 0, 1, \ldots, n$, $x_j = x_j(P_n)$, is defined as follows:

$$\Delta_j = (-1)^j \max_{x \in [x_j, x_{j+1}]} (-1)^j [P_n(x) - g_\nu(x)], \quad \text{where} \quad \nu := \frac{1 + (-1)^{j+1}}{2},$$

and the *defect* on $[a, b]$ (or simply the *defect*) is introduced as

$$\Delta = \Delta(P_n) = \max_{1 \le j \le n} |\Delta_j|.$$

Clearly, if a polynomial $\overline{P}_n$ is an upper snake, then $\overline{P}_n \in \mathcal{P}_n$ and $\Delta(\overline{P}_n) = 0$, and vice versa. Roughly speaking, the defect should be regarded as a measure of deviation of the polynomial $P_n \in \mathcal{P}_n$ from the snake.

We now prove that the upper snake exists. Indeed, since the coefficients of all polynomials $P_n \in \mathcal{P}_n$ are bounded, by virtue of the first inequality in (2.31), there exists at least one polynomial $P_n^* \in \mathcal{P}_n$ with the smallest defect Δ^*. Let us show that $\Delta^* = 0$ and, hence, P_n^* is a snake.

In fact, suppose, on the contrary, that $\Delta^* > 0$. Then, by Theorem 2.3, we can construct a T-polynomial π_n with the following properties:

1. $\operatorname{sgn} \pi_n(x) = \operatorname{sgn} \Delta_j(P_n^*)$ for all $x \in (x_j, x_{j+1})$ and all $j = 1, \ldots, n$ such that $\Delta_j(P_n^*) \neq 0$;

2. $\|\pi_n\| = \dfrac{1}{2}\Delta^*$.

At the same time, in this case, it is easy to see that $\Delta(P_n^* - \pi_n) < \Delta(P_n^*)$. This contradiction implies the existence of a snake.

II. If $\check{P}_n(x) \not\equiv 0$, then the functions g_0 and g_1 should be replaced by the following functions:

$$\tilde{g}_0(x) := g_0(x) - \check{P}_n(x) > 0, \quad x \in [a, b],$$

and

$$\tilde{g}_1(x) := g_1(x) - \check{P}_n(x) < 0, \quad x \in [a, b].$$

Further, we find the snakes $\tilde{\overline{P}}_n$ and $\tilde{\underline{P}}_n$ for these functions and set $\overline{P}_n = \tilde{\overline{P}}_n + \check{P}_n$ and $\underline{P}_n = \tilde{\underline{P}}_n + \check{P}_n$.

Finally, it remains to note that the uniqueness of a snake and the fact that the zeros of the upper and lower snakes are alternating follow from the evident fact that, otherwise, the difference between the corresponding polynomials has at least $n + 1$ zeros, which is impossible.

■

Supplement to Theorem 2.9. *If, under the conditions of Theorem 2.9, the function* g_0 *is a polynomial with respect to a system* $\{\varphi_j\}_{j=0}^n$, *i.e.,*

$$g_0(x) = \sum_{j=0}^{n} c_j^0 \varphi_j(x), \quad x \in [a, b],$$

then necessarily $\overline{x}_1 = a$ *and, in addition,* $\overline{x}_{n+1} = b$ *if* n *is even.*

If, under the conditions of Theorem 2.9, both functions g_0 *and* g_1 *are polynomials with respect to a system* $\{\varphi_j\}_{j=0}^n$, *then* $\overline{x}_{n+1} = b$ *for all odd* n. *Similar assertions also hold if* g_0 *is replaced by* g_1 *and the points* $\overline{x}_1$, $\overline{x}_{n+1}$, *and* $\underline{x}_{n+1}$ *are replaced by the points* $\underline{x}_1$, $\underline{x}_{n+1}$, *and* $\overline{x}_{n+1}$, *respectively.*

This supplement is a consequence of the following fact: If, e.g., n is even, then every point $\overline{x}_{2j-1}$, $j = 1, 2, \ldots, n/2 + 1$, satisfying the condition $a < \overline{x}_j < b$ is an even zero of the nonzero polynomial $g_0 - \overline{P}_n$ with respect to the Chebyshev system $\{\varphi_j\}_{j=0}^n$. However, if each even zero of the polynomial $g_0 - \overline{P}_n$ is taken into account twice, then, according to Lemma 2.2, this polynomial cannot have more than n zeros on $[a, b]$.

2.7. Representation of positive polynomials

Theorem 2.10 [Karlin (1963)]. *Every strictly positive polynomial* P_n^* *with respect to* *a Chebyshev system of functions* $\{\varphi_j\}_{j=0}^n$ *on a segment* $[a, b]$ *admits a unique representation in the form*

$$P_n(x) = \overline{P}_n(x) + \underline{P}_n(x),$$

where $\overline{P}_n$ *and* $\underline{P}_n$ *are polynomials with respect to a T-system* $\{\varphi_j\}_{j=0}^n$ *with the following properties:*

(a) *the polynomials* $\overline{P}_n$ *and* $\underline{P}_n$ *are nonnegative; each of these polynomials possesses exactly* n *zeros on* $[a, b]$ *counting multiplicity;*

(b) *the zeros of the polynomials* $\overline{P}_n$ *and* $\underline{P}_n$ *are alternating.*

Proof. We set $g_0(x) = P_n(x)$, $x \in [a, b]$, and $g_1(x) \equiv 0$. Let $\overline{P}_n$ be the snake for these functions. By virtue of the supplement to Theorem 2.9, the snake $\overline{P}_n(x)$ satisfies the condition $\overline{P}_n(a) = g_0(a) = P_n(a)$. Therefore, it is easy to see that the polynomial $P_n - \overline{P}_n$ satisfies all conditions imposed on a lower snake and, hence, can be regarded as $\underline{P}_n$. It is clear that the polynomials $\overline{P}_n$ and $\underline{P}_n$ are nonnegative and the condition $\underline{P}_n(a) = 0$ is satisfied. The uniqueness of these polynomials and the fact that the zeros of $\overline{P}_n$ and $\underline{P}_n$ are alternating follow from Theorem 2.9.

■

The following theorem is a corollary of Theorem 2.10:

Theorem 2.11 [Markov (1906); Lukacs (1918)]. *Every algebraic polynomial* P_n *of degree* n *positive on* $[a, b]$ *admits a representation*

$$P_n(x) = A_\nu^2(x) + (b-x)(x-a)B_{\nu-1}^2(x)$$

for any natural $\nu \geq n/2$ *and a representation*

$$P_n(x) = (x-a)C_\nu^2(x) + (b-x)D_\nu^2(x)$$

for any natural $\nu \geq (n-1)/2$; *here,* A_ν, $B_{\nu-1}$, C_ν, *and* D_ν *are polynomials of degrees equal to their subscripts.*

If the zeros of the polynomials A_ν *and* $B_{\nu-1}$ *in the first case and* C_ν *and* D_ν *in the second case are alternating and the polynomials* A_ν *and* $B_{\nu-1}$ (C_ν *and* D_ν, *respectively*) *have exactly* ν *zeros on* $[a, b]$, *then these representations are unique.*

Theorem 2.12 [Fejér (1915)]. *Assume that a trigonometric polynomial of degree n of the form*

$$T_n(t) = \sum_{k=0}^{n} \left(a_k \cos kt + b_k \sin kt\right), \quad a_n^2 + b_n^2 \neq 0, \quad b_1 := 0, \tag{2.32}$$

or, equivalently,

$$T_n(t) = \sum_{k=-n}^{n} c_k e^{ikt}, \quad c_k := \frac{a_k - ib_k}{2}, \tag{2.32$'$}$$

where a_k and b_k are real and $a_k := a_{-k}$ and $b_k := b_{-k}$ for $k < 0$, takes only nonnegative values for all $t \in \mathbb{R}$. Then

(i) this polynomial admits a (generally speaking, nonunique) representation as the squared modulus of another trigonometric polynomial t_n of the same degree n given by the formula

$$t_n(t) = \sum_{k=0}^{n} \gamma_k e^{ikt}, \tag{2.33}$$

i.e.,

$$T_n(t) = |t_n(t)|^2, \quad t \in \mathbb{R}; \tag{2.34}$$

(ii) the coefficients γ_k of the polynomial t_n can be chosen to guarantee that all zeros z_k of a polynomial

$$\Pi_n(z) := \sum_{k=0}^{n} \gamma_k z^k \tag{2.35}$$

of degree n are located in the unit disk: $|z_k| \leq 1$ (with $|z_k| < 1$ for all k if $T_n(t)$ is strictly positive for any $t \in \mathbb{R}$);

(iii) if the polynomial T_n is even, i.e., all $b_k = 0$ in relation (2.32) [or, equivalently, all c_k are real and $c_k = c_{-k}$ in relation (2.32$'$)], then all coefficients γ_k in relation (2.33) are real to within a common factor whose absolute value is equal to one for any representation of T_n in terms of t_n by relation (2.34).

Proof. (i). According to relation (2.32), we have $a_n^2 + b_n^2 \neq 0$. Thus, by virtue of Corollary 1.3, the polynomial T_n has $2n$ zeros ζ_k (real and complex) in the strip $0 \le \operatorname{Re} z < 2\pi$ and can be represented in terms of these zeros as follows:

$$T_n(t) = C\prod_{k=1}^{2n}\sin\frac{t-\zeta_k}{2} = C\prod_{k=1}^{2n}e^{-i(t+\zeta_k)/2}\,\frac{e^{it}-e^{i\zeta_k}}{2i}$$

$$= c_n e^{-int}\prod_{k=1}^{2n}\left(e^{it}-e^{i\zeta_k}\right) = \sum_{k=-n}^{n} c_k z^k$$

$$= c_n z^{-n}\prod_{k=1}^{2n}\left(z-z_k\right) = z^{-n}P_{2n}(z) = \pi_n(z), \tag{2.36}$$

$$z := e^{-it}, \quad z_k := e^{-i\zeta_k}, \quad C := (-4)^n c_n \prod_{k=1}^{2n} e^{i\zeta_k/2} = \text{const},$$

where

$$P_{2n}(z) := \sum_{j=0}^{2n} c_{j-n} z^j = z^n \sum_{j=0}^{2n} c_{j-n} z^{j-n} \quad \text{and} \quad \pi_n(z) := \sum_{k=-n}^{n} c_k z^k. \tag{2.37}$$

Hence, in view of the fact that

$$P_{2n}(0) = c_{-n} = \frac{1}{2}(a_n + ib_n) \neq 0$$

by virtue of relation (2.32′), we conclude that (see also Section 1):

— each zero ζ_k of the polynomial T_n is associated with a zero $z_k = e^{i\zeta_k}$ of the same multiplicity of the function π_n, and vice versa;

— each real zero of the polynomial T_n is associated with a zero of π_n located on the unit circle, and vice versa.

Since the polynomial T_n is nonnegative, all its zeros are even and, therefore, each zero of the function π_n located on the unit circle is also even.

By (2.32′) and (2.37), the function $\pi_n(t)$ for $z = e^{it}$ coincides with $T_n(t)$, $t \in \mathbb{R}$, and hence, takes only real values on the unit circle $|z| = 1$. By the Riemann–Schwarz symmetry principle [or by the direct substitution in (2.37)], we conclude that this function satisfies the identity

$$\pi_n(z) = \overline{\pi_n\left(\bar{z}^{-1}\right)} \tag{2.38}$$

for all $z \neq 0$. Therefore, if we denote by $z_1, z_2, \ldots, z_r$ the zeros from the collection of $2n$ zeros of the polynomial P_{2n} located inside the unit circle $(0 < |z_j| < 1,\ j = 1, 2, \ldots, r)$, then, by virtue of (2.36) and (2.38), we conclude that the points $\bar{z}_1^{-1}, \bar{z}_2^{-1}, \ldots, \bar{z}_r^{-1}$ are also zeros of the polynomial P_{2n}. The remaining $2n - 2r$ zeros of P_{2n} are, clearly, located on the unit circle. Moreover, as shown above, they are even and have the form

$$\xi_\nu = e^{it_\nu}, \quad \nu = 1, 2, \ldots, n-r, \quad t_\nu \in \mathbb{R}. \tag{2.39}$$

This means that the polynomial P_{2n} admits a representation

$$P_{2n}(z) = c_n \prod_{j=1}^{r} \left(z - z_j\right)\left(z - \bar{z}_j^{-1}\right) \prod_{\nu=1}^{n-r} \left(z - e^{it_\nu}\right)^2, \tag{2.40}$$

$$0 < |z_j| < 1, \quad t_\nu \in \mathbb{R}.$$

Therefore, in view of relation (2.36) and the equalities

$$\left|e^{it} - \bar{z}_j^{-1}\right| = \left|\bar{z}_j^{-1}\right| \left|\bar{z}_j - e^{-it}\right| = \left|\bar{z}_j^{-1}\right| \left|e^{it} - z_j\right|,$$

we find

$$T_n(t) = |T_n(t)| = \left|P_{2n}(e^{it})\right|$$

$$= \left| c_n \prod_{j=1}^{r} \left(e^{it} - z_j\right)^2 z_j^{-1} \prod_{\nu=1}^{n-r} \left(e^{it} - e^{it_\nu}\right)^2 \right| = |t_n(t)|^2, \quad t \in \mathbb{R},$$

where, for all $\alpha \in \mathbb{R}$ and $\xi_\nu = e^{it_\nu}$,

$$t_n(t) := A\, e^{i\alpha} \prod_{j=1}^{r} \left(e^{it} - z_j\right) \prod_{\nu=1}^{n-r} \left(e^{it} - \xi_\nu\right) \quad \text{and} \quad A := \left(|c_n| \prod_{j=1}^{r} |z_j|^{-1}\right)^{1/2}. \tag{2.41}$$

(ii). Representing t_n in the form (2.41), in view of relations (2.33) and (2.35), we obtain

$$\Pi_n(z) = A e^{i\alpha} \prod_{j=1}^{r} (z - z_j) \prod_{\nu=1}^{n-r} (z - \xi_\nu),$$

i.e., all zeros of Π_n are indeed located in the unit disk.

(iii). If, in addition, the polynomial T_n is even, then $b_k = 0$ and, hence, $c_{-k} = c_k$ according to (2.32). Therefore, the function π_n from relation (2.37) satisfies the identity

$$\pi_n(z) = \pi_n(1/z)$$

for for all $z \neq 0$. Thus, if z_j is a zero of the function π_n, then z_j^{-1}, $\bar{z}_j^{-1}$, and $\bar{z}_j$ are also its zeros and t_n can be indeed chosen in the form

$$t_n = A e^{i\alpha} \prod_{j=1}^{r/2} \left(e^{it} - z_j\right)\left(e^{it} - \bar{z}_j\right) \prod_{\nu=1}^{(n-r)/2} \left(e^{it} - e^{it_\nu}\right)\left(e^{it} - \bar{e}^{it_\nu}\right)$$

$$= A e^{i\alpha} \prod_{j=1}^{r/2} \left(e^{2it} - 2\operatorname{Re}(z_j)\, e^{it} + |z_j|^2\right) \prod_{\nu=1}^{(n-r)/2} \left(e^{2it} - 2\cos t_\nu\, e^{it} + 1\right),$$

i.e., in the form of the product of the number $A e^{i\alpha}$ by a polynomial of degree n with real coefficients.

■

Theorem 2.13 [Chebyshev (1859); Markov (1906); Dzyadyk (1977)].[4] *Let* $P_l(x) = a_0 x^l + a_1 x^{l-1} + \ldots + a_l$ *be an arbitrary polynomial positive on* $[-1, 1]$. *By using this polynomial* P_l, *we construct the following even trigonometric polynomial of degree* l:

$$T_l(t) = P_l(\cos t) = \sum_{k=0}^{l} \tilde{a}_k \cos kt\,.$$

By virtue of Theorem 2.12, this polynomial admits a representation

$$T_l(t) = \left(|t_l(t)|\right)^2, \qquad t_l(t) = \sum_{k=0}^{l} \gamma_k e^{ikt}, \tag{2.42}$$

where $t_l(t)$ is a polynomial all coefficients of which γ_k are real and such that all zeros of the algebraic polynomial

$$\Pi_l(z) := \sum_{k=0}^{l} \gamma_k z^k \tag{2.43}$$

are located inside the unit circle $|z| < 1$.

Then, for all $\nu \geq -l/2$, the algebraic polynomial $P^*_{l+\nu}$ defined as a linear combination of Chebyshev polynomials $T_j(x) := \cos j \arccos x$, $j = 1, 2, \ldots$,

$$P^*_{l+\nu}(x) = \sum_{k=0}^{l} \gamma_k T_{|\nu+k|}(x) \tag{2.44}$$

is a snake for the pair of functions $-\sqrt{P_l}$ and $+\sqrt{P_l}$.

Proof. For all $\nu \geq -l/2$, we set

$$\Pi_{l+\nu}(z) := z^{\nu}\, \Pi_l(z).$$

In view of relations (2.42) and (2.43), this yields

(a) $\left|\pi_{l+\nu}(e^{it})\right| := \sqrt{P_l(\cos t)}, \quad t \in \mathbb{R};$

(b) $\operatorname{Re} \pi_{l+\nu}(e^{it}) = \sum_{k=0}^{l} \gamma_k \cos(\nu+k)t, \quad t \in \mathbb{R};$

(c) on traversing the circle $|z| = 1$ counterclockwise, the argument of the function $\pi_{l+\nu}$ increases by $2\pi\,(l+\nu)$ per revolution.

By virtue of property (c), there exist $2\,(l+\nu)$ points $\xi_j = e^{i\eta_j}$ at which

$$\operatorname{Arg} \pi_{l+\nu}(\xi_j) = j\pi = \operatorname{Arg} \pi_{l+\nu}(\overline{\xi}_j).$$

For the sake of definiteness, we can set $\xi_0 = 1$ and, hence, $\xi_{\nu+l} = -1$.

Therefore,

$$\operatorname{Re} \pi_{l+\nu}(e^{i\eta_j}) = (-1)^j \sqrt{P_l(\cos \eta_j)}, \quad j = 0, 1, 2, \ldots, l+\nu,$$

and, at the same time,

$$\left|\operatorname{Re}\pi_{l+\nu}(e^{it})\right| = \sqrt{P_l(\cos t)}, \quad t \in \mathbb{R}.$$

Finally, by setting

$$P^*_{l+\nu}(x) = \sum_{k=0}^{l} \gamma_k T_{|\nu+k|}(x),$$

we conclude that

(a′) $\left|P^*_{l+\nu}(x)\right| = \sqrt{P_l(x)}, \quad x \in [-1, 1];$

(b′) there are $l + \nu + 1$ points $x_j \in [-1, 1]$ at which $P^*_{l+\nu}$ takes alternating values equal in the absolute value to $\sqrt{P_l(\eta_j)}$, respectively.

Thus, $P^*_{l+\nu}$ is indeed a snake for the pair of functions $-\sqrt{P_l}$ and $+\sqrt{P_l}$.

■

Remark 2.2. Theorem 2.13 enables one

(i) to construct the polynomials A_ν, B_ν, C_ν and D_ν appearing in the Markov–Lukacs theorem (Theorem 2.11) in the explicit form;

(ii) to find, in the explicit form, the polynomials of "close-to-the-best" approximation for numerous important functions (such as $\arctan x$, $\ln(1 + x)$, $(1 + x)^\alpha$, etc.) encountered as solutions of linear differential equations with polynomial coefficients (see [Dzyadyk (1974)] , etc.).

3. Chebyshev polynomials

The so-called polynomials least deviating from zero or the Chebyshev polynomials can be regarded as a remarkable example of the application of Theorem 2.13 (for $P_l(x) \equiv 1$) and the Chebyshev theorem (Theorem 1.2).

For the function $f(x) = x^n$, by P^*_{n-1} we denote the polynomial of its best approximation of the $(n-1)$th degree. Then, for any polynomial P_{n-1} of degree $n-1$, we can always write

$$\left\| f - P_{n-1}^* \right\|_{[-1,1]} \le \left\| f - P_{n-1} \right\|_{[-1,1]}.$$

This implies that the difference $x^n - P_{n-1}^*(x)$, i.e., an algebraic polynomial of degree n of the form

$$x^n + a_1^* x^{n-1} + \ldots + a_n^*, \tag{3.1}$$

takes the least value (in norm) on the segment $[-1, 1]$ as compared with all other algebraic polynomials of degree n whose leading coefficient is equal to 1. Therefore, this polynomial is called the *nth degree polynomial least deviating from zero on* $[-1, 1]$. It is possible to show that

$$x^n + a_1^* x^{n-1} + \ldots + a_n^* = \frac{1}{2^{n-1}} \cos n \arccos x, \qquad x \in [-1, 1], \tag{3.2}$$

and, hence,

$$E_{n-1}(f) = \left\| f - P_{n-1}^* \right\|_{[-1,1]} = \max_{x \in [-1,1]} \left| x^n + a_1^* x^{n-1} + \ldots + a_n^* \right| = \frac{1}{2^{n-1}}. \tag{3.2′}$$

Indeed, since

$$\cos nt = \frac{e^{int} + e^{-int}}{2} = \frac{\left(\cos t + \sqrt{\cos^2 t - 1}\right)^n + \left(\cos t - \sqrt{\cos^2 t - 1}\right)^n}{2},$$

by setting $\cos t = x$, $t = \arccos x$, $x \in [-1, 1]$, we find

$$\frac{1}{2^{n-1}} \cos n \arccos x = \frac{1}{2^n} \left[\left(x + \sqrt{x^2 - 1}\right)^n + \left(x - \sqrt{x^2 - 1}\right)^n \right]. \tag{3.3}$$

Expanding the binomials, we arrive at a polynomial of degree n with coefficient at the leading term equal to 1 because

$$\lim_{x \to \infty} \frac{\cos n \arccos x}{2^{n-1} x^n} = \lim_{x \to \infty} \frac{1}{2^n} \left[\left(1 + \sqrt{1 - 1/x^2}\right)^n + \left(1 - \sqrt{1 - 1/x^2}\right)^n \right] = 1. \tag{3.4}$$

By virtue of (3.2), for all $x \in [-1, 1]$, this polynomial takes absolute values not greater than $1/2^{n-1}$ and, at the same time, at the $(n-1) + 2 = n + 1$ points

$$x_0 = \cos 0, \quad x_1 = \frac{\pi}{n}, \ \dots, \ x_n = \cos \pi,$$

the polynomial alternately takes the values $\pm 1/2^{n-1}$. This means that, among all possible algebraic polynomials of degree $n-1$, the function x^n is indeed best approximated by a polynomial P_{n-1}^* such that

$$x^n - P_{n-1}^*(x) = \frac{1}{2^{n-1}} \cos n \arccos x.$$

According to Theorem 2.5, this polynomial is unique.

Definition 3.1. *A polynomial of the degree* n

$$\cos n \arccos x = \frac{1}{2}\left[\left(x+\sqrt{x^2-1}\right)^n + \left(x-\sqrt{x^2-1}\right)^n\right], \quad x \in [-1, 1], \tag{3.5}$$

is called the n*th Chebyshev polynomial and denoted by* $T_n(x)$.

Since the right-hand side of equality (3.5) is defined for all $x \in (-\infty, \infty)$, is positive for $x > 1$, and its parity coincides with the parity of the number n, in view of the identity

$$\cosh nt = \frac{e^{nt}+e^{-nt}}{2} = \frac{\left(\cosh t + \sqrt{\cosh^2 t - 1}\right)^n + \left(\cosh t - \sqrt{\cosh^2 t - 1}\right)^n}{2},$$

we conclude that the equality

$$T_n(x) = \frac{1}{2}\left[\left(x+\sqrt{x^2-1}\right)^n + \left(x-\sqrt{x^2-1}\right)^n\right] = (\operatorname{sgn} x)^n \cosh n \operatorname{arccosh} |x| \tag{3.5$'$}$$

holds for all values of x outside the segment $[-1, 1]$.

The Chebyshev polynomials have numerous remarkable properties. In what follows, we present only the most significant features of these polynomials.

1. A drawback of the presented procedure of finding the polynomial least deviating from zero is that we simply check the validity of equality (3.2). To obtain this equality in a natural way, we note that, by virtue of the Chebyshev theorem, the absolute value of polynomial (3.1)

$$\frac{T_n(x)}{2^{n-1}} = x^n + a_1^* x^{n-1} + \dots + a_n^* =: \tilde{T}_n(x) \tag{3.$\tilde{1}$}$$

must be equal to $E_{n-1} := E_{n-1}(x)$ at at least $(n-1)+2 = n+1$ different points x_j.

Since the derivative of this polynomial $\tilde{T}_n'$ is a polynomial of degree $n-1$ and, hence, has at most $n-1$ zeros in the interval $(-1, 1)$ and $\tilde{T}_n'(x_j) = 0$ for all $x_j \in (-1, 1)$, the collection of points x_j contains exactly $n-1$ points located in the interval $(-1, 1)$ and exactly two points located at the ends -1 and $+1$ of the interval $[-1, 1]$. Therefore, the polynomial $\tilde{T}_n'$ must satisfy the identity

$$-\tilde{T}_n^2(x) + E_{n-1}^2 = \frac{1}{n^2}\left[\tilde{T}_n'(x)\right]^2(-x^2+1), \tag{3.6}$$

both sides of which are polynomials of degree $2n$ with leading coefficients equal to -1, the identical zeros -1 and $+1$, and $n-1$ double zeros $x_j \in (-1, 1)$.

We now rewrite relation (3.6) in the form

$$\frac{\tilde{T}_n'(x)}{\sqrt{E_{n-1}^2 - \tilde{T}_n^2(x)}} = \pm\frac{n}{\sqrt{1-x^2}}.$$

Integrating this equality, we obtain

$$\arccos\frac{\tilde{T}_n(x)}{E_{n-1}} = \pm\arccos x + C,$$

$$\begin{aligned}\tilde{T}_n(x) &= E_{n-1}\cos\left[\pm n\arccos x + C\right]\\ &= E_{n-1}\left[\cos C\cos n\arccos x + C \mp \sin C\sin n\arccos x\right].\end{aligned}$$

Since $\tilde{T}_n'$ is a polynomial, we conclude that $\sin C = 0$[†] and, hence, $\cos C = \pm 1$. On the other hand, in view of relation (3.4), we have

$$\lim_{x\to\infty}\left(x^{-n}\cos n\arccos x\right) = 2^{n-1}.$$

[†] Since

$$\sin nt = \sin t\sum_{k=0}^{n-1}\alpha_k\cos kt = \sin t\sum_{k=0}^{n-1}\beta_k\cos t,$$

where α_k and β_k are certain numbers, the expression $\sin n\arccos x = \sqrt{1-x^2}\sum_{k=0}^{n-1}\beta_k x^k$ is not a polynomial.

Hence, by using relation $(3.\tilde{1})$ and the fact that $E_{n-1} > 0$, we conclude that $\cos C = +1$ and $E_{n-1} = 1/2^{n-1}$.

2. (a) Differentiating identity (3.6) k times $(k = 1, \dots, n-1)$, we readily get

$$(1-x^2)\tilde{T}_n^{(k+1)}(x) - (2k-1)x\tilde{T}_n^{(k)}(x) + [n^2 - (k-1)^2]\tilde{T}_n^{(k-1)}(x) = 0, \qquad (3.7)$$

$$k = 1, \dots, n-1.$$

In view of the fact that these equations are uniform, the Chebyshev polynomials $T_n(x) = \cos n \arccos x$ also satisfy these equations. Thus, in particular, for $k = 1$, we see that the polynomials T_n satisfy the following differential equation:

$$(1-x^2)y'' - xy' + n^2 y = 0. \qquad (3.8)$$

(b) By induction, relation (3.7) yields the following equality required in our subsequent presentation:

$$T_n^{(k)}(1) = \frac{n^2(n^2-1^2)(n^2-2^2)\dots[n^2-(k-1)^2]}{(2k-1)!!}, \quad k = 1, \dots, n. \qquad (3.9)$$

(c) Representing, in view of relation (3.1), the polynomial T_n in the form

$$T_n(x) = \cos n \arccos x = 2^{n-1}x^n + a_1 x^{n-1} + \dots + a_n$$

and substituting this expression in equality (3.8), we can specify all numbers a_k by the method of undetermined coefficients. As a result, after necessary transformations, we arrive at the following explicit expression for $T_n(x)$:

$$T_n(x) = \sum_{k=0}^{[n/2]} (-1)^k \frac{n}{n-k} \binom{n-k}{k} 2^{n-2k-1} x^{n-2k}. \qquad (3.10)$$

3. For all $k = 0, \dots, n$, we always have

$$\left|T_n^{(k)}(x)\right| \le T_n^{(k)}(1), \quad x \in [-1, 1]. \qquad (3.11)$$

Indeed, since $T_n(x) = \cos n\theta$, where $\theta = \arccos x$, we have

$$T_n'(x) = n\frac{\sin n\theta}{\sqrt{1-x^2}} = n\frac{\sin n\theta}{\sin\theta} = 2n\left[\cos(n-1)\theta + \cos(n-3)\theta + \ldots\right]$$

and, hence, by induction,

$$T_n^{(k)}(x) = \sum_{j=1}^{n-k} \lambda_j \cos j\theta,$$

where all $\lambda_j = \lambda_j(k) \geq 0$. This implies that

$$\left|T_n^{(k)}(x)\right| \leq \sum_{j=1}^{k} \lambda_j = T_n^{(k)}(1).$$

4. The following recurrent relation is true for the Chebyshev polynomials T_n :

$$T_n(x) = 2xT_{n-1}(x) - T_{n-2}(x), \quad n = 2, 3, \ldots . \tag{3.12}$$

Its validity immediately follows from the equality

$$\cos n\theta + \cos(n-2)\theta = 2\cos\theta\cos(n-1)\theta.$$

In view of the fact that $T_0(x) = 1$ and $T_1(x) = x$, by using (3.12), we consecutively obtain

$$T_2(x) = 2x^2 - 1, \quad T_3(x) = 4x^3 - 3x, \quad T_4(x) = 8x^4 - 8x^2 + 1,$$

$$T_5(x) = 16x^5 - 20x^3 + 5x, \quad T_6(x) = 32x^6 - 48x^4 + 18x^2 - 1,$$

$$T_7(x) = 64x^7 - 112x^5 + 56x^3 - 7x, \quad \text{etc.} \tag{3.13}$$

5. Since

$$\frac{2}{\pi}\int_{-1}^{1} \frac{T_k^2(x)}{\sqrt{1-x^2}}\,dx = \frac{2}{\pi}\int_{-1}^{1} \frac{\cos^2 k\arccos x}{\sqrt{1-x^2}}\,dx = \frac{2}{\pi}\int_{0}^{\pi} \cos^2 k\theta\,d\theta = 1 \tag{3.14}$$

and, for $j \neq k$,

$$\int_{-1}^{1} \frac{T_j(x)T_k(x)}{\sqrt{1-x^2}}\,dx = \int_{0}^{\pi} \cos j\theta \cos k\theta \, d\theta = 0, \tag{3.14'}$$

the system

$$T_0(x),\ T_1(x),\ \ldots,\ T_k(x) \tag{3.14''}$$

is orthonormalized by the weight $2(1-x^2)^{-1/2}/\pi$ on the segment $[-1, 1]$.

4. On the best uniform approximation of continuous functions of complex variable

In the present section, we show that the principal results obtained in Sections 1 and 2 admit a generalization to the case of approximation of complex-valued continuous functions f defined on a closed bounded set $\mathfrak{M}$ in the complex plane. Since the existence of the polynomial P_n^* of the best approximation for a given function f follows from Theorem 1.1″ and the problem of uniqueness of this polynomial is completely solved by Theorem 2.6, we first prove the Kolmogorov theorem which, by analogy with the Chebyshev theorem 2.4, gives necessary and sufficient conditions for a polynomial P_n^* to be the polynomial of the best approximation of degree n for a function f defined on $\mathfrak{M}$.

After this, we also briefly discuss the following problems:

1. Lower estimates of the best approximation of a given function by polynomials of a given degree.

2. Determination of the minimum subset $E_0 \subset \mathfrak{M}$ for which the value of the best uniform approximation is equal to the value of the best uniform approximation in the entire set $\mathfrak{M}$.

3. Determination of an analog of the rule of alternation of signs of the difference $f - P_n$ at the (+)- and (–)-points (in the Chebyshev theorems, etc.) in the case of approximation of functions of complex variable.

4. Approximation of abstract functions on compact sets by elements of convex sets.

Definition 4.1. *If a continuous function f and a (generally speaking, generalized) polynomial P_n are given on a closed bounded set $\mathfrak{M}$, then any point $z_0 \in \mathfrak{M}$ such that the following equality is satisfied:*

$$|f(z_0) - P_n(z_0)| = \|f - P_n\|_{\mathfrak{M}}$$

is called an $e = e(P_n)$*-point (for the difference* $f - P_n$*).*

4.1. Kolmogorov theorem

Theorem 4.1 [Kolmogorov (1948)].[5] *Let*

$$\varphi_0, \varphi_1, \dots, \varphi_n \tag{4.1}$$

be $n+1$ *fixed continuous functions given on a closed bounded set* $\mathfrak{M}$ *and let* f *be a continuous function approximated by generalized polynomials* $P_n(z) = P_n(\varphi_k; c_k; z)$, *of the form*

$$P_n(z) = \sum_{k=0}^{n} c_k \varphi_k(z). \tag{4.2}$$

Then, for a polynomial $P_n^* = \sum_{k=0}^{n} c_k^* \varphi_k$ *to be the polynomial of the best uniform approximation of the function* f *in a sense that*

$$\|f - P_n^*\|_{\mathfrak{M}} = \inf_{P_n} \|f - P_n\|_{\mathfrak{M}},$$

it is necessary and sufficient that the inequality

$$\min_{z\in E} \operatorname{Re}\left\{P_n(z)\overline{\left[f(z) - P_n^*(z)\right]}\right\} \le 0 \tag{4.3}$$

be true in the set $E = E(P_n^*)$ *of all* $e = e(P_n^*)$*-points from* $\mathfrak{M}$ *for any polynomial* $P_n(z)$ *of the form (4.2).*

Proof. *Necessity.* Let P_n^* be a polynomial of the best approximation for the function f. Assume the opposite, i.e., that there exists a polynomial P_n satisfying the following inequality opposite to (4.3):

$$\min_{z\in E} \operatorname{Re}\left\{P_n(z)\overline{\left[f(z) - P_n^*(z)\right]}\right\} > c > 0. \tag{4.4}$$

In this case, by virtue of the fact that the set E is closed and bounded, inequality (4.4) is true, for some $\delta > 0$, also in the set $E_\delta \supset E$ of all points from $\mathfrak{M}$ each of which is located at a distance smaller than δ from the set E. Thus, clearly, the set $\mathfrak{M} \setminus E_\delta$ is closed. We now write

$$\max_{z \in \mathfrak{M}} \left| f(z) - P_n^*(z) \right| = E_n, \qquad \max_{z \in \mathfrak{M} \setminus E_\delta} \left| f(z) - P_n^*(z) \right| = E_n',$$

$$E_n - E_n' = h, \tag{4.5}$$

$$\max_{z \in \mathfrak{M}} |P_n(z)| = M, \quad \min \left\{ \frac{c}{M^2}; \frac{h}{2M} \right\} = \lambda.$$

Note that the numbers h and λ are, clearly, positive. Further, we construct a polynomial

$$Q_n = P_n^* + \lambda P_n \tag{4.6}$$

and show that this polynomial approximates the function f better than the polynomial P_n^* in contradiction with our assumption. Indeed, for any $z \in E_\delta$, in view of relations (4.4) and (4.5) and the fact that the equality $w_0 + \overline{w_0} = 2 \operatorname{Re} w_0$ is true for any complex number w_0, we find

$$\begin{aligned} |f(z) - Q_n(z)|^2 &= \left[f(z) - P_n^*(z) - \lambda P_n(z) \right] \left[\overline{f(z) - P_n^*(z)} - \overline{\lambda P_n(z)} \right] \\ &= \left| f(z) - P_n^*(z) \right|^2 - 2\lambda \operatorname{Re} \left\{ P_n(z) \left[\overline{f(z) - P_n^*(z)} \right] \right\} + \lambda^2 |P_n(z)|^2 \\ &\le E_n^2 - 2\lambda c + \lambda \frac{c}{M^2} M^2 = E_n^2 - \lambda c < E_n^2; \end{aligned}$$

At the same time, if $z \in \mathfrak{M} \setminus E_\delta$, then, by virtue of relations (4.6) and (4.5), we get

$$\begin{aligned} |f(z) - Q_n(z)| &\le \left| f(z) - P_n^*(z) \right| + \lambda |P_n(z)| \\ &\le E_n' + \frac{h}{2M} M = E_n - \frac{h}{2} < E_n. \end{aligned}$$

Hence, for all $z \in \mathfrak{M}$, we obtain

$$\left|f(z) - Q_n(z)\right| < E_n = \max_{z \in \mathfrak{M}} \left|f(z) - P_n^*(z)\right|$$

in contradiction with the definition of the polynomial P_n^*.

Sufficiency. Assume that a polynomial P_n^* approximating the function f possesses a property that inequality (4.3) is true for any polynomial P_n of the form (4.2). Let us show that P_n^* is the polynomial of the best approximation for the function f. Indeed, if Q_n is an arbitrary polynomial of the form (4.2) and z_0 is a point from E for which the following inequality is true:

$$\operatorname{Re}\left\{\left[Q_n(z_0) - P_n^*(z_0)\right]\left[\overline{f(z_0) - P_n^*(z_0)}\right]\right\} \leq 0$$

(here, the role of the polynomial P_n is played by the polynomial $Q_n - P_n^*$), then we conclude that, at the indicated point,

$$\begin{aligned}
\left|f(z_0) - Q_n(z_0)\right|^2 &= \left|f(z_0) - P_n^*(z_0) - \left[Q_n(z_0) - P_n^*(z_0)\right]\right|^2 \\
&= \left|f(z_0) - P_n^*(z_0)\right|^2 \\
&\qquad - 2\operatorname{Re}\left\{\left[Q_n(z_0) - P_n^*(z_0)\right]\left[\overline{f(z_0) - P_n^*(z_0)}\right]\right\} \\
&\qquad + \left|Q_n(z_0) - P_n^*(z_0)\right|^2 \\
&\geq \max_{z \in E} \left|f(z) - P_n^*(z)\right|^2.
\end{aligned}$$

This means that the polynomial P_n^* is the polynomial of the best approximation for the function f. Theorem 4.1 is proved.

■

4.2. Examples of application of Theorem 4.1

Example 4.1. A polynomial $P_{n-1}^*(z_0) \equiv 0$ gives the best approximation of the function $f(z) = z^n$ defined in the unit disk $|z| \leq 1$ among all algebraic polynomials of degree $n-1$.

Indeed, note that the role of the set E is, in this case, played by the unit circle $E = \{z: |z| = 1\}$. Hence, for any polynomial P_{n-1} of degree $n-1$, the product

$$P_{n-1}(z)\left[\overline{f(z) - P^*_{n-1}(z)}\right] = P_{n-1}(z) \cdot \bar{z}^n \tag{4.7}$$

possesses the following property: On tracing the unit circle E by the point z in the positive direction, the argument of the factor $\bar{z}^n$ decreases by $2n\pi$ and the argument of the factor $P_{n-1}(z)$ increases by at most $2(n-1)\pi$ (since P_{n-1} has at most $n-1$ zeros inside E) and, thus, the argument of product (4.7) decreases by at least 2π. Therefore, by continuity, this implies that the real part of product (4.7) is nonpositive at at least one point $z_0 \in E$ (this happens at a point z_0, where either the argument of product (4.7) is equal to $\pi + 2k\pi$, $k = 0, 1, -1, \ldots$, or this product is equal to zero).

Example 4.2 [Al'per (1959)]. For a function

$$f(z) = \frac{1}{z-a}, \qquad |a| > 1,$$

defined in the unit disk $|z| \le 1$, we choose a polynomial P^*_n of degree n and a number γ_n such that the following equality is true:

$$\frac{1}{z-a} - P^*_n(z) = \frac{1-(z-a)P^*_n(z)}{z-a} = \gamma_n \frac{(1-\bar{a}z)z^n}{z-a}, \qquad |z| \le 1, \tag{4.8}$$

and prove that the polynomial P^*_n constructed as indicated above is the polynomial of the best approximation for the function $1/(z-a)$. To this end, we show that the following inequality is true for any polynomial P_n of degree n:

$$\min_{z\in E} \operatorname{Re}\left\{P_n(z)\left[\overline{\frac{1}{z-a} - P^*_n(z)}\right]\right\} = \min_{z\in E} \operatorname{Re}\left\{P_n(z)\left[\overline{\gamma_n \frac{(1-\bar{a}z)z^n}{z-a}}\right]\right\} \le 0, \tag{4.9}$$

where E is the set of points $|z| \le 1$ at which the difference $1/(z-a) - P^*_n(z)$ takes its maximum absolute value for $|z| \le 1$.

Since the function $(1-\bar{a}z)/(z-a)$ maps the unit circle $|z| < 1$ and its boundary $|z| = 1$ onto themselves, expression (4.8) takes its maximum value equal to $|\gamma_n|$ for $|z| = 1$, and, hence, the role of the set E is again played by the unit circle $|z| = 1$. In view of the fact that the argument of the factor $\overline{(1-\bar{a}z)z^n/(z-a)}$ decreases by $2(n+1)\pi$

on tracing the unit circle E in the positive direction [since all $n+1$ zeros of the product $(1-\bar{a}z)z^n$ lie inside the unit circle and the zero of the difference $z-a$ lies outside this circle] and the argument of the polynomial $P_n(z)$ increases by at most $2n\pi$, we conclude that the argument of the expression in braces in relation (4.9) decreases by at least 2π. After this, in exactly the same way as in Example 4.1, we conclude that inequality (4.9) is true.

Note that, as a result of the multiplication of the left- and right-hand sides of equalities (4.8) by $z-a$ and passing to the limit as $z \to a$, we obtain $\gamma_n = [a^n(1-|a|^2)]^{-1}$. Hence, for the value of the best uniform approximation of the function $(z-a)^{-1}$, we can write

$$E_n\left(\frac{1}{z-a}\right) = \min_{P_n} \max_{|z|\le 1} \left|\frac{1}{z-a} - P_n(z)\right| = |\gamma_n| = \frac{1}{|a|^n(|a|^2-1)}. \tag{4.10}$$

Example 4.3. Let us show that the Chebyshev theorem (Theorem 1.2) can be fairly simply derived from Theorem 4.1.

Necessity. In the notation used in the proof of Theorem 1.2, we now show that the segment $[a, b]$ contains at least one system of $n+2$ different e-points x_k satisfying the rule of alternation of signs, i.e.,

$$r_n(x_1) = -r_n(x_2) = \ldots = (-1)^{n+1} r_n(x_{n+2}).$$

Indeed, if the maximum number of e-points (i.e., points from E) with alternation of signs is equal to $m+1$, where $m+1 < n+2$, then we can define P_m by using relation (1.13) for $m \ge 1$ and setting $P_0(x) \equiv 1$ or -1 for $m=0$ and, as a result, construct a polynomial of degree not greater than n which has the same sign as the difference $f(x) - P_n^*(x)$ for all $x \in E$ (i.e., at all e-points) and, hence, satisfies the equality

$$\min_{x\in E} \operatorname{Re}\left\{P_m(x)\left[\overline{f(x)-P_n^*(x)}\right]\right\} = \min_{x\in E} \left\{P_m(x)\left[f(x)-P_n^*(x)\right]\right\} > 0,$$

i.e., we arrive at a contradiction.

Sufficiency. Assume that the polynomial P_n^* is such that the difference $f - P_n^*$ takes its maximum absolute value on $[a, b]$ with consecutive alternation of signs at $n+2$ points $x_k \in [a, b]$. Then, in view of the fact that any polynomial $P_n(x) \not\equiv 0$ of degree

n can consecutively take values of different signs on the segment $[a, b]$ at at most $n+1$ points x_k, we conclude that

$$\min_{x \in E} \operatorname{Re}\left\{P_n(x)\left[\overline{f(x) - P_n^*(x)}\right]\right\} \le \min_k \left\{P_n(x_k)\left[f(x_k) - P_n^*(x_k)\right]\right\} \le 0.$$

This means that P_n^* is the polynomial of the best uniform approximation for the function f, which completes the proof of the Chebyshev theorem (Theorem 1.2).

Note that if we use Theorem 2.3 and its corollary, then the Kolmogorov theorem also implies, in exactly the same way, the other Chebyshev theorem (Theorem 2.4).

4.3. De-la-Vallée-Poussin-type theorem

We now show that it is possible to obtain an analog of the de la Vallée Poussin theorem (Theorem 2.8) on the lower bound of the value of the best approximation E_n of a function by using the reasoning from the proof of sufficiency in Theorem 4.1.

Theorem 4.2. *If, for a function f continuous on $\mathfrak{M}$, a polynomial Q_n of type (4.2) is such that the inequality*

$$\min_{z \in E} \operatorname{Re}\left\{P_n(z)\left[\overline{f(z) - Q_n(z)}\right]\right\} \le 0 \tag{4.11}$$

holds on a certain subset $E \subset \mathfrak{M}$ for any polynomial P_n of type (4.2), then the value of the best uniform approximation E_n of the function f on $\mathfrak{M}$ by polynomials of degree n satisfies the inequality

$$E_n = \min_{P_n} \max_{z \in \mathfrak{M}} |f(z) - P_n(z)| \ge \min_{z \in E} |f(z) - Q_n(z)|. \tag{4.12}$$

Proof. By z_0 we denote any point from E at which the following inequality is true:

$$\operatorname{Re}\left[P_n^*(z_0) - Q_n(z_0)\right]\left[\overline{f(z_0) - Q_n(z_0)}\right] \le 0,$$

where P_n^* is the polynomial of the best approximation for the function f on $\mathfrak{M}$. This yields

$$E_n^2 \ge \left|f(z_0) - P_n^*(z_0)\right|^2 = \left|f(z_0) - Q_n(z_0) - \left[P_n^*(z_0) - Q_n(z_0)\right]\right|^2$$

$$= |f(z_0) - Q_n(z_0)|^2 - 2\,\mathrm{Re}\left\{\left[P_n^*(z_0) - Q_n(z_0)\right]\left[\overline{f(z_0) - Q_n(z_0)}\right]\right\}$$

$$+ \left|P_n^*(z_0) - Q_n(z_0)\right|^2$$

$$\geq |f(z_0) - Q_n(z_0)|^2 + \left|P_n^*(z_0) - Q_n(z_0)\right|^2$$

$$\geq |f(z_0) - Q_n(z_0)|^2 \geq \min_{z \in E} |f(z) - Q_n(z)|^2,$$

as required.

■

4.4. Characteristic sets

The theorems presented in what follows mainly without proofs (the reader is referred to the monograph [Smirnov and Lebedev (1964), Chapter 5, Section 3]) solve problems 2 and 3 formulated at the beginning of the present section.

For the the sake of simplicity, in the remaining part of the section, a closed bounded set of points of the complex plane is denoted by $\mathfrak{M}$.

Definition 4.2. *For a complex-valued function f continuous on $\mathfrak{M}$ which is not a polynomial with respect to a system of functions $\{\varphi_k\}_{k=0}^{n}$, $n = 0, 1, 2, \dots$, continuous on $\mathfrak{M}$, a characteristic set of order n with respect to this system of functions is defined as an arbitrary finite subset $E_0 \subset \mathfrak{M}$, $E_0 = E_0(n)$, with the following properties:*

(i) the value of the best uniform approximation of the function f by polynomials of the form

$$P_n = \sum_{k=0}^{n} c_k \varphi_k$$

on E_0 is equal to the value of the best uniform approximation of this function on the entire set $\mathfrak{M}$;

(ii) on any subset E_0' from E_0 that does not coincide with E_0, the value of the best uniform approximation of the function f is strictly smaller than its value on E_0 and, hence, on $\mathfrak{M}$.

Remark 4.1. The fact that, for any function f continuous on $\mathfrak{M}$, there exists at least one characteristic set is established in proving the following theorem:

Theorem 4.3 (on the existence of characteristic sets [de la Vallée Poussin (1911)]). *Assume that, for some integer $n \geq 0$, a system of $n+1$ continuous functions $\{\varphi_k\}_{k=0}^n$ is given on a set $\mathfrak{M} \subset \mathbb{C}$ formed by at least $n+2$ points. Then, for any continuous function f on $\mathfrak{M}$ different from a polynomial of degree $\leq n$ of the form $\sum_{k=0}^n c_k \varphi_k$:*

(i) there exists at least one characteristic set $E_0 = E_0(n)$ of order n;

(ii) every characteristic set E_0 of order n is formed by a finite number $m = m(E_0)$ of points such that

$$n + 2 \leq m \leq 2n + 3, \tag{4.13}$$

if $\{\varphi_k\}_{k=0}^n$ is a Chebyshev system or

$$1 \leq m \leq 2n + 3, \qquad (4.13'),$$

otherwise.

To prove this theorem, we need the following well-known Helly theorem on the intersection of convex sets. Here, we present only the statement of this theorem. A fairly simple proof of the Helly theorem can be found, e.g., in [Krein and Nudel'man (1973), p. 46).

Theorem 4.4 (on the intersection of convex sets [Helly (1936)]). *For a certain natural m, let L_m be a real m-dimensional linear normed space containing a collection K of convex closed sets formed by at least $m+1$ elements Ω_α. If any $m+1$ sets $\Omega \in K$ have at least one common point and K contains at least one system formed by a finite number n of sets $\Omega_i \in K$, $i = 1, 2, \dots, n$, whose intersection is bounded, then there exists at least one point common for all sets $\Omega \in K$.*

Proof of Theorem 4.3. We prove Theorem 4.3 only for the most important case where $\{\varphi_k\}_{k=0}^n$ is a Chebyshev system. First, we note that any polynomial given by the equality

$$P_n(z) = P_n(z; \varphi) = \sum_{j=0}^{n} (\alpha_j + i\beta_j)\,\varphi_j(z), \tag{4.14}$$

where α_j and β_j are real numbers, can be represented in the form

$$P_n(z) = \sum_{j=0}^{n} \alpha_j \varphi_j(z) + \sum_{j=0}^{n} \beta_j \psi_j(z), \quad \psi_j(z) := i\varphi_j(z) \tag{4.14'}$$

and, hence, the set of all possible polynomials of type (4.14) with ordinary norm

$$\| P_n \| = \max_{z \in \mathfrak{M}} |P_n(z)|$$

can be regarded as a real linear normed space of dimension $m = 2n + 2$ spanned by the system of functions $\varphi_0, \varphi_1, \ldots, \varphi_n, \psi_0, \psi_1, \ldots, \psi_n$, and vice versa.

To prove Theorem 4.3, we assume the opposite, i.e., that, under all conditions of the theorem, there exists a function f continuous on $\mathfrak{M}$ for which all finite sets $\{z_1, z_2, \ldots, z_p\} \subset \mathfrak{M}$ formed by at most $2n + 3$ points $(p \le 2n + 3)$ are not characteristic. To arrive at a contradiction, for any $z_1, \ldots, z_p \in \mathfrak{M}$, we set

$$E_n^0(f; z_1, z_2, \ldots, z_p) := \inf_{P_n} \max_{1 \le j \le p} \left| f(z_j) - P_n(z_j) \right|,$$

$$\sup_{z_1 \in \mathfrak{M}; \ldots; z_p \in \mathfrak{M}} E_n^0(f; z_1, \ldots, z_p) := E_{n,p}^*(f). \tag{4.15}$$

Further, for any $p \le 2n + 3$, we note that, on the one hand, the inequality

$$E_n^0(f; z_1, \ldots, z_p) < E_n(f) = \min_{P_n} \max_{z \in \mathfrak{M}} |f(z) - P_n(z)|, \tag{4.16}$$

is true according to the assumption and, on the other hand, every function $E_n^0(f; z_1, \ldots, z_p)$ is continuous on the compact set $\underbrace{\mathfrak{M} \times \mathfrak{M} \times \ldots \times \mathfrak{M}}_{p \text{ times}}$ and, hence, reaches its upper bound at a certain point $(z_1^*, \ldots, z_p^*)$, $z_j^* \in \mathfrak{M}$. This means that the following inequalities are true:

$$E_{n,j}^*(f) \le E_{n,j+1}^*(f), \quad j = 1, 2, \ldots, 2n + 2,$$

$$E_{n,2n+3}^* := E_{n,2n+3}^*(f) = E_n^0(f, z_1^*, \ldots, z_{2n+3}^*) < E_n(f). \tag{4.17}$$

For any fixed $\xi \in \mathfrak{M}$, we now introduce a set $\Omega(\xi)$ of all possible polynomials $P_n(\cdot;\varphi)$ each of which deviates in the absolute value from the function f at the point ξ by at most $E^*_{n,2n+3}$:

$$\Omega(\xi) := \{P_n(z;\varphi)\colon |f(\xi) - P_n(\xi;\varphi)| \le E^*_{n,2n+3}\}. \tag{4.18}$$

By direct verification, we can show that each set $\Omega(\xi)$ is nonempty, closed and convex. Moreover, by virtue of (4.15), we conclude that the intersection of any $2n+3$ sets $\Omega(\xi_i)$, $\xi_i \in \mathfrak{M}$, $i = 1, 2, \ldots, 2n+3$, is nonempty and contains, e.g., a polynomial P_n^0 satisfying the equality

$$\max_{1\le j\le 2n+3} \left|f(\xi_j) - P_n^0(\xi_j)\right| = E_0^n(f, \xi_1, \xi_2, \ldots, \xi_{2n+3}).$$

The existence of this polynomial follows from the Borel theorem (Theorem 1.1).

To apply the Helly theorem to the convex sets $\Omega(\xi)$ defined in a real $(2n+2)$-dimensional linear space of polynomials of the form (4.14)–(4.14′), it is necessary to show that there exists at least one finite system of sets $\Omega(\xi)$ whose intersection is bounded. To do this, we take arbitrary $n+1$ different points $\xi_1^0, \xi_2^0, \ldots, \xi_{n+1}^0$ in $\mathfrak{M}$ and consider the intersection $\Omega(\xi_1^0) \cap \ldots \cap \Omega(\xi_{n+1}^0) = \Omega^0$. Since, by virtue of (4.18), each polynomial $P_n(\cdot;\varphi) \in \Omega^0$ is bounded at all points ξ_j^0, i.e.,

$$\left|P_n(\xi_j^0, \varphi)\right| \le \left|P_n(\xi_j^0, \varphi) - f(\xi_j^0)\right| + \left|f(\xi_j^0)\right| < E_n(f) + \|f\|,$$

by using the interpolation formula (4.18), we conclude that each polynomial $P_n(\cdot;\varphi) \in \Omega^0$ and, hence, the entire set Ω^0 is indeed bounded. Therefore, by the Helly theorem, we conclude that there exists at least one polynomial (point) $P_n^*(\cdot, \varphi)$ contained in all sets $\Omega(\xi)$, $\xi \in \mathfrak{M}$, such that, according to relations (4.18) and (4.17),

$$\left|f(\xi) - P_n^*(\xi, \varphi)\right| \le E^*_{n,2n+3} < E_n(f)$$

for all $\xi \in \mathfrak{M}$, which is impossible.

The indicated contradiction proves that the characteristic set exists and consists of at most $2n+3$ elements. The fact that the characteristic set of the T-system $\{\varphi_j\}_{j=0}^n$ contains at least $n+2$ points follows from the fact that, for any set $\{\xi_1, \xi_2, \ldots, \xi_p\}$ with $p \le n+1$, according to the interpolation formula (2.11), we have

$$E_n^0(f; \xi_1, \xi_2, \dots, \xi_p) = 0 < E_n(f).$$

This proves Theorem 4.3 (in the case where $\{\varphi_j\}_{j=0}^n$ is a Chebyshev system).

Example 4.4. In the case of approximation of the function $f(z) = z^4$ by algebraic polynomials P_3 of the third degree on the disk $\mathfrak{M} = \{z : |z| \le 1\}$, the role of a characteristic set can be played, e.g., by a set of eight points $E_0 := \bigcup_{k=0}^{7} \{e^{ik\pi/4}\}$. In this case, $P_3^0(z) \equiv 0$ is the polynomial of the best approximation of the third degree for the function z^4 both on $\mathfrak{M}$ and on $E_0 \subset \mathfrak{M}$ and, hence,

$$\inf_{P_3(z)} \max_{z \in \mathfrak{M}} \left| z^4 - P_3(z) \right| = \max_{z \in \mathfrak{M}} \left| z^4 - P_3^*(z) \right|$$

$$= \max_{0 \le k \le 7} \left| \left(e^{ik\pi/4}\right)^4 - P_3^*\left(e^{ik\pi/4}\right) \right| = 1.$$

Indeed, if there exists a polynomial $P_3^*(z)\ (\not\equiv 0)$ of the third degree such that $\max_{z \in \mathfrak{M}} \left| f(z) - P_3^*(z) \right| < 1$ and, in particular,

$$\max_{0 \le k \le 7} \left| \left(e^{ik\pi/4} \right)^4 - P_3^*\left(e^{ik\pi/4} \right) \right| = \max_{0 \le k \le 7} \left| (-1)^k - P_3^*\left(e^{ik\pi/4} \right) \right| < 1,$$

then the conditions $\operatorname{sgn} T_3^*(k\pi/4) = (-1)^k$, $k = 0, 1, \dots, 7$, hold for the trigonometric polynomial $T_3^*(t) := \operatorname{Re} P_3^*(e^{it}) \not\equiv 0$. This means that the polynomial $T_3^*(t)$ of the third degree must have at least seven zeros on $[0, 2\pi)$ but, by virtue of Theorem 1.3, this is impossible since

$$T_3^*(t) = \operatorname{Re} P_3^*(e^{it}) \not\equiv 0.$$

At the same time, for any $k_0 = 0, 1, \dots, 7$ (for the sake of definiteness, we assume that $k_0 = 4$), on the subset $E_0' = E_0 \setminus \{e^{ik_0\pi/4}\} \subset E_0$, by setting

$$P_3(k_0, z) := -\varepsilon_1 \left(z e^{-ik_0\pi/4} + a + \varepsilon_2 \right)^3 e^{ik_0\pi}, \qquad a := \frac{\sqrt{6} - \sqrt{2}}{2},$$

for sufficiently small positive ε_1 and ε_2, we readily conclude that

$$\left|z^4 - P_3(k_0, z)\right| < 1, \qquad z \in E_0'.$$

Similarly, we can show that, in the case of approximation of the function $f(z) = z^3$ by algebraic polynomials P_2 of the second degree on the set $\mathfrak{M} = \{z : |z| \le 1\}$, the set of seven points $E := \left\{e^{ik2\pi/7}\right\}_{k=0}^{6}$ is characteristic and, in addition, the polynomial $P_3(k_0, \cdot)$ should be replaced (in the same notation) by a polynomial

$$P_2(k_0, z) := \varepsilon_1 \left(z e^{-ik_0 2\pi/7} + a + \varepsilon_2\right)^2,$$

where

$$a := \frac{\sin\frac{3\pi}{28}}{\sin\frac{\pi}{7}}.$$

E. Ya. Remez and V. K. Ivanov proposed necessary and sufficient conditions for the polynomial P_n^* of the form (4) to be the polynomial of the best approximation for a function f continuous on $\mathfrak{M}$ in the form different from the conditions used in the Kolmogorov theorem (Theorem 4.1) and sometimes more convenient for applications.

Theorem 4.5 [Remez (1953), (1957); Ivanov (1951), (1952)]. *In order that a polynomial P_n^* of the form (4.2) be the polynomial of the best approximation for a function f continuous on $\mathfrak{M}$, it is necessary and sufficient that there exist a set $\{z_k\}_1^m$, $1 \le m \le 2n+3$, such that $\left|f(z_k) - P_n^*(z_k)\right| = \left\|f(z) - P_n^*(z)\right\|_{\mathfrak{M}}$ and positive numbers δ_k, $k = 1, 2, \dots, m$, such that the equality*

$$\sum_{k=1}^{m} \delta_k \left[\overline{f(z_k) - P_n^*(z_k)}\right] P_n(z_k) = 0 \tag{4.19}$$

holds for any polynomial P_n of the form (4.2).

Remark 4.2. Since the equality

$$f(z_k) - P_n^*(z_k) = \rho \operatorname{sgn}\left[f(z_k) - P_n^*(z_k)\right],$$

where $\rho = \left\|f - P_n^*\right\|_{\mathfrak{M}}$, is true for any $k = 1, 2, \dots, m$, condition (4.19) can be rewritten in the following equivalent form:

$$\sum_{k=1}^{m} \delta_k \operatorname{sgn}\left[\overline{f(z_k) - P_n^*(z_k)}\right] P_n(z_k) = 0, \quad \delta_k > 0. \tag{4.19′}$$

In the case of approximation of real continuous functions f on the segment $[a, b]$ by algebraic polynomials $P_n(x)$, by virtue of the Chebyshev theorem (Theorem 1.2), the characteristic set $E_0(n)$ consists of $n + 2$ points x_k and the following equalities hold at these points:

$$\operatorname{sgn}\left[\overline{f(x_k) - P_n^*(x_k)}\right] = \operatorname{sgn}\left[f(x_k) - P_n^*(x_k)\right] = \pm(-1)^k, \quad k = 1, 2, \dots, n+2.$$

These equalities reflect the so-called *rule of alternation of signs.*

In the case of approximation of continuous functions on closed bounded sets $\mathfrak{M}$ of the complex plane, the behavior of the quantity $\operatorname{sgn}\left[\overline{f(z_k) - P_n^*(z_k)}\right]$ is described by the following theorem obtained in the process of subsequent development of Theorem 4.5:

Theorem 4.6 [Videnskii (1956)]. *In the case of approximation of a function f continuous on $\mathfrak{M}$, in order that numbers $\left\{\delta_k \operatorname{sgn}\left[\overline{f(z_k) - P_n^*(z_k)}\right]\right\}_{k=1}^{m}$, where P_n^* is the algebraic polynomial of the best approximation of the function f of degree n and $n + 2 \le m \le 2n + 3$, satisfy conditions (4.19′), it is necessary and sufficient that there exist an algebraic polynomial $U_{m-n-2}(z)$ of degree $m - n - 2$ such that*

$$\delta_k \operatorname{sgn}\left[\overline{f(z_k) - P_n^*(z_k)}\right] = \frac{U_{m-n-2}(z_k)}{\omega'(z_k)}, \quad k = 1, 2, \dots, m, \tag{4.20}$$

where

$$\omega(z) = \prod_{k=1}^{m} (z - z_k).$$

5. Approximation of functions on sets of finitely many points[6]

5.1. Best approximation of a system of linear equations

Theorems 1.2 and 2.4 (Chebyshev) and Theorem 4.3 (de la Vallée Poussin) show that the problem of approximation of a function $f \in C$ on a closed bounded set $\mathfrak{M}$ is equivalent to the problem of approximation of the same function on a certain subset $E_0 \subset \mathfrak{M}$ of finitely many points (characteristic set). This fact and the problem of approximate construction of the polynomial P_n^* of the best approximation for a given continuous function solved in Section 6 reveal the importance of the problem of finding the polynomial with least deviation from a given function at finitely many points. In particular, it is important to be able to solve, for any given $n = 1, 2, \ldots$, the following problems:

1. Find a T-polynomial of the form

$$P_n(x) = \sum_{k=0}^{n} c_k \varphi_k(x), \quad x \in [a, b],$$

least deviating from a given real function at $n + 2$ points.

2. Find a trigonometric polynomial T_n of degree n least deviating from a given real function at $2n + 2$ points.

3. Find a T-polynomial of the form

$$P_n(z) = \sum_{k=0}^{n} c_k \varphi_k(z)$$

(with complex coefficients) least deviating from a given (generally speaking, complex) function at m points of the complex plane, where $n + 2 \le m \le 2n + 3$.

Consider an incompatible system of equations

$$\begin{aligned} a_{11} z_1 + a_{12} z_2 + \ldots + a_{1n} z_n &= w_1, \\ a_{21} z_1 + a_{22} z_2 + \ldots + a_{2n} z_n &= w_2, \\ \cdots\cdots\cdots\cdots\cdots\cdots\cdots\cdots \\ a_{m1} z_1 + a_{m2} z_2 + \ldots + a_{mn} z_n &= w_m, \end{aligned} \tag{5.1}$$

all coefficients a_{ik} and free terms w_i of which are, generally speaking, complex numbers.

Definition 5.1. *In the set of points z of the form $z = (z_1, z_2, \ldots, z_n)$, a point $z^* = (z_1^*, z_2^*, \ldots, z_n^*)$ is called the best approximate solution (or simply the b.a.-solution) of* *system (5.1) if it possesses the following property:*

$$\max_i \left|(a_i, z^*) - w_i\right| = \min_z \max_i \left|(a_i, z) - w_i\right|, \quad i = 1, 2, \dots, m, \tag{5.2}$$

where $a_i = (a_{i1}, a_{i2}, \dots, a_{in})$ *and* $(a_i, z) = a_{i1} z_1 + a_{i2} z_2 + \dots + a_{in} z_n$. *The quantity*

$$\rho^* = \rho^*(z^*) := \max_i \left|(a_i, z^*) - w_i\right| \tag{5.2'}$$

is called the value of the best approximation of system (5.1).

It is easy to see that all problems of finding (generally speaking, generalized) polynomials of the best approximation of functions different from polynomials on finite sets of points are reduced to the solution of systems of the form (5.1). Thus, e.g., the problem of finding the polynomial

$$P_n^* = \sum_{k=0}^{n} c_k^* \varphi_k$$

least deviating from a given function f at points z_i, $i = 1, 2, \dots, m$, is reduced to a system of the form (5.1) if we set

$$\varphi_k(z_i) = a_{ik}, \quad c_k = z_k, \quad \text{and} \quad f(z_i) = w_i.$$

Definition 5.2. *Any point* z^0 *equidistant from the right-hand sides of system (1) in a sense that*

$$\left|(a_1, z^0) - w_1\right| = \left|(a_2, z^0) - w_2\right| = \dots = \left|(a_m, z^0) - w_m\right| \tag{5.4}$$

is called an equidistant point of system (5.1). The common value of these quantities is called the A-distance between the point z^0 *and the point* $w = (w_1, w_2, \dots, w_m)$.

It is easy to see that, for $m = 2$, $n = 1$, and $a_{11} = a_{21} = 1$, any equidistant point z is located on the perpendicular line drawn through the middle of the segment connecting the points w_1 and w_2, i.e., in the geometric locus of points equidistant from the points w_1 and w_2 in the ordinary sense.

The next assertion (Theorem 5.1) gives the complete description of all equidistant points of the simplest system (1) containing $n + 1$ equations. The importance of this theorem is explained by the facts that the b.a.-solution of system (1) coincides with an equidistant point of the system (or with one of its subsystems) and that the problem of finding equidistant points of system (1) is nonlinear.

Theorem 5.1 [Dzyadyk (1974c)]. *Let*

$$(a_i, z) = w_i, \tag{5.1'}$$

$$a_i = (a_{i1}, \dots, a_{in}), \quad z = (z_1, z_2, \dots, z_n), \quad i = 1, 2, \dots, n+1,$$

be an incompatible system of $n+1$ *equations of the form (5.1). Assume that, as a result of the removal from this system of an arbitrary equation*

$$(a_j, z) = a_{j1} z_1 + a_{j2} z_2 + \dots + a_{jn} z_n = w_j,$$

the determinant D^j *of the subsystem obtained as a result differs from zero. The solution of this subsystem is denoted by* $z^j = \left(z_1^j, z_2^j, \dots, z_n^j\right)$, *i.e.,* $(a_\nu, z^j) = w_\nu$, $\nu \neq j$.

Then

(i) for any $j = 1, 2, \dots, n+1$, *the following equality is true:*

$$(a_j, z^j) - w_j = \frac{(-1)^{j+1}}{D^j} \sum_{\nu=1}^{n+1} (-1)^\nu w_\nu D^\nu, \tag{5.5}$$

where

$$D^j = \begin{vmatrix} a_{11} & a_{12} & \dots & a_{1n} \\ \dots & \dots & \dots & \dots \\ a_{j-11} & a_{j-12} & \dots & a_{j-1n} \\ a_{j+11} & a_{j+12} & \dots & a_{j+1n} \\ \dots & \dots & \dots & \dots \\ a_{n+11} & a_{n+12} & \dots & a_{n+1n} \end{vmatrix}; \tag{5.6}$$

(ii) for all real numbers k_j, $j = 1, 2, \dots, n+1$, *such that*

$$\sum_{j=1}^{n+1} \left| D^j \right| e^{ik_j} \neq 0, \tag{5.7}$$

the point z *given by the formula*† :

† It is easy to see that each component of this point for variable k_j represents an $(n+1)$-dimensional torus.

$$z = \rho \sum_{j=1}^{n+1} \frac{z^j e^{ik_j}}{|(a_j, z^j) - w_j|} = \rho \frac{\sum_{j=1}^{n+1} |D^j| z^j e^{ik_j}}{\left|\sum_{j=1}^{n+1} (-1)^j D^j w_j\right|}, \tag{5.8}$$

where

$$\rho := \left(\sum_{j=1}^{n+1} \frac{e^{ik_j}}{|(a_j, z^j) - w_j|} \right)^{-1} = \frac{\left|\sum_{j=1}^{n+1} (-1)^j D^j w_j\right|}{\sum_{j=1}^{n+1} |D^j| e^{ik_j}}, \tag{5.9}$$

is A-equidistant for system (5.1) and the quantity $|\rho|$ is the A-distance between the point z and the point w;

(iii) vice versa, every A-equidistant point z of system (5.1′) admits representation (5.8) for some real k_j; specifically, the numbers k_j can be expressed via the point z by the formula

$$k_j = \arg[(a_j, z) - w_j] - \arg[(a_j, z^j) - w_j] \tag{5.10}$$

and the number ρ computed by using relation (5.9) for these k_j is positive.

Proof. (i) Note that the coordinates z_k^j of the solution z^j are given by the formula

$$z_k^j = \frac{\begin{vmatrix} a_{11} & \cdots & a_{1\,k-1} & w_1 & a_{1\,k+1} & \cdots & a_{1n} \\ \cdots & \cdots & \cdots & \cdots & \cdots & \cdots & \cdots \\ a_{j-1\,1} & \cdots & a_{j-1\,k-1} & w_{j-2} & a_{j-1\,k+1} & \cdots & a_{j-1\,n} \\ a_{j+1\,1} & \cdots & a_{j+1\,k-1} & w_{j+1} & a_{j+1\,k+1} & \cdots & a_{j+1\,n} \\ \cdots & \cdots & \cdots & \cdots & \cdots & \cdots & \cdots \\ a_{n+1\,1} & \cdots & a_{n+1\,k-1} & w_{n+1} & a_{n+1\,k+1} & \cdots & a_{n+1\,n} \end{vmatrix}}{\begin{vmatrix} a_{11} & \cdots & a_{1\,k-1} & a_{1k} & a_{1\,k+1} & \cdots & a_{1n} \\ \cdots & \cdots & \cdots & \cdots & \cdots & \cdots & \cdots \\ a_{j-1\,1} & \cdots & a_{j-1\,k-1} & a_{j-1\,k} & a_{j-1\,k+1} & \cdots & a_{j-1\,n} \\ a_{j+1\,1} & \cdots & a_{j+1\,k-1} & a_{j+1\,k} & a_{j+1\,k+1} & \cdots & a_{j+1\,n} \\ \cdots & \cdots & \cdots & \cdots & \cdots & \cdots & \cdots \\ a_{n+1\,1} & \cdots & a_{n+1\,k-1} & a_{n+1\,k} & a_{n+1\,k+1} & \cdots & a_{n+1\,n} \end{vmatrix}}$$

$$= \frac{\sum_{\nu=1,\,\nu\neq j}^{n+1} w_\nu A_{\nu k}^j}{D^j} = \frac{\sum_{\nu=1,\,\nu\neq j}^{n+1} w_\nu (-1)^{\nu+k} \operatorname{sgn}(j-\nu) M_{\nu k}^j}{D^j}, \tag{5.11}$$

where $A_{\nu k}^j$ and $M_{\nu k}^j$ are, respectively, the cofactor and the minor of an element $a_{\nu k}$ in the determinant D^j. Hence, in view of the facts that

$$M^j_{\nu k} = M^\nu_{jk}$$

and

$$A^j_{\nu k} = (-1)^{\nu+k}\operatorname{sgn}(j-\nu)M^j_{\nu k} = -(-1)^{\nu+k}A^\nu_{jk}$$

for any $j = 1, 2, \dots, n+1$, we find

$$\begin{aligned}(a_j, z^j) - w_j &= \sum_{k=1}^{n} a_{jk} z^j_k - w_j = \sum_{k=1}^{n} a_{jk} \frac{\sum_{\nu=1,\,\nu\neq j}^{n+1} w_\nu A^j_{\nu k}}{D^j} - w_j \\ &= \frac{1}{D^j}\sum_{k=1}^{n} a_{jk} \sum_{\nu=1,\,\nu\neq j}^{n+1} w_\nu (-1)^{\nu+j+1} A^\nu_{jk} - w_j \\ &= \frac{(-1)^{j+1}}{D^j} \sum_{\nu=1,\,\nu\neq j}^{n+1} (-1)^\nu w_\nu \sum_{k=1}^{n} a_{jk} A^\nu_{jk} - w_j \\ &= \frac{(-1)^{j+1}}{D^j} \sum_{\nu=1,\,\nu\neq j}^{n+1} (-1)^\nu w_\nu D^\nu - w_j \\ &= \frac{(-1)^{j+1}}{D^j} \sum_{\nu=1}^{n+1} (-1)^\nu w_\nu D^\nu.\end{aligned}$$

This proves equality (5.5).

Prior to proving assertion (ii) of Theorem 5.1, we establish the validity of assertion (iii).

(iii) Let z be an arbitrary point equidistant for system (5.1′). It is necessary to show that the point z admits representation (5.8) with

$$k_j = \arg\,[(a_j, z) - w_j] - \arg\,[(a_j, z^j) - w_j]. \tag{5.12}$$

Indeed, since the point z is equidistant, according to the definition, there exists a number $\rho > 0$ [this number is positive because system (1) is incompatible by the condition of the theorem] such that the following equalities hold:

$$(a_j, z) - w_j = \rho b_j, \quad j = 1, 2, \dots, n+1, \tag{5.13}$$

where $b_j = e^{i \arg[(a_j, z) - w_j]}$. By virtue of (5.12), we can also represent the numbers b_j in the form

$$b_j = \frac{[(a_j, z^j) - w_j]e^{ik_j}}{|(a_j, z^j) - w_j|}. \tag{5.14}$$

The incompatibility of system (5.13) implies that

$$\begin{vmatrix} a_{11} & a_{12} & \dots & a_{1n} & w_1 + \rho b_1 \\ \dots & \dots & \dots & \dots & \dots \\ a_{n+11} & a_{n+12} & \dots & a_{n+1n} & w_{n+1} + \rho b_{n+1} \end{vmatrix} = 0$$

and, therefore,

$$\rho = -\frac{\sum_{\nu=1}^{n+1} (-1)^{\nu} w_{\nu} D^{\nu}}{\sum_{\nu=1}^{n+1} (-1)^{\nu} b_{\nu} D^{\nu}}. \tag{5.15}$$

By virtue of (5.14) and (5.5), for any $j = 1, 2, \dots, n+1$, we can write

$$b_j D^j = D^j \frac{[(a_j, z^j) - w_j]e^{ik_j}}{|(a_j, z^j) - w_j|} = (-1)^{j+1} \sum_{\nu=1}^{n+1} (-1)^{\nu} w_{\nu} D^{\nu} \frac{e^{ik_j}}{|(a_j, z^j) - w_j|}.$$

Thus, it follows from (5.15) that

$$\rho = \left(\sum_{j=1}^{n+1} \frac{e^{ik_j}}{|(a_j, z^j) - w_j|} \right)^{-1}. \tag{5.16}$$

Finally, to show that the point z admits representation (5.8), we set

$$z' := \rho \sum_{j=1}^{n+1} \frac{z^j e^{ik_j}}{|(a_j, z^j) - w_j|} \tag{5.8'}$$

and prove that $z = z'$. Indeed, according to the definition of the solutions z^j, in view of the incompatibility of system (1′), for all $i, j = 1, 2, \dots, n+1$, we obtain

$$(a_j, z^i) = w_j, \quad j \neq i,$$
$$(a_j, z^j) - w_j \neq 0, \quad j = 1, 2, \dots, n+1. \tag{5.17}$$

Hence, the right-hand side of (5.8′) is meaningful.

By using these relations and equalities (5.16) and (5.8′) for all $j = 1, 2, \dots, n+1$, we find

$$(a_j, z') - w_j = \rho \sum_{\nu=1}^{n+1} \frac{(a_j, z^\nu)e^{ik_\nu}}{\left|(a_j, z^\nu) - w_\nu\right|} - \rho \sum_{\nu=1}^{n+1} \frac{w_j e^{ik_\nu}}{\left|(a_\nu, z^\nu) - w_\nu\right|}$$

$$= \rho \frac{\left[(a_j, z^j) - w_j\right]e^{ik_j}}{\left|(a_j, z^j) - w_j\right|} = \rho b_j. \tag{5.18}$$

Subtracting now the left- and right-hand sides of equalities (5.18) from relation (5.13) term by term, we get

$$(a_j, z - z') = a_{j1}(z_1 - z_1') + a_{j2}(z_2 - z_2') + \dots + a_{jn}(z_n - z_n') = 0,$$

$$j = 1, 2, \dots, n+1.$$

Hence, in view of the fact that, e.g., the determinant D^{n+1} differs from zero, we conclude that $z_\nu = z_\nu'$, $\nu = 1, 2, \dots, n$, and, consequently, $z = z'$.

This proves assertion (iii) of Theorem 5.1.

(ii) First, we note that the second equalities in relations (5.8) and (5.9) follow from equality (5.5). By analogy with the procedure used to deduce equalities (5.18), we can show that, under condition (5.7) (with $|\rho| < \infty$), the point z defined by equality (5.8) satisfies the conditions

$$(a_j, z) - w_j = \rho b_j, \quad b_j = \frac{\left[(a_j, z^j) - w_j\right]e^{ik_j}}{\left|(a_j, z^j) - w_j\right|}, \tag{5.18′}$$

and, hence, the conditions $\left|(a_1, z) - w_1\right| = \left|(a_2, z) - w_2\right| = \dots = \left|(a_{n+1}, z) - w_{n+1}\right|$. Therefore, this point is indeed an equidistant point of system (5.1′).

This proves assertion (ii).

The proof of Theorem 5.1 is thus completed.

Remark 5.1. Generally speaking, the set $\mathcal{E}$ of equidistant points is not convex for $n > 0$.

Indeed, for $n = 1$, by setting

$$z_1 = \rho_1 \frac{z^1}{\left|(a_1, z^1) - w_1\right|} + \rho_1 \frac{z^2}{\left|(a_2, z^2) - w_2\right|},$$

$$z_2 = \rho_2 \frac{z^1}{\left|(a_1, z^1) - w_1\right|} + \rho_2 \frac{z^2 e^{i\pi}}{\left|(a_2, z^2) - w_2\right|}$$

$$= \rho_2 \frac{z^1}{\left|(a_1, z^1) - w_1\right|} + \rho_2 \frac{z^2}{\left|(a_2, z^2) - w_2\right|},$$

$$z^* = \frac{1}{2}(z_1 + z_2),$$

$$\rho_1 = \left(\frac{1}{\left|(a_1, z^1) - w_1\right|} + \frac{1}{\left|(a_2, z^2) - w_2\right|} \right)^{-1},$$

$$\rho_2 = \left(\frac{1}{\left|(a_1, z^1) - w_1\right|} - \frac{1}{\left|(a_2, z^2) - w_2\right|} \right)^{-1}$$

we readily conclude that

$$\left|(a_1, z^*) - w_1\right| = \frac{1}{2}\left|\rho_1 + \rho_2\right| \neq \frac{1}{2}\left|\rho_1 - \rho_2\right| = \left|(a_2, z^*) - w_2\right|,$$

i.e., we see that

$$z^* = \frac{1}{2}(z_1 + z_2) \notin \mathcal{E},$$

despite the fact that, according to Theorem 5.1, $z_1 \in \mathcal{E}$ and $z_2 \in \mathcal{E}$.

Remark 5.2. Since the quantity

$$\rho = \left| \sum_{j=1}^{n+1} \frac{e^{ik_j}}{\left|(a_j, z^j) - w_j\right|} \right|^{-1}$$

takes the least value if and only if $e^{ik_1} = e^{ik_2} = \ldots = e^{ik_{n+1}}$, the best solution z^* of system (5.1) and the corresponding A-distance ρ^* of this solution from the point $(w_1, w_2, \ldots, w_{n+1})$ under the conditions of Theorem 5.1 are given by the formulas

$$z^* = \rho^* \sum_{j=1}^{n+1} \frac{z^j}{\left|(a_j, z^j) - w_j\right|} = \sum_{j=1}^{n+1} \frac{D^j}{\sum_{\nu=1}^{n+1} \left|D^\nu\right|} z^j, \tag{5.19}$$

$$\rho^* \left[\sum_{j=1}^{n+1} \frac{z^j}{\left|(a_j, z^j) - w_j\right|}\right]^{-1} = \frac{\left|\sum_{j=1}^{n+1} (-1)^j w_j D^j\right|}{\sum_{j=1}^{n+1} \left|D^j\right|}. \tag{5.20}$$

By using relation (5.20), we can represent the quantity ρ in relation (5.9) in the form

$$\rho = \rho^* \frac{\sum_{j=1}^{n+1} \left|D^j\right|}{\sum_{j=1}^{n+1} \left|D^j\right| e^{ik_j}}. \tag{5.21}$$

5.2. General case

Let us now establish the conditions under which the point z^* is the b.a.-solution of system (5.1) in the general case.

Definition 5.1′. *A point z^* is called the point of local best approximation for system (5.1) if there exists a number $\varepsilon > 0$ such that condition (5.2) holds for this point for any $z \in \overline{U(z^*; \varepsilon)} := \{z : |z - z^*| \le \varepsilon\}$.*

Definition 5.3. *System (5.1) is called irreducible if, for any its proper subsystem, the value of the best approximation is smaller than for the entire system.*

Theorem 5.2 (on the conditions of local best approximation of a system). *If, for an incompatible irreducible system of equations (in the notation of Theorem 5.1)*

$$(a_i, z) = w_i, \quad i = 1, 2, \ldots, n+1+k, \tag{5.1′}$$

where $1 \le k \le n$, the point

$$z^0 = \rho^0 \sum_{j=1}^{n+1} \frac{z_j e^{ix_j^0}}{\left|(a_j, z^j) - w_j\right|} = \rho^0 \sum_{j=1}^{n+1} C_j z^j e^{ix_j^0}, \tag{5.22}$$

$$C_j := \frac{1}{\left|(a_j, z^j) - w_j\right|}, \qquad \rho^0 = \left(\sum_{j=1}^{n+1} C_j e^{ix_j^0}\right)^{-1}, \tag{5.23}$$

is equidistant for a subsystem

$$(a_i, z) - w_i = 0, \quad i = 1, 2, \ldots, n+1, \tag{5.$\tilde{1}$}$$

then this point is the point of the best local approximation of system $(5.\tilde{1})$ *if and only if, at the point* $x^0 := \left(x_1^0, x_2^0, \ldots, x_{n+1}^0\right) \in R^{n+1}$ *specifying the point* z^0 *[according to relation (5.22)], the function*

$$f(x) = f(x_1, x_2, \ldots, x_{n+1}) := \sum_{j,k=1}^{n+1} C_j C_k \cos(x_j - x_k), \quad x_j, x_k \in [0, 2\pi), \tag{5.24}$$

possesses a local conditional maximum with the following k *constraints:*

$$\Phi_\nu(x) = \Phi_\nu(x_1, x_2, \ldots, x_{n+1}) := \sum_{j,k=1}^{n+1} [(A_{\nu j}A_{\nu k} + B_{\nu j}B_{\nu k}) \cos(x_j - x_k)$$

$$+ (A_{\nu j}B_{\nu k} - A_{\nu k}B_{\nu j}) \sin(x_j - x_k)] - 1 = 0, \tag{5.25}$$

where

$$A_{\nu j} := C_j \operatorname{Re}[(a_\nu z^j) - w_\nu] \quad \text{and} \quad B_{\nu j} := C_j \operatorname{Im}[(a_\nu z^j) - w_\nu], \tag{5.26}$$

$$\nu = n+2, \ldots, n+k+1; \quad j = 1, 2, \ldots, n+1.$$

Remark 5.3. Note that, according to Theorem 4.3 (on cleaning) reformulated for system $(5.\tilde{1})$, the characteristic set of any system of the form $(5.\tilde{1})$ consists of at most $2n+1$ points and, hence, in view of the irreducibility of system $(5.\tilde{1})$ only the case of systems of the form (5.1) with $k \le n$ is indeed important.

Proof of Theorem 5.2. According to Theorem 4.1 (Kolmogorov) reformulated for system (5.1′), the condition $\left|(a_\nu z^0) - w_\nu\right| = \text{const}$ must hold for all $\nu = 1, 2, \ldots, n+1+k$. Thus, by virtue of equalities (5.22) and (5.23) and the definition of the points z^j, we find

$$\left|(a_i z^0) - w_i\right|^2 = \left|\rho^0\right|^2, \quad i = 1, 2, \ldots, n+1,$$

and, moreover,

$$\begin{aligned}
\left|(a_\nu z^0) - w_\nu\right|^2 &= \left|\rho^0\right|^2 \left|\sum_{j=1}^{n+1} \frac{\left[(a_\nu, z^j) - w_\nu\right]}{\left|(a_j, z^j) - w_j\right|}\right|^2 \\
&= \left|\rho^0\right|^2 \sum_{k=1}^{n+1} (A_{\nu k} + iB_{\nu k}) e^{ix_k^0} \sum_{j=1}^{n+1} (A_{\nu j} - iB_{\nu j}) e^{-ix_j^0} \\
&= \left|\rho^0\right|^2 [\Phi_\nu(x^0) + 1] = \left|\rho^0\right|^2, \quad \nu = n+2, \ldots, n+k+1.
\end{aligned}$$

This immediately implies that the equations of constraints (5.25) hold at the point x^0. At the same time, in view of the fact that, according to (5.23),

$$\left|\rho^0\right|^2 = \left|\rho^0(x)\right|^2 = \left(\left|\sum_{j=1}^{n+1} C_j e^{ix_j}\right|^2\right)^{-1} = [f(x)]^{-1},$$

we conclude that, in the presence of constraints (5.25), the function $(\rho^0)^2$ (and, hence, $|\rho^0|$) possesses a local minimum at the point x^0 if and only if the function f possesses a local maximum at the same point. This proves Theorem 5.2.

Note that

(i) by virtue of the periodicity of the function f in each variable x_j, its absolute maximum is necessarily local and it is easy to see that the total number of local maxima is finite; therefore, the absolute minimum ρ^* of the function ρ and, clearly, the b.a.-solution of system $(5.\tilde{1})$ can always be found as the solution for which the function ρ attains its absolute minimum;

(ii) if we consider solely the equations of constraints (5.25), then we obtain their solution in the form of a set of points $x \in R^{n+1}$ specifying, according to relation (5.22), the set of equidistant points z of system $(5.\tilde{1})$.

Reformulating Theorem 5.1 for the case of approximation of functions by polynomials with regard for relations (5.3), we arrive at the following assertions:

Theorem 5.1′. *Assume that a Chebyshev system of functions* $\{\varphi_j\}_{j=0}^n$ *and a function* f *are given on a set* $\mathfrak{M} = \{z_0, z_1, \ldots, z_{n+1}\}$ *of different points* z_j. *The values of the function* f *on the set* $\mathfrak{M}$ *do not coincide with the corresponding values of any polynomial of the form*

$$P_n(z) = P_n(\varphi; z) = \sum_{k=0}^{n} c_k \varphi_k(z).$$

Then

(i) for any $j = 0, 1, \dots, n+1$, *the (interpolation) polynomial* $P_{nj}(z) = P_{nj}(\varphi; z)$ *whose values coincide with the corresponding values of the function* $f(z)$ *at* $n + 1$ *points* z_ν, $\nu = 0, 1, \dots, j-1, j+1, \dots, n+1$, *differs from its value* $f(z_j)$ *at the point* z_j *by the quantity*

$$P_{nj}(z_j) - f(z_j) = \frac{(-1)^{j+1}}{D^j} \sum_{\nu=0}^{n+1} (-1)^\nu f(z_\nu) D^\nu, \tag{5.5'}$$

where

$$D^\nu = \begin{vmatrix} \varphi_0(z_0) & \dots & \varphi_n(z_0) \\ \dots & \dots & \dots \\ \varphi_0(z_{\nu-1}) & \dots & \varphi_n(z_{\nu-1}) \\ \varphi_0(z_{\nu+1}) & \dots & \varphi_n(z_{\nu+1}) \\ \dots & \dots & \dots \\ \varphi_0(z_{n+1}) & \dots & \varphi_n(z_{\nu+1}) \end{vmatrix} = D\begin{pmatrix} \varphi_0 & \varphi_1 & \dots & & & \varphi_n \\ z_0 & \dots & z_{\nu-1} & z_{\nu+1} & \dots & z_{n+1} \end{pmatrix}; \tag{5.6'}$$

(ii) for all real numbers k_j, $j = 0, 1, \dots, n+1$, *satisfying the inequality*

$$\sum_{j=0}^{n+1} \left| D^j \right| e^{ik_j} \neq 0, \tag{5.7'}$$

the polynomial P_n *given by the formula*

$$P_n(z) = \rho \sum_{j=0}^{n+1} \frac{P_{nj}(z) e^{ik_j}}{\left| P_{nj}(z_j) - f(z_j) \right|} = \rho \frac{\sum_{j=0}^{n+1} \left| D^j \right| P_{nj}(z) e^{ik_j}}{\left| \sum_{\nu=0}^{n+1} (-1)^\nu D^\nu f(z_\nu) \right|}, \tag{5.8''}$$

where

$$\rho = \left(\sum_{j=0}^{n+1} \frac{e^{ik_j}}{\left| P_{nj}(z_j) - f(z_\nu) \right|} \right)^{-1} = \frac{\sum_{j=0}^{n+1} (-1)^j D^j f(z_j)}{\left| \sum_{j=0}^{n+1} \left| D^j \right| e^{ik_j} \right|}, \tag{5.9'}$$

satisfies the conditions

$$|P_n(z_0) - f(z_0)| = |P_n(z_1) - f(z_1)| = \ldots = |P_n(z_{n+1}) - f(z_{n+1})| \quad (= |\rho|); \qquad (5.4')$$

(iii) vice versa, any polynomial P_n *satisfying conditions (5.4′) can be represented in the form (5.8″) for some real numbers* k_j, $j = 0, 1, \ldots, n+1$; *the numbers* k_j *can be, in particular, expressed via the polynomial* P_n *by the formula*

$$k_j = \arg[P_n(z_j) - f(z_j)] - \arg[P_{nj}(z_j) - f(z_j)]. \qquad (5.10')$$

The number ρ *computed for these* k_j *by using relation (5.9′) is positive.*

Theorem 5.3. *Assume that a Chebyshev system of functions* $\{\varphi_j\}_{j=0}^{n}$ *and a function* f *are given on the set* $\mathfrak{M} = \{z_0, z_1, \ldots, z_{n+1}\}$ *of different points* z_j. *The values of the function* f *on the set* $\mathfrak{M}$ *do not coincide with the corresponding values of any polynomial of the form*

$$P_n(z) = P_n(\varphi; z) = \sum_{k=0}^{n} c_k \varphi_k(z). \qquad (5.22)$$

Then

(i) the polynomial

$$P_n^*(z) = \sum_{k=0}^{n} c_k^* \varphi_k(z)$$

of the best approximation of the function f *on the set* $\mathfrak{M}$ *in a sense that*

$$\max_{j=0,1,\ldots,n} |f(z_j) - P_n^*(z_j)| = \min_{P_n} \max_{j=0,1,\ldots,n} |f(z_j) - P_n(z_j)|,$$

is defined by each of the following equalities:

$$P_n^*(z) = E_n(f) \sum_{j=0}^{n+1} \frac{P_{nj}(z)}{|P_{nj}(z_j) - f(z_j)|} = \sum_{j=0}^{n+1} \frac{|D^j|}{\sum_{\nu=0}^{n+1} |D^\nu|} P_{nj}(z), \qquad (5.19')$$

where, for any $j = 0, 1, \ldots, n+1$, $P_{nj}(z)$ *is a polynomial interpolating the function* f

at $n+1$ points z_ν, $\nu = 0, 1, \dots, j-1, j+1, \dots, n+1$, and $E_n(f)$ is the value of the best approximation of the function f by polynomials of the form (5.22);

(ii) the quantity $E_n(f)$ can be found by using each of the following relations:

$$E_n(f) = \min_{P_n} \max_j \left| f(z_j) - P_n(z_j) \right| = \left[\sum_{j=0}^{n+1} \frac{1}{\left| P_{nj}(z_j) - f(z_j) \right|} \right]^{-1}$$

$$= \frac{\left| \sum_{j=0}^{n+1} (-1)^j D^j f(z_j) \right|}{\sum_{j=0}^{n+1} |D^j|} = \frac{\left| \sum_0^{n+1} (-1)^j D^j \left[f(z_j) - P_n^0(z_j) \right] \right|}{\sum_{j=0}^{n+1} |D^j|}, \qquad (5.20')$$

where P_n^0 is an arbitrary polynomial.

We now apply the results established above to a Chebyshev system of functions $\{\varphi_k\}_{k=0}^n$ on the segment $[a, b]$. According to Lemma 2.1, for this system, the function

$$D\begin{pmatrix} \varphi_0 & \varphi_1 & \dots & \varphi_n \\ \cdots & \cdots & \cdots & \cdots \\ x_0 & x_1 & \dots & x_n \end{pmatrix} = \begin{vmatrix} \varphi_0(x_0) & \varphi_0(x_1) & \dots & \varphi_0(x_n) \\ \cdots & \cdots & \cdots & \cdots \\ \varphi_n(x_0) & \varphi_n(x_1) & \dots & \varphi_n(x_0) \end{vmatrix}$$

does not change its sign under all possible changes in the locations of the points x_0, $x_1, \dots, x_n$ preserving the order of their appearance. This means that if $a \le x_0 < x_1 < \dots < x_n < x_{n+1} \le b$ and D^ν is given by relation (5.6′), then all D^ν have the same sign.
By using this remark and Theorem 5.3, we arrive at the following assertion:

Theorem 5.4. *Let $\{\varphi_k\}_{k=0}^n$ be a Chebyshev system of functions given on the segment $[a, b]$ and let f be a function defined at $n+2$ points x_j of this segment such that $a \le x_0 < x_1 < \dots < x_n < x_{n+1} \le b$. Assume that the values of this function on the set $\mathfrak{M} = \{x_0, x_1, \dots, x_{n+1}\}$ do not coincide with the corresponding values of any polynomial P_n of the form*

$$P_n = \sum_{k=0}^{n} c_k \varphi_k.$$

Then, in the notation of Theorem 5.3,

(i) the polynomial

$$P_n^*(x) = \sum_{k=0}^{n} c_k^* \varphi_k(x)$$

of the best approximation of the function f *on* $\mathfrak{M}$ *is defined by each of the following equalities:*

$$P_n^*(x) = E_n(f) \sum_{j=0}^{n+1} \frac{P_{nj}(x)}{|P_{nj}(x_j) - f(x_j)|} = \sum_{j=0}^{n+1} \frac{|D^j|}{\sum_{\nu=0}^{n+1} |D^\nu|} P_{nj}(x); \qquad (5.19'')$$

(ii) the value $E_n(f)$ *of the best approximation of the function* f *on* $\mathfrak{M}$ *can be found by using each of the following formulas*[7]:

$$E_n(f) = \left[\sum_{j=0}^{n+1} \frac{1}{|P_{nj}(x_j) - f(x_j)|} \right]^{-1} = \frac{\left|\sum_{j=0}^{n+1} (-1)^j |D^j| f(x_j)\right|}{\sum_{j=0}^{n+1} |D^j|}. \qquad (5.20'')$$

6. Algorithms of construction of polynomials of the best approximation

In most cases, the problem of finding, for a given continuous function f, the polynomial of its best approximation $P_n^*(f; \cdot)$ of the form

$$P_n(\cdot) = \sum_{k=0}^{n} c_k \varphi_k(\cdot) \qquad (6.1)$$

is unsolvable. Therefore, the problem of algorithms aimed at the construction of polynomials $P_n(f; \cdot)$ arbitrarily close to the polynomial of the best approximation $P_n^*(f; \cdot)$ for a given function $f \in C$ is of significant practical and theoretical interest. There are several algorithms of this sort.

In the present section, we describe the Remez algorithm,[8] which proves to be especially convenient in many aspects. This algorithm is generally accepted and used in practice for the approximate representation of continuous functions by polynomials. In what follows, we mainly deal with the theoretical analysis of the Remez algorithm. Various practical problems connected with the application of this algorithm are discussed in [Remez (1969)]; see also [Meinardus (1964)] and [Laurent (1975)].

We now give a description of the Remez algorithm for the case of approximation of a function $f(x)$ continuous on the segment $[a, b]$ by polynomials of the form (1) constructed according to a Chebyshev system of functions $\{\varphi_k\}_{k=0}^{n}$:

Step 1. Consider an arbitrary system[†] of $n+2$ different points $x_k^{(1)}$ on the segment $[a, b]$:

$$a \le x_0^{(1)} < x_1^{(1)} < \ldots < x_n^{(1)} < x_{n+1}^{(1)} \le b. \tag{6.2}$$

By using Theorem 5.4, we construct the polynomial $P_n^{(1)}$ of the best approximation of the function f at the points of the system $\left\{x_k^{(1)}\right\}_{k=0}^{n+1}$. Further, in view of the fact that

$$\left|f\left(x_k^{(1)}\right) - P_n^{(1)}\left(x_k^{(1)}\right)\right| = \text{const}, \quad k = 0, 1, \ldots, n+1,$$

we set

$$f(x) - P_n^{(1)}(x) = r_n^{(1)}(x), \quad x \in [a, b], \quad \left|r_n^{(1)}\left(x_k^{(1)}\right)\right| = \underline{E}_n^{(1)}, \tag{6.3}$$

$$\left\| r_n^{(1)} \right\|_{[a,b]} = \max_{a \le x \le b} \left| f(x) - P_n^{(1)}(x) \right| = \overline{E}_n^{(1)}.$$

Step 2. First, we note that the value $E_n(f)$ of the best approximation of the function f in the entire segment is not smaller than the value of the best approximation $\underline{E}_n^{(1)}$ of this function on the system of points $\left\{x_k^{(1)}\right\}_{k=0}^{n+1}$ and, therefore, $\underline{E}_n^{(1)} \le E_n(f)$. At the same time, if $E_n(f) = \overline{E}_n^{(1)}$, then $P_n^{(1)}$ is just the polynomial of the best approximation of the function f and, hence, the process of construction is completed. Thus, we assume that $E_n(f) < \overline{E}_n^{(1)}$ and, hence,

$$\underline{E}_n^{(1)} \le E_n(f) < \overline{E}_n^{(1)}. \tag{6.4}$$

According to the Weierstrass theorem, there exists a point $x^* \in [a, b]$ such that $\left|r_n^{(1)}(x^*)\right| = \overline{E}_n^{(1)}$.

We now replace the system of points (6.2) by a system $\left\{x_k^{(2)}\right\}_{k=0}^{n+1}$,

$$a \le x_0^{(2)} < x_1^{(2)} < \ldots < x_n^{(2)} < x_{n+1}^{(2)} \le b, \tag{6.2$'$}$$

[†] In practice, in the case of approximation by algebraic polynomials, it is reasonable to choose the points $x_k^{(1)} = a + 0.5(b-a)x_k^{(0)}$, where $x_k^{(0)} := 1 - \cos(k\pi/(n+1))$, $k = 0, 1, \ldots, n+1$.

such that the following three conditions are true for all $k = 0, 1, \ldots, n+1$:

$$\operatorname{sgn} r_n^{(1)}\big(x_{k+1}^{(2)}\big) = -\operatorname{sgn} r_n^{(1)}\big(x_k^{(2)}\big), \quad \big|r_n^{(1)}\big(x_k^{(2)}\big)\big| \geq \underline{E}_n^{(1)}, \quad 0 \leq k \leq n+1;$$
$$\max_k \big|r_n^{(1)}\big(x_k^{(2)}\big)\big| = \overline{E}_n^{(1)}. \tag{6.5}$$

To satisfy all these conditions, it suffices to replace one point in system (6.2) by the point x^* and preserve all other points in the system. The system obtained as a result is regarded as $\big\{x_k^{(2)}\big\}_{k=0}^{n+1}$. The process of replacement can be realized, e.g., as follows: If the point x^* is located between two points $x_k^{(1)}$ and $x_{k+1}^{(1)}$ of system (6.2), then one of these points is replaced by x^* (at this point, the difference $r_n^{(1)}$ must have the same sign as at the point x^*). If the point x^* is located to the left of all points of system (6.2) and $\operatorname{sgn} r_n^{(1)}(x^*) = \operatorname{sgn} r_n^{(1)}\big(x_0^{(1)}\big)$, then the point $x_0^{(1)}$ is replaced by x^*. If, on the contrary, $\operatorname{sgn} r_n^{(1)}(x^*) = -\operatorname{sgn} r_n^{(1)}\big(x_0^{(1)}\big)$, then the system $\big\{x_k^{(2)}\big\}_{k=0}^{n+1}$ is chosen as follows: x^*, $x_0^{(1)}$, $x_1^{(1)}, \ldots, x_n^{(1)}$. The case where the point x^* lies to the right of all points of system (6.2) is studied similarly.

Note that, in practice, it is preferable to replace more points of the system $\big\{x_k^{(1)}\big\}_{k=0}^{n+1}$ by new points (x^* is one of these points) in order that, first, all conditions (6.5) be satisfied and, second, the quantities $\big|r_n^{(1)}\big(x_k^{(2)}\big)\big|$ be as large as possible.

As soon as system (6.2′) is obtained, we construct the polynomial $\pi_n^{(1)}$ of the best approximation for the function $r_n^{(1)} = f - P_n^{(1)}$ on this system and set

$$P_n^{(2)} = P_n^{(1)} + \pi_n^{(1)},$$
$$f - P_n^{(2)} = r_n^{(1)} - \pi_n^{(1)} = r_n^{(2)}, \tag{6.3′}$$
$$r_n^{(2)}\big(x_k^{(2)}\big) = \underline{E}_n^{(2)}, \quad \big\|r_n^{(2)}\big\|_{[a,b]} = \overline{E}_n^{(2)}.$$

Step 3. As in the second step, we assume that $E_n(f) < \overline{E}_n^{(2)}$ (for $E_n(f) = \overline{E}_n^{(2)}$, the polynomial $P_n^{(2)}$ is just the required polynomial of the best approximation for the function f). Therefore,

$$\underline{E}_n^{(2)} \leq E_n(f) < \overline{E}_n^{(2)}. \tag{6.4′}$$

Then, as in the second step, we replace system (6.2′) with the system $\big\{x_k^{(3)}\big\}_{k=0}^{n+1}$,

$$a \le x_0^{(3)} < x_1^{(3)} < \ldots < x_n^{(3)} < x_{n+1}^{(3)} \le b, \tag{6.2''}$$

such that

$$\operatorname{sgn} r_n^{(2)}\big(x_{k+1}^{(3)}\big) = -\operatorname{sgn} r_n^{(2)}\big(x_k^{(3)}\big), \quad \big|r_n^{(2)}\big(x_k^{(3)}\big)\big| \ge \underline{E}_n^{(2)}, \quad 0 \le k \le n+1,$$
$$\max_{0\le k\le n+1} \big|r_n^{(2)}\big(x_k^{(3)}\big)\big| = \overline{E}_n^{(2)}. \tag{6.5'}$$

After this, we construct the polynomial $\pi_n^{(2)}$ of the best approximation for the function $r_n^{(2)}$ on system (2) and set

$$P_n^{(3)} = P_n^{(2)} + \pi_n^{(2)}, \quad f - P_n^{(3)} = r_n^{(3)},$$
$$\big|r_n^{(3)}\big(x_k^{(3)}\big)\big| = \underline{E}_n^{(3)}, \quad \big\|r_n^{(3)}\big\|_{[a,b]} = \overline{E}_n^{(3)}, \tag{6.3''}$$

etc.

Theorem 6.1 [Remez (1957)]. *The Remez algorithm converges with the rate of a geometric progression in a sense that, for any function f continuous on the segment $[a, b]$, one can find numbers $A > 0$ and $0 < q < 1$ such that the deviations $\overline{E}_n^{(k)}$ of the polynomials $P_n^{(k)}$ constructed by using this algorithm from the function f satisfy the inequalities*

$$\overline{E}_n^{(k)} - E_n(f) = \big\|f - P_n^{(k)}\big\|_{[a,b]} - E_n(f) \le A\,q^k, \quad k = 1, 2, \ldots, \tag{6.6}$$

where $E_n(f)$ is the value of the best approximation of the function $f(x)$ on $[a, b]$ by polynomials P_n of the form (6.1).

To prove Theorem 6.1, we need the following lemma:

Lemma 6.1. *Let $f \in C[a, b]$ and let $\{x_i\}_{i=0}^{n+1}$ be a certain system of points on the segment $[a, b]$. Then, for any $\varepsilon > 0$, one can find a number $\delta > 0$ such that if the minimum distance between the points of the system $\{x_i\}_{i=0}^{n+1}$ is smaller than δ, then the best approximation of the function f on this system by polynomials of the form (6.1) is smaller than ε.*

Proof. Assume the opposite, i.e., that the assertion of Lemma 6.1 is not true. Then there exists a sequence of systems $\{x_i^{(k)}\}_{i=0}^{n+1}$, $k = 1, 2, \ldots,$ in which the minimum distance between the points approaches zero as $k \to \infty$:

$$\lim_{k\to\infty} \min_{i\neq j} \left\{ \left| x_i^{(k)} - x_j^{(k)} \right| \right\} = 0 \tag{6.7}$$

but the best approximation of the function f on each of these systems is greater than ε. In the sequence $\{x_i^{(k)}\}_{i=0}^{n+1}$, we choose a subsequence of systems $\{x_i^{(k_j)}\}_{i=0}^{n+1}$ such that each of their $n+2$ components $x_i^{(k_j)}$ approaches (as $j \to \infty$) a certain point x_i^0 of the limiting system

$$x_0^0 \le x_1^0 \le x_2^0 \le \ldots \le x_{n+1}^0.$$

By virtue of condition (6.7), the system obtained as a result necessarily contains less than $n+2$ different points $\bar{x}_l^0$:

$$\bar{x}_0^0 < \bar{x}_1^0 < \ldots < \bar{x}_{m+1}^0, \qquad m < n. \tag{6.8}$$

However, in this case, it follows from Theorem 2.2 that there exists a polynomial $\overline{P}_n^{(0)}$ of the form (6.1) interpolating the function f at all $m+2 \le n+1$ points of system (6.8). By virtue of continuity, this polynomial differs from f less than by ε at all points of system $\{x_i^{(k_j)}\}_{i=0}^{n+1}$ for sufficiently large j because each point $x_i^{(k_j)}$ is arbitrarily close to a certain point of system (6.8) for large j. This contradiction implies the validity of Lemma 6.1.

Proof of Theorem 6.1. By virtue of Theorem 5.4 [see (5.20″) and equalities (6.5), (6.5′), etc.], we find

$$\underline{E}_n^{(l)} = \frac{1}{\sum_{\nu=0}^{n+1} \left| D_l^{\nu} \right|} \left| \sum_{j=0}^{n+1} (-1)^j \left| D_l^j \right| r_n^{(l-1)}\left(x_j^{(l)}\right) \right| = \sum_{j=0}^{n+1} \frac{\left| D_l^j \right|}{\sum_{\nu=0}^{n+1} \left| D_l^{\nu} \right|} \left| r_n^{(l-1)}\left(x_j^{(l)}\right) \right|, \tag{6.9}$$

$$l = 2, 3, \ldots,$$

where

$$D_l^j = D\begin{pmatrix} \varphi_0 & \varphi_1 & \dots & \varphi_j & \varphi_{j+1} & \dots & \varphi_n \\ x_0^{(l)} & x_1^{(l)} & \dots & x_{j-1}^{(l)} & x_{j+1}^{(l)} & \dots & x_{n+1}^{(l)} \end{pmatrix} \neq 0 \tag{6.10}$$

for $x_{k+1}^{(l)} = x_k^{(l)}$, $k = 0, 1, \dots, n$. Thus, in view of the fact that, by virtue of (6.5), (6.5′), ... and (6.4), (6.4′), ..., we have

$$\max_j \left| r_n^{(l)}\left(x_j^{(l+1)}\right) \right| = \overline{E}_n^{(l)} > E_n(f) \geq \underline{E}_n^{(l)},$$

it is possible to conclude that

$$\underline{E}_n^{(l+1)} - \underline{E}_n^{(l)} = \sum_{j=0}^{n+1} \frac{|D_{l+1}^j|}{\sum_{\nu=0}^{n+1} |D_{l+1}^\nu|} \left(\left| r_n^{(l)}\left(x_j^{(l+1)}\right) \right| - \underline{E}_n^{(l)} \right)$$

$$\geq \frac{\min_j |D_{l+1}^j|}{\sum_{\nu=0}^{n+1} |D_{l+1}^\nu|} \left(\overline{E}_n^{(l)} - \underline{E}_n^{(l)} \right) > 0 \tag{6.11}$$

and, hence,

$$0 \leq \underline{E}_n^{(1)} < \underline{E}_n^{(2)} < \ldots < E_n(f). \tag{6.12}$$

According to Lemma 6.1, these inequalities imply that there exists a number $c > 0$ such that the following inequalities are true for all $l \geq 2$ (and $\underline{E}_n^{(l)} \geq \underline{E}_n^{(2)} > 0$):

$$x_{k+1}^{(l)} - x_k^{(l)} > c, \quad k = 0, 1, \dots, n, \quad l = 2, 3, \dots. \tag{6.13}$$

Since all points $x_k^{(1)}$ of system (6.2) are different, we can choose a number $c > 0$ such that inequalities (6.13) remain true for $l = 1$. By using these inequalities, the continuity of the functions φ_k, and the corollary of Theorem 2.1, we conclude that there exist positive numbers m and M such that

$$m < |D_l^j| = \left| D\begin{pmatrix} \varphi_0 & \varphi_1 & \dots & \varphi_j & \varphi_{j+1} & \dots & \varphi_n \\ x_0^{(l)} & x_1^{(l)} & \dots & x_{j-1}^{(l)} & x_{j+1}^{(l)} & \dots & x_{n+1}^{(l)} \end{pmatrix} \right| < M, \tag{6.14}$$

$$j = 0, 1, \dots, n+1, \quad l = 1, 2, \dots.$$

It follows from inequalities (6.11) and (6.14) that

$$E_n(f) - \underline{E}_n^{(l)} - \left(E_n(f) - \underline{E}_n^{(l+1)}\right) = \underline{E}_n^{(l+1)} - \underline{E}_n^{(l)} \geq \frac{m}{(n+2)M}\left(\overline{E}_n^{(l)} - \underline{E}_n^{(l)}\right), \quad (6.15)$$

and, consequently,

$$E_n(f) - \underline{E}_n^{(l+1)} \leq E_n(f) - \underline{E}_n^{(l)} - \frac{m}{(n+2)M}\left(E_n(f) - \underline{E}_n^{(l)}\right) = q\left(E_n(f) - \underline{E}_n^{(l)}\right), \quad (6.16)$$

where

$$q = 1 - \frac{m}{(n+2)M}, \qquad 0 < q < 1. \quad (6.17)$$

In view of relation (6.16), we consecutively find

$$E_n(f) - \underline{E}_n^{(l)} \leq q\left(E_n(f) - \underline{E}_n^{(l-1)}\right) \leq \ldots \leq q^{l-1}\left(E_n(f) - \underline{E}_n^{(1)}\right) = aq^l, \quad (6.18)$$

where

$$a = \frac{1}{q}\left(E_n(f) - \underline{E}_n^{(1)}\right). \quad (6.19)$$

Substituting inequality (6.18) in relation (6.15), we conclude that

$$\overline{E}_n^{(l)} - \underline{E}_n^{(l)} \leq \frac{(n+2)M}{m} cq^l = Aq^l,$$
$$A = \frac{(n+2)M}{m} c, \qquad l = 1, 2, \ldots. \quad (6.20)$$

This inequality implies the required inequality (6.6).

The proof Theorem 6.1 is thus completed.

Corollary 6.1. *Note that, by virtue of inequality (6.20),*

$$\left\| f - P_n^{(l)} \right\|_{[a,b]} = \overline{E}_n^{(l)} \to E_n(f) \quad \text{as} \quad l \to \infty.$$

Thus, in view of the uniqueness of the polynomial $P_n^*(f;\cdot)$ *of the best approximation for the function* f, *we conclude that* $P_k^{(l)}(x) \to P_n^*(f;x)$ *as* $l \to \infty$ *for* $x \in [a,b]$.

7. Approximation of functions by polynomials with linear constraints imposed on their coefficients[9]

7.1. The problem of finding the polynomial least deviating from zero in the presence of linear constraints imposed on its coefficients is one of the classical (and extremely complicated) problems of the theory of approximation.

Our aim is to find a polynomial

$$P_n^*(\varphi; x) = \sum_{k=0}^{n} a_k^* \varphi_k(x)$$

with the smallest norm (in a certain space) in the family of polynomials $P_n(\varphi; \cdot)$ of the form

$$P_n(\varphi; x) = \sum_{k=0}^{n} a_k \varphi_k(x), \tag{7.1}$$

where $\{\varphi_k(x)\}_{k=0}^n$ is the Chebyshev system of functions on the set $\mathfrak{M}$, under the condition that the coefficients of the required polynomial satisfy r linear constraints

$$\sum_{k=0}^{n} c_{ik} a_k = b_i, \quad i = 1, 2, \dots, r, \tag{7.2}$$

where c_{ik} and b_i are given numbers and r is an arbitrary natural number.

Constraints (7.2) are called *compatible* if there exists at least one polynomial satisfying these constraints or, in other words, if the system of equations (7.2) is solvable with respect to the unknown a_k.

In the uniform metric, the first problem of this sort was the Chebyshev problem (1859) of finding the polynomial least deviating from zero. As shown in Section 3, this problem can be formulated as follows:

To find a polynomial P_n^* with the least norm in the metric of $C[-1,1]$ in the family of algebraic polynomials P_n of the form

$$P_n(x) = \sum_{k=0}^{n} a_k x^{n-k}$$

with a single constraint specified by the conditions

$$c_{10} = 1, \quad c_{1k} = 0, \quad k = 1, 2, \dots, n; \quad b_1 = 1$$

or, in the other words, by the condition $a_0 = 1$. As shown in Section 3,

$$P_n^*(x) = \frac{1}{2^{n-1}} T_n(x) = \frac{1}{2^{n-1}} \cos n \arccos x.$$

In 1877, Zolotarev (published in 1912) solved this problem in terms of elliptic functions under the condition that the values of the first two coefficients of the polynomial

$$\sum_{k=0}^{n} a_k x^{n-k}$$

are given. Akhiezer (1928) solved a similar problem in terms of automorphic functions under the condition that the values of the first three coefficients of the polynomial

$$P_n(x) = \sum_{k=0}^{n} a_k x^{n-k}$$

are fixed.[†]

In what follows, we show that the problem with constraints is fairly simply solved in any Hilbert space H. More exactly, it is shown that, in the space H, this problem is a consequence of some algebraic theorems.

7.2. We now establish a criterion for the solvability of a system of linear algebraic equations and a criterion for the compatibility of constraints (7.2).

To do this, we introduce the following notation: For any matrix $C = \{c_{ik}\}$, $i = 1, 2, \dots, r$, $k = 0, 1, 2, \dots, n$,

$$C = \begin{pmatrix} c_{10} & c_{11} & \dots & a_{1n} \\ c_{20} & c_{21} & \dots & a_{2n} \\ \dots & \dots & \dots & \dots \\ c_{r0} & c_{r1} & \dots & a_{rn} \end{pmatrix}, \tag{7.3}$$

by C^* we denote its adjoint matrix $C^* = \{c_{ki}^*\}$, where $c_{ki}^* = \bar{c}_{ik}$:

[†] Numerous important investigations dealing with the problem of approximation of functions by algebraic polynomials with linear constraints imposed on their coefficients and polynomials of the form (7.1), both in the uniform metric and in some other metrics, were carried out by Markov (1892), Remez (1969), Shokhat (1918), Grebenyuk (1960), Geronimus, Meiman, Voronovskaya, Gol'dshtein, Rymarenko, Grigor'eva, Koromyslichenko, Chernykh, and many others.

$$C^* = \begin{pmatrix} \bar{c}_{10} & \bar{c}_{20} & \dots & \bar{c}_{r0} \\ \bar{c}_{11} & \bar{c}_{21} & \dots & \bar{c}_{r1} \\ \dots & \dots & \dots & \dots \\ \bar{c}_{1n} & \bar{c}_{2n} & \dots & \bar{c}_{rn} \end{pmatrix}. \tag{7.3*}$$

In this case, the family of column matrices A with the same number of elements a_k forms a Hilbert space if we introduce a scalar product

$$(A, A') = \left|A^* A'\right| = \sum_{k=0}^{n} \bar{a}_k a'_k, \quad A = \begin{pmatrix} a_0 \\ a_1 \\ \vdots \\ a_n \end{pmatrix}, \quad A' = \begin{pmatrix} a'_0 \\ a'_1 \\ \vdots \\ a'_n \end{pmatrix}, \tag{7.4}$$

where the determinant of the one-element matrix $A^* A'$ is denoted by $\left|A^* A'\right|$. Thus, the column matrices are also called vectors and we set $\|A\|^2 = (A, A)$. In the next theorem, parallel with the matrices C and C^*, we also use the matrices

$$X = \begin{pmatrix} x_0 \\ x_1 \\ \vdots \\ x_n \end{pmatrix}, \quad B = \begin{pmatrix} b_1 \\ b_2 \\ \vdots \\ b_r \end{pmatrix}, \quad Y = \begin{pmatrix} y_1 \\ y_2 \\ \vdots \\ y_r \end{pmatrix}, \quad C_j = (c_{j0}, c_{j1}, \dots, c_{jn}),$$

$$A = \begin{pmatrix} a_1 \\ a_2 \\ \vdots \\ a_r \end{pmatrix}, \quad b_i = (b_i), \quad \text{and} \quad C_j^* = \begin{pmatrix} \bar{c}_{j0} \\ \bar{c}_{j1} \\ \vdots \\ \bar{c}_{jn} \end{pmatrix}. \tag{7.5}$$

Theorem 7.1. *1. In order that a system of linear equations of the form*

$$CX = B \tag{7.6}$$

be solvable with respect to X, *it is necessary and sufficient that the system of equations of the form*

$$CC^* Y = B \tag{7.7}$$

be solvable with respect to Y.

2. If system (7.6) is solvable, then

(i) the smallest (in norm) solution X^0 of this system can be expressed via an arbitrary solution Y of system (7.7) by the formula

$$X^0 = C^* Y; \tag{7.8}$$

(ii) the norm of the solution X^0 is expressed with the help of Y by the formula

$$\|X^0\|^2 = \sum_{k=1}^{r} \bar{y}_k b_k = (Y, B); \tag{7.9}$$

(iii) the solution X^0 is unique.

Proof. 1. First, we prove that the solvability of system (7.7) follows from the solvability of system (7.6). Note that if a certain solution X of system (7.6) can be represented in the form

$$X = C^* A = \sum_{j=1}^{n} a_j C_j^*, \tag{7.10}$$

then, in view of (7.6), we immediately obtain $CX = CC^*A = B$. Hence, in this case, system (7.7) is also solvable.

To prove, in the general case, that the solvability of system (7.7) follows from the solvability of system (7.6), we now show that if a vector X is a solution of system (7.6), i.e.,

$$C_i X = b_i, \quad i = 1, 2, \dots, r, \tag{7.11}$$

but cannot be represented in the form (7.10), then the vector

$$X^* = \sum_{j=1}^{r} a_j^* C_j^*$$

whose coefficients a_j^* are determined from the condition

$$\|X - X^*\|^2 = \inf_{a_j} \left\| X - \sum_{j=1}^{r} a_j C_j^* \right\|^2 \tag{7.12}$$

[i.e., X^* is the vector of the best approximation for the vector X in the family of vectors of the form (7.10)] is also a solution of system (7.6). The indicated solution also has the form (7.10) and, therefore, as shown above, this means that system (7.7) is solvable.

Indeed, assume the opposite, i.e., that X^* is not a solution of system (7.6). This means that, for some i, we get the inequality $C_i X^* = b_i' \neq b_i$. Further, by setting $\| C_i^* \|^2 = s_i$, we conclude that the vector $X^* + C_i^* \left(\overline{b_i - b_i'}\right)/s_i$ approximates the vector X better than the vector X^*. Thus, by using relation (7.11) and the equality $C_i^{**} = C_i$, we obtain

$$\left\| X - \left(X^* + \frac{\overline{b_i - b_i'}}{s_i} C_i^* \right) \right\|^2 = \left(X - X^* - \frac{\overline{b_i - b_i'}}{s_i} C_i^*,\ X - X^* - \frac{\overline{b_i - b_i'}}{s_i} C_i^* \right)$$

$$= \left\| X - X^* \right\|^2 - 2\operatorname{Re}\left\{ \frac{b_i - b_i'}{s_i} C_i^{**} \left(X - X^* \right) \right\} + \frac{|b_i - b_i'|^2}{s_i^2} s_i$$

$$= \left\| X - X^* \right\|^2 - \frac{|b_i - b_i'|^2}{s_i} < \left\| X - X^* \right\|^2,$$

i.e., arrive at a contradiction with condition (7.12).

Vice versa, if system (7.7) is solvable and a vector Y^0 satisfies equation (7.7), then the vector $C^* Y^0$ satisfies equation (7.6), i.e., system (7.6) is also solvable.

2. We denote a solution of system (7.7) by Y [by virtue of assertion 1, this solution exists because system (7.6) is solvable] and set $X^0 = C^* Y$. Then, in view of the fact that the scalar product of X^0 by any other solution X of system (7.6) satisfies the condition

$$(X^0, X) = \left| (C^* Y)^* X \right| = \left| Y^* C X \right| = \left| Y^* B \right| = (Y, B)$$

$$= \text{const} = (X^0, X^0) = \left\| X^0 \right\|^2, \tag{7.13}$$

we conclude that

$$\left\| X - X^0 \right\|^2 = (X - X^0, X - X^0)$$

$$= \left\| X^0 \right\|^2 - 2\operatorname{Re}(X^0, X) + \left\| X^0 \right\|^2 = \| X \|^2 - \left\| X^0 \right\|^2. \tag{7.14}$$

Relations (7.14) and (7.13) imply that

(i) X^0 computed according to relation (7.8) possesses the minimum norm as compared with all other solutions X of system (7.6);

(ii) by virtue of relation (7.13), the norm of X^0 is given by relation (7.9);

(iii) the solution X^0 with minimum norm is unique because, by virtue of relation (7.14), it follows from the equality $\|X_0^1\| = \|X^0\|$ that $\|X_0^1 - X^0\| = 0$ and, hence, $X_0^1 = X^0$.

7.3. Let us show that Theorem 7.1 enables to prove the following Fredholm theorem in a fairly simple way:

Theorem 7.2 (Fredholm alternative for linear systems). *In order that a system of equations*

$$C^* Y = D, \quad D = \begin{pmatrix} d_0 \\ d_1 \\ \vdots \\ d_n \end{pmatrix} \tag{7.15}$$

be compatible, it is necessary and sufficient that the vector D be orthogonal to any solution X of the homogeneous system

$$CX = 0, \tag{7.16}$$

i.e., that the condition

$$D \perp N_x \tag{7.17}$$

be satisfied, where $N_x = \{X\colon\ CX = 0\}$ is the set of zeros of system (7.16).

Proof. *Necessity.* Assume that system (7.15) is solvable and Y_0 is a solution of this system. We set $CC^* Y_0 = CD = B$ and denote the solution of the compatible system $CX = B$ with minimum norm by X^0. By virtue of relation (7.8), it follows from Theorem 7.1 that $X^0 = C^* Y = D$. Since X^0 has the minimum norm among all solutions of the equations $CX = B$, we conclude that

$$\|X^0\| = \|X^0 - 0\| = \min_{X \in N_x} \|X^0 - X\|$$

and, hence (see [Akhiezer (1965), Section 13, p. 23]), $X^0 \perp N_x$ and, in addition, $D \perp N_x$.

Sufficiency. If $D \perp N_x$, then, by setting $CD = B$, we conclude that the system $CX = B$ is compatible and, by virtue of condition (7.17), the vector D has the minimum norm among all solutions of this system. Therefore, $D = X^0$ and, hence, by virtue of statement (i) in Theorem 7.1, the system $CC^* Y = B$ is also compatible and if Y_0 is a solution of this system, then $D = X^0 = C^* Y_0$, which means that system (7.15) is compatible.

■

Remark 7.1. Let us show that, by using Theorem 7.2 (Fredholm), we can immediately get statement (i) of Theorem 7.1.

Indeed, the assumption that system (7.6) is solvable and X^0 is the solution of this system with minimum norm [the existence of this solution is guaranteed by the Borel theorem (Theorem 1.1′)] implies that $X^0 \perp N_x$. Thus, by virtue of Theorem 7.2, the equation $C^* Y = X^0$ is solvable. Hence, the equation $CC^* Y = CX^0 = B$ is also solvable, as required.

7.4. In the present subsection, we show that the problem with constraints in the general case is very simply solved in any Hilbert space H. In order to formulate the corresponding theorem, we first reformulate the problem with constraints as follows:

Consider a system of $n + 1$ orthogonal vectors $e_0, e_1, e_2, \ldots, e_n$ and r linear functionals $L_1, L_2, \ldots, L_r$ in a Hilbert space H. Also let $b_1, b_2, \ldots, b_r$ be r given, generally speaking, complex numbers. In the set E_{n+1} of elements X of the form

$$X = \sum_{k=0}^{n} x_k e_k \tag{7.1′}$$

satisfying the following r constraints:

$$L_i(x) = \sum_{k=0}^{n} x_k L_i(e_k) = b_i, \quad i = 1, 2, \ldots, r, \tag{7.2″}$$

it is necessary to find the element with minimum norm in H, i.e., in the set of elements $X \in E_{n+1}$ of the form (7.1′), it is necessary find the element (polynomial) least deviating from zero.

Note that if we set $L_i(e_k) = c_{ik}$, then condition (7.2″) immediately takes the form (7.2). Therefore, in exactly the same way as in Theorem 7.1, we arrive at the following theorem on polynomials least deviating from zero in a Hilbert space:

Theorem 7.3. *If constraints (7.2″) are compatible, then*

(i) the element $X^0 \in E_{n+1}$ *satisfying all these constraints with the minimum norm in* H *is given by formula*

$$X^0 = C^* Y = \sum_{k=0}^{n} \left(\sum_{j=1}^{r} \bar{c}_{jk} y_j \right) e_k, \tag{7.8′}$$

where Y *is a solution of the system* $CC^* Y = B$;

(ii) by using Y, *the norm of the solution* X^0 *is given by the formula*

$$\|X^0\|^2 = \sum_{k=1}^{r} \bar{y}_k b_k = (B, Y) = (Y, B), \tag{7.9′}$$

(iii) the solution X^0 *is unique.*

7.5. Snakes. In the present subsection, for fixed natural k and $n > k$, we consider systems of functions $\{\varphi_j\}_{j=0}^{n}$ continuous on the segment $[-1, 1]$ and such that both $\{\varphi_j\}_{j=0}^{n}$ and $\{\varphi_j\}_{j=0}^{n-k}$ are Chebyshev systems on $[-1, 1]$. These systems are called $TM(k, n)$ systems.

Let $g_1(x)$ and $g_2(x) > g_1(x)$ be functions continuous on $[-1, 1]$. We consider the following problem [Dzyadyk (1978a)]: In the set of generalized polynomials constructed according to a $TM(k, n)$-system with fixed leading $k \ge 1$ coefficients $A_n, A_{n-1}, \ldots,$ A_{n-k+1} [the set of these coefficients is denoted by $A(k)$] of the form

$$T_n(B; A(k); c; x) := B \sum_{j=n-k+1}^{n} A_j \varphi_j(x) + \sum_{j=0}^{n-k} c_j \varphi_j(x), \quad B = \text{const} > 0, \tag{7.18}$$

it is necessary to choose numbers $c_0^0, c_1^0, \ldots, c_{n-k}^0$ and a constant B^0 guaranteeing (if this is possible) the validity the following conditions:

1. $$g_1(x) \le T_n(B^0; A(k); c^0; x) \le g_2(x) \quad \text{for any } x \in [-1, 1], \tag{7.19}$$

where $c^0 := \{c_0^0, c_1^0, \ldots, c_{n-k}^0\}$;

2. There exist at least $n+2-k$ points x_j, $-1 \le x_1 < x_2 < x_3 < \ldots < x_{n+2-k} \le 1$ such that the following equalities are true:

$$T_n(B^0; A(k); c^0; x_j) = \begin{cases} g_1(x), & j = 2, 4, \ldots, \\ g_2(x), & j = 1, 3, \ldots, \end{cases}$$

or

$$T_n(B^0; A(k); c^0; x_j) = \begin{cases} g_1(x), & j = 1, 3, 5, \ldots, \\ g_2(x), & j = 2, 4, 6, \ldots. \end{cases} \tag{7.20}$$

The polynomials with properties (7.19) and (7.20) are called *k-snakes.* The points x_j at which the equalities $T_n(B^0; A(k); c^0; x_j) = g_2(x_j)$ or $g_1(x_j)$ are satisfied are called (+)- or (–)-*points,* respectively; the (+)- and (–)-points are also called (e)-points.

Definition 7.1. *An increasing (decreasing) branch of a k-snake* $y = T_n(B^0; A(k); c^0; x)$ *is defined as a part of its plot connecting two consecutive* (e)*-points the first of which is a* (–)*-point* [(+)*-point*] *and the second is a* (+)*-point* [(–)*-point*]*. A semibranch of the k-snake* $y = T_n(B^0; A(k); c^0; x)$ *is defined as a part of its plot located either in front the first* (e)*-point or behind the last* (e)*-point. A semibranch is called increasing (decreasing) if it is located in front of the* (+)*-point*[(–)*-point*] *or behind the* (–)*-point* [(+)*-point*].

For the case $k=1$, the problem formulated the above was solved by Karlin (see Theorem 2.9). In what follows (see the first part of Theorem 7.5), it is solved for $k=2$.

Theorem 7.4. *Let* $g_1(x)$ *and* $g_2(x) \ge g_1(x)$ *be two continuous functions given on the segment* $[-1, 1]$ *and let* $\{\varphi_j\}_{j=0}^n$ *be a Chebyshev system of functions on the same segment. If, in addition, for some* $l \le (n-1)/2$*, a polynomial* $T_n(\tilde{c}, x)$ *of the form*

$$T_n(\tilde{c}, x) := \sum_{j=0}^{n} \tilde{c}_j \varphi_j(x) \tag{7.21}$$

located between the functions $g_1(x)$ *and* $g_2(x)$*, i.e.,*

$$T_n(\tilde{c}, x_i) = y_i, \quad i = 1, 2, \ldots, l; \quad \forall x \in [-1, 1]: \; g_1(x) \le T_n(\tilde{c}, x) \le g_2(x), \tag{7.22}$$

passes through l points (x_i, y_i) located between the same functions:

$$-1 \le x_1 < x_2 < \ldots < x_l \le 1, \quad g_1(x_i) \le y_i \le g_2(x_i), \quad i = 1, 2, \ldots, l, \tag{7.23}$$

then there exists at least one $(2l+1)$-snake $T_n(\tilde{c}_0, x)$ which also passes through all points (x_i, y_i), $i = 1, 2, \ldots, l$.

Theorem 7.5.[10] *Let $g_1(x)$ and $g_2(x) \ge g_1(x)$ be two continuous functions given on the segment $[-1, 1]$ and let $\{\varphi_j\}_{j=0}^n$ be a $TM(2, n)$-system on the same segment. Then*

(a) under the assumption that, for a fixed coefficient A_n^* and all possible changes of a fixed coefficient A_{n-1}^*, there exists at least one polynomial of the form

$$T_n(x) = B \sum_{i=0}^{1} A_{n-i}^* \varphi_{n-i}(x) + \sum_{j=0}^{n-2} c_j \varphi_j(x)$$

satisfying, for all $x \in [-1, 1]$, the conditions

$$g_1(x) < T_n(x) < g_2(x), \tag{7.24}$$

the 2-snakes $T_n(B^0; A(2); c^0; x) = T_n(B^0; A_{n-1}^*; c^0; x)$ exist for all A_{n-1}^*;

(b) the functions

$$\tilde{g}_1(x) = \tilde{g}_1\left(A_{n-1}^*; x\right) = \inf_{A_{n-1}^*} T_n(B^0; A_{n-1}^*; c^0; x),$$

$$\tilde{g}_2(x) = \tilde{g}_2\left(A_{n-1}^*; x\right) = \sup_{A_{n-1}^*} T_n(B^0; A_{n-1}^*; c^0; x)$$

playing the roles of, respectively, the lower and the upper envelopes of the indicated family of 2-snakes form a "passage" in which the 2-snakes $T_n(B^0; A_{n-1}^*; c^0; x) = T_n(A_{n-1}^*; x)$ possess the following properties:

1. Exactly two 2-snakes $T_n(A_{n-1}^*; x)$ pass through every point (x_0, y_0) of the indicated passage for $x_0 \in [-1, 1]$ and $\tilde{g}_1(x_0) < y_0 < \tilde{g}_2(x_0)$. For one of these snakes, *the branch (or semibranch) passing through the point* (x_0, y_0) *is increasing and, for the second snake, the corresponding branch is decreasing. Exactly one 2-snake passes through every point* (x_0, y_0) *at which* $-1 < x_0 < 1$ *and* $y_0 = \tilde{g}_1(x_0)$ *or* $y_0 = \tilde{g}_2(x_0)$ *in the case where each function in the system* $\{\varphi_j\}_{j=0}^n$ *is twice differentiable.*

2. *Increasing (and decreasing) branches and semibranches of different 2-snakes do not intersect.*[11]

Remark 7.2. Note that if $g_1(x) = -1$ and $g_2(x) = 1$ for T-systems $1, x, x^2, \dots, x^n$ and $1, \cos x, \sin x, \dots, \cos nx, \sin nx$, then $\tilde{g}_1(x) = -1$ and $\tilde{g}_2(x) = 1$. We now present an example demonstrating that $\tilde{g}_1(x)$ and $\tilde{g}_2(x)$, generally speaking, do not coincide with $g_1(x)$ and $g_2(x)$ even if the functions $g_1(x)$ and $g_2(x)$ are sufficiently smooth.

Example 7.1. Let $r(x)$ be a function continuous on $[-1, 1]$ and satisfying the conditions

$$r(x) = 1 \quad \text{for} \quad x \in \left[-1, \frac{5}{8}\right],$$

$$r\left(\frac{3}{4}\right) = 8, \quad r(x) = 1 \quad \text{for} \quad x \in \left[\frac{7}{8}, 1\right],$$

$$r(x) \ge 1 \quad \text{and} \quad \left|r(x)^2(2x^2 - 1)\right| \le 1 \quad \text{for} \quad x \in \left[\frac{5}{8}, \frac{7}{8}\right].$$

Then the system of functions $\varphi_0(x) := r(x)$, $\varphi_1(x) := xr(x)$, and $\varphi_2(x) := (2x^2 - 1)r(x)$ is a T-system on $[-1, 1]$ and, at the same time, $\tilde{g}_2(x) \ne g_2(x)$ for $g_\nu(x) = (-1)^\nu$, $\nu = 1, 2$. This follows from the fact that the polynomial (2-snake)

$$T_2(x) := \varphi_2(x) = (2x^2 - 1)r(x)$$

has the following properties:

$$T_2(-1) = g_2(-1) = 1, \quad T_2(0) = g_2(0) = -1,$$

$$T_2\left(\frac{3}{4}\right) = g_2\left(\frac{3}{4}\right) = T_2(1) = g_2(1) = 1,$$

$$T_2\left(\frac{7}{8}\right) = \frac{17}{32} = g_2\left(\frac{7}{8}\right) = 1.$$

By virtue of these properties, if a 2-snake $\breve{T}_2$ is such that $\breve{T}_2(7/8) > 17/32$, then the difference $T_2(x) - \breve{T}_2(x)$ definitely has at least four zeros on the segment $[1, 1]$ [located in the intervals $(-1, 0)$, $(0, 3/4)$, $(3/4, 7/8)$, and $(7/8, 1)$], which is impossible. Therefore,

$$\tilde{g}_2\left(\frac{7}{8}\right) \leq \frac{17}{32} < g_2\left(\frac{7}{8}\right) \Rightarrow \tilde{g}_2(x) \neq g_2(x).$$

Proof of Theorem 7.4. Since the system of functions $\{\varphi_j\}_{j=0}^{n}$ is linearly independent, the set of all possible generalized polynomials $T_n(c, \cdot)$ satisfying the inequalities

$$g_1(x) \leq T_n(c, x) \leq g_2(x), \tag{7.25}$$

is clearly a bounded equicontinuous and, hence, compact set.

Let us now replace, for sufficiently large natural numbers $m \geq m_0$, the functions g_ν, $\nu = 1, 2$, by the functions $g_{\nu,m}$ as follows: First, we assume that the inequalities $-1 < x_1$, $x_l < 1$ take place in (7.23) and that $g_1(x_j) < y_j < g_2(x_j)$. In order to get the functions $g_{2,m}$ for each $j = 1, 2, \ldots, l$, we replace the plot of the function $g_2(x)$ on the segment $[x_j - m^{-1}, x_j + m^{-1}]$ by a continuous curve passing through the points

$$(x_j - m^{-1}, g_2(x_j - m^{-1})), \quad (x_j, y_j + m^{-1}), \quad \text{and} \quad (x_j + m^{-1}, g_2(x_j + m^{-1}))$$

whose plot to the left and right of the point $(x_j, y_j + m^{-1})$ so rapidly goes up to the points $(x_j - m^{-1}, g_2(x_j - m^{-1}))$ and $(x_j - m^{-1}, g_2(x_j + m^{-1}))$ that, by virtue of the equicontinuity of the polynomials $T_n(c, \cdot)$ satisfying inequalities (7.25), each polynomial $T_n(c, \cdot)$ passing through a point (x_j, y), $\tilde{g}_2(x_j) \leq y \leq y_j + m^{-1}$ and satisfying inequalities (7.25) can have a single common point $(x_j, y_j + 1/m)$ with the constructed curve $g_{2,m}$ on the segments $[x_j - m^{-1}, x_j + m^{-1}]$. The function $g_{1,m}$ is constructed similarly. The same is true for the functions $g_{\nu,m}$ in the case where $x_1 = -1$ or $x_l = 1$. If the functions $g_{\nu,m}$, $\nu = 1, 2$, $m = m_0, m_0 + 1, \ldots$, are already constructed, then, by using Theorem 2.9 (Karlin), for any m, we construct a 1-snake $T_n(c^{(m)}, \cdot)$ located between the curves $g_{\nu,m}$, $\nu = 1, 2$, i.e., $g_{1,m}(x) \leq T_n(c^{(m)}, x) \leq g_{2,m}(x)$. This snake possesses, in turn, $n + 1$ common points with the indicated curves and its first e-point is the (+)-point. Further, if we now choose a convergent subsequence $\{T_n(c^{(m_k)}, \cdot)\}$ of the compact sequence $T_n(c^{(m)}, \cdot)$, then, clearly, the function

$$T_n(\tilde{c}_0; x) := \lim_{k\to\infty} T_n(c^{(m_k)}, x)$$

is a $(2l+1)$-snake for the functions g_ν. This snake passes through all points (x_i, y_i), $i = 1, 2, \ldots, l$, and its first e-point is the (+)-point. If, in relations (7.23), we get the sign of equality for some y_i, then, instead of the functions g_ν, it is necessary to additionally consider the following auxiliary sequence of functions:

$$g_{\nu,m}(x) := g_\nu(x) + (-1)^\nu m^{-1}$$

and again apply the Arzelà theorem to the sequence of polynomials $T_n(c; m; \cdot)$ obtained as a result. Theorem 7.4 is proved.

■

The fact that the l-snake satisfying the conditions of Theorem 7.4 [with (+)-point as the first e-point] is, generally speaking, not unique can be illustrated by the following example:

Let a T-system $\varphi_0(x) = 1$, $\varphi_1(x) = T_1(x)$, $\varphi_2(x) = T_2(x)$, $\varphi_3(x) = T_3(x)$, and $\varphi_4(x) = T_4(x)$ be given on the segment $[-1, 1]$, let $g_\nu(x) = (-1)^\nu$, $\nu = 1, 2$, and let (x_i, y_i), $i = 1, 2, 3$, be three points of intersection of the functions $-T_3$ and T_4 from $[-1, 1]$. Then, both polynomials $-T_3$ and T_4, are, clearly, 3-snakes on the segment $[-1, 1]$ passing through the same three points (x_i, y_i), $i = 1, 2, \ldots, l$, $l = 3$, and the first e-point of each of these snakes is a (+)-point.

Proof of Theorem 7.5. (a) First, we note that, in this case, according to the condition of the theorem, the coefficient A_n^* is fixed and the coefficient A_{n-1}^* is variable. Hence, as follows from relation (7.18), the analyzed family of 2-snakes $T_n\left(A_{n-1}^*; \cdot\right)$ depends on a single parameter α (the ratio of coefficients A_n^* and A_{n-1}^*). Therefore, every 2-snake $T_n\left(A_{n-1}^*; \cdot\right)$ can be obtained as follows: Note that, according to the conditions, for any fixed A_n^* and A_{n-1}^*, there exists a polynomial $T_n(x)$ of the form (7.18) satisfying the inequalities

$$g_1(x) < \sum_{i=0}^{1} A_{n-i}^* \varphi_{n-i}(x) + \sum_{j=0}^{n-2} c_j \varphi_j(x) < g_2(x),$$

or, equivalently,

$$-\frac{g_2(x)-g_1(x)}{2} < \frac{g_1(x)+g_2(x)}{2} - T_n(x) < \frac{g_2(x)-g_1(x)}{2},$$

or (for $B = 1$)

$$-1 < \frac{g_1(x)+g_2(x)}{g_2(x)-g_1(x)} - B\sum_{i=0}^{1} A_{n-i}^{*}\,\tilde{\varphi}_{n-i}(x) - \sum_{j=0}^{n-2} c_j\,\tilde{\varphi}_j(x) < 1,$$

$$\tilde{\varphi}_j(x) := \frac{2\varphi_j(x)}{g_2(x)-g_1(x)}. \tag{7.26}$$

Thus, the quantity

$$E_{n-2} = E_{n-2}(B) = \min_{c_j}\left\| f(B;x) - \sum_{j=0}^{n-2} c_j\,\varphi_j(x)\right\|_{[-1,1]} < 1$$

is the best uniform approximation of the function

$$f(B;x) := \frac{g_1(x)+g_2(x)}{g_2(x)-g_1(x)} - B\left[A_n^{*}\,\tilde{\varphi}_n(x) + A_{n-1}^{*}\,\tilde{\varphi}_{n-1}(x)\right]$$

continuous on $[-1, 1]$ by generalized polynomials of the form $\sum_{j=0}^{n-2} c_j\,\tilde{\varphi}_j(x)$. Hence, in view of the linear independence of the functions in the T-system $\{\varphi_j\}_{j=0}^{n}$, the quantity

$$E_{n-2}(B) = \min_{c_j}\left\| \frac{g_1+g_2}{g_2-g_1} - B\left[A_n^{*}\tilde{\varphi}_n + A_{n-1}^{*}\tilde{\varphi}_{n-1}\right] - \sum_{j=0}^{n-2} c_j\tilde{\varphi}_j\right\|_{[-1,1]} \to \infty$$

as $B \to \infty$ and the variations of the functional

$$E_\nu(f) := \min_{c_j}\left\| f(x) - \sum_{j=0}^{\nu} c_j\,\tilde{\varphi}_j(x)\right\|_{[-1,1]}$$

caused by continuous variations of the function f are continuous. This enables us to conclude that there exists a number $B^0 > 0$ such that, for some c^0, we have

$$E_{n-2}(B^0) = \min_{c_j} \left\| f(B^0; x) - \sum_{j=0}^{n-1} c_j(x)\varphi_j(x) \right\|_{[-1,1]}$$

$$= \left\| \frac{g_1(x)+g_2(x)}{g_2(x)-g_1(x)} - B^0\left[A_n^*\tilde{\varphi}_n(x) + A_{n-1}^*\tilde{\varphi}_{n-1}(x)\right] - \sum_{j=0}^{n-2} c_j^0\tilde{\varphi}_j(x) \right\|_{[-1,1]}$$

$$= 1. \tag{7.27}$$

Thus, by virtue of the Chebyshev theorem on alternation, there exist at least $(n-2)+2=n$ points x_i,

$$-1 \le x_1 < x_2 < \ldots < x_n \le 1,$$

at which the following equalities are true:

$$\frac{g_1(x_i)+g_2(x_i)}{g_2(x_i)-g_1(x_i)} - B^0\left[A_n^*\tilde{\varphi}_n(x_i) + A_{n-1}^*\tilde{\varphi}_{n-1}(x_i)\right] - \sum_{j=0}^{n-2} c_j^0\tilde{\varphi}_j(x_i) = \pm(-1)^i \tag{7.28}$$

or, in other words,

$$B^0\left[A_n^*\tilde{\varphi}_n(x_i) + A_{n-1}^*\tilde{\varphi}_{n-1}(x_i)\right] - \sum_{j=0}^{n-2} c_j^0\varphi_j(x_i)$$

$$= \frac{g_1(x_i)+g_2(x_i) \pm (-1)^i[g_2(x_i)-g_1(x_i)]}{2}, \quad i = 1, 2, \ldots, n.$$

Therefore, the polynomial

$$T_n\left(B^0; A_n^*, A_{n-1}^*; x\right) := B^0\left[A_n^*\varphi_n(x) + A_{n-1}^*\varphi_{n-1}(x)\right] + \sum_{j=0}^{n-2} c_j^0\varphi_j(x) \tag{7.29}$$

is a 2-snake with respect to the functions g_1 and g_2.

This proves part (a) of Theorem 7.5.

(b) In order to establish the dependence of 2-snakes (7.29) on the ratio of the coefficients A_n^* and A_{n-1}^*, we represent these snakes in the form

$$T_\alpha(x) := \begin{cases} A_\alpha\Big[(1+\alpha)\varphi_n(x) + \big(1-(1+\alpha)^2\big)\varphi_{n-1}(x)\Big] + T_{n-2}(\alpha; c^0; x), & \alpha \in (-2, 0), \\ A_\alpha\Big[(1-\alpha)\varphi_n(x) - \big(1-(1-\alpha)^2\big)\varphi_{n-1}(x)\Big] + T_{n-2}(\alpha; c^0; x), & \alpha \in (0, 2), \end{cases} \tag{7.30}$$

where the polynomials $T_{n-2}(\alpha; c^0; x)$ are chosen so that all their coefficients coincide with the coefficients of the polynomial

$$\tilde{T}_{n-2}(\alpha; c^0; x) := \sum_{j=0}^{n-2} c_j^0\, \tilde{\varphi}_j(x)$$

of the best uniform approximation for the functions

$$f_\alpha(x) := \begin{cases} \dfrac{g_1(x)+g_2(x)}{g_2(x)-g_1(x)} - A_\alpha\Big[(1+\alpha)\tilde{\varphi}_n(x) + \big(1+(1+\alpha)^2\big)\tilde{\varphi}_{n-1}(x)\Big] & \text{for } \alpha \in (-2, 0], \\ \dfrac{g_1(x)+g_2(x)}{g_2(x)-g_1(x)} - A_\alpha\Big[(1-\alpha)\tilde{\varphi}_n(x) - \big(1-(1-\alpha)^2\big)\tilde{\varphi}_{n-1}(x)\Big] & \text{for } \alpha \in (0, 2], \end{cases}$$

in the system $\{\tilde{\varphi}_j(x)\}$ and a constant factor $A_\alpha > 0$ is chosen to guarantee the validity of the condition

$$\Big\| f_\alpha(x) - \tilde{T}_{n-2}(\alpha; c^0; x) \Big\|_{[-1,1]} = 1.$$

The proof of part (b) of Theorem 7.5 is now reduced to the analysis of the properties of the 2-snakes $T_\alpha(x)$. In what follows, we additionally establish some properties of these snakes not included in the statement of Theorem 7.5.

1. There are exactly two points $\alpha_0, \alpha_1 \in (-2, 2]$, such that, for each polynomial T_{α_0} or T_{α_1}, one can find at least one alternation formed by $n+1$ [alternating (+)- and (–)-] points x_i^0, $-1 \le x_1^0 < x_2^0 < \ldots < x_{n+1}^0 \le 1$, and there is no alternation formed by $n+2$ [alternating (+)– and (–)-] points. In particular, the polynomials T_{α_0} and T_{α_1} are 1-snakes for the functions g_1 and g_2. By virtue of the definition of the polynomials T_α, this property is a consequence of Theorem 2.9 (Karlin).

2. For any $\alpha \in (-2, 2] \setminus \{\alpha_0, \alpha_1\}$, there exists an alternation for the polynomial $T_\alpha(x)$ formed by n points and there is no alternation formed by $n+1$ points.

This property has already been established in deducing equalities (7.27)–(7.29).

3. The increasing (decreasing) branches and semibranches of the curves $y = T_\alpha(x)$ and $y = T_{\alpha'}(x)$ with $\alpha' \neq \alpha$ cannot intersect at any point (x_0, y_0) such that $x_0 \in [-1, 1]$ and $\tilde{g}_1(x_0) < y_0 < \tilde{g}_2(x_0)$. In the case of a system of twice differentiable functions $\{\varphi_j\}_{j=0}^n$, these branches, in addition, cannot intersect at the points $(x_0, \tilde{g}_\nu(x_0))$, $x_0 \in [-1, 1]$, $\nu = 1, 2$.

Indeed, assume the opposite, i.e., that some increasing branches of two polynomials T_α and $T_{\alpha'}$, $\alpha \neq \alpha'$, intersect. Then we easily find that the "less steep" increasing branch of, e.g., the polynomial T_α is crossed by the polynomial $T_{\alpha'}$ (counting double zeros) at at least three points. Since the polynomial $T_{\alpha'}$ also crosses all other $n-2$ branches of T_α at least once, the difference $T_\alpha - T_{\alpha'}$ should have $\geq n + 1$ zeros on $[-1, 1]$ (counting double zeros), which is impossible because $\{\varphi_j\}_{j=0}^n$ is a T-system.

Let us now present an example showing that the polynomials T_α and $T_{\alpha'}$ can intersect at the points $(-1, \tilde{g}_\nu(-1))$ and $(1, \tilde{g}_\nu(1))$, $\nu = 1, 2$. Indeed, by setting $g_\nu(x) = (-1)^\nu$, $\varphi_0(x) = 1$, $\varphi_1(x) = x$, and $\varphi_2(x) = 2x^2 - 1$, for $n = 2$, we conclude that

$$T_0(-1) = T_1(-1) = 1$$

and, therefore,

$$T_0(1) = T_1(1) = 1.$$

It is easy to see that the requirement of differentiability in the second case of property 3 is indeed necessary.

The following assertion is a consequence of the established properties:

Corollary 7.1. *At most two 2-snakes T_α and $T_{\alpha'}$ pass through each point (x_0, y_0) such that $x_0 \in [-1, 1]$ and $\tilde{g}_1(x_0) < y_0 < \tilde{g}_2(x_0)$; the branch of one snake passing through the point (x_0, y_0) is increasing and the branch of the second snake is decreasing.*

If the system of functions $\{\varphi_j\}_{j=0}^n$ is twice differentiable, then at most one 2-snake passes through each point (x_0, y_0) such that $x_0 \in (-1, 1)$ and $y_0 = \tilde{g}_1(x_0)$ or $y_0 = \tilde{g}_2(x_0)$.

4. Exactly two 2-snakes T_α pass through each point (x_0, y_0) such that

$$x_0 \in [-1, 1] \quad \text{and} \quad \tilde{g}_1(x_0) < y_0 < \tilde{g}_2(x_0),$$

and at least one 2-snake T_α passes through each point (x_0, y_0) such that

$$x_0 \in (-1, 1) \quad \text{and} \quad y_0 = \tilde{g}_\nu(x_0), \qquad \nu = 1 \quad \text{or} \quad \nu = 2.$$

Moreover, if the system of functions $\{\varphi_j\}_{j=0}^{n}$ is twice differentiable, then the indicated 2-snake is unique.

Proof. Since, in view of relation (7.30),

$$T_{-2}(x) := \lim_{\alpha \to 0} T_\alpha(x) \equiv T_2(x),$$

the family of 2-snakes $T_\alpha(x)$ can be continued in α onto the entire axis according to the rule

$$T_{\alpha+4}(x) \equiv T_\alpha(x). \tag{7.31}$$

By the theorem on continuous dependence of the polynomial of the best approximation on the approximated function, the obtained family $T = \{T_\alpha\}$, $-\infty < \alpha < \infty$, is continuous in α in a sense that, for any $\alpha_0 \in (-\infty, \infty)$,

$$\lim_{\alpha \to \alpha_0} \left\| T_{\alpha_0} - T_\alpha \right\|_{[-1,1]} = 0. \tag{7.32}$$

First, we prove the second part of property 4.

Assume that, e.g.,

$$y_0 = \tilde{g}_2(x_0) = \sup_{A_{n-1}^*} T_n\left(B^0; A_{n-1}^*; x_0\right) = \sup_\alpha T_\alpha(x_0) = \lim_{\nu \to \infty} T_{\alpha_\nu}(x_0),$$

$$\alpha_\nu \in (-2, 2].$$

We now select a subsequence $\{\alpha_{\nu_j}\}$ of the sequence $\{\alpha_\nu\}$ convergent to a certain number $\overline{\alpha}_0$. Then, according to relation (7.32), we get

$$\lim_{\nu \to \infty} T_{\alpha_\nu}(x) = T_{\overline{\alpha}_0}(x)$$

uniformly in all x, i.e.,

$$\lim_{\nu \to \infty} \left\| T_{\alpha_\nu}(x) - T_{\overline{\alpha}_0}(x) \right\|_{[-1,1]} = 0, \quad T_{\overline{\alpha}_0}(x_0) = y_0.$$

Hence, $T_{\overline{\alpha}_0}$ is just the required polynomial. The uniqueness in the case of a twice differentiable system of functions follows from Corollary 7.1.

We now consider the case where $x_0 \in [-1, 1]$ and $\tilde{g}_1(x_0) < y_0 < \tilde{g}_2(x_0)$. By α_1 we denote a real number for which

$$T_{\alpha_1}(x_0) = \tilde{g}_1(x_0).$$

Then, according to relation (7.30), for $\alpha_2 = \alpha_1 + 2$, we have

$$T_{\alpha_2}(x_0) = -g_1(x_0) = g_2(x_0).$$

Further, we set $y = y(\alpha) = T_\alpha(x)$. According to property 3, as α changes from α_1 to α_2, the points of the curves T_α located on g_1 and g_2 move in the same direction and, hence, the branches of the curves T_α participate in the corresponding motion. In this motion, all branches of the curves located above the point x_0 are either all increasing or all decreasing. Therefore, in view of the continuity of the function $y = y(\alpha)$ and property 3 of the polynomials, we conclude that the function $y = y(\alpha)$ homeomorphically maps the segment $[\alpha_1, \alpha_2]$ onto a certain set of numbers from the segment $[\tilde{g}_1(x_0), \tilde{g}_2(x_0)]$, the least of which is the number $\tilde{g}_1(x_0)$ and the greatest number is $\tilde{g}_2(x_0)$. However, in this case, it is known (see, e.g., [Dieudonne (1964), p. 97)] that this mapping is a strictly monotonic and homeomorphic mapping of the segment $[\alpha_1, \alpha_2]$ onto the segment $[\tilde{g}_1(x_0), \tilde{g}_2(x_0)]$. By virtue of Corollary 7.1, this yields statement (b, 1) of Theorem 7.5.

■

Remark 7.3.[12] The definition of the functions $\tilde{g}_\nu$ appearing in Theorem 7.5 for $k = 2$ is equivalent to the following more natural definition:

The functions $\tilde{g}_1$ and $\tilde{g}_2$ from Theorem 7.5 can be defined by the equalities

$$\tilde{g}_2(x) = \sup_{T \in K} \{T(x)\} \quad \text{and} \quad \tilde{g}_1(x) = \inf_{T \in K} \{T(x)\}, \tag{7.33}$$

where K is the set of polynomials of the form

$$T(x) = \sum_{k=0}^{N} c_k \varphi_k(x)$$

satisfying the condition $g_1(x) \le T(x) \le g_2(x)$ for all $x \in [-1, 1]$.

Proof. We set

$$\sup_{T\in K}\{T(x)\} = g_2^*(x) \quad\text{and}\quad \inf_{T\in K}\{T(x)\} = g_1^*(x). \tag{7.33$'$}$$

Let us show that, e.g., $g_2^*(x) = \tilde{g}_2(x)$. The inequality $\tilde{g}_2(x) \le g_2^*(x)$ is obvious. We now prove that the inverse inequality $\tilde{g}_2(x) \ge g_2^*(x)$ is also true. Indeed, assume the contrary, i.e., that the opposite inequality $\tilde{g}_2(x_0) < g_2^*(x_0)$ holds at a certain point $x_0 \in [-1, 1]$. Then, by virtue of (7.33), there exists a polynomial $\tilde{T}(x) \in K$ such that

$$\tilde{T}(x_0) > \tilde{g}_2(x_0). \tag{7.34}$$

It is clear that if, for the 1-snake $T^+(x)$ the first (or the last) point x^* of the Chebyshev alternation is a (+)-point and, hence, $T^+(x^*) = g_2(x^*)$, then, for any $x \in [-1, x^*]$ (or for any $x \in [x^*, 1]$), we have

$$g_2^*(x) = T^+(x) = \tilde{g}_2(x).$$

Indeed, the assumption that, e.g., for $x' \in [-1, x^*)$, we have $g_2^*(x') > T^+(x')$ or $\tilde{g}_2(x') > T^+(x')$ implies the existence of a T-polynomial $\tilde{\tilde{T}}$ such that $\tilde{\tilde{T}}_n(x') > T^+(x')$, which is impossible because, in this case, the polynomial $\tilde{\tilde{T}}$ must cross the increasing semibranch and n branches of the polynomial T^+ (1-snake) and, consequently, the difference $T^+ - \tilde{\tilde{T}}$ must have at least $n = 1$ zeros on $[-1, 1]$, which is impossible. Therefore, the point x_0 must belong to an interval $(\alpha, \beta) \subset (-1, 1)$ with the following properties:

(i) $\tilde{g}_2(x) < g_2(x)$,

(ii) $\tilde{g}_2(\alpha) = g_2(\alpha)$ and $\tilde{g}_2(\beta) = g_2(\beta)$.

However, in this case, by virtue of the continuity of 2-snakes T_α with respect to α in C and property 2 (or 1) of the polynomials T_α for some α^*, the polynomial (2-snake) T_{α^*} possesses the following properties:

(a) $\forall x \in (\overline{\alpha}, \overline{\beta}) \subset (\alpha, \beta)$: $T_{\alpha_0^*}(x) \le \tilde{g}_2(x) < g_2(x)$;

(b) $T_{\alpha^*}(\overline{\alpha}) = g_2(\overline{\alpha})$, $T_{\alpha^*}(\overline{\beta}) = g_2(\overline{\beta})$.

According to (7.26), this yields the inequality $\tilde{T}(x_0) > T_{\alpha_0^*}(x_0)$, which is impossible because, in this case, the difference $\tilde{T} - T_{\alpha^*}$ must have at least $n+1$ zeros on $[-1, 1]$.

■

The presented definition enables us to reformulate Theorem 7.5 as follows:

Theorem 7.5. *Let $g_1(x)$ and $g_2(x) \geq g_1(x)$ be two continuous functions given on the segment $[-1, 1]$ and let $\{\varphi_j\}_{j=0}^n$ be a T-system given on the same segment. If at least one polynomial $T_n(c; \cdot)$ such that $g_1(x) < T_n(c; x) < g_2(x)$ passes through the point (x_0, y_0) such that $x_0 \in [-1, 1]$ and $g_1(x_0) < y_0 < g_2(x_0)$, then there are exactly two 2-snakes passing through the point (x_0, y_0). Moreover, if $x_0 \in (-1, 1)$ and $y_0 = g_1(x_0)$ or $y_0 = g_2(x_0)$ and a 2-snake exists, then this snake is unique provided that all functions φ_j and the functions g_ν, $\nu = 1, 2$, are twice differentiable.*

Remarks to Chapter 1

1. In the case of real variables, this theorem was established by Haar. Kolmogorov generalized this theorem to the case of complex variables. The proof presented in our book is, for the most part, taken from [Akhiezer (1965), p. 80].

2. The scheme of the presented proof is taken from the course of lectures by S. Nikol'skii [S. Nikol'skii (1947), p. 47].

3. This theorem is correlated with a much more general Theorem 7.5 proved by Dzyadyk in analyzing the applied problems connected with the Dirichlet problem for a circle, approximate conformal mappings of domains in $\mathbb{C}$, estimation of the absolute value of the derivative of a polynomial, etc. (in this connection, the reader is referred to the works [Dzyadyk (1973)] and [L. Shevchuk (1974)]. Since, at that time, the works by Karlin and Studden were unavailable for the authors, the snakes appear in cited works under the name of φ-extremal polynomials (much less expressive). For the presentation of the results obtained by Karlin and Studden, the reader is referred to the monograph [M. Krein and Nudel'man (1973), pp. 492–504]. Note that the theory of the above-mentioned applied problems seems to be quite complicated and incomplete and, therefore, is not included in our monograph.

4. Chebyshev (1859) posed and solved the following problem:

Chebyshev problem. *Let P_l be an arbitrary polynomial positive on $[-1, 1]$. For any $n \geq l$, it is necessary to find a polynomial P_n^* of degree n with coefficient of the term*

x^n equal to one and such that

$$\left\| \frac{P_n^*}{P_l} \right\|_{[-1,1]} = \min_{P_i} \left\| \frac{x^n + P_1 x^{n-1} + \ldots + P_n}{P_l(x)} \right\|_{[-1,1]},$$

i.e., it is necessary to find, for all $n \geq l$, the polynomial least deviating from zero in the uniform metric with weight $1/P_l$.

Chebyshev constructed the polynomial P_n^* in the form of the solution of an equation of the form

$$y^2 - (x^2 - 1)\,[\,C(x - x_1)(x - x_2) \ldots (x - x_n)]^2 = L^2 P_l^2(x),$$

where $x_1, x_2, \ldots, x_n$, C and L are unknown numbers.

Since this equation possesses several solutions from which it is necessary to choose the required solution of the posed problem, Markov (1906) proposed a quite witty but fairly complicated direct solution of this (and even a somewhat more general) problem:

Markov problem. *Under the Chebyshev assumption imposed on the polynomial $P_l(x)$, for any $n \geq l/2$, it is necessary to find a polynomial $P_n^*(x)$ such that*

$$\left\| \frac{P_n^*(x)}{\sqrt{P_l(x)}} \right\|_{[-1,1]} = \min_{P_i} \left\| \frac{x^n + P_1 x^{n-1} + \ldots + P_n}{\sqrt{P_l(x)}} \right\|_{[-1,1]}.$$

In exactly the same way as in the proof of Theorem 1.2 (Chebyshev), we can show that the polynomial P_n^* is the solution of the Markov problem if an only if the polynomial

$$\tilde{P}_n^*(x) := \frac{\sqrt{P_l(1)}}{P_n^*(1)}\, P_n^*(x)$$

constructed by using the indicated polynomial is a snake for a couple of functions $-\sqrt{P_l(x)}$ and $\sqrt{P_l(x)}$.

The solution of this problem obtained by Markov is given by the formula

$$P_n^*(x) = \sqrt{P_l(x)}\,\cos(\varphi_1 + \varphi_2 + \varphi_{2n}),$$

where the numbers $\cos\varphi_k$ and $\sin\varphi_k$ are given by the formulas

$$\cos\varphi_k = \frac{\sqrt{(1+a_k)/2}\sqrt{1+x}}{\sqrt{1+a_k x}} \quad \text{and} \quad \sin\varphi_k = \frac{\sqrt{(1-a_k)/2}\sqrt{1-x}}{\sqrt{1+a_k x}},$$

where, in turn, $-\pi < \varphi_k \le \pi$ and a_k, $k = 1, 2, \ldots, 2n$, are the numbers for which the polynomial $P_l(x)$ can be represented in the form

$$P_l(x) = c(1+a_1x)(1+a_2x)\ldots(1+a_{2n}x), \quad c = \text{const}.$$

Theorem 2.13 [Dzyadyk (1977)], in addition to the Chebyshev and Markov results, establishes that the representation

$$P^*_{n+l}(x) = c_1 \sum_{k=0}^{l} \gamma_k T_{|n+k|}(x), \quad n \ge l, \quad c_1 = \text{const},$$

takes place for all $n \ge -l/2$. The coefficients γ_k of this representation are independent of n and $T_j(x) = \cos j \arccos x$.

5. For an important case $\varphi_k(z) = z^k$, $k = 0, 1, \ldots, n$, this theorem was earlier established by Tonelli (1908) and de la Vallée Poussin (1911).

6. The present section is based on the work [Dzyadyk (1974c)].

7. The second equality is well known. For the first time, it was deduced by Kirchberger (1903). Later, it was thoroughly investigated by de la Vallée Poussin (1910), (1919).

8. The problem of efficient construction of the polynomial of the best approximation of fixed degree n was apparently posed for the first time in the monograph [de la Vallée Poussin (1919)]. In the case where the set $\mathfrak{M} = \mathfrak{M}_N$ is finite and consists of N points $x_1 < x_2 < \ldots < x_N$, $N \ge n+2$, de la Vallée Poussin suggested to consider all possible subsystems

$$\check{\mathfrak{M}} := \left\{x_{k_0}, x_{k_1}, \ldots, x_{k_{n+1}}\right\} \subset \mathfrak{M}_N$$

each of which contains $n+2$ points $x_{k_i} \in \mathfrak{M}$ and find, for each subsystem, the polynomial $P^*_n(x) = P^*_n(\check{\mathfrak{M}}; x)$ of the best approximation of a given function f by using Theorem 5.4.

Thus, by virtue of the fact that, for the set $\mathfrak{M}_N$, the characteristic set E_0 consists of $n+2$ points (see Remark 4.2), it is possible to conclude that, in the family of $\binom{N}{n+2}$ subsets $\breve{\mathfrak{M}} \subset \mathfrak{M}_N$, the characteristic set $\breve{\mathfrak{M}}_0$ is defined as the subset for which the value of the deviation

$$\max_{x \in \breve{\mathfrak{M}}_0} \left| f(x) - P_n^*\left(\breve{\mathfrak{M}}_0; x\right) \right|$$

is maximum. Moreover, the polynomial $P_n^*\left(\breve{\mathfrak{M}}_0; x\right)$ is just the required polynomial of the best approximation for the function f on $\mathfrak{M}_N$. Since it is possible to find the polynomial of the best approximation on the sets $\mathfrak{M}_N$, we can also find (with any required accuracy) the polynomial of the best approximation on the segment $[a, b]$.

By using the results obtained by de la Vallée Poussin, Remez proposed a different very efficient procedure of transition from a subset $\breve{\mathfrak{M}} \subset [a, b]$ of $n+2$ points to a new "worse" subset $\breve{\mathfrak{M}}'$. On the basis of this procedure, he created a new very efficient algorithm of construction of the polynomial of the best approximation and proved (see Theorem 6.1) that the proposed algorithm is rapidly convergent.

9. The present section is based on the works [Dzyadyk (1971)] and [Dzyadyk (1973)].

10. See Remark 3.

11. This theorem, on the one hand, generalizes some results from the monograph [Voronovskaya (1963), pp. 82, 102–106, 119–126, and 158–163] obtained for $TM(n, n)$-systems $\{x^j\}_{j=0}^n$ under the assumption that $g_\nu(x) = (-1)^\nu$ and, on the other hand, generalizes Theorem 2.9 (Karlin) concerning the behavior of 1-snakes to the case $k = 2$, i.e., to the case where the constraints are absent. Moreover, the polynomials appearing in this theorem are, in fact, generalized Zolotarev polynomials.

12. This remark was proposed by Golub and Kovtunets.

Chapter 2
Weierstrass theorems

In the present chapter, the reader can find the Weierstrass theorems and important results from the theory of approximation established as a result of the analysis of these theorems.

Consider a real-valued continuous function f defined on a segment $[a, b]$. Our aim is to study whether it is possible to approximate this function by algebraic polynomials with any required accuracy, i.e., whether, for any $\varepsilon > 0$, there exists a polynomial P such that the inequality

$$|f(x) - P(x)| < \varepsilon$$

holds for all $x \in [a, b]$.

Similarly, we can pose the problem on the possibility of approximation of periodic continuous functions by trigonometric polynomials.

Both these problems are positively solved by the Weierstrass theorems.

To prove these theorems, we first present a general procedure of construction of algebraic and trigonometric polynomials convenient for the approximate representation of functions and three inequalities for the function $\sin t$.

1. Let

$$K_n(x) = \sum_{k=0}^{n} a_k x^k$$

be a polynomial of degree n. Then, for any function f integrable on $[a, b]$, the convolution

$$P(x) = \int_a^b f(t) K_n(x-t)\,dt$$

is an algebraic polynomial of degree $\leq n$. This follows from the fact that $K_n(x-t)$ is clearly a polynomial of degree n in the variable x of the form

$$K_n(x-t) = \sum_{\nu=0}^{n} c_\nu(t)x^\nu.$$

The coefficients of this polynomial $c_\nu(t)$ are also polynomials (of degree not greater than $n-\nu$).

In exactly the same way, if

$$K_n(t) = \sum_{k=0}^{n} (a_k \cos kt + b_k \sin kt)$$

is a trigonometric polynomial of degree n, then, for any function f summable on $[0, 2\pi]$, the integral

$$\int_0^{2\pi} f(u)K_n(t-u)\,du = T_n(t)$$

is a trigonometric polynomial of degree $\leq n$. It is customary to say that these polynomials are obtained with the help of polynomial kernels K_n.

2. The following important inequalities are true:

$$|\sin nt| \leq n|\sin t| \quad \text{for all} \quad t \in (-\infty, \infty), \tag{0.1}$$

$$\sin t \geq \frac{2}{\pi}t \quad \text{for all} \quad t \in \left[0, \frac{\pi}{2}\right], \tag{0.2}$$

$$\sin t \leq t \quad \text{for all} \quad t \geq 0. \tag{0.3}$$

The first of these inequalities can readily be proved by induction and the other two inequalities immediately follow from Fig. 3. It is also well known that

$$\tan t \geq t \quad \text{for all} \quad t \in \left[0, \frac{\pi}{2}\right). \tag{0.4}$$

1. First Weierstrass theorem

Theorem 1.1 [Weierstrass (1885)].[1] *For any function f continuous on the segment $[a, b]$ and any $\varepsilon > 0$, there exists an algebraic polynomial P such that*

$$\max_{x\in[a,b]} |f(x) - P(x)| < \varepsilon. \tag{1.1}$$

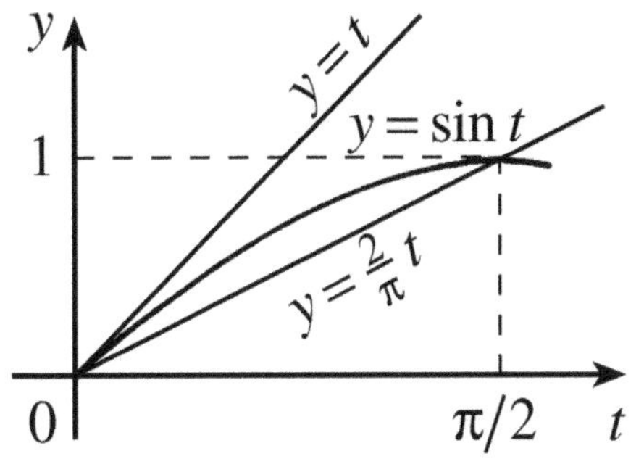

Fig. 3

There are many different proofs of Theorem 1.1. We present a proof of this theorem based on the use of the polynomial kernels K_n obtained for the Chebyshev polynomials T_n. This proof is not simpler than the other available proofs but we want the reader to get acquainted with the polynomials kernels K_n playing an important role in various problems of the theory of approximation.

Proof [Dzyadyk (1958)]. Let $T_{2n+1}(x) = \cos(2n+1)\arccos x$ be a Chebyshev polynomial of degree $2n+1$. Since $T_{2n+1}(0) = 0$, the polynomial $T_{2n+1}(x)$ can be divided by x. Thus, we consider a polynomial $K_n(x)$ of degree $4n$ of the form

$$K_n(x) = \frac{1}{\gamma_n}\left[\frac{\cos(2n+1)\arccos x}{x}\right]^2, \tag{1.2}$$

where

$$\gamma_n = \int_{-1}^{1}\left[\frac{\cos(2n+1)\arccos x}{x}\right]^2 dx. \tag{1.3}$$

This polynomial plays the role of a kernel and possesses the following properties:

1. $$\int_{-1}^{1} K_n(x)\,dx = 1. \tag{1.4}$$

This property directly follows from equalities (1.2) and (1.3).

2. Since

$$T_{2n+1}(-x) = \cos(2n+1)\arccos(-x)$$

$$= \cos(2n+1)(\pi - \arccos x) = -\cos(2n+1)\arccos x = -T_{2n+1}(x),$$

the polynomial T_{2n+1} is odd and K_n is an even polynomial of degree $4n$.

3. $\gamma_n > n, \quad n = 1, 2, \dots .$ (1.5)

Indeed, in view of relation (1.3), property 2, and inequalities (0.2) and (0.3), we find

$$\gamma_n = 2\int_0^1 \left[\frac{\cos(2n+1)(\pi/2 - \arcsin x)}{x}\right]^2 dx$$

$$= 2\int_0^1 \left[\frac{\sin(2n+1)\arcsin x}{x}\right]^2 dx = \int_0^{\pi} \left(\frac{\sin(t(2n+1)/2)}{\sin(t/2)}\right)^2 \cos\frac{t}{2}\, dt$$

$$> \int_0^{\pi/(2n+1)} \left(\frac{\sin(t(2n+1)/2)}{\sin(t/2)}\right)^2 \cos\frac{t}{2}\, dt > \int_0^{\pi/(2n+1)} \frac{(t(2/\pi)(2n+1)/2)^2}{t^2} 4\frac{1}{2} dt$$

$$= 2\frac{(2n+1)^2}{\pi^2}\frac{\pi}{2n+1} = 2\frac{2n+1}{\pi} > n.$$

It is possible to show that the exact value of γ_n is equal to

$$2\left\{2n + 1 + \sum_{k=1}^{2n} \frac{(-1)^{k-1}}{(2n-k-1/2)(2n-k+3/2)}\right\}.$$

4. For any $\delta \in (0, 1)$, $n = 1, 2, \dots,$ we have

$$\int_\delta^1 K_n(x)dx = \frac{1}{\gamma_n}\int_\delta^1 \left[\frac{\cos(2n+1)\arccos x}{x}\right]^2 dx \le \frac{1}{n}\int_\delta^1 \frac{dx}{x^2} < \frac{1}{n\delta}. \tag{1.6}$$

First, we prove the Weierstrass theorem for the case $a = -1$ and $b = 1$. The function f is continuously extended onto the interval $[-2, 2]$ by setting

$$f(x) = \begin{cases} f(-1) & \text{for } x \in [-2, -1], \\ f(1) & \text{for } x \in [1, 2]. \end{cases}$$

The function f is continuous and, hence, uniformly continuous on $[-2, 2]$. Hence, for any $\varepsilon > 0$, one can find $\delta > 0$, $0 < \delta < 1$, such that the following inequality is true for any two points x' and x'' from the interval $[-2, 2]$ such that $|x' - x''| < \delta$:

$$|f(x') - f(x'')| < \frac{\varepsilon}{2}.$$

For any $n = 1, 2, \ldots$, we define a polynomial P_n of degree $\leq 4n$ by setting

$$P_n(x) = \frac{1}{3}\int_{-2}^{2} f(t) K_n\left(\frac{t-x}{3}\right) dt.$$

By the change of variables $(t-x)/3 = \eta$, we obtain

$$P_n(x) = \int_{(-2-x)/3}^{(2-x)/3} f(3\eta + x) K_n(\eta)\, d\eta.$$

Since, by virtue of equality (1.4),

$$f(x) = \int_{-1}^{1} f(x) K_n(\eta)\, d\eta,$$

we conclude that, for $x \in [-1, 1]$,

$$|f(x) - P_n(x)| < \int_{-\delta/3}^{\delta/3} |f(x) - f(3\eta + x)| K_n(\eta)\, d\eta$$

$$+ \left(\int_{-1}^{-\delta/3} + \int_{\delta/3}^{1}\right) |f(x)| K_n(\eta)\, d\eta$$

$$+ \left(\int_{(-2-x)/3}^{-\delta/3} + \int_{\delta/3}^{(2-x)/3}\right) |f(3\eta + x)| K_n(\eta)\, d\eta,$$

whence, by using the properties of the polynomial K_n and denoting the maximum absolute value of the function $|f|$ on $[-2, 2]$ by M, we obtain

$$|f(x) - P_n(x)| \le \frac{\varepsilon}{2}\int_{-1}^{1} K_n(\eta)\,d\eta + 2M\int_{\delta/3}^{1} K_n(\eta)\,d\eta + 2M\int_{\delta/3}^{1} K_n(\eta)\,d\eta$$

$$\le \frac{\varepsilon}{2} + 4M\frac{3}{n\delta}.$$

Further, if n is sufficiently large, then for all $x \in [-1, 1]$, we can write

$$|f(x) - P_n(x)| < \varepsilon.$$

This proves the Weierstrass theorem in the case where $a = -1$ and $b = 1$.

We now prove the theorem for arbitrary a and b. To do this, we perform the change of variables

$$x = a + \frac{b-a}{2}(u+1), \tag{1.7}$$

and, hence,

$$u = \frac{2x - a - b}{b - a}. \tag{1.7'}$$

This enables us to construct a function

$$\varphi(u) = f\left[a + \frac{b-a}{2}(u+1)\right]$$

continuous on $[-1, 1]$. Further, as already shown, we can find a polynomial π_n for this function such that

$$|\varphi(u) - \pi_n(u)| < \varepsilon. \tag{1.8}$$

Thus, we set

$$P_n(x) = \pi_n\left(\frac{2x - a - b}{b - a}\right)$$

and, by virtue of relations (1.7) and (1.7′), conclude that

$$|f(x) - P_n(x)| = |\varphi(u) - \pi_n(u)| < \varepsilon.$$

This completes the proof of the Weierstrass theorem.

■

In Section 3, we present examples of different kernels very important for what follows.

2. Stone's theorem

In the present section, with an aim to generalize the Weierstrass theorem to other important cases and, in particular, to the cases of approximation of periodic functions by trigonometric polynomials and approximation of functions given on bounded sets in the space R^n, $n = 1, 2, \ldots,$ we prove a very general Stone's theorem.

2.1. Stone's theorem

We introduce the following definitions:

1. A set A of functions defined on a nonempty set E is called an *algebra* if this set is closed both with respect to the multiplication by all possible real (or complex) numbers and with respect to the summation and multiplication of the functions. This means that if A is an algebra and f and g are elements of this algebra, then

$$cf \in A, \quad f + g \in A, \quad \text{and} \quad fg \in A, \tag{2.1}$$

where c is an arbitrary constant. In what follows, all functions of the algebra A are assumed to be continuous.

2. We say that an *algebra* A defined on E *separates points of the set* E if, for any pair of different points $x_1, x_2 \in E$, there exists at least one function $f \in A$ such that

$$f(x_1) \neq f(x_2). \tag{2.2}$$

3. We say that an *algebra* A defined on E *does not vanish at any point of the set* E if, for any point $x_0 \in E$, there exists at least one function $h \in A$ such that

$$h(x_0) \neq 0. \tag{2.3}$$

An algebra A defined on a set E is called *Stone's algebra* if it separates the points and does not vanish at any point of the set E.

Examples of Stone's algebras

1. The set of all possible trigonometric polynomials T_n on a half interval $[a, b)$ such that $b - a \leq 2\pi$. In the case where $b - a > 2\pi$, the algebra of polynomials T_n is not a Stone's algebra because it does not separate the points of the half interval $[a, b)$: In this case, we always have $T(a) = T(a + 2\pi)$, where $a, a + 2\pi \in [a, b)$.

2. The set of all possible algebraic polynomials given on a bounded set in the space R^n, $n = 1, 2, \ldots$.

3. The set of all possible algebraic polynomials of the form

$$a_1 x^3 + a_2 x^6 + \ldots + a_n x^{3n},$$

where n is an arbitrary positive integer, defined on a segment $[a, 1]$, $a \in (0, 1)$.

Theorem 2.1 [Stone (1937)]. *If an algebra A defined on a compact set K is Stone's algebra, then the set of functions from A is everywhere dense in the set C of all functions continuous on K, so that $\overline{A} = C$. Therefore, any function F continuous on K can be arbitrarily well approximated by elements of the algebra A.*

Stone's theorem is proved by using geometric reasoning similar to the proof of the Weierstrass theorem (Theorem 1.1) proposed by Lebesgue (1898). More precisely, the validity of Theorem 2.1 immediately follows from Lemmas 2.1, 2.3, and 2.4 established in what follows. Lemma 2.2 in required to prove Lemmas 2.3 and 2.4.[2]

Lemma 2.1 (on the approximation of envelopes for the family of functions from A). *If a finite number of functions $f_1, f_2, \ldots, f_n$ defined on a compact set K belongs to A, then the functions*

$$M = \max_i \{f_i\} \quad \text{and} \quad m = \min_i \{f_i\}, \quad i = 1, 2, \ldots, n, \tag{2.4}$$

called the upper and lower envelopes of the functions f_i, respectively, can be arbitrarily exactly uniformly approximated by functions from the algebra A in the set K.

We split the proof of the lemma into three stages.

1. First, we prove the validity of the lemma in the following special case: If f is an arbitrary function from the algebra A, then the function

$$M = \max\{f, -f\} = |f| \tag{2.5}$$

can be approximated arbitrarily well by functions from the algebra A uniformly on K.

Indeed, we fix a number $\varepsilon > 0$ and set $\max_{x \in K} |f(x)| = a$. Further, by using the Weierstrass theorem 1.1, for a function $|\cdot|$ continuous on the segment $[-a, a]$, we find an algebraic polynomial P_N of the form

$$P_N(y) = \sum_{i=0}^{N} c_i y^i$$

such that

$$\max_{-a \le y \le a} \big||y| - P_N(y)\big| \le \frac{\varepsilon}{2}.$$

Since $c_0 = P_N(0)$, for all $y \in [-a, a]$, we get

$$\left| |y| - \sum_{i=1}^{N} c_i y^i \right| = \big||y| - P_N(y) + P_N(0)\big| < \varepsilon$$

and, hence, for $x \in K$,

$$\left| |f(x)| - \sum_{i=1}^{N} c_i [f(x)]^i \right| < \varepsilon.$$

Since the sum $\sum_{i=1}^{N} c_i [f]^i$ does not contain the term c_0, which may not belong to A and A is an algebra, the element $\sum_{i=1}^{N} c_i [f]^i \in A$. Thus, Lemma 2.1 is proved in the first case.

2. Since

$$\max\{f_1, f_2\} = \frac{f_1 + f_2}{2} + \frac{|f_1 - f_2|}{2},$$

$$\min\{f_1, f_2\} = \frac{f_1 + f_2}{2} - \frac{|f_1 - f_2|}{2},$$

in view of the first case the assumption that $f_1, f_2 \in A$ implies that the function

$$\frac{|f_1 - f_2|}{2} = \left|\frac{f_1}{2} - \frac{f_2}{2}\right|$$

and, hence, the functions $\max\{f_1, f_2\}$ and $\min\{f_1, f_2\}$ can be approximated arbitrarily well by the elements of A uniformly on K. This proves Lemma 2.1 for the case of two functions f_1 and f_2.

3. In the case of an arbitrary finite number n of functions, Lemma 2.1 is proved by induction.

Lemma 2.2 (on the interpolation property of algebras). *If an algebra A defined on K separates the points of K and does not vanish at any point of K, then, for any different points $x_1, x_2 \in K$ and any numbers c_1 and c_2, one can find a function $l \in A$ such that*

$$l(x_1) = c_1 \quad \text{and} \quad l(x_2) = c_2. \tag{2.6}$$

Proof. Since the algebra A separates the points of K, in A, there exists a function g such that $g(x_1) \neq g(x_2)$. On the other hand, since the algebra A does not vanish on K, there exists a function h such that $h(x_1) \neq 0$. Therefore, the function

$$f_1 := \frac{h}{h(x_1)} \frac{g(x_2) - g}{g(x_2) - g(x_1)} \in A \tag{2.7}$$

and, clearly, is such that $f_1(x_1) = 1$ and $f_1(x_2) = 0$.

Similarly, the algebra A contains a function f_2 such that $f_2(x_1) = 0$ and $f_2(x_2) = 1$. However, this means that the function

$$l := c_1 f_1 + c_2 f_2 \tag{2.8}$$

also belongs to the algebra A and, at the same time, satisfies conditions (2.6).

Lemma 2.2 is thus proved.

Lemma 2.3 (on the existence of dominating envelopes with junction point). *Let F be an arbitrary function continuous on a compact set K. Then, for any point $x^0 \in K$*

and any number $\varepsilon > 0$, *one can find finitely many functions* $f_1, f_2, \ldots, f_r \in A$, *whose upper envelope* M_{x^0} *possesses the following properties:*

$$M_{x^0}(x^0) = F(x^0) \quad \textit{and} \quad M_{x^0}(x) > F(x) - \varepsilon. \tag{2.9}$$

The function M_{x^0} *is called an envelope dominating over* F *with accuracy* ε *and junction point* x^0.

Proof. According to Lemma 2.2, for the point x^0 $(\in K)$ and any point $y \in K$ other than x^0, one can construct a function l_y such that

$$l_y(x^0) = F(x^0) \quad \text{and} \quad l_y(y) = F(y). \tag{2.10}$$

In view of the continuity of the function l_y, there exists an open set G_y containing the point y such that

$$l_y(x) > F(x) - \varepsilon \quad \text{for all } x \in G_y \cap K. \tag{2.11}$$

Since K is a compact set, there are finitely many points $y_1, y_2, \ldots, y_r$ such that the corresponding open sets $G_{y_1}, G_{y_2}, \ldots, G_{y_r}$ cover K: $K \subset G_{y_1} \cup \ldots \cup G_{y_r}$. Therefore, if we set $f_i = l_{y_i}$, then the upper envelope M_{x^0} of the functions $f_1 = l_{y_1}, \ldots, f_r = l_{y_r}$ satisfies all conditions of Lemma 2.3 because, by virtue of relations (2.10),

$$M_{x^0}(x^0) = \max\{f_1(x^0), \ldots, f_r(x^0)\} = \max\{l_{y_1}(x^0), \ldots, l_{y_r}(x^0)\} = F(x^0),$$

On the other hand, any point $x \in K$ belongs to a certain G_{y_i} and, hence,

$$M_{x^0}(x) = \max\{l_{y_1}(x), \ldots, l_{y_r}(x)\} \geq l_{y_i}(x) > F(x) - \varepsilon$$

by virtue of inequality (2.11). Lemma 2.3 is proved.

Lemma 2.4 (on the approximation of continuous functions by envelopes). *Let* F *be an arbitrary function continuous on a compact set* K. *Then, for any* $\varepsilon > 0$, *there exist finitely many envelopes* $M_{x_1}, M_{x_2}, \ldots, M_{x_l}$ *dominating over* F *with junction points* x_i; *moreover, the lower envelope* m *in this family satisfies the inequalities*

$$F(x) - \varepsilon < m(x) < F(x) + \varepsilon \quad \textit{for all } x \in K. \tag{2.12}$$

Proof. Every function M_{x^0} is clearly continuous and satisfies the equality in (2.9). Hence, there exists an open set D_{x^0} containing the point x^0 and such that

$$M_{x^0}(x) < F(x) + \varepsilon \quad \text{for all} \quad x \in D_{x^0} \cap K. \tag{2.13}$$

Since K is a compact set, there exists a finite set of points $x_1, x_2, \ldots, x_l$ such that the corresponding open sets $D_{x_1}, D_{x_2}, \ldots, D_{x_l}$ cover K: $K \subset D_{x_1} \cup D_{x_2} \cup \ldots \cup D_{x_l}$. Let us show that, in this case, the function

$$m = \min\{M_{x_1}, \ldots, M_{x_l}\}$$

satisfies inequalities (2.12).

Indeed, on the one hand, each function M_{x_i} satisfies inequality (2.9) and, thus,

$$m(x) > F(x) - \varepsilon \quad \text{for all} \quad x \in K. \tag{2.14}$$

On the other hand, every point $x \in K$ belongs to a certain set D_{x_i} and, hence, by virtue of inequality (2.13),

$$m(x) = \min\{M_{x_1}(x), \ldots, M_{x_l}(x)\} \le M_{x_i}(x) < F(x) + \varepsilon. \tag{2.15}$$

Inequalities (2.14) and (2.15) imply the validity of Lemma 2.4.

Proof of Stone's theorem. Since each envelope M_{x_i} dominating over F is an upper envelope of finitely many functions from A, by virtue of Lemma 2.1, each of these functions can be arbitrarily exactly approximated by a certain function f_j from A. However, on the other hand,

$$m = \min\{M_{x_1}, M_{x_2}, \ldots, M_{x_l}\}$$

and, hence, m can be arbitrarily exactly approximated by envelopes $\tilde{m}$ of the form $\tilde{m}(x) = \min\{f_1, f_2, \ldots, f_l\}$, where $f_j \in A$, and consequently, in view of Lemma 2.1, by functions g from A. Therefore, in view of (2.12), there exists a function $g \in A$ such that $F(x) - \varepsilon < g(x) < F(x) + \varepsilon$ or, in other words, such that $|F(x) - g(x)| < \varepsilon$, $x \in K$.

The proof of Stone's theorem is thus completed.

■

2.2. Corollaries of Stone's theorem

Theorem 2.2 (first Weierstrass theorem in R^k). *Let K be an arbitrary closed bounded set in R^k. Then, for any function F continuous on K and any $\varepsilon > 0$, one can find a polynomial $P_{\bar{n}}$, where $\bar{n} = (n_1, n_2, \ldots, n_k)$ are nonnegative integers, of the form*

$$P_{\bar{n}}(x) = P_{n_1, n_2, \ldots, n_k}(x_1, x_2, \ldots, x_k) = \sum_{j_1=0}^{n_1} \sum_{j_2=0}^{n_2} \ldots \sum_{j_k=0}^{n_k} c_{j_1, j_2, \ldots, j_k} x_1^{j_1} \ldots x_k^{j_k}, \tag{2.16}$$

where $x = (x_1, x_2, \ldots, x_k)$ is a point of R^k, such that the inequality

$$|F(x) - P_{\bar{n}}(x)| < \varepsilon \tag{2.17}$$

holds for all $x \in K$.

For the sake of simplicity, the next theorem is formulated for $k = 2$.

Theorem 2.3 (second Weierstrass theorem in R^k). *For any function F 2π-periodic in each variable x_j, $j = 1, 2$, and continuous in the entire space R^2 and any $\varepsilon > 0$, there exists a trigonometric polynomial $T_{\bar{n}}$, $\bar{n} = (n_1, n_2)$, of the form*

$$\begin{aligned} T_{\bar{n}}(x) &= T_{n_1, n_2}(x_1, x_2) \\ &= \sum_{j_1=0}^{n_1} \sum_{j_2=0}^{n_2} [a_{j_1, j_2} \cos j_1 x \cos j_2 y + b_{j_1, j_2} \cos j_1 x \sin j_2 y \\ &\qquad + c_{j_1, j_2} \sin j_1 x \cos j_2 y + d_{j_1, j_2} \sin j_1 x \sin j_2 y] \end{aligned} \tag{2.18}$$

such that the inequality

$$|F(x) - T_{\bar{n}}(x)| < \varepsilon \tag{2.19}$$

holds for all $x = (x_1, x_2) \in R^2$.

Theorem 2.4. *For any function F continuous on the entire real axis and such that the following finite limits exist and are equal:*

$$-\infty < \lim_{x \to -\infty} F(x) = \lim_{x \to \infty} F(x) < \infty \tag{2.20}$$

and any $\varepsilon > 0$, *there exists a rational polynomial* R_n *of the form*

$$R_n(x) = \frac{a_0 + a_1 x + \ldots + a_n x^n}{b_0 + b_1 x + \ldots + b_n x^n} \tag{2.21}$$

such that the inequality

$$|F(x) - R_n(x)| < \varepsilon \tag{2.22}$$

holds for all $x \in (-\infty, \infty)$.

Proofs of Theorems 2.2–2.4

1. The validity of Theorem 2.2 follows from the facts that, first, any closed bounded set K in a finite-dimensional space R^k is compact and, second, the set of polynomials $P_{\bar{n}}$ of the form (2.16) is a Stone's algebra on K.

2. Since the set of functions 2π-periodic in each variable and continuous in the entire space R^2 can be regarded as the set of continuous functions given on a torus and the torus is a compact set and, moreover, the set of polynomials of the form (2.18) is a Stone's algebra on the torus, Theorem 2.3 also follows Stone's theorem.

3. The validity of Theorem 2.4 follows from the facts that, first, the set of functions continuous on the entire real axis and satisfying condition (2.20) is equivalent to the set of functions continuous, e.g., on the circle $C: \rho = 1, \ \varphi \in [-\pi, \pi]$, i.e., on a compact set, and second, the set of rational polynomials of the form (2.21) regarded on the circle C [by the change of variables $x = \tan(\varphi/2), \ \varphi = 2 \arctan x$] is clearly a Stone's algebra.

3. Examples of polynomial kernels

In the present section, we consider examples of different polynomial kernels K_n defined on a symmetric interval $[-a, a]$ and establish their most important properties, including, first of all, the following properties, which are of especial interest for our presentation:

(a) normalization of the kernel, i.e., the fact that

$$\frac{1}{a}\int_{-a}^{a} K_n(x)\,dx = 1.$$

(b) Lebesgue constants of the kernels $K_n(x)$, i.e., the quantities

$$L(K_n) := \frac{1}{a}\int_{-a}^{a} |K_n(x)|\,dx.$$

A sequence of kernels $\{K_n(x)\}$ is called δ-*shaped* if the following conditions are satisfied for any $c \in (0, a)$:

$$\lim_{n\to\infty} \frac{1}{a}\int_{-c}^{c} K_n(x)\,dx = 1,$$

$$\lim_{n\to\infty} \int_{[-a,a]\setminus[-c,c]} |K_n(x)|\,dx = 0, \tag{3.1}$$

$$L(K_n) < A = \text{const}.$$

In addition, in all cases, we indicate whether the analyzed kernel K_n is even, nonnegative, etc.

3.1. Periodic kernels

The Dirichlet-, Fejér-, and Jackson-type kernels prove to be especially important among trigonometric polynomial kernels.

1. Dirichlet kernels. A function

$$D_n(t) = \frac{\sin\frac{2n+1}{2}t}{2\sin\frac{t}{2}} \tag{3.2}$$

is called the *Dirichlet kernel of order* n.

Let us show that this kernel is a trigonometric polynomial of the degree n and, in addition,

$$D_n(t) = \frac{1}{2} + \cos t + \cos 2t + \ldots + \cos nt. \qquad (3.2')$$

Indeed,

$$\frac{1}{2} + \cos t + \cos 2t + \ldots + \cos nt$$

$$= \frac{\sin\frac{t}{2} + 2\sin\frac{t}{2}\cos t + \ldots + 2\sin\frac{t}{2}\cos nt}{2\sin\frac{t}{2}}$$

$$= \frac{\sin\frac{t}{2} + \left(\sin\frac{3t}{2} - \sin\frac{t}{2}\right) + \ldots + \left(\sin\frac{(2n+1)t}{2} - \sin\frac{(2n-1)t}{2}\right)}{2\sin\frac{t}{2}}$$

$$= \frac{\sin\frac{(2n+1)t}{2}}{2\sin\frac{t}{2}} = D_n(t).$$

This kernel possesses the following properties:

(a) $D_n(t)$ is an even trigonometric polynomial of degree n;

(b) $\frac{1}{\pi}\int_{-\pi}^{\pi} D_n(t)\,dt = 1;$

(c) $L_n := L(D_n) = \frac{1}{\pi}\int_{-\pi}^{\pi} |D_n(t)|\,dt = \frac{4}{\pi^2}\ln n + R_n, \quad n = 1, 2, \ldots,$ (3.3)

where

$$|R_n| \le 3. \qquad (3.3')$$

Properties (a) and (b) are obvious. We prove property (c) for $n > 1$ [for $n = 1$, this property readily follows from (3.2′)]. Since

$$\frac{\sin\dfrac{(2n+1)t}{2}}{2\sin\dfrac{t}{2}} = \frac{\sin nt}{2\sin\dfrac{t}{2}} + \frac{\cos\dfrac{(4n+1)t}{4}}{2\cos\dfrac{t}{4}},$$

we conclude that

$$L_n = \frac{2}{\pi}\int_0^{\pi}\left|\frac{\sin\dfrac{(2n+1)t}{2}}{2\sin\dfrac{t}{2}}\right|dt = \frac{2}{\pi}\int_0^{\pi}\left|\frac{\sin nt}{2\sin\dfrac{t}{2}}\right|dt + r_1,$$

where, by virtue of the fact that $\cos\frac{t}{4} > \frac{1}{\sqrt{2}}$ for all $t \in [0, \pi]$, we have

$$|r_1| \le \frac{1}{\pi}\int_0^{\pi}\frac{1}{\cos(t/4)}\left|\cos\left(n+\frac{1}{4}\right)t\right|dt$$

$$< \frac{\sqrt{2}}{\pi}\int_0^{\pi}\left|\cos\left(n+\frac{1}{4}\right)t\right|dt = \frac{\sqrt{2}}{\pi}\,\frac{2n+\sqrt{2}/2}{n+1/4} < \frac{3}{\pi}.$$

Further, we have

$$\frac{\sin nt}{2\sin\dfrac{t}{2}} = \frac{\sin nt}{t} + \frac{\sin nt\left(t-2\sin\dfrac{t}{2}\right)}{2t\sin\dfrac{t}{2}}.$$

Thus, in view of the fact that, by virtue of (0.4),

$$0 < \frac{1}{2\sin\dfrac{t}{2}} - \frac{1}{t} < \frac{1}{2\sin\dfrac{t}{2}} - \frac{1}{4\tan\dfrac{t}{4}} < \frac{1}{4}\tan\left(\frac{t}{4}\right) < \frac{1}{4}$$

for all $t \in (0, \pi)$, we find

$$L_n = \frac{2}{\pi}\int_0^{\pi}\left|\frac{\sin nt}{t}\right|dt + r_1 + r_2,$$

where

$$|r_2| \le \frac{2}{\pi}\int_0^{\pi} \left| \frac{\sin nt\left(t - 2\sin\frac{t}{2}\right)}{2t\sin\frac{t}{2}} \right| dt < \frac{2}{\pi}\frac{1}{4}\int_0^{\pi} |\sin nt|\, dt = \frac{1}{\pi}.$$

For the integral $\int_0^{\pi} \left|\frac{\sin nt}{t}\right| dt$, we get

$$\int_0^{\pi} \left|\frac{\sin nt}{t}\right| dt = \sum_{k=0}^{n-1} \left| \int_{k\pi/n}^{(k+1)\pi/n} \frac{\sin nt}{t}\, dt \right|$$

$$\le \sum_{k=1}^{n-1} \frac{n}{k\pi} \left| \int_{k\pi/n}^{(k+1)\pi/n} \sin nt\, dt \right| + \int_0^{\pi} \frac{\sin t}{t}\, dt = \frac{2}{\pi}\sum_{k=1}^{n-1}\frac{1}{k} + \int_0^{\pi}\frac{\sin t}{t}\, dt$$

and, similarly,

$$\int_0^{\pi} \left|\frac{\sin nt}{t}\right| dt \ge \sum_{k=0}^{n-1} \frac{n}{(k+1)\pi} \left| \int_{k\pi/n}^{(k+1)\pi/n} \sin nt\, dt \right| = \frac{2}{\pi}\sum_{k=1}^{n}\frac{1}{k}.$$

Thus,

$$L_n = \frac{4}{\pi^2}\sum_{k=1}^{n-1}\frac{1}{k} + r_1 + r_2 + r_3,$$

where

$$|r_3| \le \frac{2}{\pi}\int_0^{\pi}\frac{\sin t}{t}\, dt = \frac{2}{\pi}\int_0^{\pi}\frac{2\sin\frac{t}{2}\cos\frac{t}{2}}{t}\, dt < \frac{2}{\pi}\int_0^{\pi}\cos\frac{t}{2}\, dt = \frac{4}{\pi}.$$

Since

$$\ln n < \sum_{k=1}^{n-1}\frac{1}{k} \quad \text{and} \quad \ln n > \sum_{k=2}^{n}\frac{1}{k},$$

we have

$$0 < \left(\sum_{k=1}^{n}\frac{1}{k} - \ln n\right) < 1.$$

Therefore,

$$L_n = \frac{4}{\pi^2}\ln n + R_n,$$

where

$$|R_n| \le |r_1| + |r_2| + |r_3| + \frac{4}{\pi^2}\left(\sum_{k=1}^{n-1}\frac{1}{k} - \ln n\right) < \frac{3}{\pi} + \frac{1}{\pi} + \frac{4}{\pi} + \frac{4}{\pi^2} < 3.$$

■

Remark 3.1. The following result is closely related to inequality (3.3):

It is easy to see that the polynomial $P_n(f;\cdot)$ of degree n interpolating a continuous function f on the segment $[-1,1]$ at the zeros $x_k = \cos\theta_k$, $\theta_k := (2k-1)\frac{\pi}{2n}$, of the Chebyshev polynomial T_n has the form

$$P_n(f;x) = \sum_{k=1}^{n}\frac{f(x_k)}{T_n'(\theta_k)}\frac{T_n(x)}{x - x_k} = \frac{1}{n}\sum_{k=1}^{n}(-1)^{k-1}f(x_k)\frac{\cos n\theta}{\cos\theta - \cos\theta_k}\sin\theta_k,$$

where $x := \cos\theta$, and the norm Λ_n of the polynomial operator T_n is equal to

$$\Lambda_n = \frac{1}{n}\max_{0\le\theta\le\pi}\sum_{k=1}^{n}\left|\frac{\cos n\theta}{\cos\theta - \cos\theta_k}\right|\sin\theta_k.$$

Improving the results obtained by Bernstein (1931) and Erdős and Turan (1961), Ivanov (1978) established the equalities

$$\Lambda_n = \frac{2}{\pi}\ln n + 1 - \theta_n, \quad 0 \le \theta_n < \frac{1}{4}, \quad n = 1, 2, \ldots.$$

Remark 3.2. We now present (without proof) the following general theorem obtained by Nikolaev (1948) and Lozinskii and Kharshiladze (see [Natanson (1949), pp. 642–676] and [Korovkin (1959), pp., 144–150]):

Theorem (Nikolaev–Lozinskii–Kharshiladze). *If, for some natural n, a linear polynomial operator $U_n(f;x)$ mapping the space of 2π-periodic continuous functions onto the space of trigonometric polynomials of degree n preserves all trigonometric polynomials T_n of degree n in a sense that $U_n(T_n;x) = T_n(x)$, then the norm A_n of the operator U_n (Lebesgue constant) satisfies the inequality*

$$\Lambda_n \geq \frac{1}{22} \ln n.$$

Remark 3.3. For a continuous 2π-periodic function f, by T_n^* and $E_n(f)$ we denote the polynomial and value of its best approximation of degree n, $n \geq 1$, respectively. By $S_n(f; t)$ we denote a partial sum of the Fourier series. Thus, by virtue of the equalities

$$S_n(f;t) = \frac{1}{\pi} \int_{-\pi}^{\pi} f(t-u) D_n(u)\,du \quad \text{and} \quad S_n(T_n^*;t) \equiv T_n^*(t),$$

we obtain

$$\begin{aligned} |f(t) - S_n(f;t)| &= \left| f(t) - T_n^*(t) - S_n(f - T_n^*;t) \right| \\ &\leq \left| f(t) - T_n^*(t) \right| + \frac{1}{\pi} \int_{-\pi}^{\pi} \left| f(t-u) - T_n^*(t-u) \right| |D_n(u)|\,du \\ &\leq E_n(f) + E_n(f) L_n. \end{aligned}$$

Hence, the following theorem is true:

Theorem 3.1 [Lebesgue (1909)]. *For any 2π-periodic function f continuous on the entire axis, the partial sums $S_n(f;t)$ of its Fourier series approximate this function so that the following inequality holds for any natural n:*

$$|f(t) - S_n(f;t)| \leq (1 + L_n) E_n(f) \leq \left(4 + \frac{4}{\pi^2} \ln n\right) E_n(f), \tag{3.4}$$

where L_n are constants specified by relations (3.3).

As follows from this inequality, the partial sums of the Fourier series of a continuous 2π-periodic function f deviate from this function, in fact, not greater than the polynomials $T_n^*(t)$ of its best uniform approximation (at most by a factor of $\ln n$).

By virtue of the equality

$$S_n(f;t) = \frac{1}{\pi} \int_{-\pi}^{\pi} f(t-u) D_n(u)\,du,$$

the Lebesgue constants

$$L_n = L(D_n)$$

given by relations (3.3), where $D_n(t)$ are the Dirichlet kernels, are clearly equal to the norms of the operators S_n.

In what follows, we often encounter Lebesgue constants for the other types of polynomial kernels.

Note that, as a result of the change of variables $x = \sin\frac{t}{2}$, we get

$$\frac{T_{2n+1}(x)}{x} = \frac{\cos(2n+1)(\pi/2 - \arcsin x)}{x}$$

$$= (-1)^n \frac{\sin(2n+1)\arcsin x}{x} = (-1)^n \frac{\sin\frac{(2n+1)t}{2}}{\sin\frac{t}{2}}.$$

Thus, the Dirichlet kernels are expressed via the Chebyshev polynomials $T_{2n+1}(x) = \cos(2n+1)\arccos x$ by the following identities:

$$D_n(t) = (-1)^n \frac{T_{2n+1}(x)}{2x}, \qquad x = \sin\frac{t}{2}. \tag{3.5}$$

2. Fejér kernels. A function

$$F_n(t) = \frac{\sin^2\frac{nt}{2}}{2n\sin^2\frac{t}{2}} \tag{3.6}$$

is called the *Fejér kernel of order* n (1904).

The Fejér kernel F_n is the arithmetic mean of the first n Dirichlet kernels and, hence, a trigonometric polynomial of degree $n-1$. Thus, we have

$$F_n(t) = \frac{D_0(t) + D_1(t) + \ldots + D_{n-1}(t)}{n}, \tag{3.6$'$}$$

$$F_n(t) = \frac{1}{2} + \left(1 - \frac{1}{n}\right)\cos t + \left(1 - \frac{2}{n}\right)\cos 2t + \ldots + \frac{1}{n}\cos(n-1)t, \tag{3.6$''$}$$

where $D_k(t)$ are the Dirichlet kernels.

Indeed, on the one hand, according to (3.2), we find

$$\frac{1}{n}\sum_{k=0}^{n-1} D_k(t) = \frac{\sin\frac{t}{2} + \sin\frac{3t}{2} + \ldots + \sin\frac{(2n-1)t}{2}}{2n\sin\frac{t}{2}}$$

$$= \frac{(1-\cos t) + (\cos t - \cos 2t) + \ldots + [\cos(n-1)t - \cos nt]}{4n\sin^2\frac{t}{2}}$$

$$= \frac{1-\cos nt}{4n\sin^2\frac{t}{2}} = \frac{2\sin^2\frac{nt}{2}}{4n\sin^2\frac{t}{2}} = \frac{\sin^2\frac{nt}{2}}{2n\sin^2\frac{t}{2}} = F_n(t).$$

On the other hand, according to (3.2′), we can write

$$\frac{1}{n}\sum_{k=0}^{n-1} D_k(t) = \frac{1/2 + (1/2+\cos t) + \ldots + [1/2 + \cos t + \ldots + \cos(n-1)t]}{n}$$

$$= \frac{1}{2} + \left(1-\frac{1}{n}\right)\cos t + \left(1-\frac{2}{n}\right)\cos 2t + \ldots + \frac{1}{n}\cos(n-1)t.$$

This kernel possesses the following important properties:

(i) $F_n(t)$ is an even nonnegative trigonometric polynomial of degree $n-1$;

(ii) $\frac{1}{\pi}\int_{-\pi}^{\pi} F_n(t)dt = 1;$ (3.7)

(iii) for any $\delta \in (0,\pi)$,

$$\int_{\delta}^{\pi} F_n(t)dt < \frac{2}{n\delta}. \tag{3.8}$$

Properties (i) and (ii) follow from equalities (3.6) and (3.6′). In order to prove (iii), it suffices to take into account inequality (0.4):

$$\int_{\delta}^{\pi} F_n(t)dt = \frac{1}{2n}\int_{\delta}^{\pi}\frac{\sin^2\frac{nt}{2}}{\sin^2\frac{t}{2}}dt < \frac{1}{2n}\int_{\delta}^{\pi}\frac{dt}{\sin^2\frac{t}{2}} = \frac{1}{n}\cot\frac{\delta}{2} < \frac{2}{n\delta}.$$

The Fejér kernels are expressed via the Chebyshev polynomials by the formula

$$F_n(t) = \frac{1-(-1)^n T_{2n}(x)}{4nx^2}, \qquad x = \sin\frac{t}{2}.$$

Indeed,

$$F_n(t) = \frac{\sin^2\frac{nt}{2}}{2n\sin^2\frac{t}{2}} = \frac{2\sin^2 n\left(\frac{\pi}{2} - \arccos x\right)}{4nx^2}$$

$$= \frac{1 - \cos 2n\left(\frac{\pi}{2} - \arccos x\right)}{4nx^2} = \frac{1-(-1)^n T_{2n}(x)}{4nx^2}.$$

Moreover, the kernels $K_n(x)$ [see (1.2)] are expressed via the Fejér kernels by the formula

$$K_n(x) = \frac{1}{\gamma_n}\frac{\cos^2\left[(2n+1)\left(\frac{\pi}{2} - \arcsin x\right)\right]}{\sin^2 \arcsin x}$$

$$= \frac{1}{\gamma_n}\frac{\sin^2(2n+1)\arcsin x}{\sin^2\arcsin x} = \tilde{\gamma}_n F_{2n+1}(t), \tag{3.9}$$

where $\tilde{\gamma}_n = \dfrac{4n+2}{\gamma_n}$ and $t = 2\arcsin x$. Therefore, $\cos t = 1 - 2x^2$.

We now set $\breve{F}_{2n+1}(x) := K_n(x/2)$. This yields

$$\breve{F}_{2n+1}(x) = \frac{4n+2}{\gamma_n} F_{2n+1}\left[\arccos\left(1 - \frac{x^2}{2}\right)\right], \quad x \in [-2, 2]. \tag{3.9'}$$

3. De la Vallée Poussin kernels (1919). For any nonnegative integers m and n $(n > m)$, the *de la Vallée Poussin kernel* is defined as the arithmetic mean of the Dirichlet kernels of orders from m to $n-1$, i.e., as the following trigonometric polynomial:

$$V_m^n(t) = \frac{1}{n-m}\{D_m(t) + D_{m+1}(t) + \ldots + D_{n-1}(t)\}. \tag{3.10}$$

Since $D_{m+k}(t) = D_m(t) + \cos(m+1)t + \ldots + \cos(m+k)t$, we get

$$V_m^n(t) = D_m(t) + \frac{n}{n-m} \sum_{k=m+1}^{n-1} \left(1 - \frac{k}{n}\right) \cos kt \tag{3.10'}$$

and, consequently,

$$V_n^{n+1}(t) = D_n(t) \quad \text{and} \quad V_0^n(t) = F_n(t).$$

By using (3.10′) and repeating the reasoning used to deduce inequality (3.4′), we find

$$\left| f(t) - V_m^n(f;t) \right| \le \left[1 + L(V_m^n)\right] E_m(f). \tag{3.10''}$$

In view of relations (3.10), (3.6′), and (3.6) for the Fejér kernels, we obtain

$$V_m^n(t) = \frac{1}{n-m} \left[nF_n(t) - mF_m(t) \right]$$

$$= \frac{1}{n-m} \frac{\sin^2 \frac{nt}{2} - \sin^2 \frac{mt}{2}}{2 \sin^2 \frac{t}{2}} = \frac{1}{n-m} \frac{\cos mt - \cos nt}{4 \sin^2 \frac{t}{2}}. \tag{3.11}$$

According to property (ii) of the Fejér kernels, this means that if $n - m \ge \varepsilon n$, where ε is an arbitrary fixed positive number, then the Lebesgue constants $L(V_m^n)$ of the corresponding de la Vallée Poussin kernels are bounded by the number $2/\varepsilon$.

Moreover, relation (3.11) yields the equalities

$$V_n^{2n}(t) = \frac{1}{4n} \frac{\cos nt - \cos 2nt}{\sin^2 \frac{t}{2}}, \tag{3.12}$$

$$V_n^{3n}(t) = \frac{1}{8n} \frac{\cos nt - \cos 3nt}{\sin^2 \frac{t}{2}}. \tag{3.12'}$$

In view of the fact that

$$\operatorname{sgn}(\cos t - \cos 2t) = \begin{cases} 1, & \text{for } t \in \left(0, \frac{2\pi}{3}\right), \\ -1, & \text{for } t \in \left(\frac{2\pi}{3}, \pi\right), \end{cases}$$

and

$$\operatorname{sgn}(\cos t - \cos 3t) = \operatorname{sgn}\cos t, \qquad t \in (0, \pi),$$

we arrive at the following expansions in Fourier series:

$$\operatorname{sgn}(\cos t - \cos 2t) \sim \frac{1}{3} + \frac{4}{\pi}\sum_{k=1}^{\infty}\frac{1}{k}\sin\frac{2k\pi}{3}\cos kt,$$

$$\operatorname{sgn}\cos t \sim \frac{4}{\pi}\sum_{k=0}^{\infty}(-1)^k\frac{\cos(2k+1)t}{2k+1}$$

and, therefore,

$$\operatorname{sgn} V_n^{2n}(t) \sim \frac{1}{3} + \frac{2}{\pi}\sqrt{3}\cos nt - \frac{\sqrt{3}}{\pi}\cos 2nt + \ldots,$$

$$\operatorname{sgn} V_n^{3n}(t) \sim \frac{4}{\pi}\cos nt - \frac{4}{3\pi}\cos 3nt + \ldots.$$

Thus, by using relation (3.10′) and the orthogonality of the trigonometric system of functions, we get the following values of the Lebesgue constants for these kernels[3]:

$$\frac{1}{\pi}\int_{-\pi}^{\pi}\left|V_n^{2n}(t)\right|dt = \frac{1}{\pi}\int_{-\pi}^{\pi}\left\{\frac{1}{2} + \sum_{k=1}^{n}\cos kt + 2\sum_{k=n+1}^{2n-1}\left(1 - \frac{k}{2n}\right)\cos kt\right\}\operatorname{sgn} V_n^{2n}(t)\,dt$$

$$= \frac{2}{3}\cdot\frac{1}{2} + \frac{2\sqrt{3}}{\pi} = \frac{1}{3} + \frac{2\sqrt{3}}{\pi}, \tag{3.13}$$

$$\frac{1}{\pi}\int_{-\pi}^{\pi}\left|V_n^{3n}(t)\right|dt$$

$$= \frac{1}{\pi}\int_{-\pi}^{\pi}\left\{\frac{1}{2} + \sum_{k=1}^{n}\cos kt + \frac{3}{2}\sum_{k=n+1}^{3n-1}\left(1 - \frac{k}{n}\right)\cos kt\right\}\operatorname{sgn} V_n^{3n}(t)\,dt = \frac{4}{\pi}. \tag{3.13′}$$

In conclusion, we note that, by virtue of the well-known Markov theorem (see, e.g., [Akhiezer (1965)]), the polynomials $V_n^{2n}(t)$ and $V_n^{3n}(t)$ are, respectively, the solutions of the following extreme problems of approximation of the Dirichlet kernels by the linear combinations of $\cos kt$ for $n<k<2n$ and $n<k<3n$:

$$\min_{c_k} \frac{1}{\pi} \int_{-\pi}^{\pi} \left| \frac{1}{2} + \sum_{j=1}^{n} \cos jt - \sum_{k=n+1}^{2n-1} c_k \cos kt \right| dt$$

$$= \frac{1}{\pi} \int_{-\pi}^{\pi} \left| V_n^{2n}(t) \right| dt = \frac{1}{3} + \frac{2\sqrt{3}}{\pi}, \qquad (3.14)$$

$$\min_{c_k} \frac{1}{\pi} \int_{-\pi}^{\pi} \left| \frac{1}{2} + \sum_{j=1}^{n} \cos jt - \sum_{k=n+1}^{3n-1} c_k \cos kt \right| dt = \frac{1}{\pi} \int_{-\pi}^{\pi} \left| V_n^{3n}(t) \right| dt = \frac{4}{\pi}. \qquad (3.14')$$

4. Rogosinski kernels (1925). A function

$$R_n(t) = R_n(t;\gamma_n) = \frac{1}{2}\left[D_n(t-\gamma_n) + D_n(t+\gamma_n)\right] = \frac{1}{2} + \sum_{k=1}^{n} \cos k\gamma_n \cos kt, \qquad (3.15)$$

where $\gamma_n = O(1/n)$ and D_n is the Dirichlet kernel of order n, is called the *Rogosinski kernel* of order n.

A. First, we set

$$\gamma_n = \frac{\pi}{2n}.$$

The kernel obtained as a result possesses the following properties:

(i) The kernel

$$R_n(t) = R_n\left(t; \frac{\pi}{2n}\right)$$

can be represented in the form

$$R_n(t) = \frac{1}{2} \sin\frac{\pi}{2n} \frac{\cos nt}{\cos t - \cos\frac{\pi}{2n}}. \qquad (3.16)$$

For $\gamma_n = \pi/2n$, this equality follows from the identity

$$R_n(t;\gamma_n) = \frac{1}{2}\left[\frac{\sin\frac{2n+1}{2}(t-\gamma_n)}{2\sin\frac{1}{2}(t-\gamma_n)} + \frac{\sin\frac{2n+1}{2}(t+\gamma_n)}{2\sin\frac{1}{2}(t+\gamma_n)}\right]$$

$$= \frac{\sin\frac{2n+1}{2}(t-\gamma_n)\sin\frac{t+\gamma_n}{2} + \sin\frac{2n+1}{2}(t+\gamma_n)\sin\frac{t-\gamma_n}{2}}{4\sin\frac{t+\gamma_n}{2}\sin\frac{t-\gamma_n}{2}}$$

$$= \frac{\cos[nt-(n+1)\gamma_n] - \cos[(n+1)t - n\gamma_n]}{4(\cos\gamma_n - \cos t)}$$

$$+ \frac{\cos[nt+(n+1)\gamma_n] - \cos[(n+1)t + n\gamma_n]}{4(\cos\gamma_n - \cos t)}$$

$$= \frac{\cos(n+1)t\cos n\gamma_n - \cos nt\cos(n+1)\gamma_n}{2(\cos t - \cos\gamma_n)}. \tag{3.17}$$

(ii) By virtue of equality (3.15), we conclude that

$$\frac{1}{\pi}\int_{-\pi}^{\pi} R_n(t)\,dt = 1. \tag{3.18}$$

(iii) Equalities (3.15) and (3.16) imply that the kernel R_n is an even trigonometrical polynomial of degree n whose positive zeros from the interval $(0, \pi)$ are located at the points

$$t_i = \frac{2i+1}{2n}\pi, \quad i = 1, 2, \ldots, n-1.$$

Therefore, the derivative R_n' does not vanish in the interval $(0, t_1)$. Note that, for $t_1 \le t \le \pi$, we have

$$|R_n(t)| = \frac{1}{2}\sin\frac{\pi}{2n}\frac{1}{\cos\frac{\pi}{2n} - \cos\frac{3\pi}{2n}} = \frac{1}{4\sin\frac{\pi}{n}} < \frac{1}{2}\cot\frac{\pi}{4n}.$$

This yields

$$\max_t |R_n(t)| = R_n(0) = \frac{1}{2}\sin\frac{\pi}{2n}\frac{1}{1-\cos\frac{\pi}{2n}} = \frac{1}{2}\cot\frac{\pi}{4n} < \frac{2n}{\pi}. \tag{3.19}$$

(iv) Note that, by virtue of relation (3.16),

$$\operatorname{sgn} R_n(t) = -\operatorname{sgn}\cos nt + e(t),$$

where

$$e(t) = \begin{cases} 2 & \text{for } |t| < \dfrac{\pi}{2n}, \\ 0 & \text{for } \dfrac{\pi}{2n} < |t| < \pi, \end{cases}$$

and that as it is easy to see that

$$e(t) \sim \frac{1}{2}\frac{2}{n} + \frac{4}{\pi}\sum_{k=1}^{\infty}\sin\frac{k\pi}{2n}\frac{\cos kt}{k},$$

$$\operatorname{sgn}\cos t \sim \frac{4}{\pi}\sum_{k=0}^{\infty}(-1)^k\frac{\cos(2k+1)t}{2k+1}.$$

Hence, we arrive at the following expression for the Lebesgue constant of the kernel R_n :

$$\begin{aligned}\frac{1}{\pi}\int_{-\pi}^{\pi}|R_n(t)|\,dt &= \frac{1}{\pi}\int_{-\pi}^{\pi}\left[\frac{1}{2}+\sum_{k=1}^{n-1}\cos\frac{k\pi}{2n}\cos kt\right]\left[\frac{1}{2}\frac{2}{n}+\frac{4}{\pi}\sum_{k=1}^{n-1}\sin\frac{k\pi}{2n}\frac{\cos kt}{k}\right]dt \\ &= \frac{1}{n}+\frac{2}{\pi}\sum_{k=1}^{n-1}\frac{1}{k}\sin\frac{k\pi}{n} \\ &= \frac{2}{\pi}\sum_{k=0}^{n-1}\frac{1}{2}\left[\sigma\left(\frac{k\pi}{n}\right)+\sigma\left(\frac{(k+1)\pi}{n}\right)\right]\frac{\pi}{n},\end{aligned} \tag{3.20}$$

where

$$\sigma(t) := \frac{\sin t}{t} \quad \text{for} \quad t \neq 0 \quad \text{and} \quad \sigma(0) := 1.$$

Since the right-hand side of (3.20) is an approximate value of the integral

$$\frac{2}{\pi}\int_0^{\pi}\frac{\sin t}{t}\,dt = \frac{2}{\pi}\operatorname{Si}\pi$$

computed by the formula of trapezoids, we arrive at the following natural assertion:

For all natural n, the following equalities are true[4]:

$$\frac{1}{\pi}\int_{-\pi}^{\pi}|R_n(t)|\,dt = \frac{2}{\pi}\operatorname{Si}\pi - r_n, \tag{3.21}$$

where

$$0 < r_n < \frac{1}{3n^2} \quad \text{and} \quad \frac{2}{\pi}\operatorname{Si}\pi \approx \frac{2}{\pi}1.851 \approx 1.185.$$

In order to prove this assertion, we represent the quantity r_n by using relation (3.20) in the form

$$\begin{aligned} r_n &= \frac{2}{\pi}\int_0^{\pi}\frac{\sin t}{t}\,dt - \frac{2}{\pi}\sum_{k=0}^{n-1}\frac{1}{2}\left[\sigma\left(\frac{k\pi}{n}\right) + \sigma\left(\frac{k+1}{n}\pi\right)\right]\frac{\pi}{n} \\ &= \frac{2}{\pi}\sum_{k=0}^{n-1}\left(\int_{t_k}^{t_{k+1}}\sigma(t)\,dt\right) - (t_{k+1}-t_k)(\sigma(t_k)+\sigma(t_{k+1})), \end{aligned} \tag{3.22}$$

where $t_k := k\pi/n$. We set

$$A := \frac{1}{12\pi}\sum_{k=0}^{n-1}\int_{t_k}^{t_{k+1}}(t-t_k)^2(t-t_{k+1})^2\sigma^{(4)}(t)\,dt.$$

By using the identity

$$\begin{aligned} &\int_a^b f(x)\,dx - \frac{1}{2}(b-a)(f(b)+f(a)) \\ &\qquad = \frac{1}{24}\int_a^b (x-a)^2(x-b)^2 f^{(4)}(x)\,dx - \frac{1}{12}(b-a)^2(f'(b)-f'(a)) \end{aligned}$$

valid for any function f four times continuously differentiable on $[a, b]$, we get

$$r_n = A - \frac{1}{6\pi}\left(\frac{\pi}{n}\right)^2 \sum_{k=0}^{n-1} (\sigma'(t_{k+1}) - \sigma'(t_k))$$

$$= A + \frac{\pi}{6n^2}(\sigma'(0) - \sigma'(\pi)) = A + \frac{1}{6n^2}. \tag{3.23}$$

To estimate A, we recall that

$$\sigma^{(4)}(t) = \frac{1}{t^5}\int_0^t u^4 \cos u\, du,$$

whence it follows that $\left|\sigma^{(4)}(t)\right| \le 1/5$ for all $t \in [0, \pi]$ and, therefore,

$$\left|\int_{t_k}^{t_{k+1}} (t - t_k)^2 (t - t_{k+1})^2 \sigma^{(4)}(t)\, dt\right| \le \frac{1}{5}\int_{t_k}^{t_{k+1}} (t - t_k)^2 (t - t_{k+1})^2\, dt = \frac{8}{75}\left(\frac{\pi}{n}\right)^5.$$

This enables us to conclude that

$$|A| \le \frac{n}{12\pi}\frac{8}{75}\left(\frac{\pi}{n}\right)^5 < \frac{1}{n^4} < \frac{1}{6n^2}, \qquad n \ne 1, 2.$$

Thus, for $n \ne 1, 2$, the required assertion follows from (3.22). For $n = 1, 2$, it is trivial.

(v) Since, for all natural n, we have

$$\sin^2\frac{\pi}{2n} = 4\sin^2\frac{\pi}{4n}\cos^2\frac{\pi}{4n} \ge 2\sin^2\frac{\pi}{4n},$$

for any $\delta \in (0, \pi)$, $n > \pi/\delta$, and all $t \in (\delta, \pi)$ (i.e., $t > \delta > \pi/n$), we can write

$$\left|\cos t - \cos\frac{\pi}{2n}\right| = \left|\cos t - 1 + 1 - \cos\frac{\pi}{2n}\right|$$

$$= 2\left(\sin^2\frac{t}{2} - \sin^2\frac{\pi}{4n}\right) > \sin^2\frac{t}{2} > \frac{t^2}{\pi^2}$$

and, hence, in view of relation (3.16), we find

$$\left|\sigma^{(4)}(t)\right|\int_{\delta}^{\pi}|R_n(t)|dt < \frac{1}{2}\sin\frac{\pi}{2n}\int_{\delta}^{\pi}\frac{dt}{|\cos t - \cos(\pi/2n)|} < \frac{\pi^2}{2}\sin\frac{\pi}{2n}\int_{\delta}^{\infty}\frac{dt}{t^2} < \frac{A}{n\delta}, \quad (3.24)$$

$$A = \frac{\pi^3}{2}.$$

B. If we take into account the geometric reasoning, then it is more natural to consider kernels of the form [Kharshiladze (1955), (1958)]

$$\tilde{R}_n(t) = \tilde{R}_n^{(1)}(t) = R_n\left(t, \frac{\pi}{2n+1}\right) = \frac{1}{2}\left[D_n\left(t - \frac{\pi}{2n+1}\right) + D_n\left(t + \frac{\pi}{2n+1}\right)\right].$$

In exactly the same way as in deducing equality (3.17), we represent these kernels in the form

$$\tilde{R}_n(t) = \frac{1}{2}\left[\frac{-\cos\frac{(2n+1)t}{2}}{2\sin\frac{1}{2}\left(t - \frac{\pi}{2n+1}\right)} + \frac{\cos\frac{(2n+1)t}{2}}{2\sin\frac{1}{2}\left(t + \frac{\pi}{2n+1}\right)}\right]$$

$$= \sin\frac{\pi}{2(2n+1)}\cos\frac{t}{2}\,\frac{\cos\frac{(2n+1)t}{2}}{\cos t - \cos\frac{\pi}{2n+1}}. \quad (3.25)$$

In this case, outside the segment

$$\left[-\frac{\pi}{2n+1}, \frac{\pi}{2n+1}\right],$$

the kernels

$$D_n\left(t - \frac{\pi}{2n+1}\right) \quad \text{and} \quad D_n\left(t + \frac{\pi}{2n+1}\right)$$

have the opposite signs and, to a significant extent, are mutually compensated. We recommend the reader to construct and add the plots of these kernels. The indicated kernels possess the properties completely similar to the properties of the kernels R_n studied in Subsection A.

By using the kernels $\tilde{R}_n^{(1)}$, we can, in a similar way, construct the kernels

$$\tilde{R}_n^{(2)}(t) = \frac{1}{2}\left[\tilde{R}_n^{(1)}\left(t - \frac{\pi}{2n+1}\right) + \tilde{R}_n^{(1)}\left(t + \frac{\pi}{2n+1}\right)\right]$$

$$= \frac{1}{2^2}\left[D_n\left(t - \frac{2\pi}{2n+1}\right) + 2D_n(t) + D_n\left(t + \frac{2\pi}{2n+1}\right)\right], \tag{3.25'}$$

$$\tilde{R}_n^{(3)}(t) = \frac{1}{2}\left[\tilde{R}_n^{(2)}\left(t - \frac{\pi}{2n+1}\right) + \tilde{R}_n^{(2)}\left(t + \frac{\pi}{2n+1}\right)\right]$$

$$= \frac{1}{2^3}\left[D_n\left(t - \frac{3\pi}{2n+1}\right) + 3D_n\left(t - \frac{\pi}{2n+1}\right)\right.$$

$$\left. + 3D_n\left(t + \frac{\pi}{2n+1}\right) + D_n\left(t + \frac{3\pi}{2n+1}\right)\right], \tag{3.25''}$$

etc. In this case, the kernels are mutually compensated outside the segments

$$\left[-\frac{3\pi}{2n+1}, \frac{3\pi}{2n+1}\right], \quad \left[-\frac{5\pi}{2n+1}, \frac{5\pi}{2n+1}\right],$$

etc., respectively. The kernels $\tilde{R}_n^{(2)}$, $\tilde{R}_n^{(3)}$, etc. take only positive values in the intervals

$$\left(-\frac{5\pi}{2n+1}, \frac{5\pi}{2n+1}\right), \quad \left(-\frac{7\pi}{2n+1}, \frac{7\pi}{2n+1}\right),$$

etc., respectively.

By using the kernels $R_n(t) := \tilde{R}_n^{(1)}(t)$ studied in Subsection A, one can also construct more general kernels of the form

$$\tilde{R}_n^{(i)}(t) := \gamma_n^{(i)} \cos nt\left[\left(\cos t - \cos\frac{\pi}{2n}\right)\left(\cos t - \cos\frac{3\pi}{2n}\right)\left(\cos t - \cos\frac{\pi(2i-1)}{2n}\right)\right]^1, \tag{3.26}$$

where the numbers $\gamma_n^{(i)}$ are chosen to satisfy the condition

$$\frac{1}{\pi}\int_{-\pi}^{\pi} R_n^{(i)}(t)\,dt = 1.$$

Finally, we mention the kernels $\tau_{k,n}$ quite closely related to the Rogosinski kernels. For any natural n and $k \le n$, these kernels are given by the formula

$$\tau_{k,n}(t) = \frac{1}{2^k} \sum_{\nu=0}^{k} \binom{k}{\nu} \left[D_n(t) + (-1)^{\nu+1} D_n\left(t + \frac{\nu\pi}{n}\right) \right]$$

$$= D_n(t) + \frac{1}{2^k} \sum_{\nu=0}^{k} \binom{k}{\nu} (-1)^{\nu+1} D_n\left(t + \frac{\nu\pi}{n}\right). \tag{3.27}$$

For the first time, these kernels were studied by Trigub (1965), who discovered that they play an important role in some problems connected with the approximation characteristics of the functions.

Let us show that, for any n and $k \le n$, the Lebesgue constants of the kernels $\tau_{k,n}$ satisfy the inequalities

$$L(\tau_{k,n}) = \frac{1}{\pi} \int_{-\pi}^{\pi} \left|\tau_{k,n}(t)\right| dt < 2k. \tag{3.27$'$}$$

Indeed, since, for any ν,

$$D_n(t) + (-1)^{\nu+1} D_n\left(t + \frac{\nu\pi}{n}\right) = \sum_{j=0}^{\nu-1} (-1)^j \left[D_n\left(t + \frac{j\pi}{n}\right) + D_n\left(t + \frac{j\pi}{n} + \frac{\pi}{n}\right) \right],$$

by virtue of relations (3.27), (3.15), and (3.21), we obtain

$$L(\tau_{k,n}) \le \frac{1}{2^k} \sum_{\nu=0}^{k} \binom{k}{\nu} 2\nu L(R_n) = \frac{k}{2} L(R_n) < 2k,$$

as required.

5. Jackson kernels (1911). A function

$$J_n(t) = \frac{3}{2n(2n^2+1)} \left(\frac{\sin\frac{nt}{2}}{\sin\frac{t}{2}} \right)^4, \quad n = 1, 2, \ldots, \tag{3.28}$$

is called the *Jackson kernel of order* n.

These kernels possess the following properties:

(a) for any n, the kernel $J_n(t)$ is an even nonnegative trigonometric polynomial of degree $2n-2$ of the form

$$J_n(t) = \frac{1}{2} + \sum_{k=1}^{2n-2} j_k \cos kt = \frac{1}{2} \sum_{k=-(2n-2)}^{2n-2} j_k e^{ikt}, \tag{3.28′}$$

where

$j_k = j_k(n)$ are certain numbers such that $j_0 := 1$ and $j_{-k} = j_k$ for $k \geq 0$;

(b)
$$\frac{1}{\pi} \int_{-\pi}^{\pi} J_n(t)\,dt = 1; \tag{3.29}$$

(c)
$$\frac{1}{\pi} \int_{0}^{\pi} t J_n(t)\,dt \leq \frac{5}{2n}; \tag{3.29′}$$

(d)
$$\frac{1}{\pi} \int_{0}^{\pi} t^2 J_n(t)\,dt \leq \frac{\pi^2}{n^2}. \tag{3.30}$$

Proof. (a) In view of the fact that that the following equalities are true for the Fejér kernels $F_n(t)$:

$$\left(\frac{\sin(nt/2)}{\sin(t/2)}\right)^2 = 2nF_n(t) = 2n\left[\frac{1}{2} + \sum_{k=1}^{n-1}\left(1 - \frac{k}{n}\right)\cos kt\right]$$

[see relations (3.6) and (3.6″)], we find

$$\begin{aligned} J_n(t) &= \frac{3 \cdot 2n}{2n^2+1}[F_n(t)]^2 \\ &= \frac{6n}{2n^2+1}\left[\frac{1}{2} + \sum_{j=1}^{n-1}\left(1-\frac{j}{n}\right)\cos jt\right]\left[\frac{1}{2} + \sum_{\nu=1}^{n-1}\left(1-\frac{\nu}{n}\right)\cos \nu t\right] \\ &= \frac{j_0}{2} + \sum_{\nu=1}^{2n-2} j_\nu \cos \nu t, \end{aligned}$$

where j_ν, $\nu = 0, 1, 2, \ldots, 2n-2$, are certain numbers. In particular, since the trigonometric system of functions is orthogonal, we obtain

$$j_0 = \frac{1}{\pi}\int_{-\pi}^{\pi} J_n(t)\,dt = \frac{12n}{2n^2+1}\left[\frac{1}{4} + \frac{1}{2}\sum_{j=1}^{n-1}\left(1-\frac{j}{n}\right)^2\right]$$

$$= \frac{6n}{2n^2+1}\left[\frac{1}{2} + \frac{(n-1)n(2n-1)}{6n^2}\right] = 1.$$

This proves property (a).

(b) Equality (3.29) follows from the equality established for j_0.

(c) Since $|\sin nt| \leq n|\sin t|$ for any $t \in (-\infty, \infty)$ and

$$\sin t \geq \frac{2}{\pi}t \quad \text{for} \quad t \in \left[0, \frac{\pi}{2}\right]$$

[see inequalities (0.1) and (0.2)], we conclude that

$$\frac{1}{\pi}\int_0^{\pi} tJ_n(t)\,dt = \frac{3}{2\pi n(2n^2+1)}\left\{\left(\int_0^{\pi/n} + \int_{\pi/n}^{\pi}\right)t\left(\frac{\sin(nt/2)}{\sin(t/2)}\right)^4 dt\right\}$$

$$\leq \frac{3}{4\pi n^3}\left(\int_0^{\pi/n} tn^4\,dt + \int_{\pi/n}^{\pi} t\frac{1}{(t/\pi)^4}\,dt\right)$$

$$\leq \frac{3}{4\pi n^3}\left(\frac{\pi^2 n^2}{2} + \frac{\pi^4}{2(\pi/n)^2}\right) = \frac{3\pi}{4n} < \frac{5}{2n}.$$

(d) As in case (c), we find

$$\frac{1}{\pi}\int_0^{\pi} t^2 J_n(t)\,dt = \frac{3}{4\pi n^3}\left(n^4\int_0^{\pi/n} t^2\,dt + \pi^4\int_{\pi/n}^{\pi}\frac{dt}{t^2}\right) \leq \frac{3}{4\pi n^3}\left(\frac{\pi^3 n}{3} + \pi^3 n\right) = \frac{\pi^2}{n^2},$$

as required.

6. Jackson-type kernels. A function

$$J_{k,n}(t) = \frac{1}{\gamma_{k,n}}\left(\frac{\sin\frac{nt}{2}}{\sin\frac{t}{2}}\right)^{2k}, \qquad n = 1, 2, 3, \ldots, k, \tag{3.31}$$

where k is a natural number and

$$\gamma_{k,n} = \frac{1}{\pi}\int_{-\pi}^{\pi}\left[\frac{\sin(nt/2)}{\sin(t/2)}\right]^{2k} dt, \tag{3.32}$$

is called the *Jackson-type kernel of order* n.

The Jackson-type kernels possess the following properties:

(a)
$$\frac{1}{\pi}\int_{-\pi}^{\pi} J_{k,n}(t)\,dt = 1. \tag{3.33}$$

This property follows from equalities (3.31) and (3.32).

(b) For a fixed natural k and any n, the kernel $J_{k,n}(t)$ is an even nonnegative trigonometric polynomial of degree $k(n-1)$.

This property follows from equality (3.31) and the fact that, by virtue of equalities (3.6) and (3.6''), we have

$$J_{k,n}(t) = \frac{1}{\gamma_{k,n}}[2nF_n(t)]^k = \frac{1}{\gamma_{k,n}}(2n)^k[F_n(t)]^k$$

$$= \frac{(2n)^k}{\gamma_{k,n}}\left[\frac{1}{2} + \sum_{\nu=1}^{n-1}\frac{n-\nu}{n}\cos\nu t\right]^k = \sum_{\nu=0}^{k(n-1)} j_\nu \cos\nu t, \tag{3.34}$$

where $j_\nu = j_\nu(k, n)$ are certain numbers.

(c)
$$\gamma_{k,n} \asymp n^{2k-1}, \tag{3.35}$$

i.e., there exist constants $c_1(k) > 0$ and $c_2(k) > 0$ such that, for all $n = 1, 2, \ldots,$

$$c_1 n^{2k-1} < \gamma_{k,n} < c_2 n^{2k-1}. \tag{3.35'}$$

We now prove that

$$\frac{3}{2\sqrt{k}}n^{2k-1} < \gamma_{k,n} < \frac{5}{2\sqrt{k}}n^{2k-1}, \qquad n \neq 1. \tag{3.35''}$$

For this purpose, we represent $n^{1-2k}\gamma_{k,n}$ as the sum of two integrals $a_{k,n}$ and $b_{k,n}$, where

$$a_{k,n} := \frac{2n}{\pi}\int_0^{2\pi/n}\left(\frac{\sin(nt/2)}{n\sin(t/2)}\right)^{2k}dt \quad \text{and} \quad b_{k,n} := \frac{2n}{\pi}\int_{2\pi/n}^{\pi}\left(\frac{\sin(nt/2)}{n\sin(t/2)}\right)^{2k}dt.$$

For $b_{k,n}$, it follows from estimate (0.2) that

$$0 \le b_{k,n} < \frac{2}{\pi}n^{1-2k}\int_{2\pi/n}^{\pi}\left(\frac{t}{2}\frac{2}{\pi}\right)^{-2k}dt < 2\left(\frac{\pi}{n}\right)^{2k-1}\int_{2\pi/n}^{\infty}\frac{dt}{t^{2k}} = \frac{1}{2k-1}\frac{1}{2^{2k-2}}.$$

To estimate $a_{k,n}$, we represent this integral in the form

$$a_{k,n} = \frac{4}{\pi}\int_0^{\pi}\left(\frac{\sin t}{n\sin(t/n)}\right)^{2k}dt.$$

For any fixed $t \in [0, \pi]$, the function $f(x) = x\sin(t/x)$ increases for $x \ge 2$ [since $0 < t/x < \pi/2$ and

$$f'(x) = \sin\frac{t}{x} - \frac{t}{x}\cos\frac{t}{x} > 0$$

by virtue of (0.4)]. Hence, $a_{k,n+1} < a_{k,n}$. In addition, inequality (0.1) implies that

$$n\sin\frac{t}{n} < t, \qquad t \in [0, \pi],$$

and therefore,

$$\frac{4}{\pi}\int_0^{\pi}\left(\frac{\sin t}{t}\right)^{2k}dt = a_{k,\infty} < a_{k,n} \le a_{k,2}$$

$$= \frac{4}{\pi}\int_0^{\pi}\left(\cos\frac{t}{2}\right)^{2k}dt = \frac{8}{\pi}\int_0^{\pi/2}(\cos t)^{2k}dt = 4\frac{(2k-1)!!}{(2k)!!},$$

where the last equality is verified by induction.

Since

$$a_{k,2} = 4\,\frac{\sqrt{2k-1}}{2k}\,\frac{\sqrt{(2k-1)(2k-3)}}{2k-2}\cdot\ldots\cdot\frac{\sqrt{3\cdot 1}}{2} < \frac{\sqrt{6}}{\sqrt{k}},$$

$$a_{2k,2} = \frac{4}{\sqrt{4k}}\,\frac{2k-1}{\sqrt{4k(2k-2)}}\cdot\ldots\cdot\frac{3}{\sqrt{4\cdot 2}}\cdot\frac{1}{\sqrt{2}} > \frac{3}{2\sqrt{k}},$$

and, by virtue of relation (0.4),

$$\frac{\sin t}{t} = \frac{2}{t}\sin\frac{t}{2}\cos\frac{t}{2} > \cos^2\frac{t}{2}, \qquad 0 < t < \pi,$$

we conclude that

$$a_{k,\infty} = \frac{4}{\pi}\int_0^{\pi}\left(\frac{\sin t}{t}\right)^{2k} dt > \frac{4}{\pi}\int_0^{\pi}\left(\cos\frac{t}{2}\right)^{4k} dt = a_{2k,2} > \frac{3}{2\sqrt{k}}$$

and

$$\frac{1}{\sqrt{k}} < a_{k,\infty} < a_{k,n} \le a_{k,2} \le n^{1-2k}\gamma_{k,n} = a_{k,n} + b_{k,n}$$

$$\le a_{k,2} + b_{k,n} < \frac{\sqrt{6}}{\sqrt{k}} + \frac{1}{2k-1}\,\frac{1}{2^{2k-2}} < \frac{5}{2\sqrt{k}}.$$

Inequality (3.35″) is proved.

(d) For any $\delta > 0$, the following inequality is true:

$$\int_{\delta}^{\pi} J_{k,n}(t)\,dt < \left(\frac{\pi}{n\delta}\right)^{2k-1}. \tag{3.36}$$

This property follows from inequalities (3.35″) and equality (3.31). Indeed,

$$\int_{\delta}^{\pi} J_{k,n}(t)\,dt = \frac{1}{\gamma_{k,n}}\int_{\delta}^{\pi}\left(\frac{\sin(nt/2)}{\sin(t/2)}\right)^{2k} dt$$

$$< \frac{1}{\gamma_{k,n}} \int\limits_{\delta}^{\infty} \left(\frac{\pi}{t}\right)^{2k} dt < \frac{2\sqrt{k}\,\pi^{2k}}{3n^{2k-1}} \frac{1}{(2k-1)\delta^{2k-1}} < \left(\frac{\pi}{n\delta}\right)^{2k-1}.$$

(e) For any natural $i \le 2k-2$,

$$\int\limits_{-\pi}^{\pi} J_{k,n}(t)\,|t|^i\,dt \asymp \frac{1}{n^i}. \tag{3.37}$$

Indeed, on the one hand, by virtue of relations (3.35), (0.1), and (0.2), we have

$$\int\limits_{-\pi}^{\pi} J_{k,n}(t)\,|t|^i\,dt = 2\int\limits_{0}^{\pi} t^i J_{k,n}(t)\,dt = \frac{2}{\gamma_{k,n}} \int\limits_{0}^{\pi} t^i \left(\frac{\sin(nt/2)}{\sin(t/2)}\right)^{2k} dt$$

$$\le \frac{2}{\gamma_{k,n}} \left\{ \int\limits_{0}^{\pi/n} t^i \left(\frac{n\sin(t/2)}{\sin(t/2)}\right)^{2k} dt + \int\limits_{\pi/n}^{\pi} t^i \left(\frac{\pi}{t}\right)^{2k} dt \right\} \le \frac{1}{n^i}.$$

On the other hand, in view of relation (3.35) and inequalities (0.2) and (0.3), we find

$$2\int\limits_{0}^{\pi} t^i J_{k,n}(t)\,dt \ge \frac{2}{\gamma_{k,n}} \int\limits_{0}^{\pi/n} t^i \left(\frac{2}{\pi}\cdot\frac{nt}{2}\cdot\frac{2}{t}\right)^{2k} dt \ge \frac{1}{n^i}.$$

Corollary 3.1. *If, for* $k \ge 2$, *the kernel* $J_{k,n}(t)$, *in view of property (b), admits a representation*

$$J_{k,n}(t) = \sum_{\nu=0}^{k(n-1)} j_\nu \cos\nu t,$$

then the following relations are true for the coefficients $j_\nu = j_\nu(k;n)$:

$$1 - j_\nu \asymp \frac{\nu^2}{n^2}. \tag{3.38}$$

Indeed, by virtue of relations (3.33) and (3.37), we find

$$1 - j_\nu = \frac{1}{\pi} \int\limits_{-\pi}^{\pi} J_{k,n}(t)(1-\cos\nu t)\,dt \asymp \nu^2 \int\limits_{-\pi}^{\pi} J_{k,n}(t)\,t^2\,dt \asymp \frac{\nu^2}{n^2}.$$

7. Poisson kernels. Kernels of the following type are very often encountered in various problems parallel with the polynomial kernels:

$$P_n(t) := \frac{1}{2} + r\cos t + r^2\cos 2t + \ldots = \mathrm{Re}\left\{\frac{1}{2} + re^{it} + r^2 e^{2it} + \ldots\right\}$$

$$= \mathrm{Re}\left\{\frac{1}{2}\frac{1-r^2}{1-2r\cos t + r^2} + i\frac{r\sin t}{1-2r\cos t + r^2}\right\}$$

$$= \frac{1}{2}\frac{1-r^2}{1-2r\cos t + r^2}. \tag{3.39}$$

These kernels are called *Poisson kernels.*

8. In conclusion, we note (for our subsequent presentation) that, by using Theorem 1.2.12 (Fejér), Korovkin (see [Korovkin (1959), pp. 76–78]) introduced the following remarkable even trigonometric kernels K_n of order n:

$$K_n(t) := A^{-1}\left|\sum_{j=0}^{n} a_j e^{ijt}\right|^2 = A^{-1}\left[\sum_{j=0}^{n} a_j^2 + 2\sum_{k=1}^{n}\sum_{j=0}^{k-1} a_j a_k \cos(k-j)t\right], \tag{3.40}$$

where

$$a_j := \sin\frac{(j+1)\pi}{n+2}, \qquad j = 0, 1, \ldots, n,$$

and

$$A := 2\sum_{j=0}^{n} a_j^2 = 2\sum_{k=1}^{n+1}\sin^2\frac{k\pi}{n+2}.$$

Therefore, the kernels $K_n(t)$ have the form

$$K_n(t) = \frac{1}{2} + \sum_{k=1}^{n}\lambda_k\cos kt,$$

where, in particular,

$$\begin{aligned}
\lambda_1 &= 2A^{-1}\sum_{k=1}^{n} a_{k-1}a_k \\
&= 2A^{-1}\sum_{k=1}^{n} \sin\frac{k\pi}{n+2}\left[\sin\frac{k\pi}{n+2}\cos\frac{\pi}{n+2} + \cos\frac{k\pi}{n+2}\sin\frac{\pi}{n+2}\right] \\
&= A^{-1}\left\{\cos\frac{\pi}{n+2}\,2\left[\sum_{k=1}^{n+1}\sin^2\frac{k\pi}{n+2} - \sin^2\frac{n+1}{n+2}\pi + \sin\frac{\pi}{n+2}\sum_{k=1}^{n}\sin\frac{2k\pi}{n+2}\right]\right\} \\
&= \cos\frac{\pi}{n+2} + A^{-1}\left\{-2\cos\frac{\pi}{n+2}\sin^2\frac{\pi}{n+2}\right. \\
&\qquad\qquad \left. + \frac{1}{2}\sum_{k=1}^{n}\left(\cos\frac{2k-1}{n+2}\pi - \cos\frac{2k+1}{n+2}\pi\right)\right\} \\
&= \cos\frac{\pi}{n+2} + A^{-1}\left\{-\sin\frac{2\pi}{n+2}\sin\frac{\pi}{n+2} + \frac{1}{2}\left(\cos\frac{\pi}{n+2} - \cos\frac{3\pi}{n+2}\right)\right\} \\
&= \cos\frac{\pi}{n+2}.
\end{aligned}$$

3.2. Aperiodic kernels

First, we consider a different approach to the construction of the kernels

$$K_n\left(\frac{x}{2}\right) = \check{F}_{2n+1}(x)$$

[see (1.2)] used in proving the first Weierstrass theorem (Theorem 1.1).

Since, for any j, we have

$$\cos jt = \sum_{i=0}^{j} c_i \cos^i t,$$

where c_i are coefficients, in view of relation (3.6″), the Fejér kernel $F_n(t)$ can be represented in the form

$$F_n(t) = \sum_{k=0}^{n-1} d_k \cos^k t.$$

Further, as a result of the substitution

$$x = 2\sin\frac{t}{2}, \quad t \in [-\pi, \pi], \quad x \in [-2, 2],$$

we obtain

$$\cos t = 1 - \frac{x^2}{2}, \quad t = \pm \arccos\left(1 - \frac{x^2}{2}\right)$$

$$F_n\left[\arccos\left(1 - \frac{x^2}{2}\right)\right] = \sum_{k=0}^{n-1} d_k\left(1 - \frac{x^2}{2}\right)^k. \tag{3.41}$$

This means that, by using the periodic Fejér kernel,

$$F_n(t) = \frac{\sin^2\frac{nt}{2}}{2n\sin^2\frac{t}{2}}$$

we can now define the following algebraic kernel for all natural n (but not only for odd n) [Dzyadyk (1958a)]:

$$\check{F}_n(x) = \frac{2n}{\gamma_n} F_n\left[\arccos\left(1 - \frac{x^2}{2}\right)\right]$$

$$= \frac{1}{\gamma_n} \frac{\sin^2\frac{n}{2}\arccos\left(1 - \frac{x^2}{2}\right)}{\sin^2\frac{1}{2}\arccos\left(1 - \frac{x^2}{2}\right)}, \quad x \in [-2, 2], \tag{3.41$'$}$$

where γ_n is chosen to guarantee that

$$\int_{-1}^{1} \check{F}_n(x)dx = 1,$$

i.e.,

$$\gamma_n = \int_{-1}^{1} \frac{\sin^2 \frac{n}{2} \arccos\left(1 - \frac{x^2}{2}\right)}{\sin^2 \frac{1}{2} \arccos\left(1 - \frac{x^2}{2}\right)} dx. \tag{3.41''}$$

The kernel $\check{F}_n$ possesses the following properties:

(a) $\check{F}_n$ is an even nonnegative polynomial of degree $2n-2$;

(b) $$\int_{-1}^{1} \check{F}_n(x)dx = 1; \tag{3.42}$$

(c) $$\gamma_n > n, \quad n \geq 3; \tag{3.43}$$

(d) for $0 < \delta < 1$, $n \geq 3$,

$$\int_{\delta}^{1} \check{F}_n(x)dx \leq \frac{\pi^2}{n\delta}. \tag{3.44}$$

Since properties (a) and (b) directly follow from the definition [see relations (3.40) and (3.40′)], we study only inequalities (3.43) and (3.44).

By virtue of the substitution $x = 2\sin(t/2)$ (and, hence, $1 - x^2/2 = \cos t$), in view of inequalities (0.2) and (0.3), we find

$$\gamma_n = 2\int_{0}^{1} \frac{\sin^2 \frac{n}{2} \arccos\left(1 - \frac{x^2}{2}\right)}{\sin^2 \frac{1}{2} \arccos\left(1 - \frac{x^2}{2}\right)} dx = 2\int_{0}^{\pi/3} \frac{\sin^2 \frac{nt}{2}}{\sin^2 \frac{t}{2}} \cos\frac{t}{2} dt$$

$$> 2\int_{0}^{\pi/n} \frac{1}{2} \frac{\left(\frac{2}{\pi}\frac{nt}{2}\right)^2}{\left(\frac{t}{2}\right)^2} dt = \frac{4}{n}\frac{n^2}{\pi} > n, \quad n \geq 3,$$

and, moreover,

$$\int_{\delta}^{1} \check{F}_n(x)dx \le \int_{2\sin(\delta/2)}^{1} \check{F}_n(x)dx$$

$$= \frac{1}{\gamma_n} \int_{2\sin(\delta/2)}^{1} \frac{\sin^2 \frac{n}{2} \arccos\left(1 - \frac{x^2}{2}\right)}{\sin^2 \frac{1}{2} \arccos\left(1 - \frac{x^2}{2}\right)} dx \le \frac{1}{n} \int_{\delta}^{\pi/3} \frac{\sin^2 \frac{nt}{2}}{\sin^2 \frac{t}{2}} \cos\frac{t}{2} dt$$

$$< \frac{1}{n} \int_{\delta}^{\pi/3} \left(\frac{\pi}{t}\right)^2 dt < \frac{\pi^2}{n} \int_{\delta}^{\infty} \frac{dt}{t^2} = \frac{\pi^2}{n\delta},$$

as required.

Remark 3.4. Note that, as a rule, the analyzed substitution $x = 2\sin(t/2)$ gives a possibility to obtain from an even trigonometric kernel an algebraic kernel with similar properties in the neighborhood of the origin, and vice versa.[5]

In order to deduce the explicit formulas for the polynomials $\check{F}_n$ (which can also be used as kernels in the proof of the first Weierstrass theorem), we now establish the recurrent relation for these polynomials:

$$\breve{F}_n(x) := \gamma_n \check{F}_n(x).$$

By virtue of (3.40′), for all integer $n \ge 1$, we can write

$$\breve{F}_n(x) := \gamma_n \check{F}_n(x) = \frac{1 - \cos nt}{1 - \cos t}, \quad t = \arccos\left(1 - \frac{x^2}{2}\right).$$

Hence, by setting $\breve{F}_0(x) \equiv 0$, taking into account the fact that $\breve{F}_1(x) \equiv 1$, and using the identity

$$\cos nt + \cos(n-2)t = 2\cos(n-1)t\cos t$$

for $n \ge 2$, we find

$$\breve{F}(x) = \frac{1}{1-\cos t}\left[1 - 2\cos(n-1)t\cos t + \cos(n-2)t + 2\cos t - 2\cos t - 1 + 1\right]$$

$$= 2\cos t \,\frac{1-\cos(n-1)t}{1-\cos t} - \frac{1-\cos(n-2)t}{1-\cos t} + 2$$

$$= 2\cos t\, \breve{F}_{n-1}(x) - \breve{F}_{n-2}(x) + 2.$$

Thus, we arrive at the following recurrent relation for the polynomials $\breve{F}_n$:

$$\breve{F}_n(x) = (2-x^2)\breve{F}_{n-1}(x) - \breve{F}_{n-2}(x) + 2. \tag{3.45}$$

Therefore, in view of the fact that $\breve{F}_0(x) \equiv 0$ and $\breve{F}_1(x) \equiv 1$, by virtue of relation (3.45), we immediately obtain

$$\breve{F}_2(x) = (2-x^2)1 - 0 + 2 = 4 - x^2,$$

$$\breve{F}_3(x) = (2-x^2)(-x^2+4) - 1 + 2 = x^4 - 6x^2 + 9,$$

etc.

It is also possible to prove that the polynomials $\breve{F}_n(2x)$ satisfy the following differential equation:

$$x^2(1-x^2)y'' - (5x^2-4)xy' + [4(n^2-1)x^2+2]y - 2n^2 = 0.$$

Further, starting from the Jackson-type kernels (3.31), we consider the following kernels [Dzyadyk (1958)]:

$$\breve{J}_{k,n}(x) = \frac{\gamma_{k,n}}{\breve{\gamma}_{k,n}} J_{k,n}\left[\arccos\left(1-\frac{x^2}{2}\right)\right]$$

$$= \frac{1}{\breve{\gamma}_{k,n}}\left\{\frac{\sin\left[\frac{n}{2}\arccos\left(1-\frac{x^2}{2}\right)\right]}{\sin\left[\frac{1}{2}\arccos\left(1-\frac{x^2}{2}\right)\right]}\right\}^{2k}, \qquad x \in [-2, 2], \tag{3.46}$$

where

$$\breve{\gamma}_{k,n} = \int_{-1}^{1}\left\{\frac{\sin\left[\frac{n}{2}\arccos\left(1-\frac{x^2}{2}\right)\right]}{\sin\left[\frac{1}{2}\arccos\left(1-\frac{x^2}{2}\right)\right]}\right\}^{2k} dx. \tag{3.47}$$

These kernels possess the following properties:

(a) For any natural n and k, $\breve{J}_{k,n}$ is an even nonnegative algebraic polynomial of degree $2k(n-1)$.

This property follows from the fact that, by virtue of equality (3.34), we have

$$J_{k,n}(t) = \sum_{\nu=0}^{k(n-1)} j_\nu \cos \nu t = \sum_{\nu=0}^{k(n-1)} d_\nu (\cos t)^\nu.$$

Therefore, in view of the equality $t = \arccos(1 - x^2/2)$, we find

$$\breve{J}_{k,n}(x) = \frac{\gamma_{k,n}}{\breve{\gamma}_{k,n}} \sum_{\nu=0}^{k(n-1)} d_\nu \left(1 - \frac{x^2}{2}\right)^\nu.$$

(b) $\int\limits_{-1}^{1} \breve{J}_{k,n}(x)\,dx = 1.$

This property follows from equalities (3.46) and (3.47).

(c) $\breve{\gamma}_{k,n} \asymp n^{2k-1}.$ (3.48)

Indeed, in view of relations (3.47) and (3.35), we conclude that

$$\breve{\gamma}_{k,n} = 2\int\limits_0^1 \left\{ \frac{\sin\left[\frac{n}{2}\arccos\left(1 - \frac{x^2}{2}\right)\right]}{\sin\left[\frac{1}{2}\arccos\left(1 - \frac{x^2}{2}\right)\right]} \right\}^{2k} dx = 2\int\limits_0^{\pi/3} \left(\frac{\sin^2 \frac{nt}{2}}{\sin^2 \frac{t}{2}} \right)^{2k} \cos\frac{t}{2}\,dt$$

$$\asymp \int\limits_0^{\pi} \left(\frac{\sin \frac{nt}{2}}{\sin \frac{t}{2}} \right)^{2k} dt - \int\limits_{\pi/3}^{\pi} \left(\frac{\sin \frac{nt}{2}}{\sin \frac{t}{2}} \right)^{2k} dt$$

$$= \gamma_{k,n} - \int\limits_{\pi/3}^{\pi} \left(\frac{\sin \frac{nt}{2}}{\sin \frac{t}{2}} \right)^{2k} dt \asymp n^{2k-1}.$$

(d) The following inequality is true for any $0 < \delta < 1$:

$$\int_{\delta}^{2} \breve{J}_{k,n}(x)\,dx \le \frac{c}{(n\delta)^{2k-1}}, \qquad c = c(k) = \text{const.} \tag{3.49}$$

Indeed, by using relation (3.48) and inequalities (0.2) and (0.3), for $\sin t$, we find

$$\int_{\delta}^{2} \breve{J}_{k,n}(x)\,dx = \frac{1}{\breve{\gamma}_{k,n}} \int_{\delta}^{2} \left\{ \frac{\sin\left[\frac{n}{2}\arccos\left(1-\frac{x^2}{2}\right)\right]}{\sin\left[\frac{1}{2}\arccos\left(1-\frac{x^2}{2}\right)\right]} \right\}^{2k} dx$$

$$= \frac{1}{\breve{\gamma}_{k,n}} \int_{2\arcsin(\delta/2)}^{\pi} \left[\frac{\sin\frac{nt}{2}}{\sin\frac{t}{2}}\right]^{2k} \cos\frac{t}{2}\,dt \le \frac{1}{\breve{\gamma}_{k,n}} \int_{\delta}^{\infty} \left(\frac{\pi}{t}\right)^{2k} dt \le \frac{c}{n^{2k-1}}.$$

(e) For any natural $i \le 2k-2$,

$$\int_{\delta}^{2} \breve{J}_{k,n}(x)\,x^i\,dx \asymp \frac{1}{n^i}. \tag{3.50}$$

Indeed, by using relations (3.48), (3.37), (0.2), and (0.3), we obtain

$$\int_{0}^{2} x^i \breve{J}_{k,n}(x)\,dx = \frac{1}{\gamma_{k,n}} \int_{0}^{\pi} \left(2\sin\frac{t}{2}\right)^i \left(\frac{\sin\frac{nt}{2}}{\sin\frac{t}{2}}\right)^{2k} \cos\frac{t}{2}\,dt$$

$$\asymp \frac{1}{\gamma_{k,n}} \int_{0}^{\pi} t^i \left(\frac{\sin\frac{nt}{2}}{\sin\frac{t}{2}}\right)^{2k} dt \asymp \frac{1}{n^i}.$$

4. Rational kernels and approximation of functions by rational polynomials

4.1. Newman theorem

Definition 4.1. *Any function representable as the ratio of two algebraic polynomials of degree n, i.e.,*

$$R_n(x) = \frac{a_0 x^k + a_1 x^{k-1} + \ldots + a_k}{b_0 x^l + b_1 x^{l-1} + \ldots + b_l}, \tag{4.1}$$

where $0 \le k, l \le n,\ a_0 \ne 0,$ *and* $b_0 \ne 0,$ *is called a rational polynomial* R_n *of degree* n.

For the first time, the fact that there are continuous functions f whose best approximation by rational polynomials of degree n is much better than their best approximation by algebraic polynomials was indicated by Newman. As an example of a function of this sort, he [Newman (1964)] studied the function $f(x) = |x|,\ x \in [-1, 1]$, and demonstrated that

$$\inf_{R_n} \max_x \big||x| - R_n(x)\big| \le \frac{3}{e^{\sqrt{n}}}, \qquad n = 5, 6, 7, 8, \ldots . \tag{4.2}$$

At the same time, the following inequality is true:

$$E_n(|x|) = \inf_{P_n} \max_x \big||x| - P_n(x)\big| \ge \frac{c}{n}, \qquad c = \text{const} > 0. \tag{4.3}$$

Prior to the Newman theorem, we prove two simple lemmas on the existence of polynomials of degree n whose absolute values on the segment $[-1, -e^{-\sqrt{n}}]$ are much smaller than on the segment $[e^{-\sqrt{n}}, 1]$.

Definition 4.2. *A function*

$$N_n(x) = \prod_{k=0}^{n-1} (x + \xi^k), \quad where \quad \xi = \xi(n) = e^{-1/\sqrt{n}}, \tag{4.4}$$

is called the Newman polynomial N_n *of degree* n.

Lemma 4.1. *The inequality*

$$\left|\frac{N_n(-\xi^{-1})}{N_n(\xi^{-1})}\right| = \prod_{k=1}^{n} \frac{1-\xi^k}{1+\xi^k} < e^{-\sqrt{n}} \tag{4.5}$$

holds for all $n \ge 5$.

Proof. First, we note that the following inequality holds for all $t \ge 0$:

$$\frac{1-t}{1+t} \le e^{-2t}. \tag{4.6}$$

This follows from the fact that the auxiliary function

$$\alpha(t) = (1-t)e^{2t} - (1+t)$$

is negative for all $t > 0$ since $\alpha(0) = 0$, $\alpha'(0) = 0$, and $\alpha''(t) = -4te^{2t} < 0$.

By virtue of inequality (4.6), we obtain

$$\left|\frac{N_n(-\xi^{-1})}{N_n(\xi^{-1})}\right| = \prod_{k=1}^{n} e^{-2\xi^k} = \exp\left\{-2\sum_{k=1}^{n}\xi^k\right\} = \exp\left\{-2\xi\frac{1-\xi^n}{1-\xi}\right\}. \tag{4.7}$$

Further, in view of the fact that, for $n \ge 5$,

$$\begin{aligned} 2\xi(1-\xi^n) &= 2\left(e^{-1/\sqrt{n}} - e^{-(n+1)/\sqrt{n}}\right) \\ &\ge 2\left(e^{-1/\sqrt{5}} - e^{-6/\sqrt{5}}\right) > 2\left(1 - \frac{1}{\sqrt{5}} + \frac{1}{2}\cdot\frac{1}{5} - \frac{1}{6\cdot 5\sqrt{5}}\right) - \frac{2}{e^{2.5}} > 1 \end{aligned}$$

and, in addition,

$$1-\xi = 1 - e^{-1/\sqrt{n}} = \frac{1}{\sqrt{n}} - \frac{1}{2n} + \ldots < \frac{1}{\sqrt{n}},$$

by using equality (4.6), we conclude that

$$\left|\frac{N_n(-\xi^{-1})}{N_n(\xi^{-1})}\right| < e^{-1/(1-\xi)} < e^{-\sqrt{n}}.$$

This proves Lemma 4.1.

Lemma 4.2. *The following inequality is true for all* $x \in [\, e^{-\sqrt{n}}, 1\,]$ *and* $n \ge 5$:

$$\left|\frac{N_n(-x)}{N_n(x)}\right| \le \frac{1}{e^{\sqrt{n}}}. \tag{4.8}$$

Proof. For the sake of definiteness, we assume that $x \in [\xi^{j+1}, \xi^j]$. Thus, in view of the fact that

$$\left(\frac{t-a}{t+a}\right)' > 0$$

for any $t > 0$ and $a > 0$ and Lemma 4.1, relation (4.4) implies that

$$\left|\frac{N_n(-x)}{N_n(x)}\right| = \prod_{k=0}^{j} \frac{\xi^k - x}{\xi^k + x} \prod_{k=j+1}^{n-1} \frac{x - \xi^k}{x + \xi^k} \leq \prod_{k=0}^{j} \frac{\xi^k - \xi^n}{\xi^k + \xi^n} \prod_{k=j+1}^{n-1} \frac{\xi^j - \xi^k}{\xi^j + \xi^k}$$

$$= \prod_{l=n-j}^{n} \frac{1-\xi^l}{1+\xi^l} \prod_{l=1}^{n-j-1} \frac{1-\xi^l}{1+\xi^l} = \prod_{l=1}^{n} \frac{1-\xi^l}{1+\xi^l} = \left|\frac{N_n(-\xi^{-1})}{N_n(\xi^{-1})}\right| < e^{-\sqrt{n}}.$$

This proves Lemma 4.2.

Theorem 4.1 [Newman (1964)]. *A rational polynomial* R_n *of the form*

$$R_n(x) = x \frac{N_n(x) - N_n(-x)}{N_n(x) + N_n(-x)} \tag{4.9}$$

whose degree is equal to n *if* n *is even and* $n+1$ *if* n *is odd approximates the function* $|x|$ *on the segment* $[-1, 1]$ *so that*[6]

$$||x| - R_n(x)| \leq 3e^{-\sqrt{n}} \tag{4.10}$$

for all $n \geq 5$.

Proof. Since both functions $|x|$ and $|R_n(x)|$ are even, it suffices to prove inequality (4.10) for $x \geq 0$.

In this case, by virtue of (4.4), we have $0 \leq N_n(-x) \leq N_n(x)$ for $x \in [0, \xi^n] = [0, e^{-\sqrt{n}}]$ and, consequently, $0 \leq R_n(x) \leq x$. Therefore,

$$||x| - R_n(x)| = x - R_n(x) \leq x < e^{-\sqrt{n}}.$$

If $x \in [e^{-\sqrt{n}}, 1]$, then, by using Lemma 4.2 and the inequality $n \geq 5$, we find

$$||x| - R_n(x)| = 2x\left|\frac{N_n(-x)}{N_n(x) + N_n(-x)}\right| < \frac{2}{|N_n(x)/N_n(-x)| - 1} < \frac{2}{e^{\sqrt{n}} - 1} < \frac{3}{e^{\sqrt{n}}}.$$

Theorem 4.1 is thus proved.

■

4.2. Rational kernels

All polynomial kernels and, in particular, all analyzed even kernels K_n (algebraic and trigonometric) are characterized by a relatively weakly pronounced δ-shape in the neighborhood of the origin in the sense that, for any natural n, the measure of the interval $[-\delta_n, \delta_n]$ outside which, e.g., the inequality

$$\int\limits_{[-1,1]\setminus[-\delta_n,\delta_n]} |K_n(x)|dx < \frac{1}{2}$$

$$\left(\frac{1}{\pi}\int\limits_{[-\pi,\pi]\setminus[-\delta_n,\delta_n]} |K_n(t)|dt < \frac{1}{2} \quad \text{in the trigonometric case}\right),$$

can be true is always greater than $1/n$, i.e., $2\delta_n \geq 1/n$.

We now show that, by using the rational polynomials (4.9) proposed by Newman, one can readily construct the rational kernels $K_n(x)$ which are much more δ-shaped in the neighborhood of the origin (in the sense of concentration of singularity near the point $x = 0$) than the polynomial kernels of the same degree.

Theorem 4.2 [Dzyadyk (1966)]. *The rational kernels $K_n(x)$ of degree n of the form*

$$K_n(x) = \frac{1}{2}\frac{d}{dx}\left[\frac{N_n(x) - N_n(-x)}{N_n(x) + N_n(-x)}\right] = \frac{N_n(x)N_n'(-x) + N_n(-x)N_n'(x)}{[N_n(x) + N_n(-x)]^2} \tag{4.11}$$

possess the following properties:

1. $K_n(x)$ *are even rational polynomials of degree* $4\left[\frac{n}{2}\right]$;

2. $\int\limits_{-1}^{1} K_n(x)dx = 1;$

3. $\int\limits_{e^{-\sqrt{n}}}^{1} |K_n(x)|dx \leq 3ne^{-\sqrt{n}}.$

Proof. 1. The first property is obvious.

2. The second property follows from the fact that, by virtue of relations (4.11) and (4.4), we can write

$$\int_{-1}^{1} K_n(x)\,dx = \frac{1}{2}\frac{N_n(x) - N_n(-x)}{N_n(x) + N_n(-x)}\bigg|_{-1}^{1} = \frac{1}{2}(1+1) = 1.$$

3. We introduce an auxiliary function α_n by the formula

$$\alpha_n(x) = \frac{N_n(-x)}{N_n(x)}. \tag{4.12}$$

The denominator of this function $N_n(x)$ is positive for all $x \geq 0$ and, according to relation (4.4), its numerator $N_n(-x)$ is equal to zero at n points $\xi^{n-1}, \xi^{n-2}, \ldots, \xi, 1$. Therefore, according to the Rolle theorem, the derivative

$$\alpha_n'(x) = \frac{N_n(x)N_n'(-x) + N_n(-x)N_n'(x)}{-N_n^2(x)} \tag{4.13}$$

and, hence, its numerator

$$\sigma_n(x) := N_n(x)N_n'(-x) + N_n(-x)N_n'(x)$$

(which is, clearly, an even polynomial of degree $2n-2$) is equal to zero at a system of points $\pm\eta_i$, $i = 1, 2, \ldots, n-1$ satisfying the conditions

$$\xi^{n-1} < \eta_{n-1} < \xi^{n-2} < \eta_{n-2} < \ldots < \xi < \eta_1 < 1. \tag{4.14}$$

Obviously, all these zeros are simple and the function $\sigma_n(x)$ does not have any other zeros.

Further, in view of the fact that, by virtue of equality (4.4), the sum

$$N_n(x) + N_n(-x) \geq \begin{cases} \prod_{k=0}^{n-1}(x+\xi^k) - \prod_{k=0}^{n-1}|x-\xi^k| > 0 & \text{for } x > 0, \\ 2\prod_{k=0}^{n-1}\xi^k > 0 & \text{for } x = 0, \end{cases}$$

we conclude that the polynomial $N_n(x) + N_n(-x)$ does not have any positive zeros. Thus, the kernel K_n has the same zeros as the numerator of σ_n, i.e., the points $\pm\eta_i$, and no other zeros. All these zeros are simple. Hence, by using equalities (4.11), (4.13), and (4.12) and inequality (4.8), we obtain

$$\int_{e^{-\sqrt{n}}}^{1} |K_n(x)|dx = \int_{e^{-\sqrt{n}}}^{1} \left| \frac{N_n(x)N_n'(-x) + N_n(-x)N_n'(x)}{\left[1 + \dfrac{N_n(-x)}{N_n(x)}\right]^2 N_n^2(x)} \right| dx$$

$$< \frac{1}{(1 - e^{-\sqrt{n}})^2} \int_{e^{-\sqrt{n}}}^{1} |\alpha_n'(x)|dx$$

$$< \frac{3}{2} \left| \int_{e^{-\sqrt{n}}}^{\eta_{n-1}} - \int_{\eta_{n-1}}^{\eta_{n-2}} + \ldots + (-1)^{n-1} \int_{\eta_1}^{1} \alpha_n'(x)dx \right|$$

$$\leq \frac{3}{2} |\alpha_n(e^{-\sqrt{n}})| + 3|\alpha_n(\eta_{n-1})| + \ldots + 3|\alpha_n(\eta_1)| + \frac{3}{2}\alpha_n(1).$$

Since, in view of relation (4.8), $0 \leq \alpha_n(x) \leq e^{-\sqrt{n}}$ for $x \geq e^{-\sqrt{n}}$, we find

$$\int_{e^{-\sqrt{n}}}^{1} |K_n(x)|dx \leq 3ne^{-\sqrt{n}}.$$

The theorem is proved. ■

Remark 4.1. A more comprehensive investigation [Dzyadyk (1966)] shows that the kernel K_n possesses the following properties:

$$\int_{e^{-\sqrt{n}}}^{1} |K_n(x)|dx < 3e^{-\sqrt{n}},$$

$$K_n(0) \sim \frac{\sqrt{n}}{2} e^{\sqrt{n}},$$

$$\int_{-e^{-\sqrt{n}}}^{e^{-\sqrt{n}}} |K_n(x)|dx \leq 1 + 6e^{-\sqrt{n}}, \quad \text{etc.}$$

Remark 4.2. By using the kernel $K_n(x)$, one can also construct trigonometric rational kernels of the form

$$\tilde{\gamma}_n K_n\left(\sqrt{\frac{1-\cos t}{2}}\right)\cos^2\frac{t}{2}$$

and rational kernels on the entire axis of the form

$$\gamma_{2n} K_{2n}(x)\,\frac{1}{1+(xe^{\sqrt{2n}}/2n)^2} \qquad \text{[Dzyadyk (1966)]}$$

with similar properties.

In conclusion, we note that the volume and scope of the present monograph do not allow us to dwell upon the very deep results by Gonchar, Vitushkin, Dovzhenko, Gilewicz, and others concerning the approximation of functions of complex variable by rational polynomials.

Remarks to Chapter 2

1. In proving Theorem 1.1, Weierstrass used polynomial kernels $W_{n,k}(x)$ of the form

$$W_{n,k}(x) = \frac{1}{\gamma_n}\left[1+\frac{(-nx^2)}{1!}+\frac{(-nx^2)^2}{2!}+\ldots+\frac{(-nx^2)^k}{k!}\right], \tag{4.1}$$

where, for given $\varepsilon > 0$, the natural number n and the number $c > 0$ are chosen to guarantee that, for

$$\gamma_n := \int_{-c}^{c} e^{-nt^2}\,dt,$$

the "δ-shaped function" e^{-nt^2}/γ_n (see Section 3) is such that the integral

$$\frac{1}{\gamma_n}\int_{-c}^{c} f(t)e^{-n(x-t)^2}\,dt = \frac{1}{\gamma_n}\int_{x-c}^{x+c} f(x-t)e^{-nt^2}\,dt \tag{4.2}$$

approximates the function

$$f(x) = \frac{1}{\gamma_n} \int_{-c}^{c} f(x) e^{-nt^2} \, dt$$

for all $x \in [a, b]$ with an error $< \varepsilon/2$. After this, the natural number n is chosen sufficiently large in order that the integral

$$\int_{-c}^{c} f(t) W_{n,k}(x-t) \, dt$$

(which is obviously a polynomial of degree $2k$) differ from integral (4.2) by less than $\varepsilon/2$ for all $x \in [a, b]$.

Note that the influence of this reasoning on the subsequent development of the theory of approximation could hardly be overestimated. Thus, the major part of polynomial kernels analyzed in the present chapter can be regarded as natural (although fairly nontrivial) generalizations of the kernels $W_{n,k}(x)$.

2. The proof presented in the monograph is a revised version of the proof taken from [Rudin (1966), pp. 178–186].

3. Note that, as a result of the generalization of some S. Nikol'skii's results [S. Nikol'skii (1940a)], Verbitskii (1940) proved that if the number

$$r := \frac{n+m}{n-m}$$

is odd, then the following equality is true:

$$\frac{1}{\pi} \int_{-\pi}^{\pi} \left| V_m^n(t) \right| dt = \frac{1}{r} + \frac{2}{\pi} \sum_{\nu=1}^{(r-1)/2} \frac{1}{\nu} \tan \frac{\pi \nu}{r}.$$

Stechkin (1951) showed that if the number

$$r = \frac{n+m}{n-m}$$

is even, then

$$\frac{1}{\pi} \int_{-\pi}^{\pi} \left| V_m^n(t) \right| dt = \frac{4}{\pi} \sum_{\nu=1}^{r/2} \frac{1}{2\nu - 1} \tan \frac{\pi (2\nu - 1)}{2r}.$$

4. Equality (3.21) was established by Korneichuk (1959) by using a different method. The quantity r_n was estimated as follows:

$$0 < r_n < \frac{2}{\sqrt{3}(2n+1)}.$$

5. It seems likely that this substitution was used, for the first time, by Dzyadyk (1958).

6. Much more accurate estimates are available at present.

Chapter 3
On smoothness of functions

Results of Chebyshev and Weierstrass laid the foundation for the theory of approximation of functions and had a significant impact on its further development. It follows from the Weierstrass theorem that, for any continuous function f on $[a, b]$, we have

$$\lim_{n\to\infty} E_n(f) = 0.$$

The following question now naturally arises: What properties of the function f do affect the rate of convergence of the sequence $\{E_n(f)\}$ to zero and how strong is this influence? It turns out that the higher the smoothness of the function, the higher the rate of convergence of $E_n(f)$ to zero. Roughly speaking, in the class of analytic functions, a function is assumed to be smoother, e.g., on $[-1, 1]$, if the distance from its nearest singular point to $[-1, 1]$ is larger; next to analytic functions are infinitely differentiable ones. If a function f has more derivatives than a function g, then f is assumed to be smoother than g.

In what follows, $A = A(a, b)$ denotes the class of functions analytic on an interval (a, b) and $MW[r, [a, b]]$, where r is natural, denotes the class of functions f for which all derivatives up to the $(r-1)$th order exist and are absolutely continuous on $[a, b]$ and the derivative $f^{(r)}$ satisfies the condition $|f^{(r)}(x)| \le M$ almost everywhere on $[a, b]$; if $M = 1$, then, for simplicity, we denote the corresponding class simply by $W[r, [a, b]]$:

$$W[r, [a, b]] := 1 \cdot W[r, [a, b]].$$

By $W^r([a, b])$ we denote the space of functions each of which belongs to the class $MW[r, [a, b]]$ for some M. As is customary, $C = C([a, b])$ denotes the space of all continuous functions on $[a, b]$, and $C^{(r)} = C^{(r)}([a, b])$ denotes the space of r times continuously differentiable functions on $[a, b]$.

In this chapter, to compare the smoothness of two continuous functions f and g that either have the same number of derivatives or do not have derivatives at all, we use special characteristics (*moduli of continuity*) of these functions (or of their derivatives $f^{(r)}$

and $g^{(r)}$, respectively); we assume that, of two functions, the function whose modulus of continuity converges to zero at higher rate is smoother.

As a rule, we denote classes and spaces of ω-periodic functions defined on the entire axis by a tilde over the corresponding symbol of a class or a space and indicate the period in brackets. For example, $\tilde{C} = \tilde{C}[0, 2\pi]$ and $\tilde{C}^{(r)} = \tilde{C}^{(r)}[0, 2\pi]$ are the spaces of all 2π-periodic functions, respectively, continuous and r times continuously differentiable on the entire axis. However, we omit the tilde if this does not lead to misunderstanding.

1. Modulus of continuity (of the first order)

1.1. Definition. Examples

Definition 1.1. *For a function f continuous on $[a, b]$, the first-order modulus of continuity, or, simply, the modulus of continuity, is the function $\omega(u) = \omega(u; f; [a, b])$ defined on $[0, b-a]$ by the equality*

$$\omega(u; f; [a, b]) = \sup_{\substack{a \le x \le b-h \\ 0 \le h \le u}} |f(x+h) - f(x)|, \tag{1.1}$$

or, which is the same,

$$\omega(u; f; [a, b]) = \sup_{\substack{|x_2 - x_1| \le u \\ x_1, x_2 \in [a, b]}} |f(x_1) - f(x_2)|. \tag{1.1'}$$

According to this definition, the modulus of continuity $\omega(u; f; [a, b])$ of a function f indicates, for every fixed $u \in [0, b-a]$, the amplitude of maximal oscillation of the function on an arbitrary segment of length u contained in $[a, b]$.

In particular, this yields

$$\begin{aligned} |f(x+h) - f(x)| &\le \omega(h), \qquad x, x+h \in [a, b]; \\ |f(x_2) - f(x_1)| &\le \omega(|x_2 - x_1|), \qquad x_1, x_2 \in [a, b]. \end{aligned} \tag{1.2}$$

This definition remains valid for the infinite interval $(-\infty, \infty)$, provided that the function is uniformly continuous on it.

Example 1.1. Let $f(x) = Ax + B$, $x \in (-\infty, \infty)$. Then, for any $u \ge 0$, we have

$$\omega(u) = \sup_{\substack{-\infty < x < \infty \\ 0 \le h \le u}} |A(x+h) + B - Ax - B| = \sup_{0 \le h \le u} |Ah| = |A|u.$$

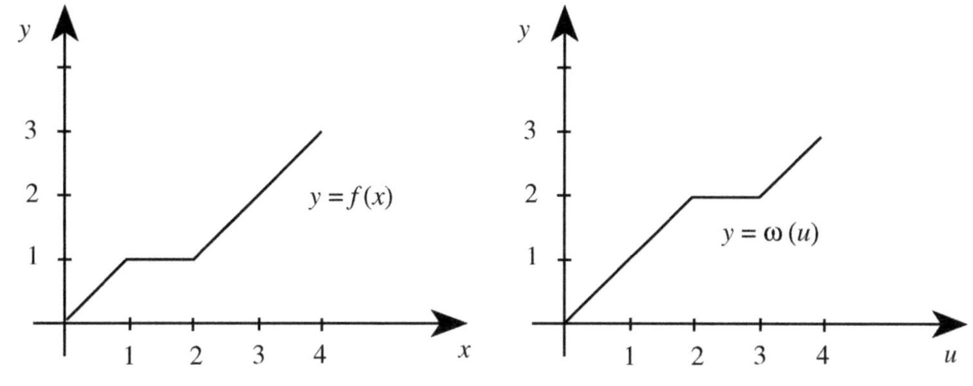

(a) (b)

Fig. 4

Example 1.2. Let the graph of a function f has the form shown in Fig. 4a. Then the graph of the function ω has the form shown in Fig. 4b.

Example 1.3. Let $f(x) = \sin x,\ x \in (-\infty, \infty)$. Then, for any $u \geq 0$, we have

$$\omega(u) = \sup_{\substack{-\infty<x<\infty \\ 0\le h\le u}} |\sin(x+h) - \sin(x)| = 2 \sup_{\substack{-\infty<x<\infty \\ 0\le h\le u}} \left| \cos\left(x + \frac{h}{2}\right) \sin\frac{h}{2} \right|$$

$$= \sup_{0\le h\le u} \left| \sin\frac{h}{2} \right| = \begin{cases} 2\sin\dfrac{u}{2} & \text{if } u \leq \pi, \\ 2 & \text{if } u \geq \pi. \end{cases}$$

Remark 1.1. Let t' and t'' be two arbitrary points on the real axis. Since, among points of the type

$$\tilde{t}'' = t'' + 2k\pi, \qquad k = 0, \pm 1, \pm 2, \pm \ldots,$$

there exists at least one point $\tilde{t}'' = t'' + 2k_0\pi$ such that $|\tilde{t}'' - t'| \leq \pi$, we conclude that, for any 2π-periodic continuous function f, the following relation holds for any $u \geq \pi$:

$$\omega(u) = \sup_{|t''-t'| \le u} |f(t'') - f(t')| = \sup_{|\tilde{t}''-t'| \le \pi} |f(\tilde{t}'') - f(t')| = \omega(\pi).$$

Therefore, for any function of this type, $\omega(u)$ is constant for all $u \geq \pi$.

Remark 1.2. If, for some $u \in [0, \pi]$, one has

$$\omega(u) = \sup_{\substack{-\infty<t<\infty \\ 0\le h\le u}} |f(t+h) - f(t)| = |f(t_0 + h_0) - f(t_0)|,$$

where $0 \le h_0 \le u \le \pi$, then, by virtue of the periodicity of the function f, we can always assume that $t_0 \in [0, 2\pi]$ and, hence, $t_0 + h_0 \in [0, 3\pi]$. Therefore, in the investigation of the modulus of continuity of a 2π-periodic function, we can restrict ourselves to the values of the argument of this function lying on the segment $[0, 3\pi]$ or on some other segment of length 3π and forget about the periodicity of f. However, as shown by Example 1.4, if the modulus of continuity is considered for the values of the argument of a 2π-periodic function f from a segment of length smaller than 3π, then, generally speaking, this modulus of continuity differs from the modulus of continuity of the function f on the entire axis.

Example 1.4. For $\varepsilon < \pi/3$, we set

$$\varphi(x) = \begin{cases} -\dfrac{x}{\pi - 2\varepsilon} & \text{for } x \in [0, \pi - 2\varepsilon], \\[2ex] \dfrac{(x-\pi)}{2\varepsilon} & \text{for } x \in [\pi - 2\varepsilon, \pi], \\[2ex] \dfrac{1}{\pi - \varepsilon}(x - \pi) & \text{for } x \in [\pi, 2\pi - \varepsilon], \\[2ex] -\dfrac{1}{\varepsilon}(x - 2\pi) & \text{for } x \in [2\pi - \varepsilon, 2\pi]. \end{cases}$$

Let

$$f(x) := \varphi\left(x - \left[\frac{x}{2\pi}\right]2\pi\right)$$

be the 2π-periodic extension of the function φ onto the entire axis. If the function f is considered on the segment $[0, 3\pi - 3\varepsilon]$ of length $3\pi - 3\varepsilon < 3\pi$, so that (see Fig. 5a)

$$f(x) = \begin{cases} \varphi(x) & \text{for } x \in [0, 2\pi], \\[2ex] -\dfrac{x - 2\pi}{\pi - 2\varepsilon} & \text{for } x \in [2\pi, 3\pi - 3\varepsilon], \end{cases}$$

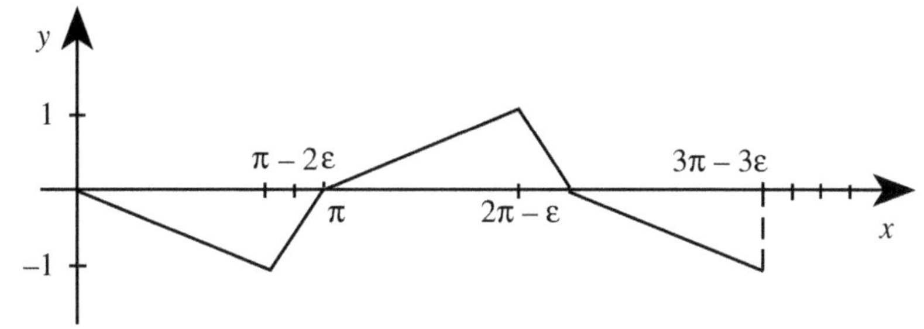

(a)

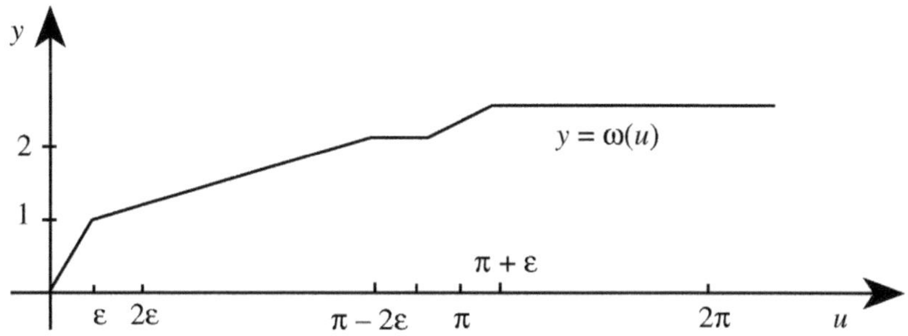

(b)

Fig. 5

then one can easily see (Fig. 5b) that

$$\omega(u;f) = \begin{cases} \dfrac{1}{\varepsilon}u & \text{for } u \in [0, \varepsilon], \\[2ex] 1 + \dfrac{u-\varepsilon}{\pi - 2\varepsilon} & \text{for } u \in [\varepsilon, \pi - 2\varepsilon], \\[2ex] 1 + \dfrac{\pi - 3\varepsilon}{\pi - 2\varepsilon} & \text{for } u \in \left[\pi - 2\varepsilon, \pi - \dfrac{\varepsilon^2}{\pi - 2\varepsilon}\right], \\[2ex] 1 + \dfrac{u - 2\varepsilon}{\pi - \varepsilon} & \text{for } u \in \left[\pi - \dfrac{\varepsilon^2}{\pi - 2\varepsilon}, \pi + \varepsilon\right], \\[2ex] 2 & \text{for } u \in [\pi + \varepsilon, 3\pi - 3\varepsilon], \end{cases}$$

i.e., the modulus of continuity of the function $f\colon [0, 3\pi - 3\varepsilon] \to R$ does not attain its greatest value at the point π and, hence, differs from the modulus of continuity of the function on the entire axis.

1.2. Properties of the modulus of continuity

The modulus of continuity possesses the following properties:

(i) $\omega(0) = 0$;

(ii) ω is a nondecreasing function;

(iii) ω is a continuous function;

(iv) ω is a semiadditive function in the sense that, for any $u_1 \geq 0$ and $u_2 \geq 0$, one has

$$\omega(u_1 + u_2) \leq \omega(u_1) + \omega(u_2). \tag{1.3}$$

Proof. Property (i) follows from the definition of modulus of continuity. Property (ii) follows from the fact that, for large u, we must consider the supremum on a wider set of values of h. Property (iv) follows from the fact that, representing a number $h \in [0, u_1 + u_2] \subset [0, b-a]$ in the form $h = h_1 + h_2$, $h_1 \in [0, u_1]$, $h_2 \in [0, u_2]$, one obtains

$$\begin{aligned}&\| f(\cdot + h_1 + h_2) - f(\cdot)\|_{[a,b-h]}\\&\quad\leq \| f(\cdot + h_1 + h_2) - f(\cdot + h_2)\|_{[a,b-h]} + \sup_{0 \leq h_2 \leq u_2} \| f(\cdot + h_2) - f(\cdot)\|_{[a,b-h]}\\&\quad\leq \omega(u_1) + \omega(u_2),\end{aligned}$$

which yields inequality (1.3). It follows from (1.3) that if $0 \leq u_1 \leq u_2$, then

$$\omega(u_2) = \omega(u_2 - u_1 + u_1) \leq \omega(u_2 - u_1) + \omega(u_1),$$

i.e.,

$$\omega(u_2) - \omega(u_1) \leq \omega(u_2 - u_1). \tag{1.4}$$

Let us prove property (iii). Since the function f is uniformly continuous on $[a, b]$, we have $\omega(\Delta) \to 0$ as $\Delta \to 0$ and, hence, for any u, $u + \Delta u \in [0, b-a]$, we get

$$|\omega(u + \Delta u) - \omega(u)| \le \omega(|\Delta u|) \to 0 \quad \text{as} \quad \Delta u \to 0,$$

which implies that the function ω is continuous.

In conclusion, we make the following remarks:

Remark 1.3. Properties (i)–(iv) completely determine a modulus of continuity in the sense that any function f that possesses these properties is the modulus of continuity for a continuous function, namely for itself; thus, for such a function, we have

$$\omega(u; f, [0, b-a]) \equiv f(u).$$

Indeed, if $f(x)$, $x \in [0, b-a]$, possesses properties (i)–(iv), then, for any x and $x + h \in [0, b-a]$, $h > 0$, we have $f(x+h) \le f(x) + f(h)$, and, for any $u \in [0, b-a]$, we get

$$\begin{aligned} f(u) &= f(u) - f(0) \le \omega(u; f, [0, b-a]) \\ &\le \sup_{\substack{0 \le x \le b-a-h \\ 0 \le h \le u}} |f(x+h) - f(x)| \le \sup_{0 \le h \le u} f(h) = f(u), \end{aligned}$$

i.e.,

$$\omega(u; f, [0, b-a]) = f(u).$$

For this reason, in what follows, any function ω that possesses properties (i)–(iv) is called a modulus of continuity.

Remark 1.4. Property (iv) (the semiadditivity of ω) is not always easy to verify. Therefore, of interest is the following sufficient condition for the semiadditivity of a function $\alpha(u)$, $u > 0$:

If $\dfrac{\alpha(u)}{u}$ *is a nonincreasing function, then the function* $\alpha(u)$ *is semiadditive.*

Indeed, if $\dfrac{\alpha(u)}{u} \downarrow$, then

$$\frac{\alpha(u_2 + u_1)}{u_2 + u_1} \le \frac{\alpha(u)}{u} \quad \text{for} \quad 0 < u < u_1 + u_2,$$

whence

$$\alpha(u_2+u_1) = u_2\frac{\alpha(u_2+u_1)}{u_2+u_1} + u_1\frac{\alpha(u_2+u_1)}{u_2+u_1}$$

$$\leq u_2\frac{\alpha(u_2)}{u_2} + u_1\frac{\alpha(u_1)}{u_1} = \alpha(u_2)+\alpha(u_1),$$

which was to be proved.

The remarks presented above enable one to obtain the following important examples of a modulus of continuity:

Example 1.5. All functions of the form $M u^{\alpha}$ $(u \geq 0)$, where $M = \text{const} \geq 0$ and $0 < \alpha \leq 1$, are moduli of continuity.

Example 1.6. For $0 < \alpha \leq 1$, the function

$$\omega(u) = \begin{cases} 0 & \text{for } u = 0, \\ u^{\alpha} \ln\dfrac{1}{u} & \text{for } u \in (0, e^{-1/\alpha}), \\ \dfrac{1}{\alpha e} & \text{for } u \geq e^{-1/\alpha} \end{cases}$$

is a modulus of continuity.

Example 1.7. For $0 \leq \alpha < 1$, the function

$$\omega(u) = \begin{cases} 0 & \text{for } u = 0, \\ \dfrac{u^{\alpha}}{\ln\dfrac{1}{u}} & \text{for } u \in \left(0, e^{-1/(1-\alpha)}\right), \\ (1-\alpha)e^{-\alpha/(1-\alpha)} & \text{for } u \geq e^{-1/(1-\alpha)} \end{cases}$$

is a modulus of continuity.

The modulus of continuity also possesses the following properties, which are consequences of properties (i)–(iv):

Property (v). For any natural n and $nu \in [0, b-a]$, one has

$$\omega(nu) \le n\omega(u), \tag{1.5}$$

and, for any $\lambda > 0$, $(\lambda+1)u \in [0, b-a]$, one has

$$\omega(\lambda u) \le [\lambda+1]\omega(u) \le (\lambda+1)\omega(u). \tag{1.6}$$

Indeed, for $n = 1$, inequality (1.5) is trivial. Assuming that it is true for some $k \ge 1$, by virtue of property (iv) we get

$$\begin{aligned}\omega((k+1)u) &= \omega(ku+u)\\ &\le \omega(ku)+\omega(u) \le k\omega(u)+\omega(u) = (k+1)\omega(u).\end{aligned}$$

In turn, this implies that inequality (1.5) holds for all natural n.

If λ is an arbitrary positive number, then, according to property (ii) and inequality (1.5), we have

$$\omega(\lambda u) \le \omega([\lambda+1]u) \le [\lambda+1]\omega(u) \le (\lambda+1)\omega(u).$$

Note that the inequality $\omega(\lambda u) \le [\lambda+1]\omega(u)$, i.e., the first part of inequality (1.6), is exact in the sense that the multiplier $[\lambda+1]$ cannot be decreased. To this end, we show that, for any noninteger $\lambda > 0$, there exist moduli of continuity $\omega(u)$ for which this inequality turns into an equality for a certain u.

For definiteness, we restrict ourselves to the case $1 < \lambda < 2$. We set (see Fig. 6)

$$f(x) = \begin{cases} \dfrac{1}{\lambda-1}x & \text{for } 0 \le x \le \lambda-1,\\[2mm] 1 & \text{for } \lambda-1 \le x \le 1,\\[2mm] 1+\dfrac{1}{\lambda-1}(x-1) & \text{for } 1 \le x \le \lambda,\\[2mm] 2 & \text{for } x \ge \lambda. \end{cases}$$

Then, obviously, $\omega(u;f) = f(u)$, and, consequently, we obtain the following relation for $u = 1$:

$$\omega(\lambda u;f) = \omega(\lambda;f) = f(\lambda) = 2 = [1+\lambda]f(\lambda) = [1+\lambda]\omega(u;f).$$

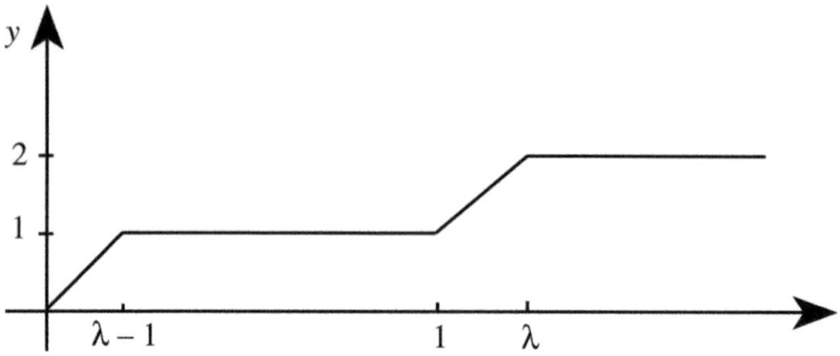

Fig. 6

Property (vi). For any ω and all $u \in [0, b-a]$, we have

$$\omega(u) \geq \frac{\omega(b-a)}{2(b-a)} u . \tag{1.7}$$

Indeed, for any $u \in [0, b-a]$, we get

$$\omega(b-a) = \omega\left(u\frac{b-a}{u}\right) \leq \left(\frac{b-a}{u}+1\right)\omega(u)$$

$$= \frac{b-a}{u}\left(1+\frac{u}{b-a}\right)\omega(u) \leq \frac{b-a}{u} 2\omega(u),$$

which yields inequality (1.7).

As is shown by the example of the function

$$f(x) = \begin{cases} \frac{1}{\varepsilon} x, & x \in [0, \varepsilon], \\ 1, & x \in [\varepsilon, 1-\varepsilon], \\ 1+\frac{1}{\varepsilon}(x-1+\varepsilon), & x \in [1-\varepsilon, 1], \end{cases}$$

the constant $\dfrac{\omega(b-a)}{2(b-a)}$ on the right hand side of (1.7) cannot be increased.

1.3. Concave majorant

Definition 1.2. *We say that a function* α *defined on a segment* $[a, b]$ *is concave if the relation*

$$\alpha(\mu t_1 + (1-\mu)t_2) \geq \mu\alpha(t_1) + (1-\mu)\alpha(t_2) \tag{1.8}$$

holds for any $t_1, t_2 \in [a, b]$ *and* $\mu \in (0, 1)$.

Definition 1.3. *Let* ω *be an arbitrary modulus of continuity defined on a segment* $[0, a]$. *The function* ω_* *defined by the formula*

$$\omega_*(t) = \sup_{0\le x<t<y\le a} \frac{(t-x)\omega(y) + (y-t)\omega(x)}{y-x} \tag{1.9}$$

is called the least concave majorant of the modulus of continuity ω.

It is obvious that the graph of this function is a curve that bounds from above the smallest convex figure containing the curvilinear trapezoid bounded by the curve $y = \omega(t)$ from above, by the abscissa axis from below, and by the straight line $y = a$ from the right. It possesses the following property:

Property (vii). The function ω_* possesses the following properties:

(a) the function ω_* is a modulus of continuity, so that

$$\omega(0) = 0, \qquad \omega_* \uparrow, \qquad \omega_* \in \boldsymbol{C}[0, a],$$

and

$$\omega_*(t_1+t_2) \leq \omega_*(t_1) + \omega_*(t_2), \qquad t_1, t_2 > 0, \quad t_1+t_2 \leq a;$$

(b) the modulus of continuity ω_* is concave, and, therefore,

$$\begin{gathered}\omega_*(\mu t_1 + (1-\mu)t_2) \geq \mu\omega_*(t_1) + (1-\mu)\omega_*(t_2),\\ \omega_*(\lambda t) \leq \lambda\omega_*(t) \quad (\lambda > 1,\ 0 < t_1, t_2 < a,\ \mu \in (0, 1));\end{gathered} \tag{1.10}$$

(c) for any $t \in [0, a]$, the following inequalities established by Stechkin (see [Efimov (1961), p. 78]) are true:

$$\omega(t) \leq \omega_*(t) \leq 2\omega(t). \tag{1.11}$$

The first inequality in (1.11) is obvious. The second inequality follows from the fact that, by virtue of (1.9), for any $0 < x < t < y \leq a$ we have

$$\omega_*(t) \le \sup \frac{1}{y-x}\left((t-x)\left(\frac{y}{t}+1\right) + (y-t) \right)\omega(t)$$

$$\le \sup \omega(t)\left(1 + \frac{y(t-x)}{t(y-x)} \right) = \sup \omega(t)\left(1 + \frac{1-\frac{x}{t}}{1-\frac{x}{y}} \right) \le 2\omega(t),$$

where the least upper bounds are determined under the condition that $0 \le x < t < y \le a$.

2. Classes of functions defined by the first modulus of continuity

2.1. Hölder (Lipschitz) classes and spaces

Using the notion of modulus of continuity, we introduce several classes of functions. Let $J \subset [a, b]$ be a closed interval.

Definition 2.1. *For any fixed* $\alpha \in (0, 1]$, *the Hölder (or Lipschitz) space of order* α *is the set of all functions* $f \in C(J) = C([a, b])$ *whose moduli of continuity satisfy the condition*

$$\omega(u; f; J) \le M u^{\alpha}, \tag{2.1}$$

where M *is a positive constant that does not depend on* u *and, generally speaking, is different for different functions. This space is denoted by* $H^{\alpha}(J)$ *or* $\mathrm{Lip}\,\alpha(J)$.

By $MH[\alpha, J]$ or $M\mathrm{Lip}[\alpha, J]$ (or $\mathrm{Lip}_{\alpha} M$) we denote the class of all functions from the Hölder (or Lipschitz) space of order α that satisfy condition (2.1) for the same value M. It is obvious that if $f \in MH[\alpha, J]$, then necessarily $f \in H^{\alpha}(J)$.

Conversely, if $f \in H^{\alpha}(J)$, then, for some M (possibly fairly large), we have $f \in MH[\alpha, J]$. Denote $H[\alpha, J] := 1H[\alpha, J]$.

Remark 2.1. By virtue of property (vi), no modulus of continuity $\omega(u) \not\equiv 0$ can be an infinitesimal of order higher than u as $u \to 0$. Therefore, inequality (2.1) is impossible for $\alpha > 1$. Hence, there is no sense in considering the classes $H[\alpha, J]$ for $\alpha > 1$.

The examples presented above show that, for all $0 < \alpha \le 1$, the following assertions are true:

(i) $Mx^{\alpha} \in MH[\alpha, [0, c]]$, $c = \text{const}$, and, hence, $Mx^{\alpha} \in H^{\alpha}([0, c])$;

(ii) $\dfrac{x^{\alpha}}{\ln \frac{1}{x}} \in H^{\alpha}([0, c])$;

(iii) $x^{\alpha} \ln \frac{1}{x} \overline{\in} H^{\alpha}([0, c])$ and, at the same time, for any $\varepsilon \in (0, \alpha)$ one has $x^{\alpha} \ln \frac{1}{x} \in MH[\alpha - \varepsilon, [0, e^{-1/\alpha}]]$, where $M \geq \dfrac{1}{e\varepsilon}$.

In approximation theory, the modulus of continuity plays an important role only for sufficiently small values u. Therefore, we assume that $u \in [0, 1]$. Then, for any $0 < \alpha < \beta \leq 1$, we get $u^{\alpha} \geq u^{\beta}$.

This implies that $H^{\beta}(J) \subset H^{\alpha}(J)$ for $\alpha < \beta$. In other words, the less the order α, the broader the class $H^{\alpha}(J)$.

2.2. Class Lip 1

Among all Hölder classes of order α, the most important is the class $H[1, J]$. This class is often called the Lipschitz class and is denoted by Lip 1.

Theorem 2.1. *In order that a function f belong to $MH[1, J]$, it is necessary and sufficient that this function be absolutely continuous and such that f' satisfies the following inequality almost everywhere on $[a, b]$:*

$$|f'(x)| \leq M.$$

Proof. *Necessity.* Let $f \in MH[1, J]$, i.e., $\omega(u) \leq Mu$. We prove that, in this case, the function f is absolutely continuous. Indeed, consider an arbitrary $\varepsilon > 0$ and an arbitrary set of disjoint elementary intervals (a_k, b_k) such that

$$\sum_{k=1}^{n} (b_k - a_k) < \frac{\varepsilon}{M}.$$

Since

$$\sum_{k=1}^{n} |f(b_k) - f(a_k)| \leq \sum_{k=1}^{n} M(b_k - a_k) < M\frac{\varepsilon}{M} = \varepsilon,$$

the function f is indeed absolutely continuous on $[a, b]$, and, hence, it has the almost-everywhere finite derivative f', which satisfies (at the points where it exists) the following inequality:

$$|f'(x)| = \left|\lim_{h\to 0}\frac{f(x+h) - f(x)}{h}\right| \le \lim_{h\to 0}\left|\frac{Mh}{h}\right| = M.$$

Sufficiency. Let f be an absolutely continuous function and let $|f'(x)| \le M$ a.e. We prove that, in this case, one has $f \in M[1, J]$. Indeed, since f is absolutely continuous, it is the indefinite integral of its own derivative, and, hence, for all $x \in J$ and $h > 0$ such that $x + h \in J$, we obtain

$$|f(x+h) - f(x)| = \left|\int_x^{x+h} f'(t)\,dt\right| \le \int_x^{x+h} |f'(t)|\,dt \le M\int_x^{x+h} dt = Mh.$$

This means that $\omega(u; f; J) \le Mu$, i.e., $f \in MH[1, J]$.

Corollary 2.1. *For any $M > 0$ and $J = [a, b]$, one has*

$$MW[1, J] = MH[1, J], \tag{2.2}$$

whence

$$W^1(J) = H^1(J). \tag{2.3}$$

2.3. Dini–Lipschitz condition

Definition 2.2. *We say that a continuous function f satisfies the Dini–Lipschitz condition if*

$$\omega(u; f; J) = o\left(\frac{1}{\ln(1/u)}\right) \quad as \quad u \to 0. \tag{2.4}$$

It is easy to see that if $f \in H^{\alpha}(J)$ for some $\alpha > 0$, then $\omega(u; f; J) = o\left(\frac{1}{\ln(1/u)}\right)$, i.e., every function f that belongs to a Hölder space also necessarily satisfies the Dini–Lipschitz condition. The fact that the converse statement is not true can be established by considering the function

$$f(t) = \frac{1}{\ln^2 t},$$

for which

$$\omega(u; f[0, e^{-2}]) = \frac{1}{\ln^2 u}$$

and which satisfies the Dini–Lipschitz condition. Nevertheless, this function does not belong to any space $H^{\alpha}(J)$ $(\alpha > 0)$.

2.4. Classes $H[\varphi, J]$

A natural generalization of the Hölder classes is provided by so called classes $H[\varphi, J]$.

Definition 2.3. *Let φ be a function that is a modulus of continuity and let M be a constant. Then $MH[\varphi, J]$ denotes the class of all continuous functions f for which*

$$\omega(u; f; J) \leq M\varphi(u), \tag{2.5}$$

and $H^{\varphi}(J)$ is the set of all functions each of which belongs to $MH[\varphi, J]$ for some M.

In the set of all differentiable functions, an important role is played by classes of functions defined as follows:

Definition 2.4. *For any fixed natural r, we denote by*

$$W^r H[\alpha, J], \quad MW^r H[\alpha, J], \quad W^r H[\varphi, J], \quad MW^r H[\varphi, J], \quad etc.$$

the classes of functions f that have continuous derivatives up to the order r and whose rth derivatives belong to the classes $H[\alpha, J]$, $MH[\alpha, J]$, $H[\varphi, J]$, and $MH[\alpha, J]$, respectively. Furthermore, for $r = 0$, we set

$$W^0 H[\alpha, J] = H[\alpha, J], \quad MW^0 H[\alpha, J] = MH[\alpha, J],$$

$$W^0 H[\varphi, J] = H[\varphi, J], \quad MW^0 MH[\varphi, J] = MH[\varphi, J].$$

If, in addition, the functions of a given class are 2π-periodic and we want to emphasize this fact, then we denote this class by $\tilde{H}[\alpha]$, $\tilde{H}[\varphi]$, etc.

3. Lagrange polynomials. Divided and finite differences

Let a number $m \in \mathbb{N}$ and points $x_i \in \mathbb{R}$, $i = 0, \dots, m$, be given. The points x_i are assumed to be different, i.e., $x_i \neq x_j$ if $i \neq j$. Let a function $f\colon \{x_i\}_{i=0}^m \to \mathbb{R}$ be defined at points x_i, $i = 0, \dots, m$.

3.1. Lagrange polynomials

Denote $L(x; f; x_0) := f(x_0)$.

Definition 3.1. *A Lagrange polynomial*

$$L(x; f) := L(x; f; x_0, x_1, \dots, x_m) \tag{3.1}$$

that interpolates a function f at points $x_0, x_1, \dots, x_m$ (interpolation nodes) is defined as an algebraic polynomial of at most mth order that takes the same values at these points as the function f, i.e.,

$$L(x_i; f) = f(x_i), \quad i = 0, \dots, m. \tag{3.2}$$

For example, for $m = 1$ we have

$$\begin{aligned} L(x; f; x_0, x_1) &= \frac{x - x_1}{x_0 - x_1} f(x_0) + \frac{x - x_0}{x_1 - x_0} f(x_1) \\ &= f(x_0) + \frac{f(x_1) - f(x_0)}{x_1 - x_0}(x - x_0). \end{aligned} \tag{3.3}$$

Definition 3.2. *The polynomials*

$$l_j(x) := l_j(x; x_0, \dots, x_m) := \prod_{i=0,\, i \neq j}^{m} \frac{x - x_i}{x_j - x_i}, \quad j = 0, \dots, m, \tag{3.4}$$

are called the fundamental Lagrange polynomials.

We set $p(x) := (x - x_0)(x - x_1) \dots (x - x_m)$ and note that

$$p'(x_j) = \lim_{x \to x_j} \frac{p(x)}{(x - x_j)} = \lim_{x \to x_j} \prod_{i=0,\, i \neq j}^{m} (x - x_i) = \prod_{i=0,\, i \neq j}^{m} (x_j - x_i).$$

Therefore, for any $j = 0, \dots, m,$ the fundamental Lagrange polynomials are representable in the form

$$l_j(x\,; x_0, x_1, \dots, x_m) = \frac{p(x)}{(x - x_j)p'(x_j)}, \quad x \neq x_j.$$

Let $\delta_{i,j}$ denote the Kronecker symbol, which is equal to 1 for $i = j$ and to 0 otherwise.

It follows from the obvious equality $l_j(x_i) = \delta_{i,j},\ i, j = 0, \dots, m,$ that the Lagrange polynomial exists and is representable by the relation

$$L(x\,; f; x_0, x_1, \dots, x_m) = \sum_{j=0}^{m} f(x_j)\, l_j(x\,; x_0, x_1, \dots, x_m). \tag{3.5}$$

By using the main theorem of algebra on the number of zeros of an algebraic polynomial, we can prove the uniqueness of the Lagrange polynomial and the validity of the following identity for any algebraic polynomial P_m of degree $\leq m$:

$$L(x\,; P_m\,; x_0, x_1, \dots, x_m) \equiv P_m(x). \tag{3.6}$$

Also note that, considered as an operator, the Lagrange polynomial is linear. Indeed, if we have another function g defined at points x_i along with the function $f,$ then

$$\begin{aligned} L(x\,; af + bg) &= \sum_{j=0}^{m} l_j(x)\big(af(x_j) + bg(x_j)\big) \\ &= a \sum_{j=0}^{m} l_j(x) f(x_j) + a \sum_{j=0}^{m} l_j(x) g(x_j) \\ &= aL(x\,; f) + bL(x\,; g), \qquad a, b = \text{const.} \end{aligned} \tag{3.7}$$

3.2. Divided differences

Let us divide the difference

$$f(x) - L(x\,; f; x_0, \dots, x_{m-1})$$

by the product $(x-x_0)\dots(x-x_{m-1})$. Using (3.4) and (3.5), we represent the quotient at the point $x=x_m$ as follows:

$$\frac{f(x_m)-L(x_m;f;x_0,\dots,x_{m-1})}{\prod_{j=0}^{m-1}(x_m-x_j)} = \sum_{j=0}^{m}\frac{f(x_j)}{\prod_{i=0,\,i\neq j}^{m}(x_j-x_i)}$$

$$=: [x_0,x_1,\dots,x_m;f]. \tag{3.8}$$

Definition 3.3. *The expression $[x_0,x_1,\dots,x_m;f]$ is called the divided difference of order m for the function f at the points $x_0,x_1,\dots,x_m$ (nodes of the divided difference).*

For example,

$$[x_0,x_1;f] = \frac{f(x_0)}{x_0-x_1} + \frac{f(x_1)}{x_1-x_0} = \frac{f(x_0)-f(x_1)}{x_0-x_1}, \tag{3.9}$$

$$[x_0,x_1,x_2;f] = \frac{f(x_0)}{(x_0-x_1)(x_0-x_2)} + \frac{f(x_1)}{(x_1-x_0)(x_1-x_2)} + \frac{f(x_2)}{(x_2-x_0)(x_2-x_1)}.$$

Let

$$[x_0;f] := f(x_0). \tag{3.10}$$

Note that the divided difference is symmetric with respect to the points x_i, i.e., the value of the divided difference does not change if the points x_i on the right hand side of (3.8) are interchanged. For example, for divided differences of order 1 we obtain the equality $[x_0,x_1;f]=[x_1,x_0;f]$, for divided differences of order 2 we get

$$[x_0,x_1,x_2;f] = [x_0,x_2,x_1;f] = [x_1,x_0,x_2;f]$$

$$= [x_1,x_2,x_0;f] = [x_2,x_0,x_1;f] = [x_2,x_1,x_0;f],$$

and so on.

Theorem 3.1. *The Lagrange polynomial $L(x;f;x_0,\dots,x_m)$ can be represented by the following Newton formula:*

$$L(x;f;x_0,x_1,\dots,x_m)$$

$$= [x_0;f] + [x_0,x_1;f](x-x_0)$$

$$+\dots+[x_0,x_1,\dots,x_m;f](x-x_0)(x-x_1)\dots(x-x_{m-1}). \qquad (3.11)$$

Proof. For $m=1$, formula (3.11) follows from (3.3), (3.9), and (3.10). Assume that (3.11) is true for a number $m-1$. By induction, let us prove that this formula is true for the number m, i.e.,

$$L(x;f;x_0,\dots,x_m) = L(x;f;x_0,\dots,x_{m-1}) + [x_0,\dots,x_m;f](x-x_0)\dots(x-x_{m-1}).$$

Since both parts of this equality are polynomials of degree $\le m$, it suffices to prove that this equality holds at all points x_i, $i = 0,\dots,m$. By the definition of Lagrange polynomial (Definition 3.1), for all $i = 0,\dots,m-1$ we have

$$L(x_i;f;x_0,\dots,x_{m-1}) + [x_0,\dots,x_m;f](x_i-x_0)\dots(x_i-x_{m-1})$$

$$= f(x_i)+0 = L(x_i;f;x_0,\dots,x_m);$$

for $i=m$, according to (3.8) we obtain

$$L(x_m;f;x_0,\dots,x_{m-1}) + [x_0,\dots,x_m;f]\dots(x_m-x_0)\dots(x_m-x_{m-1})$$

$$= f(x_m) = L(x_m;f;x_0,\dots,x_m).$$

■

Corollary 3.1. *If $x_i\in[a,b]$, $i=0,\dots,m$, and the function f has the mth derivative on (a,b), then there exists a point $\theta\in(a,b)$ such that*

$$[x_0,\dots,x_m;f] = \frac{1}{m!}f^{(m)}(\theta). \qquad (3.12)$$

Indeed, the function $f(x)-L(x;f;x_0,\dots,x_m)$ vanishes at at least $m+1$ points x_i. Therefore, according to the Rolle theorem, there exists a point $\theta\in(a,b)$ at which

$$f^{(m)}(\theta) - L^{(m)}(\theta;f;x_0,\dots,x_m) = 0.$$

On the other hand, according to (3.11), we have

$$L^{(m)}(\theta;f;x_0,\dots,x_m)\equiv m![x_0,\dots,x_m;f].$$

Hence, relation (3.12) is true.

It will be shown in Section 8 that if $f^{(m)}(x_0)$ exists, then

$$\lim_{x_i\to x_0,\, i=1,\dots,m}[x_0,\dots,x_m;f] = \frac{1}{m!}f^{(m)}(x_0),$$

i.e., the divided difference of order m is approximately equal to the coefficient of $(x-x_0)^m$ in the Taylor formula for the function f.

Corollary 3.2. *If* P_{m-1} *is a polynomial of degree* $\le m-1$, *then*

$$[x_0,\dots,x_m;P_{m-1}] = 0. \tag{3.13}$$

If $f(x)=x^m$, *then*

$$[x_0,\dots,x_m;f] = 1. \tag{3.14}$$

Note that relation (3.13) can also be derived directly from (3.8) and (3.6), and relation (3.14) follows from (3.8) and the obvious identity

$$P_m(x) - L(x;P_m;x_0,\dots,x_{m-1}) = a_0(x-x_0)\dots(x-x_{m-1}), \tag{3.15}$$

where $P_m(x)=a_0x^m+\dots+a_m$ is a polynomial of degree m.

Lemma 3.1. *The following identity is true:*

$$(x_0-x_m)[x_0,\dots,x_m;f] = [x_0,\dots,x_{m-1};f] - [x_1,\dots,x_m;f]. \tag{3.16}$$

Proof. Let $L(x):=L(x;f;x_0,\dots,x_m)$. It follows from (3.11) that

$$L^{(m-1)}(x) = [x_0,\dots,x_{m-1};f](m-1)!$$

$$+ [x_0,\dots,x_m;f](m!x-(m-1)!)(x_0+\dots+x_{m-1})).$$

Interchanging the points x_0 and x_m in (3.11), we get

$$L^{(m-1)}(x) = [x_m, x_1, \dots, x_{m-1}](m-1)!$$

$$+ [x_m, x_1, \dots, x_{m-1}, x_0](m!x - (m-1)!(x_m + x_1 + \dots + x_{m-1}))$$

$$= [x_1, \dots, x_m](m-1)! + [x_0, \dots, x_m](m!x - (m-1)!(x_1 + \dots + x_m)).$$

Subtracting the obtained equalities, we get (3.16).

■

Corollary 3.3. *Assume that a function f is defined not only at points x_i but also at $p+1$ points y_i, $i = 0, \dots, p$, $p \le m$, and, moreover, all $m+p+2$ points x_i and y_i are different. Then*

$$[x_0, \dots, x_m; f] - [y_0, \dots, y_p, x_{p+1}, \dots, x_m; f]$$

$$= \sum_{i=0}^{p} [x_i, \dots, x_m, y_0, \dots, y_i; f](x_i - y_i). \tag{3.17}$$

Note that identity (3.16) is often used as a definition of divided difference.

Let $x_0, x_1 \in [a, b]$ and let a function f be absolutely continuous on $[a, b]$. Then, according to the Lebesgue theorem, we have

$$f(x_1) - f(x_0) = \int_{x_0}^{x_1} f'(t)\,dt,$$

Performing the change of variables $t = x_0 + (x_1 - x_0)t_1$, we obtain

$$[x_0, x_1; f] = \frac{1}{x_1 - x_0} \int_{x_0}^{x_1} f'(t)\,dt = \int_0^1 f'(x_0 + (x_1 - x_0)t_1)\,dt_1. \tag{3.18}$$

A similar representation is true for any m by virtue of the following theorem:

Theorem 3.2. *Let $x_i \in [a, b]$ for all $i = 0, \dots, m$. If the function f has the absolute continuous $(m-1)$th derivative on $[a, b]$, then*

$$[x_0, x_1, \dots, x_m; f]$$

$$= \int_0^1 \int_0^{t_1} \dots \int_0^{t_{m-1}} f^{(m)}(x_0 + (x_1 - x_0)t_1 + \dots + (x_m - x_{m-1})t_m)\,dt_m \dots dt_1. \tag{3.19}$$

Proof. Assume that representation (3.19) is true for a number $m-1$. By induction, let us prove that (3.19) is also true for the number m. Denote $t_0 := 1$. According to relation (3.16) and the induction hypothesis, we have

$$(x_m - x_{m-1})[x_0, \dots, x_m; f] = [x_0, \dots, x_{m-2}, x_m; f] - [x_0, \dots, x_{m-1}; f]$$

$$= \int_0^{t_0} \dots \int_0^{t_{m-2}} \left(\int_u^v f^{(m)}(t)dt \right) dt_{m-1} \dots dt_1,$$

where

$$v = x_0 + \dots + (x_{m-2} - x_{m-3})t_{m-2} + (x_m - x_{m-2})t_{m-1},$$

$$u = x_0 + \dots + (x_{m-1} - x_{m-2})t_{m-1}.$$

It remains to introduce a new integration variable t_m instead of t in the last integral by using the change of variables

$$t = x_0 + (x_1 - x_0)t_1 + \dots + (x_{m-1} - x_{m-2})t_{m-1} + (x_m - x_{m-1})t_m$$

and then note that this linear change of variables transforms the segment $[0, t_m]$ into the segment that connects the points u and v.

■

Corollary 3.4. *If, under the conditions of Theorem 3.2, one has* $\left|f^{(m)}(x)\right| \le 1$ *for almost all* $x \in [a, b]$, *then*

$$|[x_0, \dots, x_m; f]| \le \frac{1}{m!}, \tag{3.20}$$

$$\left|f(x) - L(x; f; x_0, \dots, x_{m-1})\right| \le \left|\frac{1}{m!}(x - x_0)\dots(x - x_{m-1})\right|. \tag{3.21}$$

For what follows, we need the following relations for the divided differences of functions f and g:

$$[x_0, \dots, x_m; af + bg] = a[x_0, \dots, x_m; f] + b[x_0, \dots, x_m; g], \tag{3.22}$$

$$a, b = \text{const};$$

$$[x_0, \dots, x_m; fg] = \sum_{i=0}^{m} [x_0, \dots, x_i; f][x_i, \dots, x_m; g]; \tag{3.23}$$

if $g(x) = (x - x_i) \dots (x - x_m)$, $i = 1, \dots, m$, then

$$[x_0, \dots, x_m; fg] = [x_0, \dots, x_{i-1}; f]; \tag{3.24}$$

if $g(x) = (x - x_0) \dots (x - x_m)$, then

$$[x_0, \dots, x_m; fg] = 0; \tag{3.24'}$$

and

$$[x_0, \dots, x_m; f] = [x_i, \dots, x_m; f_i], \tag{3.25}$$

where

$$f_i(x) = [x_0, \dots, x_{i-1}, x; f].$$

Relation (3.22), which means that the divided difference is linear, and relations (3.24) and (3.24′) follow directly from (3.8). Relations (3.23) and (3.25) can easily be proved by induction with the use of (3.16).

Simple corollaries of (3.25), (3.19), and (3.18) are identifies (3.25′) and (3.25″) presented below. Let $j \in \mathbb{N}$, $j \le m$, and let $x_i \in [a, b]$ for all $i = 0, \dots, m$. If a function f is j times continuously differentiable on $[a, b]$ or f has the $(j-1)$th absolutely continuous derivative on $[a, b]$, then

$$[x_0, \dots, x_m; f] = \int_0^1 \dots \int_0^{t_{j-1}} [x_j, \dots, x_m; f_{t_1, \dots, t_j}] dt_j \dots dt_1, \tag{3.25'}$$

where

$$f_{t_1, \dots, t_j}(x) := f^{(j)}(x_0 + (x_1 - x_0)t_1 + \dots + (x_{j-1} - x_{j-2})t_{j-1} + (x - x_{j-1})t_j);$$

in particular, if $j = 1$, then

$$[x_0, \dots, x_m; f] = \int_0^1 [x_1, \dots, x_m; f_{t_1}] dt_1, \tag{3.25''}$$

where

$$f_{t_1}(x) := f'(x_0 + (x - x_0)t_1).$$

Below we establish one more representation [representation (3.26)], which we call the Hermite preformula.

Let q nonnegative integer numbers $p_1, p_2, \dots, p_q$ be given and let

$$\sum_{s=1}^{q} (p_s + 1) = m + 1.$$

We fix an arbitrary one-to-one correspondence between $m + 1$ numbers $i = 0, \dots, m$ and $m + 1$ pairs (s, j), where $s = 1, \dots, q$ and $j = 0, \dots, p_s$. Denote $x_{s,j} := x_i$ if $i \leftrightarrow (s, j)$. We set

$$A_s := A_s(x) := \prod_{\nu=1, \nu \neq s}^{q} \prod_{j=0}^{p_\nu} (x - x_{\nu, j})^{-1};$$

$$f_s(x) := f(x) A_s(x).$$

Using (3.16), we establish by induction that

$$[x_0, \dots, x_m; f] = \sum_{s=1}^{q} [x_{s,0}, \dots, x_{s,p_s}; f_s].$$

Taking (3.23) into account, we obtain the representation

$$[x_0, \dots, x_m; f] = \sum_{s=1}^{q} \sum_{j=1}^{p_s} [x_{s,0}, \dots, x_{s,j}; A_s]\,[x_{s,j}, \dots, x_{s,p_s}; f]. \qquad (3.26)$$

Finally, we generalize the trivial identity

$$[x_0, x_N; f] = \sum_{n=0}^{N-1} (x_{n+1} - x_n)[x_0, x_{n+1}; f] \frac{1}{x_N - x_0}, \qquad N \in \mathbb{N},$$

to the case of divided differences of order $m \geq 2$.

We fix $N + 1$ points $y_0, y_1, \dots, y_N$, $N \geq m \geq 2$, among which $m + 1$ points coincide with the points $x_0, x_1, \dots, x_m$, i.e., $y_{n_0} = x_0, \dots, y_{n_m} = x_m$. Then we obtain the Popoviciu identity [Popoviciu (1934)] (see also [Tamrazov (1975)], [De Boor (1976)], and [Ciesielski (1979)]):

$$[x_0, \ldots, x_m; f] = \sum_{n=0}^{N-m} (y_{n+m} - y_n)[y_n, \ldots, y_{n+m}; f][x_0, \ldots, x_m; p_{n,m}], \tag{3.27}$$

where

$$p_{n,m} := p_{n,m}(x_s) := \prod_{j=1}^{m-1} (x_s - y_{n+j})_+,$$

$(x_s - y_{n+j})_+ := x_s - y_{n+j}$ if $n_s \geq n + j$, and $(x_s - y_{n+j})_+ := 0$ if $n_s < n + j$.

Indeed, for $m = 2$, identity (3.27) can be proved by direct verification. By induction, we assume that (3.27) is true for a number $m - 1$. Denote

$$F_m(x_s) := \sum_{n=0}^{N-m} (y_{n+m} - y_n)[y_n, \ldots, y_{n+m}; f]\, p_{n,m}(x_s).$$

It follows from relation (3.16) and the equality

$$p_{N+m-1,m}(x_s) = 0$$

that

$$F_m(x_s) = -[y_0, \ldots, y_{m-1}; f]p_{0,m}(x_s)$$

$$+ \sum_{n=1}^{N-m+1} [y_n, \ldots, y_{n+m-1}; f](p_{n-1,m}(x_s) - p_{n,m}(x_s)).$$

For all $n = 1, \ldots, N - m + 1$ and $s = 0, \ldots, m$, we have

$$p_{n-1,m}(x_s) - p_{n,m}(x_s) = (y_{n+m-1} - y_n)p_{n,m-1}(x_s).$$

Therefore,

$$F_m(x_s) = F_{m-1}(x_s) - [y_0, \ldots, y_{m-1}; f](p_{0,m}(x_s) + (y_{m-1} - y_0)p_{0,m-1}(x_s))$$

$$= F_{m-1}(x_s) - [y_0, \ldots, y_{m-1}; f](x_s - y_0)(x_s - y_1) \cdot \ldots \cdot (x_s - y_{m-2}).$$

It follows from the induction hypothesis and the equality

$$F_m(x_s) = F_{m-1}(x_s) + P_{m-1}(x_s),$$

where P_{m-1} is an algebraic polynomial of degree $\leq m-1$, that

$$[x_0, \dots, x_m; f] = \frac{[x_0, \dots, x_{m-1}; f] - [x_1, \dots, x_m; f]}{x_0 - x_m}$$

$$= \frac{[x_0, \dots, x_{m-1}; F_{m-1}] - [x_1, \dots, x_m; F_{m-1}]}{x_0 - x_m}$$

$$= [x_0, \dots, x_m; F_{m-1}] = [x_0, \dots, x_m; F_m - P_{m-1}]$$

$$= [x_0, \dots, x_m; F_m].$$

■

By setting $f(x) = x^m$ in (3.27), we obtain

$$\sum_{n=0}^{N-m} (y_{n+m} - y_n)[x_0, \dots, x_m; p_{n,m}] = 1.$$

Let $x_0 = y_0$ and $x_m = y_N$ in (3.27) and let the points x_i and y_j be enumerated in the increasing order, i.e., $x_0 < x_1 < \dots < x_m$ and $y_0 < y_1 < \dots < y_N$ (recall that each point x_i must coincide with a certain point y_j). Then

$$[x_0, \dots, x_m; p_{n,m}] > 0,$$

and, therefore, the following representation holds in this case:

$$[x_0, \dots, x_m; f] = \sum_{n=0}^{N-m} \alpha_n [y_n, \dots, y_{n+m}; f],$$

where

$$\sum_{n=0}^{N-m} \alpha_n = 1$$

and $\alpha_n > 0$ for all $n = 0, \dots, N-m$.

Remark 3.1. It is obvious that

$$f(x) - L(x;f;x_0,\dots,x_{m-1})$$

$$= \begin{vmatrix} 1 & x_0 & \dots & x_0^{m-1} & f(x_0) \\ \dots & \dots & \dots & \dots & \dots \\ 1 & x_{m-1} & \dots & x_{m-1}^{m-1} & f(x_{m-1}) \\ 1 & x & \dots & x^{m-1} & f(x) \end{vmatrix} \Bigg/ \begin{vmatrix} 1 & x_0 & \dots & x_0^{m-1} \\ \dots & \dots & \dots & \dots \\ 1 & x_m & \dots & x_m^{m-1} \end{vmatrix}.$$

Hence, by virtue of (3.8) and (3.15), we get

$$[x_0,\dots,x_m;f] = \begin{vmatrix} 1 & x_0 & \dots & x_0^{m-1} & f(x_0) \\ \dots & \dots & \dots & \dots & \dots \\ 1 & x_m & \dots & x_m^{m-1} & f(x_m) \end{vmatrix} \Bigg/ \begin{vmatrix} 1 & x_0 & \dots & x_0^{m} \\ \dots & \dots & \dots & \dots \\ 1 & x_m & \dots & x_m^{m} \end{vmatrix}.$$

Wronicz (1984) established several properties, which are analogous to properties of divided differences, for the expression

$$[x_0,\dots,x_m;f]_\Phi = \begin{vmatrix} \varphi_0(x_0) & \dots & \varphi_{m-1}(x_0) & f(x_0) \\ \dots & \dots & \dots & \dots \\ \varphi_0(x_m) & \dots & \varphi_{m-1}(x_m) & f(x_m) \end{vmatrix} \Bigg/ \begin{vmatrix} \varphi_0(x_0) & \dots & \varphi_m(x_0) \\ \dots & \dots & \dots \\ \varphi_0(x_m) & \dots & \varphi_m(x_m) \end{vmatrix},$$

where $\Phi = \{\varphi_0(x),\dots,\varphi_m(x)\}$ is a Chebyshev system of functions.

3.3. Finite differences

In this subsection, we assume that the points x_i are equidistant, i.e., for all $i = 0,\dots,m$, we have

$$x_i = x_0 + ih, \quad h \in \mathbb{R}, \quad h \neq 0.$$

For the Lagrange interpolation polynomial

$$L(x) = L(x;f;x_0,\dots,x_{m-1}) = \sum_{j=0}^{m-1} f(x_j)\, l_j(x, x_0,\dots,x_{m-1}),$$

we determine the values of the fundamental Lagrange polynomials l_j at the point $x = x_m$. According to (3.4), we have

$$l_j(x_m, x_0, \dots, x_{m-1}) = \prod_{i=0,\, i\neq j}^{m-1} \frac{(x_m - x_i)}{(x_j - x_i)} = -(-1)^{m-j}\binom{m}{j}.$$

We represent the difference $f(x_m) - L(x_m)$ in the form

$$f(x_m) - L(x_m) = \sum_{j=0}^{m} (-1)^{m-j}\binom{m}{j} f(x_0 + jh). \tag{3.28}$$

Definition 3.4. *The expression*

$$\Delta_h^m(f; x_0) := \sum_{j=0}^{m} (-1)^{m-j}\binom{m}{j} f(x_0 + jh) \tag{3.29}$$

is called the m*th difference of the function* f *at the point* x_0 *with step* h.
For example,

$$\Delta_h^1(f; x_0) = -f(x_0) + f(x_0 + h),$$

$$\Delta_h^2(f; x_0) = f(x_0) - 2f(x_0 + h) + f(x_0 + 2h),$$

$$\Delta_h^3(f; x_0) = -f(x_0) + 3f(x_0 + h) - 3f(x_0 + 2h) + f(x_0 + 3h), \dots .$$

Denote $\Delta_h^0(f; x_0) = f(x_0)$ and $\Delta_0^m(f; x_0) := 0$.
The identities

$$\Delta_h^m(f; x_0) = \sum_{i=0}^{m-j} (-1)^{m-j-i}\binom{m-j}{i} \Delta_h^j(f; x_0 + ih), \quad j = 0, \dots, m-1; \tag{3.30}$$

$$\Delta_{hn}^m(f; x_0) = \sum_{i_1=0}^{n-1} \dots \sum_{i_m=0}^{n-1} \Delta_h^m(f; x_0 + h(i_1 + \dots + i_m)), \quad n \in \mathbb{N}, \tag{3.31}$$

can easily be proved by induction.

In what follows, for definiteness, we assume that $h > 0$. Relations (3.28) and (3.8) yield

$$\Delta_h^m(f; x_0) = h^m m!\,[x_0, x_0 + h, \dots, x_0 + mh; f]. \tag{3.32}$$

Hence, the mth differences inherit the properties of divided differences. In particular, the following assertions are true:

(i) the Newton formula (3.11) takes the form

$$L(x;f;x_0,\dots,x_m) = \Delta_h^0(f;x_0) + \sum_{j=1}^{m} \frac{\Delta_h^j(f;x_0)}{j!h^j} \prod_{i=0}^{j-1}(x-x_0-ih)\,; \qquad (3.33)$$

(ii) if the function f has the mth derivative on (x_0, x_0+mh), then

$$\Delta_h^m(f;x_0) = h^m f^{(m)}(\theta), \quad \theta \in (x_0, x_0+mh); \qquad (3.34)$$

if $f(x)=x^m$, then

$$\Delta_h^m(f;x_0) = h^m m!; \qquad (3.35)$$

and if $f(x)=P_{m-1}(x)$ is a polynomial of degree $\le m-1$, then

$$\Delta_h^m(P_{m-1}\,;x_0) = 0; \qquad (3.36)$$

(iii) if the function f has the jth derivative on (x_0, x_0+mh), $j=0,\dots,m$, then

$$\Delta_h^m(f;x_0) = h^j\Delta_h^{m-j}(f^{(j)};\theta), \quad \theta \in (x_0, x_0+(m-j)h); \qquad (3.37)$$

indeed, denoting $g(x) := \Delta_h^{m-j}(f,x)$, we establish that

$$g^{(j)}(x) = \Delta_h^{m-j}(f^{(j)};x), \qquad x \in (x_0, x_0+(m-j)h),$$

and, using (3.30) and (3.34), we get

$$\Delta_h^m(f;x_0) = \Delta_h^j(g,x_0)h^j g^{(j)}(\theta) = h^j\Delta_h^{m-j}(f^{(j)};\theta);$$

(iv) the following identity is true:

$$\Delta_h^{m-1}(f;x_0+h) - \Delta_h^{m-1}(f;x_0) = \Delta_h^m(f;x_0); \qquad (3.38)$$

for $p \in \mathbb{N}$, this identity yields

$$\Delta_h^{m-1}(f;x_0+ph) - \Delta_h^{m-1}(f;x_0) = \sum_{i=0}^{p-1} \Delta_h^m(f;x_0+ih); \qquad (3.39)$$

(v) if, along with the function f, a function g is also defined at the points $x_0 + ih$, $i = 0, \dots, m$, then

$$\Delta_h^m(af + bg; x_0) = a\,\Delta_h^m(f; x_0) + b\,\Delta_h^m(g; x_0), \qquad a, b = \text{const}; \tag{3.40}$$

$$\Delta_h^m(fg; x_0) = \sum_{i=0}^{m} \binom{m}{i} \Delta_h^i(f; x_0)\, \Delta_h^{m-i}(g; x_0 + ih); \tag{3.41}$$

(vi) if $g(x) = (x - x_i) \dots (x - x_m)$, $i = 1, \dots, m$, then

$$\Delta_h^m(fg; x_0) = h^{m+1-i} \frac{m!}{(i-1)!} \Delta_h^{i-1}(f; x_0); \tag{3.42}$$

in particular, if $g(x) = x - x_0$, then

$$\Delta_h^m(fg; x_0) = mh\Delta_h^{m-1}(f; x_0 + h); \tag{3.43}$$

(vii) if the function f has the $(m-1)$th absolute continuous derivative $f^{(m-1)}(x)$ on $[x_0, x_0 + mh]$, then

$$\Delta_h^m(f; x_0) = h^m m! \int_0^1 \int_0^{t_1} \dots \int_0^{t_{m-1}} f^{(m)}(x_0 + h(t_1 + \dots + t_m))\, dt_m \dots dt_1; \tag{3.44}$$

note one more useful representation, namely,

$$\Delta_h^m(f; x_0) = \int_0^h \dots \int_0^h f^{(m)}(x_0 + t_1 + \dots + t_m) dt_m \dots dt_1, \tag{3.45}$$

which can easily be proved by induction with the use of (3.38).

Note that identity (3.31) is implied by (3.26).

Lemma 3.2. *If a function f continuous on a segment $[a, b]$ has the nonincreasing derivative $f^{(m)}(x) \geq 0$ on the half-open interval $(a, b]$, then, for $h \leq t \leq (b-a)/m$, the following estimate holds for any $x_0 \in [a, b - mh]$:*

$$\Delta_h^m(f; x_0) \leq \Delta_t^m(f; a). \tag{3.46}$$

Proof. By assumption, for any positive $\varepsilon < b-a$ the derivative $f^{(m)}(x)$ is bounded on $[a+\varepsilon, b]$. Hence, the $(m-1)$th derivative $f^{(m-1)}(x)$ is absolutely continuous on $[a+\varepsilon, b]$. Assuming that $x_0 \neq a$, $t \neq (b-a)/m$, and $\varepsilon < \min\{x_0 - a, b - mt\}$ and using (3.44) and (3.45), we obtain

$$\Delta_h^m(f; x_0) \le \Delta_h^m(f; a+\varepsilon) \le \Delta_t^m(f; a+\varepsilon).$$

It remains to use the fact that ε is arbitrary and the mth difference $\Delta_h^m(f; x)$ is continuous in x and h.

■

For $x \in [x_0 - h, x_0 + mh]$, we often use the inequality

$$\left|L(x; f; x_0, x_1, \ldots, x_{m-1})\right| \le (2^m - 1) \max_{j=0,\ldots,m-1} \left|f(x_j)\right|, \tag{3.47}$$

which follows from the estimate

$$\left|l_j(x; x_0, \ldots, x_{m-1})\right| \le \left|l_j(x_m; x_0, \ldots, x_m)\right| = \binom{m}{j},$$

and the inequality

$$\left|L^{(p)}(x; f; x_0, \ldots, x_{m-1})\right| \le ch^{-p} \max_{j=0,\ldots,m-1} \left|f(x_j)\right|, \quad p \in \mathbb{N}, \tag{3.48}$$

where $c = c(m) = \text{const}$.

4. Moduli of continuity of higher order

4.1. Modulus of continuity of order k

In what follows, we assume that $k \in \mathbb{N}$.

Definition 4.1. *The modulus of continuity of order k (the kth modulus of continuity) of a function $f \in C([a, b])$ is defined as follows:*

$$\omega_k(t) := \omega_k(t; f; [a, b]) := \sup_{h \in [0,t]} \|\Delta_h^k(f; \cdot)\|_{[a, b-kh]} \quad \textit{for} \quad t \in [0, (b-a)/k],$$

$$\omega_k(t) := \omega_k((b-a)/k) \quad \textit{for} \quad t > (b-a)/k.$$

Remark 4.1. The modulus of continuity of order k, $k \geq 2$, is often called the modulus of smoothness of order k.

For example, if a function f has the kth derivative on (a, b) and $\left|f^{(k)}(x)\right| \leq 1$ for all $x \in (a, b)$, then, according to (3.34), we have

$$\omega_k(t; f; [a, b]) \leq t^k; \tag{4.1}$$

furthermore, if $f(x) = P_{k-1}(x)$ is a polynomial of degree $\leq k-1$, then

$$\omega_k(t; P_{k-1}; [a, b]) = 0, \tag{4.2}$$

if $f(x) = x^k$, then

$$\omega_k(t; f; [a, b]) = k!\min\{t^k, ((b-a)/k)^k\}, \tag{4.3}$$

and if $f(x) = \sin x$, then

$$\omega_k(t; f; [0, \pi/2]) = 2^k \sin^k \frac{t}{2} \sin kt/2, \qquad t \leq \pi/2k,$$

$$\omega_k(t; f; [0, \pi]) = 2^k \sin^k \frac{t}{2}, \qquad t \leq \pi/k. \tag{4.4}$$

Note that $\omega(t; f; [a, b]) = \omega_1(t; f; [a, b])$, i.e., the modulus of continuity and the modulus of continuity of order 1 have the same meaning. For convenience, we denote

$$\omega_0(t; f; [a, b]) = \|f\|_{[a,b]}.$$

In what follows, if it is clear what a segment $[a, b]$ is considered, then we write $\omega_k(t; f)$ instead of $\omega_k(t; f; [a, b])$; if, in addition, it is clear what a function f is considered, then we write $\omega_k(t)$ instead of $\omega_k(t; f)$.

It follows immediately from (3.40), (3.41), (3.45), and (3.46) that the kth moduli of continuity possess the following properties:

$$\omega_k(t; Af) = |A|\omega_k(t; f), \qquad A = \text{const};$$
$$\omega_k(t; f+g) \le \omega_k(t; f) + \omega_k(t; g); \tag{4.5}$$

$$\omega_k(t; fg) \le \sum_{i=0}^{k} \binom{k}{i} \omega_i(t; f)\omega_{k-i}(t; g); \tag{4.6}$$

if the function f has the $(k-1)$th absolutely continuous derivative and $\left|f^{(k)}(x)\right| \le 1$ almost everywhere on $[a, b]$, then

$$\omega_k(t) \le t^k; \tag{4.7}$$

and if the function f has the nonnegative nonincreasing kth derivative on the half-open interval $(a, b]$, then

$$\omega_k(t; f) = \Delta_t^k(f; a), \qquad t \in [0, (b-a)/k]. \tag{4.8}$$

Using (4.8), one can easily calculate the kth modulus of continuity of the functions $f_0(x) = x^\alpha$, $0 < \alpha < k$, $\alpha \in \mathbb{N}$, and $f_j(x) := x^j \ln x$, $j = 1, \dots, k$, $f_j(0) = 0$. Denote $A_j := |\Delta_1^k(f_j; 0)| = \text{const}$. Then, on the segment $[0, b]$, we obtain the following relations for $t \le b/k$:

$$\omega_k(t; f_0) = A_0 t^\alpha, \tag{4.9}$$

$$\omega_k(t; f_j) = A_j t^j, \qquad j = 1, \dots, k-1, \tag{4.10}$$

$$\omega_k(t; f_k) = A_k t^k + k!\, t^k \ln(1/t), \tag{4.11}$$

[in (4.11) b is sufficiently small].

In particular, for $f(x) = x \ln x$ we have

$$\omega_2(t; f; [0, b]) = (2 \ln 2)t, \qquad t \le \frac{b}{2}. \tag{4.10′}$$

Relations (3.30), (3.31), and (3.37) mean that

$$\omega_k(t) \le 2^{k-j}\omega_j(t), \qquad j = 1, \dots, k, \tag{4.12}$$

and, in particular,

$$\omega_k(t) \le 2^k \| f \|_{[a,b]}, \tag{4.13}$$

$$\omega_k(nt) \le n^k \omega_k(t), \qquad n \in \mathbb{N}, \tag{4.14}$$

$$\omega_k(t;f) \le t^j \omega_{k-j}(t;f^{(j)}). \tag{4.15}$$

Lemma 4.1. *The kth modulus of continuity* $\omega_k(t) = \omega_k(t; f; [a,b])$ *is equal to zero at the point* $t = 0$, *is a nondecreasing continuous function on the interval* $[0, \infty)$, *i.e.,*

$$\omega_k(0) = 0, \tag{4.16}$$

$$\omega_k(t_1) \le \omega_k(t_2), \qquad 0 \le t_1 \le t_2, \tag{4.17}$$

$$\omega_k \in C([0,\infty)),$$

and satisfies the inequality

$$t_2^{-k} \omega_k(t_2) \le 2^k t_1^{-k} \omega_k(t_1), \qquad 0 < t_1 \le t_2. \tag{4.18}$$

Proof. Relations (4.16) and (4.17) are obvious. Inequality (4.18) follows from (4.14). Indeed, we have

$$\omega_k(t_2) = \omega_k\left(\frac{t_2}{t_1} t_1\right) \le \omega_k\left(\left[\frac{t_2}{t_1} + 1\right] t_1\right) \le \left[\frac{t_2}{t_1} + 1\right]^k \omega_k(t_1) \le \left(\frac{2t_2}{t_1}\right)^k \omega_k(t_1).$$

It remains to verify the continuity of ω_k. To this end, we prove the inequality

$$\omega_k(t_2) \le \omega_k(t_1) + k\,2^{k-1} \omega_1(t_2 - t_1), \qquad 0 \le t_1 \le t_2. \tag{4.19}$$

Since the case $t_2 > (b-a)/k$ can be reduced to the case $t_2 = (b-a)/k$ by using (4.17), we assume that $t_2 \le (b-a)/k$. Let us represent an arbitrary number $h \in [0, t_2]$ as a sum of two terms $(h = h_1 + h_2)$ such that $h_1 \in [0, t_1]$ and $h_2 \in [0, t_2 - t_1]$. It follows from the equality

$$\Delta_h^k(f;x) = \Delta_{h_1}^k(f;x) + \sum_{i=0}^{k} \binom{k}{i} (-1)^{k-i} \Delta_{ih_2}^1(f, x + ih_1), \qquad x \in [a, b - kh],$$

that

$$\left|\Delta_h^k(f;x)\right| \le \omega_k(h_1) + k\,2^{k-1}\omega_1(h_2) \le \omega_k(t_1) + k\,2^{k-1}\omega_1(t_2-t_1).$$

By virtue of the arbitrariness of h, this yields (4.19). Thus, for any $t \ge 0$ and $t_0 \ge 0$, we have

$$\left|\omega_k(t) - \omega_k(t_0)\right| \le k\,2^{k-1}\,\omega_1(|t-t_0|) \to 0 \quad \text{as} \quad |t-t_0| \to 0,$$

which yields the continuity of the function ω_k.

■

It follows from inequality (4.18) that

$$\omega_k(t) \ge t^k \frac{\omega_k((b-a)/k)}{(2(b-a)/k)^k}, \qquad t \le (b-a)/k, \tag{4.20}$$

and if $\omega_k(t_0)=0$ for $t_0 \ne 0$, then $\omega_k(t) \equiv 0$.

It will be shown later that if $\omega_k(t) \equiv 0$, then $f(x) = P_{k-1}(x)$ is a polynomial of degree $\le k-1$.

If $f^{(k)} \in C[a,b]$, then, with regard for (4.1) or (4.7), we get

$$\omega_k(t) \le t^k \left\| f^{(k)} \right\|_{[a,b]}. \tag{4.21}$$

4.2. Approach of Stechkin

Definition 4.2. *A continuous function φ nondecreasing on $[0,\infty)$ is called a majorant if $\varphi(0)=0$. The set of all majorants is denoted by Φ.*

Definition 4.3. *A majorant ω is called a function of the kth-modulus-of-continuity type if $t_2^{-k}\omega(t_2) \le 2^k t_1^{-k}\omega(t_1)$ for $0 < t_1 \le t_2$.*

Definition 4.4. *A majorant φ is called a k-majorant if the function $t^{-k}\varphi(t)$ does not increase on $(0,\infty)$, i.e.,*

$$t_2^{-k}\omega(t_2) \le t_1^{-k}\varphi(t_1), \qquad 0 < t_1 \le t_2. \tag{4.22}$$

The set of all k-majorants is denoted by Φ^k.

It is clear that if $\varphi \in \Phi^k$, then φ is a function of the kth-modulus-of-continuity type. Generally speaking, the converse statement is not true. Nevertheless, Theorem 4.1 presented below is true.

Definition 4.5 (Stechkin; see [Efimov (1961)]). *For a function ω of the kth-modulus-of-continuity type, we denote*

$$\omega^*(t) := \sup_{u>t} \frac{t^k \omega(u)}{u^k}, \qquad t \ge 0. \tag{4.23}$$

Theorem 4.1 (Stechkin). *The following relations are true:*

$$\omega(t) \le \omega^*(t) \le 2^k \omega(t), \qquad \omega^* \in \Phi^k. \tag{4.24}$$

Proof. If $t = 0$, then $\omega^*(0) = \omega(0) = 0$. If $t \ne 0$, then

$$\omega(t) = \sup_{u>t} \frac{t^k \omega(t)}{u^k} \le \sup_{u>t} \frac{t^k \omega(u)}{u^k} = \omega^*(t) \le \sup_{u>t} t^k 2^k \frac{\omega(t)}{t^k} = 2^k \omega(t).$$

It is obvious that the function $t^{-k}\omega^*(t)$ does not increase on $(0, \infty)$. Let us prove that the function ω^* is monotone. Let $0 < t_1 < t_2$. Consider a point $t_3 \ge t_1$ at which we have

$$\sup \{ u^{-k} \omega(u) \mid u > t_1 \} = t_3^{-k} \omega(t_3).$$

If $t_3 \ge t_2$, then

$$\frac{\omega^*(t_1)}{\omega^*(t_2)} = \left(\frac{t_1}{t_2} \right)^k < 1.$$

For $t_3 < t_2$, we have

$$\omega^*(t_1) = \left(\frac{t_1}{t_3} \right)^k \omega(t_3) \le \omega(t_2) \le \omega^*(t_2).$$

The second inequality in (4.24) implies the continuity of the function ω^* at the point $t = 0$. Finally, the continuity of the function ω^* at the point $t > 0$ follows from the fact that, for any pair of points t_1 and t_2 such that $2t \ge t_1 > t_2 \ge t/2$, we have

$$0 \le \omega^*(t_2) - \omega^*(t_1) \le \left(\frac{t_2}{t_1}\right)^k \omega^*(t_1) - \omega^*(t_1)$$

$$= t_1^{-k}\omega^*(t_1)(t_2^k - t_1^k) \le \left(\frac{t}{2}\right)^{-k} \omega^*(2t)\left(t_2^k - t_1^k\right) \to 0 \qquad (4.25)$$

as $(t_2 - t_1) \to 0$.

■

Note that the reasoning used above means the equivalence of Definition 4.4 and the following definition:

Definition 4.4′. *A function* φ *nondecreasing on* $[0, \infty)$ *is called a k-majorant if the function* $t^{-k}\varphi(t)$ *does not increase on* $(0, \infty)$, $\varphi(0) = 0$, *and* $\varphi(t) \to 0$ *as* $t \to 0$.

4.3. Extremal function

It was noted by S. Nikol'skii (1946b) that, in order that a majorant be the modulus of continuity of a continuous function, it is necessary and sufficient that the majorant be semi-additive (see Remark 1.3).

In the case $k > 1$, a necessary and sufficient condition for a majorant to be the kth modulus of continuity of a continuous function is not found as yet.

Nevertheless, by using the approaches of Bari and Stechkin [Bari (1955), (1961); Bari and Stechkin (1956); Stechkin (1961)], Geit (1972) proved that, for any majorant $\varphi \in \Phi^k$, there exists a function $f \in C([a, b])$ such that

$$\varphi(t) \le \omega_k(t; f; [a, b]) \le c\varphi(t), \qquad (4.26)$$

$$0 \le t \le (b-a)/k, \quad c = c(k) = \text{const}.$$

In this subsection, we introduce a so-called extremal function and, using it, prove relation (4.26).

Definition 4.6 [Shevchuk (1976)]. *Let* $k \in \mathbb{N}$ *and let* φ *be a majorant. The extremal function is defined as follows:*

$$F(x) = F(x; \varphi; k) := \frac{1}{(k-2)!} \int_1^x \frac{x(x-u)^{k-2}}{u^k} \varphi(u)\, du, \quad x \ge 0, \quad \textit{if} \quad k \ne 1,$$

$$F(x) = F(x; \varphi; 1) := \varphi(x), \quad x \ge 0, \quad \textit{if} \quad k = 1. \qquad (4.27)$$

Lemma 4.2. *Let $k \in \mathbb{N}$, $\varphi \in \Phi$, $x > 0$, $h \geq 0$, and $p = 1, \ldots, k-2$. The extremal function $F(x) = F(x, \varphi, k)$ possesses the following properties:*

(i)
$$\lim_{x \to 0} F(x) = F(0) = 0, \tag{4.28}$$

i.e., $F \in C([0, \infty))$;

(ii)
$$F^{(p)}(x) = \frac{1}{(k-p-1)!} \int_1^x ((k-1)x - pu) \frac{(x-u)^{k-2-p}}{u^k} \varphi(u)\, du, \tag{4.29}$$

$$F^{(k-1)}(x) = x^{1-k}\varphi(x) + (k-1) \int_1^x u^{-k}\varphi(u)\, du, \tag{4.30}$$

and

$$F^{(k)}(x) = x^{1-k}\varphi'(x) \quad \textit{for almost all} \quad x > 0; \tag{4.31}$$

(iii)
$$\Delta_h^k(F; x) \geq 0; \tag{4.32}$$

(iv)
$$\Delta_h^k(F; 0) = k! \int_0^1 \int_0^{s_1} \ldots \int_0^{s_{k-2}} \frac{\varphi\big(h(1 + s_1 + \ldots + s_{k-1})\big)}{(1+s)^k} ds_{k-1} \ldots ds_1, \tag{4.33}$$

$$\Delta_h^k(F; 0) = kh \int_0^h \ldots \int_0^h \frac{\varphi(h + s_1 + \ldots + s_{k-1})}{(h + s_1 + \ldots + s_{k-1})^k} ds_{k-1} \ldots ds_1, \tag{4.34}$$

$$\Delta_h^k(F; 0) = k h^k \theta^{-k} \varphi(\theta), \qquad h < \theta < kh; \tag{4.35}$$

(v)
$$\varphi(h) \leq \Delta_h^k(F; 0) \leq \varphi(kh) \tag{4.36}$$

and if $\varphi \in \Phi^k$, *then*

$$\varphi(h) \leq \Delta_h^k(F; 0) \leq k\varphi(h); \tag{4.37}$$

(vi) if $0 \leq x \leq h$, $m \in \mathbb{N}$, *and* $m \geq k$, *then*

$$|\Delta_h^m(F; x)| \leq 2^m \varphi\big((m+1)h\big). \tag{4.38}$$

Proof. For $k = 1$, the lemma is trivial. Therefore, in what follows, we assume that $k > 1$.

(i) The equality $F(0) = 0$ is obvious. The equality $F(0+) = 0$ follows from the estimates

$$
\begin{aligned}
(k-2)!\,|F(x;\varphi;k)| &\le x \int_x^1 u^{-2}\varphi(u)\,du \\
&= x\left(\int_x^{\sqrt{x}} + \int_{\sqrt{x}}^1\right) u^{-2}\varphi(u)\,du \\
&\le \varphi(\sqrt{x})\,x \int_x^{\sqrt{x}} u^{-2}\,du + \varphi(1)\,x \int_{\sqrt{x}}^1 u^{-2}\,du \\
&\le \varphi(\sqrt{x}) + \sqrt{x}\,\varphi(1) \to 0 \quad \text{as} \quad x \to 0.
\end{aligned}
$$

(ii) Equalities (4.29)–(4.31) can be verified by simple differentiation. In (4.31), one should take into account that the majorant $\varphi = \varphi(t)$ is monotone.

(iii) Let us prove that $F^{(k-1)}$ does not decrease (generally speaking, the fact that $F^{(k)}(x) \ge 0$ a.e. does not imply that $F^{(k-1)}$ does not decrease). Indeed, if $x_1 > x_2$, then, by virtue of (4.30), we get

$$
F^{(k-1)}(x_1) - F^{(k-1)}(x_2) = \frac{\varphi(x_1) - \varphi(x_2)}{x_1^{k-1}} + (k-1)\int_{x_2}^{x_1} \frac{\varphi(u) - \varphi(x_2)}{u^k}\,du \ge 0.
$$

Using (3.38) and (3.45), we obtain

$$
\begin{aligned}
\Delta_h^k(F;x) &= \Delta_h^{k-1}(F;x+h) - \Delta_h^{k-1}(F;x) \\
&= \int_0^h \ldots \int_0^h \left(F^{(k-1)}(x+h+s) - F^{(k-1)}(x+s)\right) ds_{k-1} \ldots ds_1 \ge 0,
\end{aligned}
$$

$$
s = s_1 + \ldots + s_{k-1}.
$$

(iv) Denoting $E(x) := x^{-1}F(x)$, we note that $E^{(k-1)}(x) = x^{-k}\varphi(x)$. By virtue of (3.43), we have $\Delta_h^k(F;0) = kh\,\Delta_h^{k-1}(E;h)$. Identities (4.33)–(4.35) now follow from (3.44), (3.45), and (3.34), respectively.

(v) Inequalities (4.36) follow from relation (4.33), the estimate $\varphi(h) \le \varphi(h(1+s)) \le \varphi(kh)$, $0 \le s \le k-1$, and the identity

$$k!\int_0^1\int_0^{s_1}\dots\int_0^{s_{k-2}}(1+s_1+\dots+s_{k-1})^{-k}ds_{k-1}\dots ds_1 \equiv k(-1)^{k-1}\Delta_1^{k-1}(\mu;1) = 1,$$

where $\mu(y) = 1/y$. However, if $\varphi \in \Phi^k$, then

$$(1+s)^{-k}\varphi(h(1+s)) = h^k(h(1+s))^{-k}\varphi(h(1+s)) \le \varphi(h).$$

(vi) Denote

$$F_h(y) := \frac{1}{(k-2)!}\int_{(m+1)h}^{y} y(y-u)^{k-2}u^{-k}\varphi(u)\,du.$$

Note that $F(y) - F_h(y)$ is an algebraic polynomial of degree $\le k-1$ and, for $0 \le y \le (m+1)h$, we have

$$|F_h(y)| \le \frac{\varphi((m+1)h)}{(k-2)!}y\int_y^{(m+1)h}u^{-2}\,du < \varphi((m+1)h).$$

Hence,

$$|\Delta_h^m(F;x)| \equiv |\Delta_h^m(F_h;x)| \le 2^m\|F_h\|_{[0,(m+1)h]} \le 2^m\varphi((m+1)h).$$

■

Theorem 4.2 (see [S. Nikol'skii (1946b)] for $k = 1$ and [Shevchuk (1976)] for $k > 1$). *Let $F(x) = F(x, \varphi, k)$ be the extremal function (4.27) and let $b > 0$. If $\varphi \in \Phi^k$, then the following relations hold for $0 \le t \le b/k$:*

$$\omega_k(t;F;[0,b]) \equiv \Delta_t^k(F;0), \tag{4.39}$$

$$\varphi(t) \le \omega_k(t;F;[0,b]) \le k\varphi(t). \tag{4.40}$$

Proof. For $k = 1$, the theorem has already been proved (see Remark 1.3). Therefore, we assume in what follows that $k > 1$. Denote $E = E(x) = x^{-1}F(x)$. Note that $E^{(k-1)}(x) = x^{-k}\varphi(x)$. Since $\varphi \in \Phi^k$, we conclude that $E^{(k-1)}(x)$ does not increase. Reasoning by analogy with the proof of assertion (iii) of the previous lemma, we get $\Delta_h^k(E; x) \le 0$, $x > 0$, $h \ge 0$. We take $t > 0$, $h \in [0, t]$, $x_0 > 0$, and denote $E_0(x) := (x - x_0)E(x)$. Using (4.32), (3.44), and (4.33), we obtain

$$\begin{aligned}
0 \le \Delta_h^k(F; x_0) &= \Delta_h^k(E_0; x_0) + x_0\Delta_h^k(E; x_0) \\
&\le \Delta_h^k(E_0; x_0) = k\,h\Delta_h^{k-1}(E; x_0 + h) \\
&= k!h^k \int_0^1\int_0^{s_1} \ldots \int_0^{s_{k-2}} (x_0 + h(s+1))^{-k}\varphi(x_0 + h(s+1))\, ds_{k-1} \ldots ds_1 \\
&\le k! \int_0^1\int_0^{s_1} \ldots \int_0^{s_{k-2}} (s+1)^{-k}\varphi(h(s+1))\, ds_{k-1} \ldots ds_1 \\
&\le k! \int_0^1\int_0^{s_1} \ldots \int_0^{s_{k-2}} (s+1)^{-k}\varphi(t(1+s))\, ds_{k-1} \ldots ds_1 = \Delta_t^k(F; 0),
\end{aligned}$$

$$s := s_1 + \ldots + s_{m-1},$$

which yields (4.39). Inequality (4.40) follows from (4.39) and (4.37).

■

4.4. Smoothing of a majorant

It is sometimes convenient to assume that a majorant φ is differentiable. This possibility can be provided by the following statement:

Lemma 4.3 [Shevchuk (1984a)]. *Let* $m \in \mathbb{N}$, $m \ne 1$. *If* $\varphi \in \Phi$, *then the function*

$$\varphi_m(t) := \varphi_m(t, \varphi) = m! \int_0^1\int_0^{s_1} \ldots \int_0^{s_{m-2}} (1+s)^{-m}\varphi(t(1+s))\, ds_{m-1} \ldots ds_1 \tag{4.41}$$

where $s := s_1 + \ldots + s_{m-1}$, *possesses the following properties:*

$$\text{(i)} \quad \varphi(t) \le \varphi_m(t) \le \varphi(mt), \ t \ge 0; \tag{4.42}$$

(ii) $\varphi_m \in \Phi$, *and if* $\varphi \in \Phi^k$, *then*

$$\varphi_m \in \Phi^k; \tag{4.43}$$

(iii) for $j = 1, \ldots, m-2$, $\varphi_m^{(j)}$ *belongs to* $C((0, \infty))$ *and*

$$|\varphi_m^{(j)}(t)| \le m 2^m t^{-j} \varphi(mt), \quad t > 0. \tag{4.44}$$

Proof. (i) Inequalities (4.42) are proved in the same way as (4.36).

(ii) It is obvious that φ_m does not decrease. Relation (4.43) follows the identity

$$t^{-k}\varphi_m(t) = m! \int_0^1 \int_0^{s_1} \ldots \int_0^{s_{m-2}} (t(1+s))^{-k} \varphi(t(1+s))(1+s)^{k-m} \, ds_{m-1} \ldots ds_1.$$

(iii) Denote

$$E_m := E_m(x) := (1/(m-2)!)(x-u)^{m-2} u^{-m} \varphi(u)\, du, \quad E_{m,j} := x^j \varphi_m^{(j)}(x).$$

Equalities (2.33) and (3.43) yield $\varphi_m(t) = mt\Delta_t^{m-1}(E_m; t)$. Hence,

$$\varphi_m^{(j)}(t) = mt^{1-j}\left(\Delta_t^{m-1}(E_{m,j}; t) + j\Delta_t^{m-1}(E_{m,j-1}; t)\right).$$

Relation (4.44) now follows from the estimate

$$\left|\Delta_t^{m-1}(E_{m,j-1}; t)\right| < 2^{m-1}(jt)^{-1} \varphi(mt).$$

■

Corollary 4.1. *Let* $m \in \mathbb{N}$, $m \ge k$, *let* $\varphi \in \Phi^m$, *let* $\varphi_{m-k+2} := \varphi_{m-k+2}(t, \varphi)$ *be defined by (4.41), and let* $F = F(x) = F(x, \varphi_{m-k+2}, k)$ *be the extremal function (4.27). Then*

$$|\Delta_h^m(F; x)| \le c\varphi(h), \quad x \ge 0, \quad h \ge 0, \tag{4.45}$$

where $c = c(m) = \text{const}$; *in particular, for any fixed* $b > 0$, *one has*

$$\omega_m(t\,;F\,;[0,b]) \le c\varphi(t), \qquad 0 \le t \le b/k. \tag{4.46}$$

Indeed, if $x \le h$, then relation (4.45) follows from (4.38) and (4.42); if $x > h$, then, with regard for (4.31) and (4.44), we obtain

$$|\Delta_h^m(F;x)| = h^m\left|F^{(m)}(\theta)\right| = h^m\,(F^{(k)}(\theta))^{(m-k)}$$

$$= h^m\,(\theta^{1-k}\varphi_{m-k+2}(\theta))^{(m-k)} \le c h^m \theta^{-m}\varphi(\theta) \le c\varphi(h), \qquad \theta > h.$$

4.5. Is the relation $\omega_k(t\,;f) = O(\omega_m(t\,;f))$ true for $m > k$?

Here and in what follows, the expressions $\alpha(t) = O(\beta(t))$ and $\alpha(t) = o(\beta(t))$, where $\alpha(t) \ge 0$ and $\beta(t) \ge 0$, $t > 0$, mean that

$$\limsup_{t\to 0}\,(\alpha(t)/\beta(t)) < \infty \qquad \text{and} \qquad \lim_{t\to 0}\,(\alpha(t)/\beta(t)) = 0,$$

respectively.

Thus, if $f(x) = x^\alpha$, $0 < \alpha < k$, then the relation

$$\omega_k(t\,;f) = O(\omega_m(t\,;f)) \tag{4.47}$$

is true [see (4.9)]. If $f \in C^k([0,1])$, then, according to (4.20) and (4.15), we have

$$\omega_k(t\,;f) \ge (k/2)^k\,\omega_k(1/k\,;f)\,\omega_m(t\,;f)/\,\omega_{m-k}(t\,;f^{(k)})$$

i.e., relation (4.47) is not true. If $f(x) = x\ln(e/x) \notin C^1([0,1])$, then

$$\omega_1(t\,;f) = (\ln 4)\,\omega_2(t\,;f)\ln(e/t),$$

i.e., relation (4.47) is not true again. D. Galan (1973) proved that, for "bad" functions whose smoothness is "strictly less" than k, one has

$$\liminf_{t\to 0}\,\omega_k(t\,;f)/\,\omega_m(t\,;f) < \infty.$$

At the same time, D. Galan and V. Galan [V. Galan (1991); V. Galan and D. Galan (1987)] and Nessel and van Wickeren (1984) constructed examples of arbitrarily "bad" functions for which relation (4.47) is not valid. In particular, the following theorem is true:

Theorem 4.3 [Nessel and van Wickeren (1984)]. *Let* $k \in \mathbb{N}$, $m \in \mathbb{N}$, $m > k$, *and let* α *be a function positive on* $(0, \infty)$ *and such that*

$$\lim_{t \to 0} \alpha(t) = \infty, \qquad \lim_{t \to 0} t^r \alpha(t) = 0, \qquad r := m - n. \tag{4.48}$$

If $\varphi \in \Phi^k$, *then there exists a function* $f_{\varphi,\alpha} \in C([0, 1])$ *such that*

$$\omega_k(t; f_{\varphi,\alpha}; ([0, 1])) \begin{cases} = O(\varphi(t)), & (4.49) \\ \neq o(\varphi(t)), & (4.50) \end{cases}$$

$$\omega_k(t; f_{\varphi,\alpha}; ([0, 1])) \neq O(\alpha(t))\, \omega_m(t; f_{\varphi,\alpha}; ([0, 1])). \tag{4.51}$$

Proof. Let us construct a sequence of points $t_n \to 0$. Denote $t_1 := 1$. Assume that the first n points of the sequence are chosen. If n is odd, then, as t_{n+1}, we take an arbitrary point satisfying the conditions

$$0 < 2t_{n+1} < t_n, \qquad \alpha(t_{n+1}) t_{n+1}^r < t_n^r.$$

If n is even, then, as t_{n+1}, we take an arbitrary point satisfying the condition $\varphi(t_{n+1}) = \varphi(t_{n-1})(t_n / t_{n-1})^m$. Denote $\varphi^*(t) = \varphi(t_{n+1})$ if $t_{n+1} \le t \le t_n$ and n is even, and $\varphi^*(t) := \varphi(t_n)(t/t_n)^m$ if $t_{n+1} \le t \le t_n$ and n is odd. It is obvious that $\varphi^* \in \Phi^m$. Starting from the majorant φ^* and using relation (4.41), we define the majorant $\varphi_* = \varphi_*(t) := \varphi_{m-k+2}(t; \varphi^*)$. By $F = F(x) := F(x, \varphi_*, k)$ we denote the extremal function (4.27). Let us show that the function $f_{\varphi,\alpha} = f_{\varphi,\alpha}(x) := F(x)$ can be taken as that indicated in the theorem.

Indeed, relations (4.36) and (4.42) yield

$$\omega_k(t; F) \ge \Delta_t^k(F; 0) \ge \varphi^*(t).$$

Hence, for odd n, we have $\omega_k(t_n; F) \ge \varphi^*(t_n) = \varphi(t_n)$, i.e., relation (4.50) is proved.

Reasoning as in the previous corollary and using (4.31) and (4.44), we obtain the following relation for $x > h$:

$$|\Delta_h^k(F;x)| = h^k\theta^{1-k}\varphi_*'(\theta) \le 2^{r+2}(r+2)h^k\theta^{-k}\varphi^*(\theta)$$

$$\le 2^{r+2}(r+2)h^k\theta^{-k}\varphi(\theta) \le 2^{r+2}(r+2)^{k+2}\varphi(h), \qquad \theta > x.$$

For $0 < x \le h$, relations (4.38) and (4.42) yield

$$|\Delta_h^k(F;x)| = 2^k\varphi_*((k+1)h) \le 2^k(r+2)^k\varphi^*((k+1)h)$$

$$\le 2^k(r+2)^k\varphi((k+1)h) \le 2^k(r+2)^k(k+1)^k\varphi(h),$$

which proves (4.49).

Finally, by virtue of (4.46), we have $\omega_m(t;F) \le c\varphi^*(t)$. Denoting $a_n := t_n/t_{n-1}$ and using (4.40) and (4.42), we obtain the following relation for even n:

$$2^k\omega_k(t_n;F) \ge a_n^k\,\omega_k(t_{n-1};F) \ge a_n^k\,\varphi(t_{n-1}) \ge a_n^{-r}\,\varphi^*(t_n)$$

$$\ge (1/c)a_n^{-r}\omega_m(t_n;F) \ge (1/c)\alpha(t_n)\omega_m(t_n;F).$$

This proves (4.51).

■

The "regular" estimate for $\omega_k(t;f)$ in terms of $\omega_m(t;f)$ with $k < m$ is the Marchaud inequality, which is considered in the next section.

4.6. Gliding-hump method

Nessel and van Wickeren (1984) deduced Theorem 4.3 from a general theorem (see Theorem 4.4 below), which is proved by the gliding-hump method. Theorem 4.4 and its analogs enabled Nessel and other authors to construct a series of counterexamples in various parts of approximation theory, including integral metrics (see [Dickmeis and Nessel (1982); Dickmeis, Nessel, and van Wickeren (1984); Nessel and van Wickeren (1984)]).

We write $\varphi \in \Phi_0^k$ if $\varphi \in \Phi^k$ and $t^{-k}\varphi(t) \to \infty$ as $t \to 0$.

Let T be a functional on $C([a,b])$. We write $T \in X^*$ if T is a sublinear functional, i.e.,

$$|T(f+g)| \le |Tf| + |Tg|,$$

$$|T(\lambda f)| = |\lambda||Tf| \tag{4.52}$$

for all $f, g \in C([a,b])$ and $\lambda \in \mathbb{R}$, and T is a bounded functional, i.e.,

$$\|T\|_{X^*} := \sup\{|Tf| : \|f\|_{[a,b]} = 1\} < \infty. \tag{4.53}$$

Theorem 4.4 [Nessel and van Wickeren (1984)]. *Suppose that* $k \in \mathbb{N}$, $M = \text{const} > 0$, $n \in \mathbb{N}$, $T_n, R_n, V_n \in X^*$, $h_n \in C([a,b])$, $\|h_h\|_{[a,b]} \le 1$, *and* $M_n = \text{const} > 0$. *If*

$$\limsup_{n\to\infty} |T_n h_n| \ge M, \qquad \limsup_{n\to\infty} |R_n h_n| \ge M, \tag{4.54}$$

and, for all $n \in \mathbb{N}$, *one has*

$$\omega_k(t; h_n; [a,b]) \le \min\{1, (tn)^k\}, \qquad 0 \le t \le 1, \tag{4.55}$$

$$|V_n h_n| \le 1, \qquad \sup_{p\in\mathbb{N}} p^k |V_p h_n| \le M_n, \tag{4.56}$$

then, for any majorant $\varphi \in \Phi_0^k$, *there exists a sequence* $\{\delta_i\} \subset \{0, 1\}$ *(i.e.,* $\delta_j = 0$ *or* $\delta_j = 1$*) such that the function*

$$f_\varphi(x) := \sum_{j=1}^{\infty} \delta_j \varphi(1/j) h_j(x)$$

possesses the following properties:

$$\omega_k(t; f_\varphi; [a,b]) = O(\varphi(t)), \qquad t \to 0, \tag{4.57}$$

$$|T_n f_\varphi| \ne o(\varphi(1/n)), \quad |R_n f_\varphi| \ne o(|V_n f_\varphi|), \qquad n \to \infty. \tag{4.58}$$

Proof. Since $\varphi \in \Phi_0^k$, we conclude that, for any $q \in \mathbb{N}$, there exists a number $N(q)$ such that, for all $n \ge N(q)$, the following inequalities are true:

$$\varphi(1/n) < (1/2)\varphi(1/q), \tag{4.59}$$

$$\|T_q\|_{X^*} \varphi(1/n) < (M/16)\varphi(1/q), \tag{4.60}$$

$$\|R_q\|_{X^*} \varphi(1/n) < (M/16)\varphi(1/q), \tag{4.61}$$

$$\|V_q\|_{X^*}\,\varphi(1/n) < \varphi(1/q), \tag{4.62}$$

$$q^{k+1}\varphi(1/q) < n^k\varphi(1/n), \tag{4.63}$$

$$q \max_{j=1,\ldots,q} |V_n h_j| \le q n^{-k} \max_{j=1,\ldots,q} |M_j| < \varphi(1/n). \tag{4.64}$$

Using $N(q)$, we construct two sequences of numbers $\{n_j\} \subset \mathbb{N}$ and $\{\delta_{n_j}\} \subset \{0, 1\}$ and the sequence of functions

$$g_p = g_p(x) := \sum_{j=0}^{p} \delta_{n_j}\varphi(1/n_j)h_{n_j}(x).$$

Denote $n_1 := 1$ and $\delta_{n_1} = \delta_1 := 1$. Assume that the first $p-1$ terms of these sequences have already been chosen and p is even. Let n_p be an arbitrary number such that $n_p > N(n_{p-1})$ and $|T_{n_p}h_{n_p}| > M/2$. If $|T_{n_p}g_{p-1}| > (M/4)\varphi(1/n_p)$, then we set $\delta_{n_p} := 0$; if $|T_{n_p}g_{p-1}| \le (M/4)\varphi(1/n_p)$, then we set $\delta_{n_p} := 1$. Thus, in both cases, we have

$$|T_{n_p}g_p| > (M/4)\varphi(1/n_p), \qquad p \text{ is even.} \tag{4.65}$$

Similarly, in the case of odd p, we denote by n_p any number for which $n_p > N(n_{p-1})$ and $|R_{n_p}h_{n_p}| > M/2$ and choose the number $\delta_{n_p} \in \{0, 1\}$ from the condition

$$|R_{n_p}g_p| > (M/4)\varphi(1/n_p), \qquad p \text{ is odd.} \tag{4.66}$$

Let us prove that the function f_φ indicated in the theorem can be defined as follows:

$$f_\varphi(x) := \sum_{j=1}^{\infty} \delta_{n_j}\varphi(1/n_j)\,h_{n_j}(x).$$

Indeed, let $1/n_{p+1} < t \le 1/n_p$. By using (4.55), (4.59), and (4.63), we obtain

$$\omega_k(t; f_\varphi; [a,b])$$

$$\le \left(\sum_{j=1}^{p-1} + \sum_{j=p}^{p+1} + \sum_{j=p+2}^{\infty}\right)\varphi(1/n_j)\,\omega_k(t; h_{n_j})$$

$$\leq t^k \sum_{j=1}^{p-1} \varphi(1/n_j) n_j^k + (tn_p)^k \varphi(1/n_p) + \varphi(1/n_{p+1}) + \sum_{j=p+2}^{\infty} \varphi(1/n_j)$$

$$\leq t^k (p-1) n_{p-1}^k \varphi(1/n_{p-1}) + 2\varphi(t) + \varphi(1/n_{p+1}) \sum_{\nu=1}^{\infty} 2^{-\nu}$$

$$\leq (tn_p)^k \varphi(1/n_p) + 3\varphi(t) = 4\varphi(t). \tag{4.67}$$

Furthermore, for even p, relations (4.53), (4.59), and (4.60) yield

$$\left| T_{n_p}(f_\varphi - g_p) \right| \leq \sum_{j=p+1}^{\infty} \varphi(1/n_j) \left| T_{n_p} h_{n_j} \right|$$

$$\leq \left\| T_{n_p} \right\|_{X^*} \sum_{j=p+1}^{\infty} \varphi(1/n_j) \leq \left\| T_{n_p} \right\|_{X^*} \varphi(1/n_{p+1}) \sum_{\nu=1}^{\infty} 2^{-\nu}$$

$$= 2 \left\| T_{n_p} \right\|_{X^*} \varphi(1/n_{p+1}) < (M/8) \varphi(1/n_p).$$

Hence, according to (4.52) and (4.65), we have

$$\left| T_n f_\varphi \right| \geq \left| T_n g_p \right| - \left| T_{n_p}(f_\varphi - g_p) \right|$$

$$\geq (M/4)\varphi(1/n_p) - (M/8)\varphi(1/n_p) \geq (M/8)\varphi(1/n_p). \tag{4.68}$$

By analogy, for odd p, relations (4.53), (4.59), (4.61), (4.52), and (4.66) yield

$$\left| R_n f \right| \geq (M/8)\varphi(1/n_p). \tag{4.69}$$

Finally, by virtue of (4.52), (4.53), (4.56), (4.62), (4.59), and (4.64), we obtain

$$\left| V_{n_p} f_\omega \right| \leq \left| V_{n_p} h_{n_p} \right| \varphi(1/n_p) + \left(\sum_{j=1}^{p-1} + \sum_{j=p+1}^{\infty} \right) \left| V_{n_p} h_{n_j} \right| \varphi(1/n_j)$$

$$\leq \varphi(1/n_p) + (p-1)\varphi(1) \max_{j=1,\ldots,p-1} \left| V_{n_p} h_{n_j} \right|$$

$$+ \left\| V_{n_p} \right\|_{X^*} \varphi(1/n_{p+1}) \sum_{\nu=0}^{\infty} 2^{-\nu}$$

$$\leq \varphi(1/n_p) + \varphi(1)\varphi(1/n_p)$$

$$+ 2\varphi(1/n_p) + (3+\varphi(1))\varphi(1/n_p). \tag{4.70}$$

Estimates (4.67)–(4.70) prove the theorem.

■

4.7. Remarks

Note that the behavior of moduli of continuity in integral metrics was thoroughly studied by Besov and Stechkin (1975), Kolyada [(1975), (1988), (1989)], Konyagin (1983), Oskolkov (1976), Radoslavova (1979), Ul'yanov [(1967), (1968)], Yudin (1979), and others.

5. Marchaud inequality

In the present section, we generalize the well-known Kolmogorov-type inequality

$$\left\| f^{(j)} \right\|_{[a,b]} \leq A_{j,r}\big((b-a)^{r-j} \left\| f^{(r)} \right\|_{[a,b]} + (b-a)^{-j} \| f \|_{[a,b]}\big), \tag{5.1}$$

where $j \in \mathbb{N}$, $0 < j < r$, $f \in C^r([a,b])$, $A_{j,r} = \text{const}$, and $0 < A_{j,r} < c = c(r)$, and the inequality (see [Besov (1965)])

$$\left\| f^{(j)} \right\|_{[a,b]} \leq A_{j,r,k}\big((b-a)^{r-j}\, \omega_k(b-a; f^{(r)}; [a,b]) + (b-a)^{-j} \| f \|_{[a,b]}\big), \tag{5.2}$$

where $j \in \mathbb{N}$, $0 < j \leq r$, $f \in C^r([a,b])$, and $0 < A_{j,r,k} < c = c(k,r)$.

In what follows, c and c_i always denote various positive numbers (constants) that can depend only on k and r.

5.1. Marchaud inequality

Theorem 5.1 formulated below gives an upper bound for the modulus of continuity of order $j < k$ in terms of the kth modulus of continuity and the norm of the function. Recall that the corresponding lower bound is given by inequality (4.12). A part of the proof of

Theorem 5.1 is presented as a separate lemma (Lemma 5.1) because this result will also be used in other cases.

Lemma 5.1. *Let*

$$k \neq 1, \qquad h > 0, \qquad l \in \mathbb{N}, \qquad H = 2^l h,$$

$$f \in C([x_0, x_0 + H(k-1)]), \qquad \omega_k(t) = \omega_k(t; f; [x_0, x_0 + H(k-1)]).$$

Then

$$\left|\Delta_h^{k-1}(f; x_0) - 2^{l(1-k)}\Delta_H^{k-1}(f; x_0)\right| \leq (k-1)^2 h^{k-1} \int_h^H u^{-k}\omega_k(u)\,du, \tag{5.3}$$

$$\left|\Delta_h^{k-1}(f; x_0 + (k-1)(H-h)) - 2^{l(1-k)}\Delta_H^{k-1}(f; x_0)\right|$$
$$\leq (k-1)^2 h^{k-1} \int_h^H u^{-k}\omega_k(u)\,du. \tag{5.4}$$

Proof. According to (3.31) and (3.39), we have

$$\begin{aligned}
&\left|\Delta_{2h}^{k-1}(f; x_0) - 2^{k-1}\Delta_h^{k-1}(f; x_0)\right| \\
&\quad= \left|\sum_{\nu=0}^{k-1}\binom{k-1}{\nu}\left(\Delta_h^{k-1}(f; x_0 + \nu h) - \Delta_h^{k-1}(f; x_0)\right)\right| \\
&\quad= \left|\sum_{\nu=0}^{k-1}\binom{k-1}{\nu}\sum_{i=0}^{\nu-1}\Delta_h^k(f; x_0 + ih)\right| \\
&\quad\leq \sum_{\nu=0}^{k-1}\binom{k-1}{\nu}\sum_{i=0}^{\nu-1}\omega_k(h) = (k-1)2^{k-2}\omega_k(h) \\
&\quad= (k-1)^2 2^{k-2} 2^{k-1}(2^{k-1}-1)^{-1} h^{k-1}\int_h^{2h} u^{-k}\omega_k(h)\,du \\
&\quad\leq (k-1)^2 2^{k-1} h^{k-1}\int_h^{2h} u^{-k}\omega_k(u)\,du.
\end{aligned}$$

Therefore,

$$\begin{aligned}
&\left|\Delta_h^{k-1}(f;x_0) - 2^{l(1-k)}\Delta_H^{k-1}(f;x_0)\right| \\
&\quad = \left|\sum_{i=1}^{l} 2^{i(1-k)}\left(2^{k-1}\Delta_{2^{i-1}h}^{k-1}(f;x_0) - \Delta_{2^i h}^{k-1}(f;x_0)\right)\right| \\
&\quad \le \sum_{i=1}^{l} 2^{i(1-k)}(k-1)^2 2^{k-1}(2^{i-1}h)^{k-1}\int_{2^{i-1}h}^{2^i h} u^{-k}\omega_k(u)\,du \\
&\quad = h^{k-1}(k-1)^2\int_h^H u^{-k}\omega_k(u)\,du.
\end{aligned}$$

Inequality (5.3) is proved. Inequality (5.4) can be proved by analogy.

■

Remark 5.1. If $\omega_k(u) \le u^k$, then, under the conditions of Lemma 5.1, we have

$$\left|\Delta_h^{k-1}(f;x_0) - 2^{l(1-k)}\Delta_H^{k-1}(f;x_0)\right| < (1/2)(k-1)Hh^{k-1}. \tag{5.5}$$

Indeed, since

$$\left|\Delta_{2h}^{k-1}(f;x_0) - 2^{k-1}\Delta_h^{k-1}(f;x_0)\right| \le (k-1)2^{k-2}h^k,$$

we also have

$$\begin{aligned}
&\left|\Delta_h^{k-1}(f;x_0) - 2^{l(1-k)}\Delta_H^{k-1}(f;x_0)\right| \\
&\quad \le \sum_{i=1}^{l} 2^{i(1-k)}(k-1)2^{k-2}(2^{i-1}h)^{k-1} < (1/2)(k-1)Hh^{k-1}.
\end{aligned}$$

Theorem 5.1 (Marchaud inequality [Marchaud (1927)]). *If* $f \in C([a,b])$, *then, for all* $j = 0, \dots, k$, *the following inequality is true:*

$$\omega_j(t) \le A_j t^j\left(\int_t^{b-a} u^{-j-1}\omega_k(u)\,du + (b-a)^{-j}\|f\|\right), \tag{5.6}$$

where $0 < t \le b-a$, $\omega_j(t) = \omega_j(t;f;[a,b])$, $\|f\| = \|f\|_{[a,b]}$, *and* $0 < A_j < c$.

Proof. For $j=0$ and $j=k$, inequality (5.6) is obvious. Let us prove (5.6) for $j=k-1$. We take a point $x_0 \in [a,(a+b)/2]$ and a number $h \in (0,(b-x_0)/(k-1)]$ and estimate the $(k-1)$th difference $\Delta_h^{k-1}(f;x_0)$. To this end, we choose a number $(l+1) \in \mathbb{N}$ from the condition

$$(k-1)2^l h \le b-x_0 < (k-1)2^{l+1}h.$$

Using inequality (5.3), we obtain

$$\left|\Delta_h^{k-1}(f;x_0)\right| \le (k-1)^2 h^{k-1} \int_h^{2^l h} u^{-k}\omega_k(u)\,du + 2^{l(1-k)}\left|\Delta_{2^l h}^{k-1}(f;x_0)\right|.$$

Since

$$2^l h \le (b-x_0)/(k-1) \le (b-a)/(k-1) \le b-a,$$

$$2^{l(1-k)} \le (4(k-1))^{k-1}(h/(b-a))^{k-1},$$

$$\left|\Delta_{2^l h}^{k-1}(f;x_0)\right| \le 2^{k-1}\|f\|,$$

we get

$$\begin{aligned}&\left|\Delta_h^{k-1}(f;x_0)\right| \\ &\quad\le (k-1)^2 h^{k-1}\int_h^{b-a} u^{-k}\omega_k(u)\,du + (8(k-1))^{k-1}(h/(b-a))^{k-1}\|f\|. \qquad (5.7)\end{aligned}$$

If $x_0 \in ((a+b)/2, b)$, then we obtain (5.7) by using estimate (5.4). Thus, inequality (5.6) is proved for $j=k-1$. Now assume that relation (5.6) holds for a number $j+1$. Then, by induction, we obtain the following relation for the number j:

$$\begin{aligned}\omega_j(t) &\le j^2 t^j \int_t^{b-a} u^{-j-1}\omega_{j+1}(u)\,du + (8j)^j\,(t/(b-a)^j)\,\|f\| \\ &\le j^2 A_{j+1} t^j \int_t^{b-a} u^{-j-1}\left(u^{j+1}\int_u^{b-a} v^{-j-2}\omega_k(v)\,dv + u^{j+1}(b-a)^{-j-1}\|f\|\right)du \\ &\quad + (8j)^j t^j (b-a)^{-j}\|f\|\end{aligned}$$

$$= j^2 A_{j+1} t^j \left(\int_t^{b-a} u^{-j-1} \omega_k(u) du + t^{j+1} \int_t^{b-a} u^{-j-2} \omega_k(u) du + (b-a-t)(b-a)^{-j-1} \| f \| \right)$$

$$+ (8j)^j t^j (b-a)^{-j} \| f \|$$

$$\leq j^2 A_{j+1} t^j \int_t^{b-a} u^{-j-1} \omega_k(u) du + (j^2 A_{j+1} + (8j)^j) t^j (b-a)^{-j} \| f \|.$$

Note that we have proved an inequality that is even stronger than (5.6), namely

$$\omega_j(t) \leq ct^j \left(\int_t^{b-a} \frac{\omega_k(u)}{u^{j+1}} du + (b-a)^{-j} \omega_j\left(\frac{b-a}{j} \right) \right), \quad j \neq 0. \tag{5.6'}$$

■

Note that if $f \in C^j([a, b])$, then, by virtue of (3.34), we have

$$\| f^{(j)} \|_{[a,b]} \leq \limsup_{t \to 0} t^{-j} \omega_j(t; f; [a, b]).$$

Hence, inequality (5.2) follows from (5.6). Inequality (5.1) follows from (5.2) and (4.13).

5.2. Some simple but important facts for the second modulus of continuity

Let $f \in C([a, b])$, $\omega_j(t) = \omega_j(t; f; [a, b])$, $j = 1, 2$, and $\| f \| = \| f \|_{[a,b]}$.

Theorem 5.1′. *The following inequality is true:*

$$\omega_1(t) \leq t \int_t^{b-a} u^{-2} \omega_2(u) du + 8t(b-a)^{-1} \| f \|. \tag{5.8}$$

Theorem 5.2 [Burkill (1952)]. *If $f(a) = f(b) = 0$, then*

$$\| f \| \leq \omega_2((b-a)/2). \tag{5.9}$$

Proof. Let $x^* \in [a, b]$ be an arbitrary point at which $|f(x^*)| = \| f \|$. For definiteness, we assume that $x^* \in [a, (a+b)/2]$ and $f(x^*) \geq 0$. Denoting $h = x^* - a$, we obtain

$$\|f\| = f(x^*) \le 2f(x^*) - f(x^*+h)$$

$$= -f(x^*+h) + 2f(x^*) - f(x^*+h) \le \omega_2(h) \le \omega_2((b-a)/2).$$

■

Theorem 5.3 (see, e.g., [Dzyadyk (1975c)]). *Let* $x \in [a, b]$ *and let* $h := \min\{x-a, b-x\}$. *Then*

$$|f(x) - L(x; f; a, b)| \le 9h \int_h^{b-a} u^{-2} \omega_2(u)\, du\,. \tag{5.10}$$

Proof. Denote $g(x) := f(x) - L(x; f; a, b)$. Note that $g(a) = g(b) = 0$. Using (5.8) and (5.9), we obtain

$$|f(x) - L(x; f; a, b)| = |g(x)| \le \omega_1(h; g)$$

$$\le h \int_h^{b-a} u^{-2} \omega_2(u; g)\, du + 8h\,(b-a)^{-1}\,\|g\|$$

$$\le h \int_h^{b-a} u^{-2} \omega_2(u; g)\, du + 8h\,(b-a)^{-1}\,\omega_2((b-a)/2; g)$$

$$\le 9h \int_h^{b-a} u^{-2} \omega_2(u; g)\, du.$$

Relation (5.10) now follows from the equality $\omega_2(t; f) = \omega_2(t; g)$.

■

Corollary 5.1. *The following inequality is true:*

$$|f(x) - L(x; f; a, b)| \le 45\,\omega_2\left(\sqrt{(x-a)(b-x)}\right), \quad x \in [a, b]. \tag{5.11}$$

Indeed, denoting

$$h^* = \sqrt{(x-a)(b-x)}, \qquad h = \min\{x-a, b-x\},$$

we obtain

$$h \int_h^{b-a} u^{-2}\omega_2(u)\,du$$

$$= h \int_h^{h^*} u^{-2}\omega_2(u)\,du + h \int_{h^*}^{b-a} u^{-2}\omega_2(u)\,du$$

$$\le h \int_h^{\infty} u^{-2}\omega_2(h^*)\,du + h \int_{h^*}^{b-a} 4(h^*)^{-2}\omega_2(h^*)\,du < 5\omega_2(h^*).$$

Theorem 5.4. *If*

$$\int_0^1 u^{-2}\omega_2(u)\,du < \infty,$$

then $f \in C^1([a,b])$ *and*

$$\omega_1(t;f';[a,b]) \le 2\int_0^t u^{-2}\omega_2(u)\,du. \tag{5.12}$$

Proof. First, we prove that the derivative $f' = f'(x)$ exists at all points $x \in [a,b]$. Using the Cauchy criterion, for an arbitrary $\varepsilon > 0$ we choose $\delta > 0$ from the condition

$$18\int_0^{2\delta} u^{-2}\omega_2(u)\,du < \varepsilon.$$

We fix arbitrary points $x_1, x_2 \in [a,b]$ from the δ-neighborhood of the point x. Among the points x, x_1, and x_2, let y_0 denote the leftmost one, y_2 the rightmost one, and y_1 the middle one. Note that

$$[x, x_1; f] - [x, x_2; f] = (x_1 - x_2)[x, x_1, x_2; f]$$

$$= (x_1 - x_2)[y_0, y_1, y_2; f]$$

$$= (x_1 - x_2)\big(f(y_1) - L(y_1; f; y_0, y_2)(y_1 - y_0)^{-1}(y_1 - y_2)^{-1}\big).$$

Denote $h = \min\{y_1 - y_0, y_2 - y_1\}$. Using (5.10), we obtain

$$\big|[x, x_1; f]\big| - \big|[x, x_2; f]\big|$$

$$\le 18\,|x_1 - x_2| \int_h^{y_2 - y_0} u^{-2}\omega_2(u)\,du\,(y_2 - y_0)^{-1}$$

$$\le 18 \int_h^{y_2 - y_0} u^{-2}\omega_2(u)\,du \le 18 \int_0^{2\delta} u^{-2}\omega_2(u)\,du < \varepsilon,$$

According to the Cauchy criterion, this means the existence of the derivative $f'(x)$.

Let us prove that the derivative $f'(x)$ is continuous and relation (5.12) is true. Consider two points $x_0, x_1 \in [a, b]$, $x_0 < x_1$, and denote $H = x_1 - x_0$, $\varepsilon_l = H2^{-l}$, $l \in \mathbb{N}$. According to Lemma 5.1, we have

$$\big|\Delta^1_{\varepsilon_l}(f; x_0) - \Delta^1_{\varepsilon_l}(f; x_1 - \varepsilon_l)\big|$$

$$\le \big|\Delta^1_{\varepsilon_l}(f; x_0) - 2^{-l}\Delta^1_H(f; x_0)\big| + \big|\Delta^1_{\varepsilon_l}(f; x_1 - \varepsilon_l) - 2^{-l}\Delta^1_H(f; x_0)\big|$$

$$\le 2\varepsilon_l \int_{\varepsilon_l}^{H} u^{-2}\omega_2(u; f; [x_0, x_1])\,du,$$

whence

$$|f'(x_0) - f'(x_1)| = \lim_{l \to \infty} \frac{1}{\varepsilon_l}\big|\Delta^1_{\varepsilon_l}(f; x_0) - \Delta^1_{\varepsilon_l}(f; x_1 - \varepsilon_l)\big|$$

$$\le 2 \int_0^{x_1 - x_0} u^{-2}\omega_2(u; f; [x_0, x_1])\,du.$$

■

5.3. Estimates for moduli of continuity of derivatives

Lemma 5.2. *Let* $k \ne 1$. *If* $f \in C([a, b])$ *and*

$$\int_0^{b-a} u^{-2}\omega_k(u; f; [a, b])\,du < \infty,$$

then $f \in C^1([a,b])$ *and*

$$\omega_{k-1}(u;f';[a,b]) \le c\int_0^t u^{-2}\omega_k(u;f;[a,b])\,du. \tag{5.13}$$

Proof. According to the Marchaud inequality (5.6), we have

$$\int_0^{b-a} u^{-2}\omega_2(u;f;[a,b])\,du$$

$$\le c_1\int_0^{b-a} u^{-2}\left(u^2\int_u^{b-a} v^{-3}\omega_k(v;f;[a,b])\,dv + u^2(b-a)^{-2}\|f\|_{[a,b]}\right)$$

$$\le c_1\int_0^{b-a} u^{-2}\omega_k(u;f;[a,b])\,du + c_1(b-a)^{-1}\|f\|_{[a,b]} < \infty.$$

Therefore, by virtue of Theorem 5.4, we also have $f \in C^1([a,b])$. Let us prove (5.13). We fix a point $x_0 \in [a,b]$ and a number $h \in (0,(b-x_0)/(k-1)]$, and denote $x_1 = x_0 + (k-1)h$. Using (2.12), (5.6), and (5.12), we get

$$|\Delta_h^{k-1}(f';x_0)| \le \omega_{k-1}(h;f';[x_0,x_1])$$

$$\le 2^{k-1}\omega_1(h;f';[x_0,x_1]) \le 2^k\int_0^h u^{-2}\omega_2(u;f;[x_0,x_1])\,du$$

$$\le 2^k c_1\int_0^{x_1-x_0} u^{-2}\omega_k(u;f;[x_0,x_1])\,du + 2^k c_1(x_1-x_0)^{-1}\|f\|_{[x_0,x_1]}$$

$$\le c_2\int_0^h u^{-2}\omega_k(u;f;[x_0,x_1])\,du + c_2h^{-1}\|f\|_{[x_0,x_1]}. \tag{5.14}$$

We set

$$g(x) = f(x) - L(x;f;x_0,x_0+h,\dots,x_0+(k-1)h)$$

and note that $\Delta_h^{k-1}(f'; x_0) = \Delta_h^{k-1}(g'; x_0)$ and $\omega_k(t; f) \equiv \omega_k(t; g)$. We also use the inequality $\|g\|_{[x_0,x_1]} \le c_3 \omega_k(h; f; [x_0, x_1])$ (Whitney inequality), which will be proved in the next section. It follows from (5.14) that

$$\begin{aligned}
\left|\Delta_h^{k-1}(f'; x_0)\right| &= \left|\Delta_h^{k-1}(g'; x_0)\right| \\
&\le c_2 \int_0^h u^{-2} \omega_k(u; g; [x_0, x_1])\, du + c_2 h^{-1} \|g\|_{[x_0,x_1]} \\
&\le c_2 \int_0^h \frac{\omega_k(u; f; [a, b])}{u^2}\, du + \frac{c_2 c_3}{h} \omega_k(h; f; [x_0, x_1]) \\
&\le c_4 \int_0^h \frac{\omega_k(u; f; [a, b])}{u^2}\, du.
\end{aligned}$$

■

Theorem 5.5. *Let $m \in \mathbb{N}$, $k \le m$, $r := m - k$, and $f \in C([a, b])$. If*

$$\int_0^{b-a} r u^{-r-1} \omega_m(u; f)\, du < \infty, \tag{5.15}$$

then $f \in C^r([a, b])$ and, for every $j = 0, \dots, k$, $j \ne m$, the following inequality is true:

$$c\,\omega_j(t; f^{(r)}) \le \int_0^t \frac{r\omega_m(u; f)}{u^{r+1}} du + (k-j)t^j \left(\int_t^{b-a} \frac{\omega_m(u; f)}{u^{j+r+1}} du + \frac{\|f\|}{(b-a)^{j+r}} \right), \tag{5.16}$$

where $0 \le t \le b - a$, $\|f\| = \|f\|_{[a,b]}$, and $\omega_j(t; f^{(s)}) = \omega_j(t; f^{(s)}; [a, b])$.

Proof. In the case $r = 0$, condition (5.15) is obviously satisfied, and inequality (5.16) is the Marchaud inequality (5.6). We assume that the theorem is valid for a number $r - 1$ and prove it by induction for the number r. Since, according to the induction hypothesis, we have

$$\int_0^t \frac{\omega_{k+1}(u; f^{(r-1)})}{u^2} du \le c_1 \int_0^t u^{-2} \int_0^u v^{-r} \omega_m(v; f) dv \le c_1 \int_0^t \frac{\omega_m(u; f)}{u^{r+1}} du < \infty,$$

it follows from Lemma 5.2 that $f \in C^r([a,b])$ and relation (5.16) is true for $j=k$. For $j=0$, taking (3.34) into account, we obtain

$$\omega_0(t;f^{(r)}) = \left\| f^{(r)} \right\| \le \limsup_{t\to\infty} \frac{\omega_r(t;f)}{t^r} \le c_2 \int_0^{b-a} \frac{\omega_m(u;f)}{u^{r+1}}\,du + \frac{c_2\|f\|}{(b-a)^r},$$

i.e., relation (5.16) is also true for $j=0$. For the other $j=1,\dots,k-1$, it remains to use the Marchaud inequality (5.6).

■

Remark 5.2. If $\omega_m(t;f;[a,b]) \le t^m$, then

$$\omega_1(t;f^{(m-1)};[a,b]) \le t. \tag{5.17}$$

Indeed, by virtue of (5.15), we have $f \in C^{m-1}([a,b])$. We take points $x_0 \in [a,b]$ and $x_1 \in [a,b]$, $x_1 > x_0$, and denote $H := (x_1 - x_0)/(m-1)$ and $h_l = 2^{-l}H$, $l \in \mathbb{N}$. Using (3.34) and (5.5), we obtain

$$\begin{aligned}
&\left| f^{(m-1)}(x_1) - f^{(m-1)}(x_0) \right| \\
&\quad= \lim_{l\to\infty} h_l^{1-m} \left| \Delta_{h_l}^{m-1}(f; x_1 - (m-1)h_l) - \Delta_{h_l}^{m-1}(f; x_0) \right| \\
&\quad\le \lim_{l\to\infty} h_l^{1-m} \left| \Delta_{h_l}^{m-1}(f; x_1 - (m-1)h_l) - 2^{l(1-m)}\Delta_{H}^{m-1}(f; x_0) \right| \\
&\qquad\qquad + \left| 2^{l(1-m)}\Delta_{H}^{m-1}(f; x_0) - \Delta_{h_l}^{k-1}(f; x_0) \right| \\
&\quad\le 2(1/2)(m-1)H = x_1 - x_0.
\end{aligned}$$

The results of Subsection 5.3 are contained in [Brudnyi and Gopengauz (1960)].

5.4. On the exactness of Theorem 5.5

The result of Theorem 5.5 is exact because the following statement is true:

Theorem 5.6. *Let* $m := r + k$, $m \ne 1$, $\varphi \in \Phi^m$, *and*

$$F(x) = F(x; \varphi; m) = \frac{1}{(m-2)!} \int_0^x x(x-u)^{m-2} u^{-m} \varphi(u)\, du, \quad x \ge 0.$$

Then the following assertions are true:

(i) if

$$\int_0^1 r u^{-r-1} \varphi(u)\, du = \infty,$$

then $F \notin C^r([0, 1])$;

(ii) if

$$\int_0^1 r u^{-r-1} \varphi(u)\, du < \infty,$$

then $F \in C^r([0, 1])$ *and, for all* $j = 0, \ldots, k$, *the following estimate is valid:*

$$\omega_j(t; F^{(r)}; [0, 1])$$

$$\ge c \int_0^t r u^{-r-1} \varphi(u)\, du + c(k-j) t^j \int_t^1 u^{-r-1} \varphi(u)\, du, \quad 0 \le tm \le 1. \tag{5.18}$$

Recall that $\varphi(t) \le \omega_m(t; F; [0, 1]) \le m\varphi(t)$ [see (4.25)].
Theorem 5.6 is a corollary of Lemma 5.3 and Remark 5.3 (see below).

Lemma 5.3 [Shevchuk (1976), (1989a)]. *Let* $m \in \mathbb{N}$ *and* $m \ne 1$. *Assume that a function* $f = f(u)$ *is given that is continuous and nonnegative on the interval* $(0, \infty)$ *and such that the integral*

$$G(x) := \frac{1}{(m-2)!} \int_1^x x(x-u)^{m-2} f(u)\, du \tag{5.19}$$

is continuous at the point $x = 0$ (*i.e., it is continuous on* $[0, \infty)$). *Then the following assertions are true:*

(i) *for* $r = 1, \dots, m-2$, *the condition*

$$\int_0^1 u^{m-r-1} f(u)\, du < \infty \tag{5.20}$$

is necessary and sufficient for G *to belong to* $C^r([0, \infty])$; *in the case* $r = m-1$, *the following condition is necessary and sufficient for* G *to belong to* $C^{m-1}([0, \infty])$:

$$\begin{cases} \int_0^1 f(u) du < \infty, & (5.21) \\ \lim_{x\to\infty} x f(x) = 0; & (5.22) \end{cases}$$

(ii) *if* $G \in C^r([0, \infty])$, $r = 0, \dots, m-1$, *then the following relation holds for* $j = 1, \dots, m-r$:

$$|\Delta_h^j(G^{(r)}; 0)| \geq c \int_0^h r u^{m-r-1} f(u)\, du + c(m-r-j) h^j \int_{mh}^1 u^{m-r-1-j} f(u)\, du, \tag{5.23}$$

where $0 \leq mh \leq 1$ *and* $c = c(m) = \text{const}$.

Proof. Denote

$$G_s(x) := (1/(s-1)!) \int_1^x (x-u)^{s-1} f(u)\, du, \qquad s \in \mathbb{N},$$

$$G_0(x) := f(x), \qquad k := m - r.$$

Note that, for all $r = 0, \dots, m-1$, the following identity is true:

$$G^{(r)}(x) = x G_{k-1}(x) + r G_k(x). \tag{5.24}$$

(i) *Necessity.* Let $G \in C^r([0, \infty))$, $r = 1, \dots, m-1$. For any $\varepsilon \in (0, 1/m)$, using (3.42) and (3.34), we obtain

$$\Delta_\varepsilon^r(G;0) = r\varepsilon\Delta_\varepsilon^{r-1}(G_{m-1};\varepsilon) = r\varepsilon^r G_{m-1}^{(r-1)}(\theta) = r\varepsilon^r G_k(\theta), \quad \theta\in[\varepsilon, r\varepsilon],$$

which yields

$$|\Delta_\varepsilon^r(G;0)| = (-1)^k r\varepsilon^r G_k(\theta)$$

$$\geq \frac{r\varepsilon^r}{(k-1)!}\int_{r\varepsilon}^1 (u-\theta)^{k-1} f(u)\,du \geq A\varepsilon^r \int_{m\varepsilon}^1 u^{k-1} f(u)\,du,$$

where $A = rk^l m^{-1}((k-1)!)^{-1}$. Therefore, with regard for (3.34), we get

$$\int_{m\varepsilon}^1 u^{k-1} f(u)\,du \leq A^{-1}\varepsilon^{-r}|\Delta_\varepsilon^r(G;0)| \leq A^{-1}\left\|G^{(r)}\right\|_{[0,1]},$$

i.e., the necessity of condition (5.20) is proved. In the case $r = m-1$, the necessity of (5.22) is an immediate consequence of the identity

$$G^{m-1}(x) = xf(x) + (m-1)\int_1^x f(u)\,du.$$

(i) *Sufficiency.* It follows from condition (5.20) that

$$x\int_x^1 u^{k-2} f(u)\,du = x\int_x^{\sqrt{x}} u^{k-2} f(u)\,du + x\int_{\sqrt{x}}^1 u^{k-2} f(u)\,du$$

$$\leq \int_0^{\sqrt{x}} u^{k-1} f(u)\,du + \sqrt{x}\int_0^1 u^{k-1} f(u)\,du \to 0 \tag{5.25}$$

as $x\to 0$. In view of (5.24), this implies that $G^{(r)}(x)\to rG_k(0) = G^{(r)}(0)$.

(ii) As proved above, if $G\in C^r([0,\infty))$, then relation (5.24) holds for $x = 0$. Let us estimate the jth difference for the first term $i_1(x) := xG_{k-1}(x)$ in (5.24). By virtue of (3.42) and (3.34), we get

$$\Delta_h^j(i_1;0) = jh\,\Delta_h^{j-1}(G_{k-1};h) = jh^j G_{k-j}(\theta), \quad \theta\in[h, jh],$$

whence

$$(-1)^{k-j}\Delta_h^j(i_1;0) \;=\; (-1)^{k-j}jh^jG_{k-j}(\theta) \;\geq\; c^{(k-j)}\int_{mh}^{1} u^{k-j}f(u)\,du.$$

It is clear that the second term $rG_k(x)$ in (5.24) is equal to zero for $r=0$. In the case where $r \neq 0$, we denote $i_2(x) := 0$ for $x > h$,

$$i_2(x) \;:=\; (1/(k-1)!)\int_{h}^{x} (x-u)^{k-1}f(u)\,du \qquad \text{for } x \in [0,h],$$

and $i_3(x) = G_k(x) - i_2(x)$. As a result, we get

$$(-1)^{k-j}\Delta_h^j(i_2;0) \;=\; (-1)^k i_2(0) \;=\; (1/(k-1)!)\int_{0}^{h} u^{k-1}f(u)\,du,$$

$$(-1)^{k-j}\Delta_h^j(i_3;0) \;=\; (-1)^{k-j}i_3^{(j)}(\theta)h^j, \quad \theta \in (0, jh).$$

If $\theta \in (h, jh]$, then

$$(-1)^{k-j}i_3^{(j)}(\theta) \;=\; (-1)^{k-j}G_{k-j}(\theta) \;\geq\; 0,$$

and if $\theta \in (0,h]$, then

$$\begin{aligned}(-1)^{k-j}i_3^{(j)}(\theta) \;&=\; (-1)^{k-j}\big(G_{k-j}(\theta) - i_2^{(j)}(\theta)\big)\\ &=\; (-1)^{k-j}(1/(k-j-1)!)\int_{1}^{h}(\theta-u)^{k-j-1}f(u)\,du.\end{aligned}$$

Thus,

$$(-1)^{k-j}\Delta_h^j(G_k;0) \;\geq\; (-1)^{k-j}\Delta_h^j(i_2;0) \;\geq\; (1/(k-1)!)\int_{0}^{h} u^{k-1}f(u)\,du.$$

■

Remark 5.3. If $\varphi \in \Phi^m$ and $f(u) = u^{-m}\varphi(u)$, then

$$\int_0^x f(u)du \geq \int_0^x x^{-m}\varphi(x)du = xf(x),$$

i.e., relation (5.22) follows in this case from (5.21).

6. Whitney inequality

Recall that the value of the best uniform approximation of a function $f \in C([a,b])$ by algebraic polynomials P_n of degree $\leq n$ is defined as the number

$$E_n(f)_{[a,b]} = \inf_{P_n} \|f - P_n\|_{[a,b]}.$$

In the case where $[a,b] = [0,1]$, we write $E_n(f)$ instead of $E_n(f)_{[a,b]}$ and $\|f\|$ instead of $\|f\|_{[a,b]}$.

It follows from Theorem 5.2 that

$$E_1(f)_{[a,b]} \leq \omega_2((b-a)/2; f; [a,b]),$$

and one can easily see that

$$E_1(f)_{[a,b]} \leq \frac{1}{2}\omega_2\left(\frac{b-a}{2}; f; [a,b]\right). \tag{6.1}$$

A generalization of this inequality to the case $k \in \mathbb{N}$, i.e., the inequality

$$E_{k-1}(f)_{[a,b]} \leq c\omega_k\left(\frac{b-a}{k}; f; [a,b]\right), \tag{6.2}$$

is called the Whitney inequality; it is proved in Subsection 6.1.

Recall that we use the notation c for different constants that may depend only on k and r (or some of these parameters). Also recall that $L(x, f; x_0, \dots, x_k)$ denotes the Lagrange polynomial of degree $\leq k$ that interpolates a function f at the points $x_0, \dots, x_k$.

6.1. Whitney inequality

First, in Lemma 6.1 and Theorem 6.1, we obtain a weak version of the Whitney inequality (for differentiable functions).

Lemma 6.1 (see, e.g., [Zhuk and Natanson (1983)]). *Let* $x_0 \in [a, b]$, $h > 0$, $x_j := x_0 + jh$, *and* $x_k \in [a, b]$. *If* $F \in C^1([a, b])$, *then, for every* $x \in [a, b]$, *the following inequality is true:*

$$|F(x) - L(x; F; x_0, \ldots, x_k)| \leq \frac{|(x - x_0) \ldots (x - x_k)|}{k! h^k} \omega_k(h), \tag{6.3}$$

where

$$\omega_k(t) := \omega_k(t; F'; [a, b]).$$

Proof. For every $t_1 \in [0, 1]$, we set $F_{t_1}(u) := F'(x + (u - x)t_1)$, $u \in [a, b]$. Then relations (3.25″) and (3.32) yield

$$[x, x_0, \ldots, x_k; F] = \int_0^1 [x_0, \ldots, x_k; F_{t_1}] dt_1 = \frac{1}{k! h^k} \int_0^1 \Delta_h^k(F_{t_1}; x_0) dt_1.$$

Since

$$\Delta_h^k(F_{t_1}; u) = \left|\Delta_{ht_1}^k(F'; x + (u - x)t_1)\right| \leq \omega_k(ht_1) \leq \omega_k(h),$$

relation (6.3) follows from (3.8).

■

Lemma 6.1 readily yields the following statement:

Theorem 6.1. *If* $F \in C^1([a, b])$, *then*

$$E_k(F)_{[a, b]} \leq \frac{b - a}{k} \omega_k\left(\frac{b - a}{k}; F'; [a, b]\right). \tag{6.4}$$

Remark 6.1. It is easy to see that the factor $\dfrac{b - a}{k}$ in (6.4) (but not the step) can be replaced by $\dfrac{b - a}{e \sigma_k k}$, $\sigma_k := 1 + \ldots + \dfrac{1}{k}$.

We are now ready to prove the main theorem of this section.

Theorem 6.2 [Whitney (1957), (1959)]. *If* $f \in C([a, b])$, *then*

$$E_{k-1}(f)_{[a,b]} \le W_k \omega_k\left(\frac{b-a}{k}; f; [a, b]\right), \tag{6.5}$$

where $W_k = \text{const}$ *depends only on* k.

Proof. We set

$$x_0 := a, \quad h := \frac{b-a}{k}, \quad x_j := x_0 + jh, \quad F(x) := \int_a^x f(u)du,$$

$$G(x) := F(x) - L(x; F; x_0, \dots, x_k), \quad g(x) := G'(x),$$

$$\omega_k(t) := \omega_k(t; f; [a, b]) \equiv \omega_k(t; g; [a, b]).$$

We fix $x \in [a, b]$ and choose δ for which $(x + k\delta) \in [a, b]$. As a result, we get

$$\int_0^1 \Delta_{t\delta}^k(g; x)dt = (-1)^k g(x) + \sum_{j=1}^k (-1)^{k-j}\binom{k}{j}\int_0^1 g(x + jt\delta)dt$$

$$= (-1)^k g(x) + \sum_{j=1}^k (-1)^{k-j}\binom{k}{j}\frac{1}{j\delta}(G(x + j\delta) - G(x)), \tag{6.6}$$

whence

$$|g(x)| \le \int_0^1 \left|\Delta_{t\delta}^k(g; x)\right|dt + \frac{2}{|\delta|}\|G\|_{[a,b]}\sum_{j=1}^k\binom{k}{j}\frac{1}{j} \le \omega_k(|\delta|) + \frac{1}{|\delta|}2^{k+1}\|G\|_{[a,b]}. \tag{6.7}$$

By virtue of Lemma 6.1, we have $\|G\|_{[a,b]} \le h\omega_k(h)$. Therefore,

$$E_{k-1}(f)_{[a,b]} \le \|g\|_{[a,b]} \le \omega_k(|\delta|) + |\delta|^{-1}h2^{k+1}\omega_k(h).$$

To complete the proof, note that δ can always be chosen so that $h \ge |\delta| \ge h/2$.

■

We denote the smallest possible constant in the Whitney inequality (6.5) by $W(k)$ and call it the Whitney constant.

6.2. Sendov's conjecture

For practical applications of the Whitney inequality, it is important to have good estimates for the Whitney constant $W(k)$. As mentioned above, it was first proved by Burkill (1952) that $W(2) \le 1$. He also formulated the conjecture that $W(k)$ is finite for each $k \ge 3$ (the case $k = 1$ is trivial: $W(1) = 1/2$). Burkill's conjecture was proved by Whitney (1957), who also proved that

$$W(2) = \frac{1}{2}, \quad \frac{8}{15} \le W(3) \le 0.7, \quad W(4) \le 3.3, \quad W(5) \le 10.4,$$

and

$$W(k) \ge \frac{1}{2}, \quad k \in \mathbb{N}. \tag{6.8}$$

Later, Brudnyi (1964) obtained the estimate $W(k) = O(k^{2k})$. Sendov (1982) proved that

$$W(4) \le 1.26, \quad W(5) \le 1.31, \quad W(6) \le 1.67, \quad W(k) \le (k+1)k^k,$$

and formulated the conjecture that

$$W(k) \le 1. \tag{6.9}$$

Ivanov and Takev (1985) proved that $W(k) = O(k \ln k)$, Binev (1985) proved that $W(k) = O(k)$, and, finally, Sendov [(1985), (1986)] established that the Whitney constant $W(k)$ is bounded by an absolute constant, namely, $W(k) \le 6$. Kryakin (1989) modified the method used by the authors cited above and obtained the estimate $W(k) \le 3$. This inequality was also announced by Brudnyi (1983) and Sendov [Sendov and Popov (1989), p. 37]. Later, Kryakin proved a more precise estimate. A modification of his proof allows one to obtain the estimate

$$W(k) \le 2 + \frac{1}{e^2}. \tag{6.10}$$

The monograph [Sendov and Popov (1988)] contains proofs of the inequality $E_{k-1}(f) \le 6\omega_k(1/k)$ and similar inequalities as well as applications of the Whitney inequality to the numerical integration, approximate solution of integral and differential equations, etc.

An analog of the Whitney inequality is also true in L_p [see Burkill (1952) ($k = 2$, $p = \infty$), Whitney [(1957), (1959)] ($p = \infty$), Brudnyi (1964) ($1 \le p \le \infty$), Storozhenko

(1977) ($0<p<1$), Sendov and Takev (1986) ($W_1(k)\le 30$), Kryakin (1989) ($W_p(k)\le 11$, $1\le p<\infty$), and Kryakin and Kovalenko ($W_p(k)<9$, $W_1(k)<6.5$)]. Multidimensional analogs of the Whitney inequality can be found in works of Brudnyi, Jonen, Scherer, Takev, Storozhenko, Osvald, Dahmen, DeVore, Binev, Ivanov, etc. (see [Brudnyi (1970); Takev (1988); Storozhenko and Oswald (1978); Dahmen, DeVore, and Scherer (1980); Binev and Ivanov (1985)]).

6.3. Corollaries of the Marchaud inequality and Whitney inequality

Let $f\in C([a,b])$, $k\in\mathbb{N}$, and $\omega_k(t):=\omega_k(t;f;[a,b])$. We fix a point $x_0\in[a,b]$ and a number $h>0$ such that $(x_0+(k-1)h)\in[a,b]$. Denote

$$J := [x_0-h, x_0+kh]\cap[a,b] \quad\text{and}\quad x_i := x_0+ih,$$

and let

$$L(x,f) := L(x,f;x_0,\dots,x_{k-1})$$

be the Lagrange polynomial that interpolates the function f at the points x_i, $i=0,\dots,k-1$. We set $g(x):=f(x)-L(x,f)$, $x\in[a,b]$, i.e. [see (3.8)],

$$g(x) = [x,x_0,\dots,x_{k-1};f]\prod_{i=0}^{k-1}(x-x_i). \tag{6.11}$$

Note that $\omega_k(t;g;[a,b])\equiv\omega_k(t)$.

Lemma 6.2. *The following inequalities are true:*

(i)
$$\|g\|_J \le c\,\omega_k(h); \tag{6.12}$$

(ii)
$$\omega_j(t;g;J)\le c t^j\int_t^{kh}u^{-j-1}\omega_k(u)\,du,\quad j=1,\dots,k-1,\ 0\le t\le h; \tag{6.13}$$

(iii)
$$|g(x)| \le c\,|x-x_0|^{k-1}\int_h^{|x-x_0|}u^{-k}\omega_k(u)\,du,\quad x\in[a,b]\setminus J; \tag{6.14}$$

(iv)
$$|g(x)| \le c(1+|x-x_0|/h)^k\omega_k(h)\,du,\quad x\in[a,b]; \tag{6.15}$$

(v) $$|f(x)| \le c|x-x_0|^{k-1} \int_h^{|x-x_0|} \frac{\omega_k(u)}{u^k}\,du + c\frac{|x-x_0|^{k-1}}{h^{k-1}}\|f\|_J, \quad x \in [a,b]\setminus J; \tag{6.16}$$

(vi) $$|f(x)| \le c(1+|x-x_0|/h)^k \times \left(\omega_k(h) + \|f\|_{[x_0,x_0+(k-1)h]}\right), \quad x \in [a,b]; \tag{6.17}$$

Proof. By virtue of the Whitney inequality (6.5), there exists an algebraic polynomial $P_{k-1} = P_{k-1}(x)$ of degree $\le k-1$ for which $\|f - P_{k-1}\|_J \le c\,\omega_k(h)$. Taking into account that

$$g(x) = f(x) - L(x,f) = (f(x) - P_{k-1}(x)) - L(x, f - P_{k-1})$$

and using estimate (1.47), we obtain

$$\|g\|_J \le c\,\omega_k(h) + (2^k - 1)\max_{i=0,\dots,k-1}|f(x_i) - P_{k-1}(x_i)| \le c2^k\omega_k(h),$$

i.e., inequality (6.12) is proved. Inequality (6.13) follows from relation (6.12) and the Marchaud inequality (5.6).

Let us prove inequality (6.14). Assume, for definiteness, that $x \ge x_0 + kh$. Let L^* denote the Lagrange polynomial that interpolates the function f at the points $x_0 + i(x-x_0)/(k-1)$, $i = 0,\dots,k-1$. We set $g^*(y) = f(y) - L^*(y)$ and note that $g^*(x_0) = g^*(x) = 0$ and $g(x) = -L(x, g^*)$. Estimate (6.13) yields

$$|\Delta_h^j(g^*, x_0)| \le c_1 h^j \int_h^{\delta} u^{-j-1}\omega_k(u)\,du, \qquad j = 1,\dots,k-1,$$

where $\delta := x - x_0$. Using the Newton formula (3.33), we obtain

$$|g(x)| = |L(x, g^*)| \le \sum_{j=1}^{k-1}\frac{c_1}{j!}\prod_{i=0}^{j-1}(x - x_i)\int_h^{\delta} u^{-j-1}\omega_k(u)\,du$$

$$\le c_2\sum_{j=1}^{k-1}\delta^j\int_h^{\delta} u^{-j-1}\omega_k(u)\,du \le (k-1)c_2\delta^{k-1}\int_h^{\delta} u^{-k}\omega_k(u)\,du.$$

Inequality (6.14) is proved. Inequalities (6.16) and (6.17) follow from the equality $f(x) = g(x) + L(x, f)$, the estimate

$$|L(x, f)| \le c(1 + |x - x_0| / h)^{k-1} (\omega_k(h) + \|f\|_{[x_0, x_0+(k-1)h]})$$

and inequalities (6.14) and (6.15), respectively.

■

Lemma 6.2′. *If* $f \in C^p([a, b])$, $p \in \mathbb{N}$, $p < k$, *then*

$$\|g^{(p)}\|_J \le c\omega_{k-p}(h; f^{(p)}; [a, b]). \tag{6.18}$$

Proof. Let

$$L_{k-p-1} := L_{k-p-1}(x, f^{(p)}) := L_{k-p-1}(x, f^{(p)}; y_0, \dots, y_{k-p-1})$$

be the Lagrange polynomial of degree $\le p - k - 1$ that interpolates the derivative $g^{(p)} = g^{(p)}(x)$ at the points $y_i := x_0 + ikh/(k - p - 1)$, $i = 0, \dots, k - p - 1$ $(L_0 \equiv g^{(p)}(x_0)$ in the case where $k = p + 1)$. We set

$$\mathcal{L}(x) := \sum_{\nu=0}^{p-1} \frac{1}{\nu!}(x_0 - x)^\nu f^{(p)}(x_0) + \frac{1}{(p-1)!} \int_{x_0}^{x} (x - u)^{p-1} L_{k-p-1}(u, f^{(p)})\, du$$

and note that

$$g(x) \equiv f(x) - L(x, f) = f(x) - \mathcal{L}(x, f) - L(x, f - \mathcal{L}),$$

i.e., $g^{(p)}(x) = f^{(p)}(x) - \mathcal{L}^{(p)}(x) + L^{(p)}(x, f - \mathcal{L})$. According to (6.12), we have

$$\|f^{(p)} - L_{k-p-1}\|_J \le c_1 \omega_{k-p}(h; f^{(p)}; J) =: c_1 \omega(h).$$

Therefore,

$$|f(x) - \mathcal{L}(x; f)|$$

$$\equiv \frac{1}{(p-1)!} \left| \int_{x_0}^{x} (x - u)^{p-1} (f^{(p)}(u) - L_{k-p-1}(u; f^{(p)}))\, du \right| \le c_2 h^p \omega(h).$$

Hence, by virtue of (3.48), we obtain

$$\left|L^{(p)}(x, f - \mathcal{L})\right| \le c_3 h^{-p} \|f - \mathcal{L}\|_J \le c_3 c_2 \omega(h), \quad x \in J.$$

Thus,

$$\begin{aligned}\left\|g^{(p)}\right\|_J &= \left\|f^{(p)} - L^{(p)}\right\|_J \le \left\|f^{(p)} - \mathcal{L}^{(p)}\right\|_J + c_3 c_2 \omega(h) \\ &= \left\|f^{(p)} - L_{k-p-1}\right\|_J + c_3 c_2 \omega(h) \le c_1 \omega(h) + c_3 c_2 \omega(h) = c\omega(h).\end{aligned}$$

■

Remark 6.2. Let x^* denote the point closest to x among the points x_i, $i = 0, \ldots, k-1$. It follows from (6.11) and (6.14) that

$$\left|[x, x_0, \ldots, x_{k-1}; f]\right| \le \frac{c}{|x - x^*|} \int_h^{|x-x^*|} \frac{\omega_k(u)}{u^k} du, \qquad x \in [a, b] \setminus J. \tag{6.19}$$

Relation (6.11) and inequality (6.13) applied to $j = 1$ yield the following estimate:

$$\left|[x, x_0, \ldots, x_{k-1}; f]\right| \le c h^{1-k} \int_{|x-x^*|}^{kh} u^{-k} \omega_k(u)\, du, \qquad x \in J, \quad k \ne 1. \tag{6.20}$$

6.4. Estimate for a divided difference for arbitrary nodes

Inequalities (6.19) and (6.20) give an estimate for the kth divided difference of a function f in terms of the kth modulus of continuity in the case where k nodes are equidistant and one node (the point x) may be arbitrary. In the present subsection, we prove an estimate for the kth divided difference in the case of arbitrary location of all $k + 1$ nodes of the divided difference. For $k = 1$, we obviously have

$$|[x_0, x_1; f]| \le (x_1 - x_0)^{-1} \omega_1((x_1 - x_0); f; [x_0, x_1]), \qquad x_0 < x_1. \tag{6.21}$$

For $k = 2$, inequality (3.10) yields

$$|[x_0, x_1, x_2; f]| \le \frac{18}{x_2 - x_0} \int_{\min\{x_2 - x_1, x_1 - x_0\}}^{x_2 - x_0} \frac{\omega_2(u; f; [x_0, x_2])}{u^2} du, \quad x_0 < x_1 < x_2. \tag{6.22}$$

If $\omega_k(t;f;[a,b]) \le t^k$, $k \in \mathbb{N}$, then, for any different points $x_i \in [a,b]$, $i = 0,\dots,k$, according to (3.17), (2.10), and (1.20) we have

$$|[x, x_0, \dots, x_k; f]| \le c. \tag{6.23}$$

Definition 6.1 ([Shevchuk (1984a)]). *Let a natural number k, a k-majorant $\varphi = \varphi(t)$ (see Definition 4.4), and $k+1$ points $x_i \in \mathbb{R}$, $x_0 < x_1 < \dots < x_k$, be given. For every pair of numbers $p = 0,\dots,k-1$ and $q = p+1,\dots,k$, we denote*

$$\Lambda_{p,q}(x_0,\dots,x_k;\varphi) := \frac{\int_{(x_q-x_p)/2}^{\min\{x_{q+1}-x_p;\, x_q-x_{p-1}\}} \frac{\varphi(u)}{u^{q-p+1}}\,du}{\prod_{i=0}^{p-1}(x_q-x_i)\prod_{i=q+1}^{k}(x_i-x_p)}, \tag{6.24}$$

where $x_{-1} := x_0-(x_k-x_0)$ and $x_{k+1} := x_k+(x_k-x_0)$. We set

$$\Lambda(x_0,\dots,x_k;\varphi) := \max_{p=0,\dots,k-1}\ \max_{q=p+1,\dots,k} \Lambda_{p,q}(x_0,\dots,x_k;\varphi). \tag{6.25}$$

We write $A \asymp B$ if $c_1 \le A/B \le c_2$. One can easily verify that

$$\Lambda(x_0,x_1;\varphi) \asymp (x_1-x_0)^{-1}\varphi(x_1-x_0), \tag{6.26}$$

$$\Lambda(x_0,x_1,x_2;\varphi) \asymp \frac{1}{x_2-x_0}\int_{\min\{x_2-x_1;\, x_1-x_0\}}^{x_2-x_0} \frac{\varphi(u)}{u^2}\,du, \tag{6.27}$$

if $\varphi(t) = t^{k-1}$, $k \ne 1$, then

$$\Lambda(x_0,x_1,\dots,x_k;\varphi) \asymp \frac{1}{x_k-x_0}\left(\left|\ln\frac{x_k-x_1}{x_{k-1}-x_0}\right|+1\right), \tag{6.28}$$

and if $\varphi(t) = t^k$, then

$$\Lambda(x_0,x_1,\dots,x_k;\varphi) \asymp 1. \tag{6.29}$$

Theorem 6.3 [Shevchuk (1984a)]. *Suppose that $\varphi \in \Phi^k$ and $x_0 < x_1 < \dots < x_k$. If $\omega_k(t;f;[a,b]) \le \varphi(t)$, then*

$$|[x_0,\dots,x_k; f]| \le c\Lambda(x_0,x_1,\dots,x_k;\varphi). \tag{6.30}$$

Theorem 6.3 is a corollary of the more general Theorem 6.4 below.

Definition 6.2 [Shevchuk (1984)]. *Let a nonnegative integer number* r, *a natural number* k, *a* k*-majorant* $\varphi = \varphi(t)$, *and* $k+r+1$ *points* x_i, $x_0 < x_1 < \ldots < x_{k+r}$, *be given. Denote* $m := r+k$, $x_{-1} := x_0 - (x_m - x_0)$, *and* $x_{m+1} := x_m + (x_m - x_0)$. *For every pair of numbers* $p = 0, \ldots, k-1$ *and* $q = p+r+1, \ldots, m$, *we set*

$$d(p, q) := \min\{x_{q+1} - x_p; x_q - x_{p-1}\},$$

$$\Lambda_{p,q,r}(x_0, \ldots, x_m; \varphi) := \frac{\int_{x_q - x_p}^{d(p,q)} u^{p+r-q-1} \varphi(u)\, du}{\prod_{i=0}^{p-1} (x_q - x_i) \prod_{i=q+1}^{m} (x_i - x_p)}. \tag{6.31}$$

The expression

$$\Lambda_r(x_0, \ldots, x_m; \varphi) := \max_{p=0,\ldots,k-1} \max_{q=p+r+1,\ldots,m} \Lambda_{p,q,r}(x_0, \ldots, x_m; \varphi) \tag{6.32}$$

is called the r*th divided majorant.*

One can easily verify that

$$\Lambda_0(x_0, \ldots, x_k; \varphi) = \Lambda(x_0, \ldots, x_k; \varphi), \tag{6.33}$$

$$\Lambda_r(x_0, \ldots, x_{r+1}; \varphi) \asymp (x_{r+1} - x_0)^{-1} \varphi(x_{r+1} - x_0), \tag{6.34}$$

and if $\varphi(t) = t^k$, *then*

$$\Lambda_r(x_0, \ldots, x_m; \varphi) \asymp 1. \tag{6.35}$$

Theorem 6.4 [Shevchuk (1984), (1984a)]. *Let* $\varphi \in \Phi^k$, $f \in C^r([x_0, x_m])$, $m := r+k$, *and* $x_0 < x_1 < \ldots < x_m$. *If* $\omega_k(t; f^{(r)}; [a, b]) \le \varphi(t)$, *then*

$$\left|[x_0, \ldots, x_k; f]\right| \le c\Lambda_r(x_0, x_1, \ldots, x_m; \varphi). \tag{6.36}$$

Remark 6.3. It will be proved in Section 8 that inequality (6.36) is also meaningful and true in the case where points x_i may "coincide" (but at most $r+1$ points at once). Inequality (6.36) is exact in the sense that, for any collection of nodes $x_0, \ldots, x_m$ and any kth majorant φ, one can find a function $f \in C^r([x_0, x_m])$ such that

$$\omega_k(t; f^{(r)}; [x_0, x_m]) \le \varphi(t),$$

but

$$\left|[x_0, \dots, x_m; f]\right| > c_0 \Lambda_r(x_0, \dots x_m; \varphi).$$

This will be proved in the next chapter.

6.5. Proof of Theorem 6.4

For $r = 0$ and $k = 1$, Theorem 6.4 is obvious [see (6.26) and (6.21)]. First, we prove this theorem in the case where $r \in \mathbb{N}$ and $k = 1$. Using (6.16) and (6.19), we can represent the divided difference in the form

$$\begin{aligned}
&[x_0, \dots, x_{r+1}; f] \\
&\quad = (x_{r+1} - x_0)^{-1}([x_1, \dots, x_{r+1}; f]) - ([x_0, \dots, x_r; f]) \\
&\quad = (x_{r+1} - x_0)^{-1} \int_0^1 \int_0^{t_1} \dots \int_0^{t_{r-1}} (f^{(r)}(x_1 + (x_2 - x_1)t_1 \\
&\qquad + \dots + (x_r - x_{r-1})t_{r-1} + (x_{r+1} - x_r)t_r) \\
&\qquad - f^{(r)}(x_1 + (x_2 - x_1)t_1 + \dots + (x_r - x_{r-1})t_{r-1} + (x_0 - x_r)t_r)\, dt_r \dots dt_1,
\end{aligned}$$

whence

$$\left|[x_0, \dots, x_{r+1}; f]\right| \le (1/r!)(x_{r+1} - x_0)^{-1} \varphi(x_{r+1} - x_0).$$

By induction, we assume that Theorem 6.4 is true for a number $k-1$ and prove it for k. We set $H := x_m - x_0$ and denote by $L(x)$ the Lagrange polynomial of degree $\le k-1$ that interpolates the rth derivative $f^{(r)} = f^{(r)}(x)$ at the points $x_0 + iH/(k-1)$, $i = 0, \dots, k-1$. We set $g(x) := f^{(r)}(x) - L(x)$, $\omega(t) := \omega_{k-1}(t; g; [x_0, x_m])$, $G(x) :=$ $g(x)$ if $r = 0$, and

$$G(x) := (1/(r-1)!) \int_{x_0}^{x} (x-t)^{r-1} g(t)\, dt$$

if $r \in \mathbb{N}$, i.e., $G^{(r)}(x) \equiv g(x)$. Since $[x_0, \dots, x_m; f] = [x_0, \dots, x_m; G]$, by the induction hypothesis we get

$$\begin{aligned} \left|[x_0, \dots, x_m; f]\right| &= \left|[x_0, \dots, x_m; G]\right| \\ &= H^{-1}\left|[x_0, \dots, x_{m-1}; G] - [x_1, \dots, x_m; G]\right| \\ &\le c_1 H^{-1}\Lambda_r(x_0, \dots, x_{m-1}; \omega) + c_1 H^{-1}\Lambda_r(y_0, \dots, y_{m-1}; \omega), \end{aligned}$$

where $y_i := x_{i+1}$. Therefore, to prove Theorem 6.4, it suffices to prove the estimates

$$\Lambda_r(x_0, \dots, x_{m-1}; \omega) \le c_2 H \Lambda_r(x_0, \dots, x_m; \varphi), \tag{6.37}$$

$$\Lambda_r(y_0, \dots, y_{m-1}; \omega) \le c_2 H \Lambda_r(x_0, \dots, x_m; \varphi). \tag{6.38}$$

We prove estimate (6.37) [estimate (6.38) can be proved by analogy]. To this end, we fix numbers $p = 0, \dots, k-2$ and $q = p + r + 1, \dots, m-1$ and define integer numbers p_s and q_s for all $s = 0, \dots, m + p - q + 1$ as follows: $p_0 := p$, $q_0 := q$; if $\min\{x_{q+1} - x_p; x_q - x_{p-1}\} = x_q - x_{p-1}$, then $p_1 := p - 1$, $q_1 = q$, otherwise $p_1 = p$, $q_1 = q + 1$; … ; if $\min\{x_{q_s+1} - x_{p_s}; x_{q_s} - x_{p_s-1}\} = x_{q_s} - x_{p_s-1}$, then $p_{s+1} = p_s - 1$, $q_{s+1} = q_s$, otherwise $p_{s+1} = p_s$, $q_{s+1} = q_s + 1$; … ; $p_{m+p-q} = 0$, $q_{m+p-q} = m$; and $p_{m+p-q+1} = -1$, $q_{m+p-q+1} = m$. Denote $d_s := x_{q_s} - x_{p_s}$; in particular $d_0 = x_q - x_p$, $d_{m+p-q} = H$, and $d_{m+p-q+1} = 2H$.

The proof of estimate (6.37) and Theorem 6.4 is completed by the following lemma:

Lemma 6.3. *The following inequality is true:*

$$\Lambda_{p,q,r}(x_0, \dots, x_{m-1}; \omega) \le c_3 H \sum_{\nu=0}^{m+p-q} \Lambda_{p_\nu, q_\nu, r}(x_0, \dots, x_m; \varphi). \tag{6.39}$$

Proof. We use inequality (6.13) for $j = k - 1$, i.e., the inequality

$$\omega(t) \le c_4 t^{k-1} \int_t^{kH/(k-1)} u^{-k}\varphi(u)\,du \le c_4 t^{k-1} \int_t^{2H} u^{-k}\varphi(u)\,du, \tag{6.40}$$

where $0 \le t \le H$. Consider three cases.

1. Let $p=0$ and $q=m-1$. Then

$$\Lambda_{p,q,r}(x_0, \dots, x_{m-1}; \omega)$$

$$= \int_{d_0}^{2d_0} u^{-k}\varphi(u)\,du \le 2^{k-1}(\ln 2)\, d_0^{1-k}\omega(d_0)$$

$$\le 2^{k-1}(\ln 2)\, c_4 \int_{d_0}^{2H} u^{-k}\varphi(u)\,du$$

$$= 2^{k-1}(\ln 2)\, c_4 H \Lambda_{p,q,r}(x_0, \dots, x_m; \varphi) + 2^{k-1}(\ln 2)\, c_4 \int_{H}^{2H} u^{-k}\varphi(u)\,du$$

$$< 2^{k-1}(\ln 2)\, c_4 H(\Lambda_{p,q,r}(x_0, \dots, x_m; \varphi) + 2\Lambda_{0,m,r}(x_0, \dots, x_m; \varphi))$$

$$\le c_3 H \sum_{\nu=0}^{1} \Lambda_{p_\nu, q_\nu, r}(x_0, \dots, x_m; \varphi).$$

2. Let $p \ne 0$, $q=m-1$, and $x_m - x_p < x_q - x_{p-1}$, i.e.,

$$d_1 = x_m - x_p,\ d_2 = x_m - x_{p-1}, \dots, d_{p+1} = x_m - x_0 = H,\ d_{p+2} = x_m - x_{-1} = 2H.$$

Relation (6.40) yields

$$\int_{d_0}^{d_2} u^{p-k}\omega(u)\,du \le c_4 \int_{d_0}^{d_2} u^{p-1}\left(\int_{u}^{2H} v^{-k}\varphi(v)\,dv\right)du$$

$$= (c_4/p)\left(u^p \int_{u}^{2H} v^{-k}\varphi(v)\,dv\right)\Bigg|_{d_0}^{d_2} + (c_4/p)\int_{d_0}^{d_2} u^{p-k}\varphi(u)\,du$$

$$\le c_4 d_2^p \int_{d_2}^{2H} u^{-k}\varphi(u)\,du + c_4 \int_{d_0}^{d_2} u^{p-k}\varphi(u)\,du.$$

Taking into account that $d_{p-i+1}/2 < x_q - x_i < d_{p-i+1}$ for all $i = 0, \dots, p-1$, we get

$$(2^p c_4)^{-1} \Lambda_{p,q,r}(x_0, \dots, x_{m-1}; \omega)$$

$$= (2^p c_4)^{-1} \prod_{i=0}^{p-1} (x_q - x_i) \int_{d_0}^{x_q - x_{p-1}} u^{p-k} \omega(u)\, du$$

$$\le c_4^{-1} \prod_{i=2}^{p+1} d_i^{-1} \int_{d_0}^{d_2} u^{p-k} \omega(u)\, du$$

$$\le d_2^p \prod_{i=2}^{p+1} d_i^{-1} \int_{d_2}^{2H} u^{-k} \varphi(u)\, du + \prod_{i=2}^{p+1} d_i^{-1} \int_{d_0}^{d_2} u^{p-k} \varphi(u)\, du$$

$$= 2H \sum_{s=2}^{P+1} \left[d_2^{s-1} \prod_{i=2}^{s} d_i^{-1} \right] \prod_{i=s+1}^{p+1} d_i^{-1} \int_{d_s}^{d_{s+1}} u^{p-k-s} \varphi(u) \Big[(d_2/u)^{p-s+1} (u/2H) \Big] du$$

$$+ H \prod_{i=2}^{p+1} d_i^{-1} \int_{d_1}^{d_2} u^{p-k-1} \varphi(u) [u/H]\, du$$

$$+ H \left[d_1 H^{-1} \prod_{i=0}^{p-1} (x_q - x_i) \prod_{i=2}^{p+1} d_i^{-1} \right] d_1^{-1} \prod_{i=0}^{p-1} (x_q - x_i)^{-1} \int_{d_0}^{d_1} u^{p-k} \varphi(u)\, du.$$

Taking into account that all expressions in brackets do not exceed unity, we obtain relation (6.39) with $c_3 < 2^{k-1} c_4$.

3. The other situations are contained in the case where $x_q - x_{p-1} \le x_m - x_p$ It follows from (6.40) that

$$\int_{d_0}^{d_1} u^{r+p-q-1} \omega(u)\, du \le c_4 d_1^{m+p-q-1} \int_{d_1}^{2H} u^{-k} \varphi(u)\, du + c_4 \int_{d_0}^{d_1} u^{r+p-q-1} \varphi(u)\, du.$$

By construction, we have

$$x_{q_s} - x_{q_{s-1}} \le x_{p_{s-1}} - x_{p_{s-1}-1} < x_q - x_{p_{s-1}},$$

$$x_{p_{s-1}} - x_{p_s} \le x_{q_{s-1}+1} - x_{q_{s-1}} \le x_{q_{s+1}} - x_p.$$

Therefore,

$$x_{q_s} - x_{p_s - 1} < (s+1)(x_q - x_{p_s - 1}), \quad x_{q_s + 1} - x_{p_s} < (s+1)(x_{q_s + 1} - x_p).$$

Hence,

$$c_4^{-1} \Lambda_{p,q,r}(x_0, \dots, x_{m-1}; \omega)$$

$$= \frac{c_4^{-1}(x_m - x_p) \int_{d_0}^{d_1} u^{r+p-q-1} \omega(u)\, du}{\prod_{i=0}^{p-1} (x_q - x_i) \prod_{i=q+1}^{m} (x_i - x_p)}$$

$$\leq \frac{(x_m - x_p) d_1^{m+p-q-1}}{\prod_{i=0}^{p-1} (x_q - x_i) \prod_{i=q+1}^{m} (x_i - x_p)} \int_{d_1}^{2H} \frac{\varphi(u)}{u^k} du + (x_m - x_p) \Lambda_{p,q,r}(x_0, \dots, x_m; \varphi)$$

$$\leq \int_{H}^{2H} \frac{\varphi(u)}{u^k} du + H \sum_{s=1}^{m+p-q-1} \frac{d_1^{m+p-q-s-1}}{\prod_{i=0}^{p_s - 1} (x_q - x_i) \prod_{i=q_s+1}^{m} (x_i - x_p)} \int_{d_s}^{d_{s+1}} \frac{\varphi(u)}{u^k} du$$

$$+ H \Lambda_{p,q,r}(x_0, \dots, x_m; \varphi)$$

$$\leq 2H \Lambda_{0,m,r}(x_0, \dots, x_m; \varphi)$$

$$+ H \sum_{s=1}^{m+p-q-1} \frac{(s+1)^{m+p-q-s} d_1^{m+p-q-s-1}}{\prod_{i=0}^{p_s - 1} (x_{q_s} - x_i) \prod_{i=q_s+1}^{m} (x_i - x_{p_s})} \int_{d_s}^{d_{s+1}} \frac{\varphi(u)}{u^k} du$$

$$+ H \Lambda_{p,q,r}(x_0, \dots, x_m; \varphi) < c_5 H \sum_{\nu=0}^{m+p-q} \Lambda_{p_\nu, q_\nu, r}(x_0 \dots, x_m; \varphi).$$

Lemma 6.3 and Theorem 6.4 are proved.

■

7. Classes and spaces of functions defined by the kth modulus of continuity

7.1. Definitions

In the second section of this chapter, we have already introduced Lipschitz and Hölder classes and spaces defined by the first modulus of continuity. In the present section, we extend this definition to the case where k is an arbitrary natural number. The important special case of Zygmund spaces and Zygmund classes is provided by Definition 7.1 below.

In the present section, we denote

$$J := [a,b]. \tag{7.1}$$

Definition 7.1. *I. Let* $M = \text{const} > 0$. *A Zygmund class* $MZ[J]$ *is the set of functions* $f \in C(J)$ *satisfying the following inequality for every pair of points* $x_1, x_2 \in J$:

$$\left| f(x_1) - 2f\left(\frac{x_1+x_2}{2}\right) + f(x_2)\right| \le \frac{M}{2}|x_1 - x_2|. \tag{7.2}$$

II. We define

$$Z[J] := 1Z[J]. \tag{7.3}$$

III. A Zygmund space $Z(J)$ *is the set of functions* $f \in C(J)$ *for which there exists a number* $M = M(f) > 0$ *such that* $\omega_2(t; f; J) \le Mt$, *i.e.,*

$$Z(J) := U_{M>0} MZ[J]. \tag{7.4}$$

IV. Let $r \in \mathbb{N}$. *We write* $f \in MW^rZ[J]$, $f \in W^rZ[J]$, *and* $f \in W^rZ(J)$ *if* $f \in C^r(J)$ *and, respectively,* $f^{(r)} \in MZ[J]$, $f^{(r)} \in Z[J]$, *and* $f^{(r)} \in Z(J)$.

V. We denote $MW^0Z[J] := WZ[J]$, $W^0Z[J] := Z[J]$, *and* $W^0Z(J) := Z(J)$.

Definition 7.2. *Let* $k \in \mathbb{N}$ *and let* φ *be a* k*-majorant.*

I. The class $MH[k; \varphi; J]$, $M = \text{const} > 0$, *is the set of functions* $f \in C(J)$ *for which*

$$\omega_k(t; f; J) \le M\varphi(t). \tag{7.5}$$

II. We denote $H[k;\varphi;J] := 1H[k;\varphi;J]$.

III. The space $H_k^{\varphi}(J)$ *is the set of functions* $f \in C(J)$ *for which there exists a number* $M = M(f)$ *such that* $\omega_k(t;f;J) \le M\varphi(t)$, *i.e.,*

$$H_k^{\varphi}(J) := U_{M>0} MH[k;\varphi;J]. \tag{7.6}$$

IV. Let $r \in \mathbb{N}$. *We write* $f \in MW^rH[k;\varphi;J]$, $f \in W^rH[k;\varphi;J]$, *and* $f \in W^rH_k^{\varphi}(J)$ *if* $f \in C^r(J)$ *and, respectively,* $f^{(r)} \in MH[k;\varphi;J]$, $f^{(r)} \in H[k;\varphi;J]$, *and* $f^{(r)} \in H_k^{\varphi}(J)$.

V. We denote $MW^0H[k;\varphi;J] := MH[k;\varphi;J]$, $W^0H[k;\varphi;J] := H[k;\varphi;J]$, *and* $W^0H_k^{\varphi}(J) := H_k^{\varphi}(J)$.

Note that $H[\alpha;J] = H[1;\varphi;J]$ for $\varphi(t) = t^{\alpha}$ and $Z[J] = H[2;\varphi;J]$ for $\varphi(t) = t$. Theorem 2.1 and relation (5.17) imply that

$$W[r;J] = W^{r-j}H[j;\varphi_j;J] \quad \text{for} \quad \varphi_j(t) = t^j, \quad j = 1,\dots,r. \tag{7.7}$$

Note that the space $W^rH_k^{\varphi}(J)$ is an algebra by virtue of the following lemma:

Lemma 7.1 (see, e.g., [Trigub (1960)]). *Let functions* f *and* g *be given. If* $g \in M_1H[k;\varphi;J]$ *and* $f \in MH[k;\varphi;J]$, *then* $fg \in M_2H[k;\varphi;J]$, *where*

$$M_2 \le c(M_1\|f\| + M\|g\|) + c(k-1)(MM_1\varphi(b-a) + \|f\|\,\|g\|/\varphi(b-a)).$$

Proof. We fix $t \in (0,(b-a]$, denote

$$t_* := t\sqrt[k]{\varphi(b-a)/\varphi(t)},$$

and note that $t \le t_* \le b-a$. For every $j = 1,\dots,k-1$, the following inequalities are true:

$$\int_t^{b-a} u^{-j-1}\varphi(u)\,du \le t^{-k}\varphi(t)(b-a)^{k-j},$$

$$\int_t^{b-a} u^{-j-1}\varphi(u)\,du = \int_t^{t_*} u^{-j-1}\varphi(u)\,du + \int_{t_*}^{b-a} u^{-j-1}\varphi(u)\,du$$

$$\le t_*^{k-j}t^{-k}\varphi(t) + t_*^{-j}\varphi(b-a) \le 2t^{-j}(\varphi(t))^{j/k}(\varphi(b-a))^{1-j/k}.$$

It follows from (3.39) that

$$\omega_k(t;fg;J) \le \sum_{i=0}^{k} \binom{k}{i} \omega_i(t;f)\omega_{k-i}(t;g).$$

Therefore, by virtue of the Marchaud inequality (5.6), we obtain

$$\omega_k(t;fg;J)$$

$$\le \|f\| M_1 \varphi(t) + c_1 t^k \sum_{i=1}^{k-1} \left(M \int_t^{b-a} u^{-i-1} \varphi(u)\,du + (b-a)^{-i} \|f\| \right)$$

$$\times M_1 \left(\int_t^{b-a} u^{i-k-1} \varphi(u)\,du + (b-a)^{i-k} \|g\| \right) + M \|g\| \varphi(t)$$

$$\le (1+(k-1)c_1)(M_1 \|f\| + M \|g\|)\varphi(t) + (k-1)c_1 t^k (b-a)^{-k} \|f\| \|g\|$$

$$+ MM_1 4(k-1)c_1 \varphi(b-a)\varphi(t) \le M_2 \varphi(t).$$

■

7.2. Relations between spaces of continuous functions

Numerous relations between various classes $MW^rH[k;\varphi;J]$ and spaces $W^rH_k^{\varphi}(J)$ were established by S. Nikol'skii (1946b), Stechkin [(1951a), (1952)], Brudnyi (1959), Brudnyi and Gopengauz (1960), Geit (1972), Guseinov and Il'yasov (1977), and others. The most general result is given by the following theorem:

Theorem 7.1 [Guseinov (1979)]. *Let $k \in \mathbb{N}$, $j \in \mathbb{N}$, $\varphi \in \Phi^k$, $\omega \in \Phi^j$, $(r+1) \in \mathbb{N}$, and $r < j$. The following assertions are true:*

1. The inclusion

$$W^rH_k^{\varphi}(J) \subset H_j^{\omega}(J) \tag{7.8}$$

is true if and only if the following conditions are satisfied:

$$(a)\quad t^r\varphi(t)=O(\omega(t)) \quad \textit{if} \quad j\ge k+r; \tag{7.9}$$

$$(b)\quad t^j\int_t^1 u^{r-j-1}\varphi(u)\,du \;=\; O(\omega(t)) \quad \textit{if} \quad j<k+r. \tag{7.10}$$

2. *Let* $r\neq 0$. *The inclusion*

$$H_j^{\omega}(J) \;\subset\; W^r H_k^{\varphi}(J) \tag{7.11}$$

is true if and only if the following conditions are satisfied:

$$(a)\quad t^j\int_t^1 u^{-r-k-1}\omega(u)\,du + \int_0^t u^{-r-1}\omega(u)\,du \;=\; O(\varphi(t)) \quad \textit{if} \quad j>k+r; \tag{7.12}$$

$$(b)\quad \int_0^t u^{-r-1}\omega(u)\,du \;=\; O(\varphi(t)) \quad \textit{if} \quad j\le k+r. \tag{7.13}$$

Proof. 1a. *Sufficiency.* If $f\in W^r H_k^{\varphi}(J)$, then, by virtue of (4.15), we can conclude that $f\in H_{k+r}^{\varphi_1}(J)$, where $\varphi_1=\varphi_1(t)=t^r\varphi(t)$. Therefore, according to (4.12), we have $f\in H_j^{\varphi_1}(J)$, Hence, taking (7.9) into account, we establish that $f\in H_j^{\omega}(J)$.

1a. *Necessity.* We set $f(x):=0$ if $x<0$ and

$$f(x) \;=\; u^{2k+r}u^{-3k-r-1}\varphi(u)(u-x)^{k+r}\,du \qquad \text{if} \quad x\ge 0.$$

For $x\in(0,1]$, we have

$$\left|f^{(r)}(x)\right| \;\le\; c_1\varphi(x), \qquad \left|f^{(r+k)}(x)\right| \;\le\; c_2u^{-k}\varphi(x).$$

Therefore, $f\in W^r H_k^{\varphi}([-1,1])$. On the other hand, if $f\in H_j^{\varphi}([-1,1])$, then, for all $t\in(0,1/j)$, we obtain

$$O(\omega(t)) \;=\; \omega_j(t;f;[-1,1]) \;\ge\; \left|\Delta_t^j(f;(1-j)t)\right| \;=\; f(t) \;\ge\; c_3t^r\varphi(t),$$

i.e., the necessity is also proved.

1b. *Sufficiency.* If $f \in W^r H_k^{\varphi}(J)$, then, by virtue of (4.15), we have $f \in H_{k+r}^{\varphi_1}(J)$, where $\varphi_1(t) = t^r \varphi(t)$. Therefore, according to the Marchaud inequality (5.6), we get

$$\omega_k(t; f; J) = O\left(\int_t^1 u^{r-j} \varphi(u)\, du\right).$$

Taking (7.10) into account, we conclude that $f \in H_j^{\varphi}(J)$.

1b. *Necessity.* Let $F(x; \varphi; k)$ be the extremal function defined by (4.27). We set

$$f(x) := \frac{1}{(r-1)!} \int_1^x F(u; \varphi; k)(x-u)^{r+1}\, du \quad \text{if} \quad r \neq 0$$

and $f(x) := F(x; \varphi; k)$ if $r = 0$. According to (4.40), we have $f \in W^r H_k^{\varphi}([0,1])$. Denote

$$\varphi_2(t) = t^r \varphi(t) - \int_0^t r u^{r-1} \varphi(u)\, du$$

and note that $f(x) = F(x; \varphi_2; r+k)$. Therefore, by virtue of (5.18), we obtain

$$\omega_j(t; f; [0,1]) \geq c_4 t^j \int_t^1 u^{j-1} \varphi_2(u)\, du$$

$$= c_4\Big(t^j (1 - r/j) \int_t^1 u^{r-j-1} \varphi(u)\, du + (1/j) t^j \int_0^1 r u^{r-1} \varphi(u)\, du$$

$$- (1/j) \int_0^t r u^{r-1} \varphi(u)\, du,$$

i.e.,

$$t^j \int_t^1 u^{r-j-1} \varphi(u)\, du \leq c_5\big(t^r \varphi(t) + \omega_j(t; f; [0,1])\big) = O(\omega(t)).$$

2a. The necessity and the sufficiency of (7.12) for (7.11) follow from Theorems 5.6 and 5.5, respectively.

2b. *Sufficiency.* If $f \in H_j^{\omega}(J)$, then, taking (7.13) and (5.15) into account, we establish that $f \in C^r(J)$ and $\omega_{j-r}(t; f^{(r)}; J) = O(\varphi(t))$, whence, according to (4.12), we get $\omega_{k+r}(t; f^{(r)}; J) = O(\varphi(t))$, i.e., $f \in W^r H_k^{\varphi}(J)$.

2b. *Necessity.* In view of Lemma 4.3, we can assume, without loss of generality, that $|\omega'(t)| \le \omega(t)/t$. Denote $f(x) := F(x; \omega; j)$. According to (4.40), we have $f \in H_j^{\varphi}([0,1])$. We set

$$i_1 := (-1)^{k+r-j} \Delta_h^{j-r}(f^{(r)}; 0)$$

and

$$i_2 := \sum_{i=1}^{k+r-j} (-1)^{k+r-j-i} \binom{k+r-j}{i} \Delta_h^{j-r}(f^{(r)}; ih)$$

so that [see (3.30)] $\Delta_h^k(f^{(r)}; 0) = i_1 + i_2$. By virtue of (5.18), we obtain

$$|i_1| \ge c_6 \int_0^t u^{-r-1} \omega(u)\, du.$$

According to (3.37), we get

$$\left|\Delta_h^{j-r}(f^{(r)}; ih)\right| = h^{j-r} \left|f^{(r)}(\theta)\right| = h^{j-r} \theta^{-j+1} \omega'(\theta) \le c_7 h^{-r} \omega(h), \quad \theta \ge h,$$

whence

$$|i_2| \le (2^{k+r-j}) c_7 h^{-r} \omega(h) = c_8 h^{-r} \omega(h).$$

Therefore, the assumption $f \in W^r H_k^{\varphi}([0,1])$ yields

$$O(\varphi(t)) = \omega_k(t; f^{(r)}; [0,1]) \ge c_6 \int_0^t u^{-r-1} \omega(u)\, du - c_8 t^{-r} \omega(t).$$

Finally, since $f \in W^r H_k^{\varphi}([0,1])$, we obviously have $f \in H_{k+r}^{\varphi_1}([0,1])$, where $\varphi_1 = \varphi_1(t) = t^r \varphi(t)$. Consequently, according to assertion 1a, we can conclude that $t^{-r}\varphi(t) = t^{-r} O(\varphi_1(t)) = O(\varphi(t))$.

■

Remark 7.1. If $j \le r$, then inclusion (7.11) is not true for any $\varphi \in \Phi^k$ and $\omega \in \Phi^j$.

The following corollary of Theorem 7.1 is true:

Corollary 7.1. *Let* $k \in \mathbb{N}$, $j \in \mathbb{N}$, $\varphi \in \Phi^k$, $\omega \in \Phi^j$, $(r+1) \in \mathbb{N}$, $(p+1) \in \mathbb{N}$, $p \le r$, *and* $q := \min\{k, j+p-r\}$. *The condition*

$$\begin{cases} t^p\omega(t) \sim t^r\varphi(t), & (7.14) \\ \int_0^t (r-p)u^{-1}\varphi(t)du + |k+r-j-p|\,t^q\int_t^1 u^{-q-1}\varphi(u)du = O(\varphi(t)) & (7.15) \end{cases}$$

is necessary and sufficient for the equality

$$W^pH_j^\omega(J) = W^rH_k^\varphi(J). \tag{7.16}$$

The notation $t^p\omega(t) \sim t^r\varphi(t)$ means that $t^p\omega(t) = O(t^r\varphi(t))$ and $t^r\varphi(t) = O(t^p\omega(t))$.

The following corollary of Theorems 5.5 and 5.6 is true:

Corollary 7.2. *The condition*

$$\int_0^1 (p-r)u^{r-p-1}\varphi(u)du < \infty \tag{7.17}$$

is necessary and sufficient for the inclusion

$$W^rH_k^\varphi(J) \subset C^p(J), \quad (r+1) \in \mathbb{N},\ (p+1) \in \mathbb{N}. \tag{7.18}$$

Let us formulate some corollaries of relations (7.8)–(7.18). As above, we assume that $k \in \mathbb{N}$, $\varphi \in \Phi^k$, $(r+1) \in \mathbb{N}$, and $p \in \mathbb{N}$. Then

$$t^{p-r} = O(\varphi(t)) \Leftrightarrow W^p(J) \subset W^rH_k^\varphi(J), \tag{7.19}$$

$$\int_0^1 (p-r)u^{r-p-1}\varphi(u)du < \infty \Leftrightarrow W^rH_k^\varphi(J) \subset W^p(J), \tag{7.20}$$

$$p = r+k,\ t^{p-r} \sim \varphi(t) \Leftrightarrow W^rH_k^\varphi(J) = W^p(J), \tag{7.21}$$

$$p \geq r+k \Leftrightarrow C^p(J) \subset W^r H_k^{\varphi}(J). \tag{7.22}$$

It follows from relations (7.17), (7.18), (7.22), and (4.22) that the equality $C^p(J) = W^r H_k^{\varphi}(J)$ is impossible for any p, r, k, and φ.

Taking (7.8)–(7.22) into account and using Theorems 5.5 and 5.6 and inequalities (4.12) and (4.15), one can easily establish various relations between different classes $MW^rH[k; \varphi; J]$.

7.3. Space $H(\bar{\varepsilon})$

Definition 7.3. *Let $\bar{\varepsilon} = \{\varepsilon_n\}$ be a decreasing sequence of positive numbers ε_n, $n \in \mathbb{N}$. By $H[\bar{\varepsilon}]_J$ we denote the class of functions $f \in C(J)$ such that*

$$E_n(f)_J \leq \varepsilon_n, \qquad n \in \mathbb{N}. \tag{7.23}$$

In the case where $J = [0, 1]$, we set $H[\bar{\varepsilon}] := H[\bar{\varepsilon}]_{[0,1]}$. By $H(\bar{\varepsilon})_J$ we denote the class of functions $f \in C(J)$ for which there exists a number $M = M(f)$ such that $E_n(f)_J \leq M\varepsilon_n$ for all $n \in \mathbb{N}$.

It will be proved in Chapter 7 that the condition

$$\sum_{j=n+1}^{\infty} rj^{2r-1} + n^{-2k} \sum_{j=1}^{n} j^{2(r+k)-1} \varepsilon_j = O(\varphi(1/n^2)), \tag{7.24}$$

$$k \in \mathbb{N}, \qquad \varphi \in \Phi^k, \qquad (r+1) \in \mathbb{N},$$

is sufficient for the inclusion

$$H(\bar{\varepsilon})_J \subset W^r H_k^{\varphi}(J). \tag{7.25}$$

In the present subsection, we prove that condition (7.24) is also necessary for (7.25). Dolzhenko gave an example of a function $f \in H[\bar{\varepsilon}]$ such that

$$\omega_1(1/n^2; f^{(r)}; [0, 1]) \geq c n^{-2} \sum_{j=1}^{n} j\varepsilon_j, \qquad n \in \mathbb{N}, \tag{7.26}$$

i.e., he proved the necessity of (7.24) for (7.25) in the case where $r = 0$ and $k = 1$. In the general case, the following theorem is true:

Theorem 7.2 [Shevchuk (1989a)]. *Let* $k \in \mathbb{N}$ *and* $(r+1) \in \mathbb{N}$. *Then the following assertions are true:*

(i) if

$$\sum_{j=1}^{\infty} rj^{2r-1}\varepsilon_j = \infty, \tag{7.27}$$

then there exists a function $f \in H[\bar{\varepsilon}]$ *such that* $f \,\overline{\in}\, C^r([0, 1])$;

(ii) if

$$\sum_{j=1}^{\infty} rj^{2r-1}\varepsilon_j < \infty, \tag{7.28}$$

then there exists a function $f \in H[\bar{\varepsilon}]$ *such that* $f \in C^r([0, 1])$, *but, for all* $n \in \mathbb{N}$, $n \geq m := r + k$, *one has*

$$\omega_k(1/n^2; f; [0, 1]) \geq c \sum_{j=n+1}^{\infty} rj^{2r-1}\varepsilon_j + cn^{-2k} \sum_{j=m}^{n} j^{2(r+k)-1}\varepsilon_j. \tag{7.29}$$

Proof. 1. Let $x \in [0, 1]$. Note that, for $l \in \mathbb{N}$, the function $\sin^2(l \arcsin\sqrt{x})$ is an algebraic polynomial of degree l. For all $n \in \mathbb{N}$, we define polynomials T_n by the formula $T_n(x) := \sin^{2(m+2)}([n/(m+2)] \arcsin\sqrt{x})$, where $[n/(m+2)]$ is the integer part. The polynomials T_n possess the following obvious properties: T_n is a polynomial of degree n, $0 \leq T_n(x) \leq 1$,

$$T_n(x) \leq c_1 x^{m+2} n^{2(m+2)} \quad \text{if} \quad x \leq 1/n^2, \tag{7.30}$$

$$T_n(x) \geq c_2 x^{m+2} n^{2(m+2)}, \quad x \leq 1/n^2, \quad n \geq m+2, \tag{7.31}$$

$$T_n^{(j)}(0) = 0, \quad j = 0, \dots, m+1. \tag{7.32}$$

We define the function

$$\beta(x) := \sum_{j=1}^{\infty} j^{-3}\varepsilon_{j-2} T_j(x) \equiv \sum_{j=m+2}^{\infty} j^{-3}\varepsilon_{j-2} T_j(x)$$

and the polynomials

$$P_n(x) := \sum_{j=1}^{n} j^{-3}\varepsilon_{j-2}T_j(x).$$

We prove that if $n \ge m$ and $(n+1)^{-2} \le x \le n^{-2}$, then

$$\beta(x) = c_2\varepsilon_n x. \tag{7.33}$$

Indeed, if $n \ge m+2$, then, according to (7.31), we have

$$\beta(x) \ge \sum_{j=m+2}^{\infty} j^{-3}\varepsilon_{j-2}T_j(x) \ge c_2\varepsilon_{n-2}x^{m+2}\sum_{j=m+2}^{\infty} j^{-3}j^{2(m+2)} = c_3\varepsilon_n x.$$

If $n = m,\ m+1$, then, for all $x \ge (m+2)^{-2}$, we get

$$\beta(x) \ge (m+2)^{-3}\varepsilon_m T_{m+2}(x) = (m+2)^{-3}\varepsilon_m x^{m+2} \ge c_3\varepsilon_n x \ge c_3\varepsilon_{m+1}x.$$

We now prove the inequality

$$0 \le \beta(x) - P_n(x) \le c_4 x\varepsilon_{n-1}, \quad n \in \mathbb{N},\ n \ne 1. \tag{7.34}$$

First, let $x > (n+1)^{-2}$. Then

$$0 \le \beta(x) - P_n(x) = \sum_{j=n+1}^{\infty} j^{-3}\varepsilon_{j-2}T_j(x)$$

$$\le \varepsilon_{n-1}\sum_{j=n+1}^{\infty} j^{-3} < \frac{1}{2}\varepsilon_{n-1}n^{-2} < 2x\varepsilon_{n-1}.$$

Now let $x \le (n+1)^{-2}$. We choose a number $n_0 \in \mathbb{N}$ from the condition $(n_0+1)^{-2} < x \le n_0^{-2}$ and note that, according to (7.30), we have

$$\sum_{j=m+1}^{n_0} j^{-3}T_j(x) \le c_1 x^{m+2} n_0^{2(m+1)} \le c_1 x.$$

Therefore,

$$0 \le \beta(x) - P_n(x) \le \varepsilon_{n-1} \sum_{j=n+1}^{n_0} j^{-3} T_j(x) + \varepsilon_{n-1} \sum_{j=n_0+1}^{\infty} j^{-3} T_j(x)$$

$$< c_1 \varepsilon_{n-1} x + (1/2) \varepsilon_{n-1} n^{-2} n_0^{-2} < c_4 x \varepsilon_{n-1} x.$$

We set [cf. (5.18)]

$$F(x) := ((m-1)!)^{-2} \int_1^x u^{-m-2} \beta(u)\, x(x-u)^{m-1}\, du,$$

$$f(x) := c_4^{-1} F(x),$$

$$Q_n(x) := ((m-1)!)^{-2} \int_1^x u^{-m-2} P_{n+1}(x)\, x(x-u)^{m-1}\, du.$$

It follows from (7.32) that Q_n is a polynomial of degree n. Taking (7.34) into account, we get

$$|F(x) - Q_n(x)| \le c_4 \varepsilon_n \int_x^1 u^{-m-2} u x (x-u)^{m-1}\, du$$

$$\le c_4 \varepsilon_n \int_x^1 x u^{-2}\, du \le c_4 \varepsilon_n,$$

i.e.,

$$E_n(f)_{[0,1]} \le \varepsilon_n, \quad n \in \mathbb{N}. \tag{7.35}$$

2. Using (7.33), we obtain the following relation for all $n > m$:

$$\int_{1/n^2}^{1/m^2} u^{-r-2} \beta(u)\, du = \sum_{j=m}^{n-1} \int_{1/(j+1)^2}^{1/j^2} u^{-r-2} \beta(u)\, du$$

$$\ge c_3 \sum_{j=m}^{n-1} \varepsilon_j \int_{1/(j+1)^2}^{1/j^2} u^{-r-1}\, du \ge c_3 \sum_{j=m}^{n-1} \varepsilon_j j^{2r-1} \varepsilon_j. \tag{7.36}$$

Therefore, relation (7.27) yields

$$\int_0^1 r u^{-r-2} \beta(u)\, du = \infty.$$

According to assertion (i) of Lemma 5.3, this means that $F \notin C^r([0, 1])$, i.e.,

$$\sum_{j=1}^{\infty} r j^{2r-1} \varepsilon_j = \infty \;\Rightarrow\; F \notin C^r([0, 1]). \tag{7.36}$$

3. By virtue of (7.30), we have

$$\int_0^1 u^{-r-2} T_n(u)\, du = \left(\int_0^{1/n^2} + \int_{1/n^2}^1 \right) u^{-r-2} T_n(u)\, du$$

$$\leq c_1 n^{2(m+2)} \int_0^{1/n^2} u^{m-r}\, du + \int_{1/n^2}^{\infty} u^{-r-2}\, du = c_5 n^{2(r+2)}.$$

Therefore, relation (7.28) yields

$$\int_0^1 r u^{-r-2} \beta(u)\, du \leq \sum_{j=m+2}^{\infty} r \varepsilon_{j-2} j^{-3} \int_0^1 u^{-r-2} T_j(u)\, du$$

$$\leq c_6 \sum_{j=m+2}^{\infty} r \varepsilon_{j-2} j^{2r-1} < \infty.$$

According to assertion (ii) of Lemma 5.3, this means that $F \in C^r([0, 1])$, i.e.,

$$\sum_{j=1}^{\infty} r j^{2r-1} \varepsilon_j < \infty \;\Rightarrow\; F \in C^r([0, 1]). \tag{7.37}$$

Thus, it remains to prove inequality (7.29). By analogy with step 2, we obtain the following relation for all $n \geq m$:

$$\int_0^{n^{-2}} r u^{-r-2} \beta(u)\, du \geq c_3 \sum_{j=n}^{\infty} \varepsilon_j\, j^{2r-1},$$

$$\int_{(n+1)^{-2}}^{1} u^{-m-2}\beta(u)\,du \ge \int_{(n+1)^{-2}}^{m^{-2}} u^{-m-2}\beta(u)\,du \ge c_3 \sum_{j=m}^{n} \varepsilon_j\, j^{2m-1}.$$

Taking into account relation (5.23) and the fact that, for any majorant ω of the kth-modulus-of-continuity type, one has

$$\omega\left(\frac{1}{n^2}\right) \ge \frac{1}{2}\omega\left(\frac{1}{n^2}\right) + c_6\omega\left(\frac{1}{k(n+1)^2}\right),$$

we get

$$\omega_k(n^{-2}; F; [0, 1])$$

$$\ge \frac{1}{2}\int_0^{n^{-2}} r u^{-r-2}\beta(u)\,du + c_7 n^{-2k} \int_{(n+1)^{-2}}^{1} u^{-m-2}\beta(u)\,du$$

$$\ge \frac{c_3}{2}\sum_{j=n}^{\infty} r\varepsilon_j\, j^{2r-1} + c_7 c_3 n^{-2k} \sum_{j=m}^{\infty} \varepsilon_j\, j^{2m-1},$$

i.e., relation (7.29) is proved with

$$c = \min\left\{\frac{c_3}{2c_4}, \frac{c_7 c_3}{c_4}\right\}. \tag{7.38}$$

Relations (7.35)–(7.38) are equivalent to the statement of Theorem 7.2.

Below, we present several corollaries of Theorem 7.2.

Theorem 7.3. *If* $H[\bar{\varepsilon}]_J \subset C^r(J)$, *then*

$$\sum_{j=1}^{\infty} \varepsilon_j\, j^{2r-1} < \infty.$$

Theorem 7.4. *If* $H[\bar{\varepsilon}]_J \subset W^r(J)$, *then*

$$\sum_{j=1}^{\infty} \varepsilon_j\, j^{2r-1} < \infty.$$

For $r = 1$, Theorems 7.3 and 7.4 follow immediately from the Dolzhenko inequality (7.26). For $r > 1$, under the additional assumption that $\varepsilon_n n^{2r}$ decreases, these theorems were proved by Xie (1985) (see also [Shevchuk (1986)]) earlier than the general theorem (Theorem 7.2) was proved; this was a positive answer to Hasson's conjecture [Hasson (1982)]. It follows from fairly old results of Ibragimov (1946) that if $\varepsilon_n n^{2r} > \text{const} > 0$, then $H[\bar{\varepsilon}]_J \not\subset W^r(J)$.

The following corollary of relations (7.24) and (7.25) and Theorem 7.2 is true:

Corollary 7.3. *Let* $k \in \mathbb{N}$, $\varphi \in \Phi^k$, *and* $(r+1) \in \mathbb{N}$. *Then*

$$H(\bar{\varepsilon})_J \subset W^r H_k^{\varphi}(J)$$

$$\Leftrightarrow \quad \sum_{j=n+1}^{\infty} r j^{2r-1} \varepsilon_j + n^{-2k} \sum_{j=1}^{n} j^{2(r+k)-1} \varepsilon_j = O(\varphi(1/n^2)); \tag{7.39}$$

in particular,

$$H(\bar{\varepsilon})_J \subset C^r(J) \Leftrightarrow H(\bar{\varepsilon})_J \subset W^r(J)$$

$$\Leftrightarrow \quad \sum_{j=n+1}^{\infty} j^{2r-1} \varepsilon_j < \infty, \qquad r \in \mathbb{N}. \tag{7.40}$$

7.4. Peetre K-functional

In approximation theory, the idea of the replacement of an arbitrary function f by a sufficiently smooth function g is often used. One of the most efficient realizations of this idea is based on the method of the Peetre K-functional from the theory of interpolation spaces (see [Peetre (1968)], [Bergh and Löfström (1976)], and others).

In the case of interpolation between $C(J)$ and $W^r(J)$, $r \in \mathbb{N}$, the K-functional has the form

$$K_r(t, f, J) := \inf_{g \in W^r(J)} \left(\|f - g\|_J + t \operatorname*{ess\,sup}_{x \in J} \left| g^{(r)}(x) \right| \right) \tag{7.41}$$

It is clear that $K_r(t, f, J)$ is a function nondecreasing with respect to t.

Let $x \in J$, $h > 0$, and $(x + rh) \in J$. Relation (3.44), or (3.45), implies that the following estimate holds for $g \in W^r(J)$:

$$\left|\Delta_h^r(g,x)\right| \le h^r \operatorname*{ess\,sup}_{x\in J}\left|g^{(r)}(x)\right|.$$

Moreover, it is obvious that

$$\left|\Delta_h^r(f-g,x)\right| \le 2^r \|f-g\|_J.$$

Hence,

$$\left|\Delta_h^r(f,x)\right| \le \left|\Delta_h^r(f-g,x)\right| + \left|\Delta_h^r(g,x)\right| \le 2^r K_r(h,f,J),$$

which yields

$$\omega_r(t,f,J) \le 2^r K_r(t^r,f,J), \qquad t>0. \tag{7.42}$$

It will be proved in Chapter 7 (Lemma 7.3.5) that the following estimate is true for $t\in[0,1/r]$:

$$K_r(t^r,f,[-1;1]) \le c\omega_r(t;f;[-1;1]). \tag{7.43}$$

For $t\ge 1/r$, estimate (7.43) is a trivial consequence of the Whitney inequality (6.5). By the linear change of variables that transforms the segment $I=[-1;1]$ into the segment $J=[a;b]$, we obtain

$$K_r(t^r,f,J) \le c\omega_r(t,f,J).$$

Thus,

$$2^{-r}\omega_r(t,f,J) \le K_r(t^r,f,J) \le c\omega_r(t,f,J), \qquad t\ge 0. \tag{7.44}$$

Freud and Popov (1972) proved relation (7.43) by using a modification of Steklov means. For applications of (7.44) to the polynomial approximation, see, e.g., DeVore (1977), DeVore and Yu (1985), and Brudnyi, Krein, and Semenov (1987).

8. Hermite formula

8.1. Introduction

Until now, the Lagrange interpolation polynomial $L(x;f;x_0,\dots,x_m)$ and the divided difference $[x_0,\dots,x_m;f]$ have been considered in the case where all points x_i are different. In the present section, we generalize this notion to the case where some (fixed) points x_i may coincide. Namely, we define and study an Hermite–Lagrange interpolation polynomial $L(x; f; \bar{x})$ and a generalized divided difference $[\bar{x}; f]$, where $\bar{x} = (x_0, x_1, \dots, x_m)$. The corresponding definitions will be introduced so that

$$L(x; f; \bar{x}) \equiv L(x; f; x_0, \dots, x_m)$$

and

$$[\bar{x}; f] \equiv [x_0, \dots, x_m; f]$$

in the case where all points x_i are different. In contrast to $L(x; f; \bar{x})$ and $[\bar{x}; f]$, the expressions $L(x; f; x_0, \dots, x_m)$ and $[x_0, \dots, x_m; f]$ are used in this book only in the cases where all points x_i are different.

In this section, we use the following notation and assumptions: $\bar{x} := (x_0, \dots, x_m)$ is a collection of points $x_i \in \mathbb{R}$, $i = 0, \dots, m$, and $X := \{x_0\} \cup \{x_1\} \cup \dots \cup \{x_m\}$ is a subset of the real axis. The set X consists of q different points y_s, $s = 1, \dots, q$, i.e., $X = \{y_1\} \cup \dots \cup \{y_q\}$, where $y_s \neq y_\nu$ for $s \neq \nu$. The points x_i, $i = 0, \dots, m$, are called the coordinates of the collection $\bar{x}$, and the points y_s, $s = 1, \dots, q$ are called the nodes of the collection $\bar{x}$. If the coordinates x_i of the collection $\bar{x}$ are different, then we obviously have $q = m + 1$. However, if, conversely, at least two coordinates of the collection $\bar{x}$ are equal, then, obviously, $q < m + 1$. If a node y_s coincides with exactly $p_s + 1$ coordinates of the collection $\bar{x}$, then the number p_s is called the multiplicity of the node y_s. If $p_s = 0$, then the node y_s is called a simple node. If $p_s \in \mathbb{N}$, then the node y_s is called a multiple node. It is clear that

$$\sum_{s=1}^{q} (1 + p_s) = m + 1.$$

For every $s = 1, \dots, q$, we denote

$$l_s(x) := l_s(x, \bar{x}) := \prod_{\nu=1, \nu \neq s}^{q} (x - y_\nu)^{p_\nu + 1},$$

which is a polynomial of degree $m - p_s$, and

$$B_s(x) := B_s(x, \bar{x}) := \frac{1}{l_s(x, \bar{x})}. \tag{8.1}$$

8.2. Hermite–Lagrange polynomial

Definition 8.1. *For every* $s = 1, \dots, q$ *and* $i = 0, \dots, p_s$, *the fundamental Hermite–Lagrange polynomial is defined as the polynomial of degree* m *that has the following form:*

$$l_{s,i}(x) := l_{s,i}(x, \bar{x}) := \frac{1}{i!} l_s(x, \bar{x}) \sum_{\mu=0}^{p_s - i} \frac{1}{\mu!} B_s^{(\mu)}(y_s)(x - y_s)^{\mu+i}. \tag{8.2}$$

Let $\nu = 1, \dots, q$ and $j = 0, \dots, p_\nu$. It is easy to verify that $l_{s,i}^{(j)}(y_\nu) = 1$ for $i = j$ and $\nu = s$, and $l_{s,i}^{(j)}(y_\nu) = 0$ otherwise. Therefore, for any collection of numbers $f_{s,i}$, the polynomial (of degree $\leq m$)

$$\mathcal{L}(x) = \sum_{s=1}^{q} \sum_{i=0}^{p_s} f_{s,i} l_{s,i}(x) \tag{8.3}$$

possesses the following property:

$$\mathcal{L}^{(j)}(y_\nu) = f_{\nu,j}, \quad \nu = 1, \dots, q, \ j = 0, \dots, p_\nu. \tag{8.4}$$

In Subsections 8.2–8.4, the symbol $f = f(x) = f^{(0)}(x)$ always denotes a function defined on $\mathbb{R}$ and having p_s derivatives $f^{(i)}(x)$, $i = 1, \dots, p_s$, at each multiple node y_s.

Definition 8.2. *An Hermite–Lagrange polynomial*

$$L(x) := L(x; f) := L(x; f; \bar{x}) \tag{8.5}$$

that interpolates a function $f = f(x)$ *and its derivatives at the nodes* $y_1, \dots, y_q$ *of a collection* $\bar{x}$ *is defined as an algebraic polynomial of at most* m *th degree that satisfies the following equalities for all* $\nu = 1, \dots, q$ *and* $j = 0, \dots, p_s$:

$$L^{(j)}(y_\nu) = f^{(j)}(y_\nu). \tag{8.6}$$

For example, if $q = m + 1$, i.e., all nodes are simple, then the Hermite–Lagrange polynomial is the Lagrange polynomial (3.1). If $q = 1$, i.e., $x_0 = x_1 = \dots = x_m = y_1$, then the Hermite–Lagrange polynomial is the Taylor polynomial, i.e., in this case, one has

$$L(x; f; \bar{x}) = \sum_{j=0}^{m} (x - x_0)^j f^{(j)}(x_0) / j!.$$

In the case where $m = 2$ and $x_0 = x_1 \neq x_2$, we have

$$L(x;f;\bar{x}) = f'(x_0)\frac{(x-x_0)(x-x_2)}{(x_0-x_2)}$$

$$+ f(x_0)\left(\frac{x-x_2}{x_0-x_2} + \frac{(x_0-x)(x-x_2)}{(x_0-x_2)^2}\right) + f(x_2)\frac{(x-x_2)^2}{(x_0-x_2)^2}$$

$$= f'(x_0)(x-x_2)B_1(y_1) + f'(x_0)((x-x_2)B_1(y_1)$$

$$+ (x-x_0)(x-x_2)B_1'(y_1)) + f(x_0)(x-x_2)^2 B(y_2)$$

$$y_1 := x_0, \qquad y_2 := x_2.$$

By virtue of (8.3) and (8.4), the Hermite–Lagrange polynomial exists and is representable in the form

$$L(x;f;\bar{x}) = \sum_{s=1}^{q}\sum_{i=0}^{p_s} f^{(i)}(y_s)\, l_{i,s}(x). \tag{8.7}$$

Using the main theorem of algebra on the number of zeros of an algebraic polynomial, one can prove that the Hermite–Lagrange polynomial is unique and

$$L(x;P_m;\bar{x}) \equiv P_m(x) \tag{8.8}$$

for any algebraic polynomial P_m of degree $\leq m$.

Similarly to Lagrange polynomials, Hermite–Lagrange polynomials are linear operators.

8.3. Generalized divided difference

Let s^* be a number for which

$$x_m = y_{s^*}, \qquad \bar{x}_{m-1} := (x_0, \dots, x_{m-1}), \qquad \text{and} \qquad L_{m-1} := L(x;f;\bar{x}_{m-1}).$$

Dividing the difference $f^{(p_{s^*})}(x_m) - L_{m-1}^{(p_{s^*})}(x_m)$ by $(p_{s^*})!\, l_{s^*}(x_m)$ and using (8.7), (8.2), and (8.1), we obtain

$$\frac{f^{(p_{s*})}(x_m) - L_{m-1}^{(p_{s*})}(x_m)}{(p_{s*})!\,l_{s*}(x_m)} = \sum_{s=1}^{q}\sum_{i=0}^{p_s} \frac{f^{(p_s-i)}(y_s)B_s^{(i)}(y_s)}{(p_s-i)!\,i}$$

$$= \sum_{s=1}^{q} f_s^{(p_s)}(y_s)/(p_s)! =: [\bar{x}, f], \tag{8.9}$$

where

$$f_s(x) := f(x)B_s(x) = f(x)\sum_{s=1}^{q} f_s^{(p_s)}(y_s)/(p_s)! \prod_{\nu=1,\, \nu\neq s}^{q} (x-y_n)^{-p_\nu-1}.$$

Definition 8.3 (Hermite formula). *The expression* $[\bar{x}, f]$ *is called the generalized divided difference of order* m *for a function* f *in a collection* $\bar{x}$.

For example, for $q = 1$, i.e., in the case where $x_0 = \ldots = x_m = y_1$, we have

$$[\bar{x}, f] = f^{(m)}(y_1)/m!; \tag{8.10}$$

for $q = m+1$, we have $[\bar{x}, f] = [x_0, \ldots, x_m; f]$. In contrast to the case $q = 1$, for $q = 2, \ldots, m$ expression (8.9) contains the values of the function f at all nodes, as well as the values of all derivatives $f^{(i)}(y_s)$, $j = 1, \ldots, p_s$, at each multiple node y_s.

Note that the generalized divided difference $[x, f]$ is symmetric in the same sense as the divided difference for different points defined by (3.8).

For Hermite–Lagrange polynomials and generalized divided differences, complete analogs of all relations (3.11)–(3.27) are valid. Namely, the following assertions are true:

(i)
$$L(x; f; \bar{x}) = \sum_{i=0}^{m} [\bar{x}_i, f] \prod_{j=0}^{i-1} (x-x_j), \tag{8.11}$$

where $\bar{x}_i := (x_0, \ldots, x_i)$;

(ii) if $x_i \in [a, b]$, $i = 0, \ldots, m$, and the function f has the mth derivative $f^{(m)}$ on (a, b), then

$$[\bar{x}, f] = f^{(m)}(\theta)/m!, \quad \theta \in (a, b); \tag{8.12}$$

(iii) if P_{m-1} is a polynomial of degree $\le m-1$, then

$$[\bar{x}, P_{m-1}] = 0; \tag{8.13}$$

(iv) if $f(x) = x^m$, then

$$[\bar{x}, f] = 1; \tag{8.14}$$

(v) if $P_m(x) = a_0 x^m + \ldots + a_m$ is a polynomial of degree m, then

$$P_m(x) - L(x; P_m; \bar{x}_{m-1}) = a_0(x - x_0) \ldots (x - x_{m-1}), \tag{8.15}$$

where $x_{m-1} = (x_0, \ldots, x_{m-1})$;

(vi)
$$(x_0 - x_m)[\bar{x}, f] = [\bar{x}_{m-1}, f] - [\bar{x}_m^0, f], \tag{8.16}$$

where $\bar{x}_{m-1} = (x_0, \ldots, x_{m-1})$ and $\bar{x}_m^0 = (x_1, \ldots, x_m)$, etc.

The proofs of relations (8.11)–(8.16) are similar to those of relations (3.11)–(3.27), respectively. For example, let us prove (8.11). For $m - 1$, relation (8.11) follows from (8.10) and (3.11). By induction, we assume that (8.11) is true for a number $m - 1$ and prove it for the number m, i.e., we must prove the equality

$$L(x) = L_{m-1}(x) + [\bar{x}, f]\Pi(x), \tag{8.17}$$

where

$$L(x) = L(x; f; \bar{x}), \quad L_{m-1}(x) = L(x; f; \bar{x}_{m-1}),$$

$$\bar{x}_{m-1} = (x_0, \ldots, x_{m-1}), \quad \Pi(x) := \prod_{j=0}^{m-1} (x - x_j).$$

Since both sides of equality (8.17) are polynomials of degree $\le m$, it suffices to prove that, for all $s = 1, \ldots, q$ and $j = 0, \ldots, p_s$, one has

$$L^{(j)}(y_s) = L_{m-1}^{(j)}(y_s) + [\bar{x}; f]\Pi^{(j)}(y_s). \tag{8.18}$$

Let s^* be the number for which $x_m = y_{s^*}$. Since $\Pi^{(p_{s^*})}(x_m) = (p_{s^*})!\, l_{s^*}(x_m)$ we conclude that, for $s = s^*$ and $j = p_{s^*}$, equality (8.18) follows immediately from (8.9). For the other $s = 1, \ldots, q$ and $j = 0, \ldots, p_s$, it follows from Definition 8.2 that

$$L_{m-1}^{(j)}(y_s) + [\bar{x}, f]\Pi^{(j)}(y_s) = f^{(j)}(y_s) + 0 = L_m^{(j)}(y_s).$$

Thus, equality (8.18) is proved, which proves (8.11).

■

8.4. Convergence

Assume the following:

(i) $\mathbb{R}^{m+1}$ is the $(m+1)$-dimensional space of points $\bar{t} = (t_0, \dots, t_m)$;

(ii) $\mathbb{R}_*^{m+1}$ is the set of points $\bar{t} \in \mathbb{R}^{m+1}$ all coordinates t_i of which are different;

(iii) $\mathbb{R}^{m+1}(\bar{x})$ is the set of points $\bar{t} \in \mathbb{R}^{m+1}$ such that each node y_s of the collection $\bar{x}$ coincides with at least one coordinate t_i of the point $\bar{t}$;

(iv) $\mathbb{R}_*^{m+1}(\bar{x}) := \mathbb{R}_*^{m+1} \cap \mathbb{R}^{m+1}(\bar{x})$.

We say that, in the collection $\bar{x}$,

(i) a generalized divided difference converges,

(ii) a divided difference converges,

(iii) a generalized divided difference converges weakly,

(iv) a divided difference converges weakly,

and write, respectively,

(i)
$$\lim_{\bar{t}\to\bar{x}} [\bar{t}; f] = A, \tag{8.19}$$

(ii)
$$\lim_{\bar{t}\to\bar{x}}{}_{*} [t_0, \dots, t_m; f] = A, \tag{8.19$_*$}$$

(iii) $$\lim_{\bar{t}\to\bar{x}}{}^{0}\,[\bar{t};f] = A, \tag{8.19^0}$$

(iv) $$\lim_{\bar{t}\to\bar{x}}{}^{0}_{*}\,[t_0,\ldots,t_m;f] = A \tag{8.19$^0_*$}$$

if, for any $\varepsilon > 0$, there exists $\delta > 0$ such that $|A - [\bar{t}; f]| < \varepsilon$ whenever $|t_i - x_i| < \delta$, $i = 0, \ldots, m$, $\bar{t} \neq \bar{x}$, and, respectively,

(i) $\bar{t} \in \mathbb{R}^{m+1}$,

(ii) $\bar{t} \in \mathbb{R}_*^{m+1}$,

(iii) $\bar{t} \in \mathbb{R}^{m+1}(\bar{x})$,

(iv) $\bar{t} \in \mathbb{R}_*^{m+1}(\bar{x})$.

Remark 8.1. A necessary condition for the convergence of the generalized divided difference in the collection $\bar{x}$ [see (i)] is the existence of the derivative $f^{(p_s)}(y)$ not only at each multiple node y_s but also in a certain neighborhood of it. For weak convergence, this condition is not necessary because, in this case, the derivative does not appear in the expression $y \neq y_s$ for $[\bar{t}; f]$, and the existence of the derivatives $f^{(j)}(y)$, $j = 0, \ldots, p_s - 1$, in the neighborhood of the node y_s is guaranteed by the assumption of the existence of $f^{(p_s)}(y)$. In the case where all coordinates x_i are different (i.e., $q = m + 1$), and only in this case, we have $\mathbb{R}_*^{m+1}(\bar{x}) = \mathbb{R}^{m+1}(\bar{x}) = \{\bar{x}\}$. Therefore, in the case where $q = m + 1$, we set

$$\lim_{\bar{t}\to\bar{x}}{}^{0}_{*}\,[t_0,\ldots,t_m;f] := [x_0,\ldots,x_m;f] =: \lim_{\bar{t}\to\bar{x}}{}^{0}\,[\bar{t};f].$$

Lemma 8.1. *The following relations are true:*

$$\lim_{\bar{t}\to\bar{x}}{}^{0}\,[\bar{t};f] = [\bar{x};f]; \tag{8.20}$$

$$\lim_{\bar{t}\to\bar{x}}{}^{0}_{*}\,[t_0,\ldots,t_m;f] = [x_0,\ldots,x_m;f]. \tag{8.21}$$

In particular, if $q = 1$, i.e., there exists $f^{(m)}(x_0)$ and $x_i = x_0$, $i = 1, \ldots, m$, then

$$\lim_{\bar{t}\to\bar{x}}{}^{0}\,[\bar{t};f] \;=\; \frac{f^{(m)}(x_0)}{m!}, \tag{8.22}$$

$$\lim_{\bar{t}\to\bar{x}}{}^{0}_{*}\,[t_0,\dots,t_m;f] = \frac{f^{(m)}(x_0)}{m!}. \tag{8.23}$$

Proof. Relation (8.21) is a corollary of (8.23) and (3.26). Relation (8.20) is a corollary of (8.22) and of an analog of (3.26) for generalized divided differences. Moreover, it is obvious that (8.23) is a corollary of (8.22). Therefore, it suffices to prove (8.22).

Since $f^{(m)}(x_0)$ exists, we conclude that the $(m-1)$th derivative $f^{(m-1)}$ exists in a certain neighborhood $(x_0-\delta,\, x_0+\delta)$ of the point x_0. For every $\bar{t}=(t_0,\dots,t_m)$ with $t_i\in(x_0-\delta,\, x_0+\delta)$, we denote the least coordinate and the greatest coordinate among t_i by $t_*=t_*(\bar{t})$ and $t^*=t^*(\bar{t})$, respectively; thus, $t^*-t_*\ge\left|t_i-t_j\right|$, $i,j=0,\dots,m$. We set $g:=f-T$, where

$$T(x) \;=\; \sum_{j=0}^{m}\frac{f^{(j)}(x_0)(x-x_0)^j}{j!}$$

is the Taylor polynomial, and take the equalities $g^{(m)}(x_0)=g^{(m-1)}(x_0)=0$ into account. By virtue of (8.16) and (8.12), there exist points $\theta_1\in(t_*,t^*)$ and $\theta_2\in(t_*,t^*)$ such that

$$(m-1)!\,[\bar{t};g] \;=\; \frac{g^{(m-1)}(\theta_1)-g^{(m-1)}(\theta_2)}{t^*-t_*}.$$

Hence,

$$(m-1)!\,\left|[\bar{t};g]\right| \;\le\; \left|\frac{g^{(m-1)}(x_0)\;-\;g^{(m-1)}(\theta_1)}{x_0-\theta_1}\right| \;+\; \left|\frac{g^{(m-1)}(x_0)\;-\;g^{(m-1)}(\theta_2)}{x_0-\theta_2}\right|$$

(if $\theta_1=x_0$, then the first term in this inequality should be replaced by zero; the case $\theta_2=x_0$ should be treated by analogy). Therefore,

$$(m-1)!\,\lim_{\bar{t}\to\bar{x}}{}^{0}\,\left|[\bar{t};g]\right| \;\le\; 2\left|g^{(m)}(x_0)\right| \;=\; 0.$$

By virtue of (8.14), we get

$$\lim_{\bar{t}\to\bar{x}}{}^{0}\,[\bar{t}; f] \;=\; \frac{f^{(m)}(x_0)}{m!} \;+\; \lim_{\bar{t}\to\bar{x}}{}^{0}\,[\bar{t}; g].$$

Relation (8.22) is proved, which completes the proof of Lemma 8.1.

■

Remark 8.2. Generally speaking, weak convergence cannot be replaced by strong convergence in (8.22). For example, for the function $f(x) = x^2 \sin(1/x)$, $f(0) := 0$, we have $f'(0) = 0$, but $\lim\limits_{(t_0, t_1)\to(0,0)} [t_0, t_1; f]$ does not exist. Nevertheless, the following lemma is true:

Lemma 8.2. *Suppose that $q = 1$, i.e., there exists $f^{(m)}(x_0)$ and $x_0 = x_1 = \ldots = x_m$. If the m th derivative $f^{(m)}(x)$ exists and is continuous in a certain neighborhood $J_\delta := (x_0 - \delta,\, x_0 + \delta)$ of the point x_0, then*

$$\lim_{\bar{t}\to\bar{x}} [\bar{t}; f] \;=\; \frac{f^{(m)}(x_0)}{m!}; \tag{8.24}$$

in particular,

$$\lim_{\bar{t}\to\bar{x}}{}_{*}\,[t_0, \ldots, t_m; f] = \frac{f^{(m)}(x_0)}{m!}. \tag{8.25}$$

Proof. For any positive $\delta_1 < \delta$, we consider $t_i \in (x_0 - \delta_1,\, x_0 + \delta_1)$, $i = 0, \ldots, m$. Using relation (8.12) and the assumption that $f^{(m)}(x)$ is continuous, we establish that

$$\left| [\bar{t}; f] - \frac{f^{(m)}(x_0)}{m!} \right| \;=\; \frac{1}{m!} \left| f^{(m)}(\theta) - f^{(m)}(x_0) \right| \;\to 0 \quad \text{as} \quad \delta_1 \to 0,$$

where $\theta \in (x_0 - \delta_1,\, x_0 + \delta_1)$.

■

Definition 8.4 [Whitney (1934a)]. *Suppose that a set $E \subset \mathbb{R}$ and a limit point x_0 of E are given. We say that divided differences $[t_0, \ldots, t_m; f]$ converge at the point x_0 if, for any $\varepsilon > 0$, there exists $\delta > 0$ such that*

$$\left| [t'_0, \ldots, t'_m; f] \;-\; [t''_0, \ldots, t''_m; f] \right| \;<\; \varepsilon \tag{8.26}$$

whenever $|t_i' - x_0| < \delta$, $|t_i'' - x_0| < \delta$, $i = 0, \dots, m$, *and the coordinates* t_i' *and* t_i'' *in each of the collections* $\bar{t}' = (t_0', \dots, t_m')$ *and* $\bar{t}'' = (t_0'', \dots, t_m'')$ *are different.*

We say that divided differences $[t_0, \dots, t_m; f]$ *converge on the set* E *if they converge at every limit point of* E.

Lemma 8.3. *If* $f \in C^m(\mathbb{R})$, *then divided differences converge on every set* $E \subset \mathbb{R}$.

Lemma 8.3 is a corollary of (8.25).

8.5. Further generalization of the Whitney inequality

Let $k \in \mathbb{N}$, $(r+1) \in \mathbb{N}$, $m = k + r$, $\varphi \in \Phi^k$, $a \le x_0 \le x_1 \le \dots \le x_m \le b$, and $p(\bar{x}) := \max\{p_s \mid s = 1, q\}$. In the case $p(\bar{x}) = 1$, the rth divided majorant $\Lambda_r(x_0, \dots, x_m; \varphi)$ was defined in Section 6 by formula (6.32). If $p(\bar{x}) \le r$, then the right-hand side of (6.32) is obviously a finite number; denote it by $\Lambda_r(\bar{x}; \varphi)$. Furthermore, we have $\Lambda_r(t_0, \dots, t_m; \varphi) \to \Lambda_r(\bar{x}; \varphi)$ as $(t_0, \dots, t_m) \to \bar{x}$. Therefore, relations (8.21) and (6.36) yield the following statement:

Lemma 8.4 [V. Galan (1991)]. *Let* $p(\bar{x}) \le r$. *If* $f \in W^r H([k; \varphi; \mathbb{R}])$, *then*

$$|[\bar{x}; f]| \le c\Lambda_r(\bar{x}; \varphi). \tag{8.27}$$

Corollary 8.1. *Let* $p(\bar{x}) \ne m$. *If* $f \in W[m; \mathbb{R}]$, *then*

$$|[\bar{x}; f]| \le c. \tag{8.28}$$

Remark 8.3. If $p(\bar{x}) = m$, i.e., $x_0 = \dots = x_m$, then, under the assumption that $f \in W[m; \mathbb{R}]$, representation (8.10) yields inequality (8.28) for almost all $x_0 \in \mathbb{R}$.

Remark 8.4. Estimate (8.27), as well as estimate (6.36), is sharp in the sense that, for any collection $\bar{x}$ with $p(\bar{x}) \le r$, there exists a function $g \in W^r H[k; \varphi; \mathbb{R}]$ for which $|[\bar{x}; g]| \ge c\Lambda_r(\bar{x}; \varphi)$ [see Lemma 4.3.1].

8.6. Existence of the derivative

Let $x_0 = x_1 = \dots = x_m$. If $f^{(m)}(x_0)$ exists, then, according to (8.23), the following limit also exists:

$$\lim_{\bar{t}\to\bar{x}}{}^{0}_{*}\,[t_0, \dots, t_m; f] = \frac{A}{m!};$$

moreover, $A = f^{(m)}(x_0)$. Generally speaking, the converse statement is not true. For example, for the function $f(x) = x^4 \sin|1/x|$, $f(0) := 0$, we have

$$\lim_{(t_0,t_1,t_2)\to(0,0,0)}{}^{0}_{*}\,[t_0, t_1, t_2; f] = 0;$$

at the same time, the derivative $f'(1/\pi n)$ does not exist at the points $(\pi n)^{-1}$, $n \in \mathbb{N}$, and, hence, the second derivative $f''(0)$ does not exist as well. Nevertheless, one can easily prove by using (3.17) that if a function f is defined in a certain neighborhood of the point x_0 $(= x_1 = \dots = x_m)$ and the limit

$$\lim_{\bar{t}\to\bar{x}}{}_{*}\,[t_0, \dots, t_m; f] = \frac{A}{m!}$$

exists, then the function f has the mth derivative at the point x_0 and, moreover, $f^{(m)}(x_0) = A$.

9. *DT*-Moduli of smoothness

9.1. Introduction

It will be shown later that, for any function $f \in H[k; \varphi]$ and every $n \geq k-1$, there exists a polynomial P_n that guarantees, generally speaking, the same order of uniform approximation on $I = [-1, 1]$ as the polynomial of the best approximation and "much better" approximates the function f near the endpoints ± 1. Therefore, it is natural to assume that the class $H[k; \varphi]$ may be "worsened" near the endpoints without worsening the order of the value $\|f - P_n\|$.

At the beginning of the 1960s, Dzyadyk and Alibekov (1968) and Volkov (1965), and later Dyn'kin (1974) and Andrievskii (1985), described such classes on sets of the complex plane by using the mapping function. Such a description can also be applied to a segment because a segment is a set of the complex plane. Anyway, the investigation on a segment reduces to the substitution $x = \cos t$ (see [Fuksman (1965)]). In this direction, the corresponding classes were constructed by Potapov (1981); one should also mention the results of Sendov and Ivanov.

In this section, we present the construction of Ditzian and Totik (1987) that generalizes the spaces B^r introduced by Babenko (1985) (see also [Boikov (1987)]).

Definition 9.1. *Let* B^r, $r \in \mathbb{N}$, *denote the space of functions* f *that have the locally absolutely continuous* $(r-1)$*th derivative on* $(-1, 1)$ *and satisfy the following inequality almost everywhere on* $(-1, 1)$:

$$\left|(1-x^2)^{r/2} f^{(r)}(x)\right| \le M, \tag{9.1}$$

where $M = M(f) = \text{const} < \infty$.

Typical examples of functions $f \in B^r$ are $f_r(x) := (1+x)^{r/2}$ for odd r and $f_r(x) := (1+x)^{r/2} \ln(1+x)$ for even r.

Remark 9.1. Bernstein (1930) and Ibragimov (1946) proved that

$$E_n(f_r)_I \sim n^{-r}.$$

Without loss of generality, we can assume that a function $f \in B^r$ is continuous on I together with its derivatives $f^{(p)}$, $p < r/2$. Indeed, we have

$$f^{(p)}(x) = ((r-p-1)!)^{-1} \int_0^x (x-t)^{r-p-1} f^{(r)}(t)\,dt + P_{r-p-1}(x), \quad x \in (-1, 1). \tag{9.2}$$

Since

$$\int_0^x (1 \pm t)^{r-p-1} (1-t^2)^{-r/2}\,dt < \infty, \qquad p < \frac{r}{2},$$

the integral in (9.2) defines a function continuous on I for $p < r/2$. For $p \ge r/2$, the derivative $f^{(p)}$ of a function $f \in B^r$ does not need to be bounded (and, all the more, continuous) on I (e.g., f_r). Nevertheless, it is convenient to assume that the function $f^{(p)}$ is defined at the points ± 1. For $p \ge r/2$, we set $f^{(p)}(\pm 1) := 0$ if the limits $f^{(p)}(\pm 1 \mp 0)$ do not exist.

By $f(a+)$ $(f(a-))$ we denote the one-sided limit

$$\lim_{x \to a,\, x > a} f(x) \qquad \left(\lim_{x \to a,\, x < a} f(x) \right),$$

provided that it exists.

9.2. Definition of *DT*-modulus of continuity (smoothness)

We set

$$w(x) := \sqrt{1-x^2}, \tag{9.3}$$

and denote by

$$\dot{\Delta}_h^m(f; x) := \sum_{j=0}^{m} (-1)^{m-j} \binom{m}{j} f\left(x - \frac{m}{2} h + jh\right) \tag{9.4}$$

the mth symmetric difference of a function f at a point x with step h. Note that

$$\dot{\Delta}_h^m(f; x) = \Delta_h^m\left(f; x - \frac{m}{2} h\right). \tag{9.5}$$

Definition 9.2. *The DT-modulus of continuity (smoothness) of order $k \in \mathbb{N}$ of $f \in C(I)$ is defined as the function*

$$\overline{\omega}_k(t; f) := \sup_{0 \le h \le t} \max_x \left| \dot{\Delta}_{hw(x)}(f; x) \right|, \quad t \ge 0, \tag{9.6}$$

where the maximum is taken over all x such that

$$\left[x - \frac{1}{2} khw(x);\ x + \frac{1}{2} khw(x)\right] \subset I. \tag{9.7}$$

Note that condition (9.7) is equivalent to the inequality

$$\frac{1}{2} khw(x) \le 1 - |x|,$$

which yields

$$\frac{1}{2} kh \le w(x) \le 1. \tag{9.8}$$

Therefore,

$$\frac{1}{2} \omega_k(t^2; f) \le \overline{\omega}_k(t; f) \le \omega_k(t; f), \quad t \ge 0, \tag{9.9}$$

and

$$\overline{\omega}_k(t; f) \equiv \overline{\omega}_k\left(\frac{2}{k}; f\right), \qquad t \geq \frac{2}{k}. \tag{9.10}$$

It will be shown below that the DT-modulus of continuity $\overline{\omega}_k(t; f)$ possesses properties similar to properties of the "ordinary" modulus of continuity $\omega_k(t; f)$; in particular, for every function $f \in C(I)$, there exists a k-majorant φ (see Definition 4.4) such that

$$\varphi(t) \leq \overline{\omega}_k(t; f) \leq c\varphi(t), \qquad 0 \leq t \leq \frac{2}{k}. \tag{9.11}$$

In Example 9.1, we prove that, conversely, for every k-majorant φ there exists a function $f \in C(I)$ such that relation (9.11) is true.

Example 9.1. Assume that $\varphi \in \Phi^k$, $\overline{\varphi}(t) := 2\varphi(\sqrt{t})$,

$$F(y) := F(y; \overline{\varphi}; k) = \begin{cases} \overline{\varphi}(y) & \text{if } k = 1, \\ \dfrac{1}{(k-2)!}\displaystyle\int_1^x \frac{x(x-u)^{k-2}}{u^k}\overline{\varphi}(u)\,du & \text{if } k > 1 \end{cases}$$

is the extremal function [see (4.27)], and

$$f(x) := F\left(\frac{x+1}{2}\right), \qquad x \in I.$$

Then relation (9.11) is true.

Proof. Taking Lemma 4.3 into account, we can assume, without loss of generality, that

$$t\varphi'(t) \leq c_1\varphi(t), \qquad t \geq 0. \tag{9.12}$$

We set

$$w_*(y) := \sqrt{y(1-y)}.$$

For x and h satisfying (9.7), we have

$$w(x) = 2w_*(y), \tag{9.13}$$

$$\dot{\Delta}^k_{hw(x)}(f;\, x) \;=\; \dot{\Delta}^k_{hw_*(y)}(F;\, y), \tag{9.14}$$

where $x = 2y - 1$. We set $\bar{h} := hw_*(y)$ and $y_0 := y - \frac{1}{2}k\bar{h}$. It follows from (9.8) and (9.13) that

$$\bar{h} \;\geq\; \frac{1}{4}kh^2, \tag{9.15}$$

which, together with the condition $\varphi \in \Phi^k$, yields

$$\theta^{-k/2}\bar{\varphi}(\theta) \;\leq\; 4h^{-k}\varphi(h), \qquad \theta \geq \bar{h}. \tag{9.16}$$

If $y_0 \geq \bar{h}$, then, successively using relations (9.5), (3.34), (4.31), (9.12), and (9.16) and the estimate

$$w_*^2(y) \;<\; y \;=\; y_0 + \frac{1}{2}k\bar{h} \;\leq\; \left(1 + \frac{1}{2}k\right)y_0,$$

we get

$$\begin{aligned}\left|\dot{\Delta}^k_{hw_*(x)}(F;\, y)\right| \;&=\; \left|\Delta^k_{\bar{h}}(F;\, y_0)\right| \;=\; \bar{h}^k\left|F^{(k)}(\theta)\right| \;=\; \bar{h}^k\theta^{1-k}\bar{\varphi}'(\theta)\\ &\leq\; \frac{1}{2}c_1\bar{h}^k\theta^{-k}\bar{\varphi}(\theta) \;\leq\; 2c_1\bar{h}^k h^{-k}y_0^{-k/2}\varphi(h)\\ &\leq\; 2c_1\left(\frac{w_*^2(y)}{y_0}\right)^{k/2}\varphi(h) \;\leq\; c\varphi(h),\end{aligned} \tag{9.17}$$

where $\theta > y_0$. If $y_0 < \bar{h}$, then

$$w_*^2(y) \;<\; y \;=\; y_0 + \frac{1}{2}khw_*(y) \;<\; \left(1 + \frac{1}{2}k\right)hw_*(y),$$

whence

$$\bar{h} \;\leq\; \frac{1}{2}(k+2)h^2.$$

Therefore, relation (4.38) yields

$$\left|\Delta^k_{\bar{h}}(F;\, y_0)\right| \;\le\; 2^k\overline{\varphi}((k+1)\bar{h}) \;\le\; 2^{k+1}\varphi\left(\frac{1}{\sqrt{2}}\,(k+2)h\right) \;\le\; c\varphi(h). \tag{9.18}$$

The estimate of $\overline{\omega}_k(t;\, f)$ from above now follows from (9.17), (9.18), and (9.14).

For the estimation of $\overline{\omega}_k(t;\, f)$ from below, we fix $h \le 2/k$, choose y from the condition $y_0 = 0$, and successively use (9.6), (9.14), (9.5), (4.37), and (9.15). As a result, we get

$$\overline{\omega}_k(h;\, f) \;\ge\; \left|\dot{\Delta}^k_{hw(x)}(f;\, x)\right| \;=\; \left|\dot{\Delta}^k_{hw_*(x)}(F;\, y)\right| \;=\; \left|\Delta^k_{\bar{h}}(F;\, 0)\right|$$

$$\ge\; \overline{\varphi}(\bar{h}) \;\ge\; 2\varphi\left(\frac{\sqrt{k}h}{2}\right) \;\ge\; \varphi(h).$$

■

Example 9.2. For $\alpha > 0$, we set

$$f_\alpha(x) \;=\; \begin{cases} (x+1)^{\alpha/2} & \text{if } \dfrac{\alpha}{2} \text{ is not an integer,} \\ (x+1)^{\alpha/2}\ln(x+1) & \text{if } \dfrac{\alpha}{2} \text{ is an integer,} \end{cases} \tag{9.19}$$

$$F_\alpha(y) := f_\alpha(2y-1), \qquad \overline{\varphi}(t) \;:=\; 2t^{\alpha/2}.$$

Let $k \ge \alpha$. Since $F^{(k)}_\alpha(y) = a_{k,\alpha}y^{1-k}\overline{\varphi}'(y)$, $y > 0$, $a_{k,\alpha} = \text{const} > 0$, it follows from Example 9.1 that

$$a_{k,\alpha}t^\alpha \;\le\; \overline{\omega}_k(t;\, f_\alpha) \;\le\; ca_{k,\alpha}t^\alpha, \qquad 0 \le t \le \frac{2}{k}. \tag{9.20}$$

For what follows, we also need the definition of the *DT*-modulus of continuity for a closed interval

$$J \;:=\; [a, b] \subset I.$$

Definition 9.3. *The DT-modulus of continuity (smoothness) of order* $k \in \mathbb{N}$ *of a function* $f \in C(J)$ *on* J *is defined as the function*

$$\overline{\omega}_k(t;\, f;\, J) \;:=\; \sup_{0\le h\le t}\, \max_x \left|\dot{\Delta}_{hw(x)}(f;\, x)\right|, \qquad t \ge 0, \tag{9.21}$$

where the maximum is taken over all x *such that*

$$\left[x-\frac{1}{2}\,khw(x);\ x+\frac{1}{2}\,khw(x)\right]\subset J. \tag{9.22}$$

If $f\in C(I)$, then, obviously,

$$\overline{\omega}_k(t;\,f;\,I)\ \equiv\ \overline{\omega}_k(t;\,f). \tag{9.23}$$

Inequalities (9.8) yield

$$\frac{1}{2}\,\omega_k(t^2;\,f;\,J)\ \le\ \overline{\omega}_k(t;\,f;\,J)\ \le\ \omega_k(t;\,f;\,J),\qquad t\ge 0, \tag{9.24}$$

where ω_k is the "ordinary" modulus of continuity.

Denote[†]

$$|J|\ :=\ b-a,\qquad /J/\ :=\ \frac{|J|}{w((a+b)/2)}. \tag{9.25}$$

Note that condition (9.22) is equivalent to the inequality

$$\frac{1}{2}\,khw(x)\ \le\ \frac{b-a}{2}-\left|x-\frac{a+b}{2}\right|.$$

Since w is a concave function, we have

$$\frac{b-a}{2}-\left|x-\frac{a+b}{2}\right|\ \le\ \frac{1}{2}\,/J/\,w(x),\qquad x\in J.$$

Therefore,

$$h\le\frac{1}{k}\,/J/ \tag{9.26}$$

for h satisfying (9.22). It follows from estimate (9.26) that

$$\omega_k\left(\frac{1}{k}|J|;\,f;\,J\right)\ =\ \overline{\omega}_k\left(\frac{1}{k}\,/J/;\,f;\,J\right)\ =\ \overline{\omega}_k(/J/;\,f;\,J). \tag{9.27}$$

[†] The notation $/J/$ has been introduced in recent papers of Leviatan and Shevchuk. We use arguments from these papers in this and the next section.

It follows from the Whitney inequality (6.1) and relation (9.27) that

$$E_{k-1}(f)_J \le c\overline{\omega}_k(/J/; f; J), \tag{9.28}$$

whence

$$\left\| f - L_{k-1}(\cdot; f; J) \right\|_J \le c\overline{\omega}_k(/J/; f; J), \tag{9.29}$$

where, as usual,

$$L_{k-1}(\cdot; f; J)$$

denotes the Lagrange polynomial of degree $\le k-1$ that interpolates a function f at k equidistant points of a closed interval J, including its endpoints, for $k \ne 1$; $L_0(x; f; J) = f((a+b)/2)$. In the case of a divided difference for an arbitrary collection of points $x_j \in J$, $x_0 < x_1 < \ldots < x_k$, estimate (9.29) yields

$$|[x_0, \ldots, x_k; f]| \le c\overline{\omega}_k\left(/J/; f; J\right) \max_{j=1,\ldots,k} |x_j - x_{j-1}|^{-k}. \tag{9.30}$$

Finally, note that

$$1 \le \frac{/J/}{/J_1/} < 2\frac{|J|}{|J_1|} \quad \text{if} \quad J_1 \subset J. \tag{9.31}$$

9.3. Properties of *DT*-modulus of continuity

The properties of the function $\overline{\omega}_k(t) := \overline{\omega}_k(t; f; J)$ are similar to those of the "ordinary" modulus of continuity $\omega_k(t) = \omega_k(t; f; J)$. In particular, it follows directly from Definition 9.3 that $\overline{\omega}_k$ does not decrease on $[0, \infty]$ and $\overline{\omega}_k(0) = 0$. Moreover, it follows from (9.24) that $\overline{\omega}_k$ is continuous at the point $t = 0$, i.e., $\overline{\omega}_k(0+) = 0$. Now let us prove an analog of the inequality $\omega_k(nt) \le n^k \omega_k(t)$, $n \in \mathbb{N}$.

Lemma 9.1. *If* $n \in \mathbb{N}$, *then*

$$\overline{\omega}_k(nt; f; J) \le cn^k\overline{\omega}_k(t; f; J), \qquad t \ge 0; \tag{9.32}$$

in particular,

$$\overline{\omega}_k(nt; f) \le cn^k\overline{\omega}_k(t; f), \qquad t \ge 0. \tag{9.33}$$

Proof. Without loss of generality, we can assume that n is odd. Let

$$x \in J, \quad h > 0, \quad \delta := hw(x),$$

$$x_0 := x - \frac{1}{2}kn\delta, \quad x^0 := x + \frac{1}{2}kn\delta, \quad [x_0, x^0] \subset J.$$

We set

$$x_j := \begin{cases} x_0 + \frac{1}{n}2\delta j^2 & \text{if } j = 0, \dots, n/2, \\ x_0 + \delta j & \text{if } j = n/2, \dots, (k-1/2)n, \\ x_0 - \frac{1}{n}2\delta(kn-j)^2 & \text{if } j = (k-1/2)n, \dots, kn, \end{cases}$$

$$J_j := [x_j, x_{j+k}].$$

One can easily verify that

$$/J_j/ \le c_1 h, \quad j = 0, \dots, k(n-1). \tag{9.34}$$

Therefore, relation (9.30) yields

$$|\Delta_j| := |[x_j, \dots, x_{j+k}; f]| < c_2|J_j|^{-k}\overline{\omega}_k(c_1h; f; J). \tag{9.35}$$

To use the Popoviciu identity (3.27), we introduce the notation

$$\Pi_{j,k}(t) := \prod_{\mu=1}^{k-1}(t - x_{j+\mu})_+, \quad y_i := x_{in}, \quad i = 0, \dots, k,$$

$$s_j := |[y_0, \dots, y_k; \Pi_{j,k}]|, \quad A := \prod_{\nu=1}^{k}(y_\nu - y_0).$$

Then identity (3.27), estimate (9.35), and the inequality

$$s_j < c_3A^{-1}n^{k-1}|J_j|^{k-1}$$

yield

$$\left|\dot{\Delta}^k_{hw(x)}(f;x)\right| = A\left|[y_0,\dots,y_k;f]\right| = A\left|\sum_{j=0}^{k(n-1)} |J_j|\Delta_j s_j\right| < c_2c_3n^k\omega_k(c_1h;f;J),$$

which leads to (9.32) with $c = (c_1+1)^k c_2c_3$.

■

The Marchaud-type inequality

$$\overline{\omega}_j(t;f;J) \le ct^j\int_t^{/J/}\frac{\overline{\omega}_k(u;f;J)}{u^{j+1}}\,du + c\left(\frac{t}{/J/}\right)^j\|f\|_J, \qquad t\ge 0, \tag{9.36}$$

$$j = 1,\dots,k-1,$$

is also true. This inequality immediately follows from Lemma 9.2 and the second inequality in (9.31).

Lemma 9.2. *Let* $j = 1,\dots,k-1,$ $x\in J,$ $h>0,$ *and*

$$J_h := \left[x - \frac{1}{2}\,jhw(x);\ x + \frac{1}{2}\,jhw(x)\right].$$

If $J_h\subset J,$ *then*

$$\left|\dot{\Delta}^j_{hw(x)}(f;x)\right| \le ct^j\int_t^{/J/}\frac{\overline{\omega}_k(u;f;J)}{u^{j+1}}\left(\frac{w(x)}{u+w(x)}\right)^j du + c\left(\frac{hw(x)}{|J|}\right)^j\|f\|_J. \tag{9.37}$$

Proof. Denote

$$\overline{\omega}_k(t) := \overline{\omega}_k(t;f;J),$$

$$x_* := x - \frac{1}{4}\,w^2(x), \qquad x^* := x + \frac{1}{4}\,w^2(x), \qquad J^* := J\cap[x_*,x^*].$$

Estimate (9.8) yields

$$\frac{1}{4}\,hw(x) \le \frac{1}{4}\,jhw(x) \le |J^*| \le \frac{1}{2}\,w^2(x). \tag{9.38}$$

Since

$$\frac{1}{2}\,w(u) \;<\; w(x) \;<\; 2w(x), \qquad u \in [x_*, x^*],$$

we have

$$\overline{\omega}_k(tw(x);\, f;\, J^*) \;\le\; \overline{\omega}_k(2t;\, f;\, J^*) \;\le\; \overline{\omega}_k(2t), \qquad t \ge 0.$$

Taking (9.31) into account, we obtain

$$\frac{|J^*|}{2w(x)} \;<\; /J^*/ \;\le\; /J/.$$

Therefore,

$$(w(x))^j \int\limits_{hw(x)}^{4/J^*/} \frac{\omega_k(u;\, f;\, J^*)}{u^{j+1}}\,du$$

$$\le\; 2^k (w(x))^j \int\limits_{hw(x)}^{4|J^*|} \frac{\overline{\omega}_k(u/w(x))}{u^{j+1}}\,du \;=\; 2^k \int\limits_{h}^{4|J^*|(w(x))^{-1}} \frac{\overline{\omega}_k(u)}{u^{j+1}}\,du$$

$$\le\; 2^k 3^j \int\limits_{h}^{8/J/} \frac{\overline{\omega}_k(u)}{u^{j+1}} \left(\frac{w(x)}{u + w(x)}\right)^j du. \tag{9.39}$$

The Marchaud inequality (5.6) and relation (9.39) yield (9.37) in the case $J = J^*$.

If $J \ne J^*$, then, obviously,

$$\frac{1}{4}\,w^2(x) \;<\; |J^*| \;\le\; \frac{1}{2}\,w^2(x),$$

and the trivial inequality

$$w^2(x_1) \;\le\; w^2(x_2) + 2|x_1 - x_2|, \qquad x_1, x_2 \in I,$$

yields

$$w^2\left(\frac{a+b}{2}\right) \;\le\; w^2(x_2) + 2\left|x - \frac{a+b}{2}\right| \;<\; 4|J^*| + |J| < 5|J|.$$

In other words,

$$\sqrt{/J/} < \sqrt{5}\,/J/ \le 3/J/.$$

Hence, taking (9.24) and (9.8) into account, we get

$$\int_{|J^*|}^{|J|} \frac{\omega_k(u; f; J)}{u^{j+1}}\,du \le 2 \int_{|J^*|}^{|J|} \frac{\overline{\omega}_k(\sqrt{u})}{u^{j+1}}\,du = 4 \int_{\sqrt{|J^*|}}^{\sqrt{|J|}} \frac{\overline{\omega}_k(u)}{u^{2j+1}}\,du \le 4 \int_{w(x)/2}^{3/J/} \frac{\overline{\omega}_k(u)}{u^{2j+1}}\,du$$

$$\le 3^j 4 \int_{h/4}^{3/J/} \frac{\overline{\omega}_k(u)}{u^{j+1}(u + w(x))^j}\,du. \tag{9.40}$$

In the case where $J_h \subset J^* \neq J$, we denote $L_{k-1}(x) := L_{k-1}(x; f; J^*)$. Inequalities (5.6), (9.39), and (9.29) yield

$$\left|\dot{\Delta}^j_{hw(x)}(f - L_{k-1}; x)\right|$$

$$\le c(hw(x))^j \int_{hw(x)}^{|J^*|} \frac{\omega_k(u; f; J^*)}{u^{j+1}}\,du + c(hw(x))^j \frac{\left\|f - L_{k-1}\right\|_{J^*}}{|J^*|}$$

$$\le ch^j \int_{h}^{8/J/} \frac{\omega_k(u)}{u^{j+1}} \left(\frac{w(x)}{u + w(x)}\right)^j du. \tag{9.41}$$

Moreover, it follows from the Newton formulas (3.11), the Marchaud inequality (5.6), and relation (9.40) that

$$\left|\dot{\Delta}_{hw(x)}(x; L_{k-1})\right| \le c(hw(x))^j |J^*|^{-j} \omega_j(|J^*|; f; J^*)$$

$$\le c(hw(x))^j \int_{|J^*|}^{|J|} \frac{\omega_k(u; f; J)}{u^{j+1}}\,du + c\left(\frac{hw(x)}{|J|}\right)^j \|f\|_J. \tag{9.42}$$

Inequality (9.37) now follows from (9.40)–(9.42) in the case under consideration.

In the case where $J_h \subset J^*$, we have

$$\frac{1}{4}\, jhw(x) \le |J^*| < jhw(x).$$

Therefore, relation (9.37) follows directly from (5.6) and (9.40).

■

By analogy with (6.15), one can prove the inequality

$$\left|f(x) - L_{k-1}(x; f; J)\right| \le c\left(\frac{|J_x|}{|J|}\right)^k \overline{\omega}_k(/J/; f; J_x), \qquad x \in I, \tag{9.43}$$

where $J_x := J \cup [a, x]$ if $x \ge a$ and $J_x := J \cup [a, b]$ if $x < a$.

Indeed, reasoning as in the proof of Lemma 6.2 and using (9.36) instead of the Marchaud inequality (5.6), and (9.28) instead of the Whitney inequality (6.1), we get

$$\left|f(x) - L_{k-1}(x; f; J)\right| \le c\left(\frac{|J_x|}{|J|}\right)^{k-1} /J/^{k-1} \int_{/J/}^{2/J/} \frac{\omega_k(t; f; J_x)}{t^k}\, dt. \tag{9.44}$$

Relation (9.43) now follows from (9.31).

9.4. Relationship between the *DT*-modulus of smoothness and the space B^r

Lemma 9.3. *If $\overline{\omega}_r(t; f) \le t^r$, then $f \in B^r$ and*

$$\left|w^r(x) f^{(r)}(x)\right| \le 1 \quad a.e., \qquad x \in I. \tag{9.45}$$

If $f \in B^r$ and relation (9.45) is true, then

$$\overline{\omega}_r(t; f) \le ct^r. \tag{9.46}$$

Proof. Assertion (9.45) follows from relation (7.7) and the inequality

$$\omega_r(t; f; J) \le \overline{\omega}_r\left(\frac{t}{\min\{w(a); w(b)\}}; f; J\right) \le \left(\min\{w(a); w(b)\}\right)^{-r} t^r.$$

To prove assertion (9.46), we take x and h satisfying (9.7), set $H := hw(x)$ and $x_* := x - kH/2$, and use (9.4) and (3.45). As a result, we get

$$\left|\dot{\Delta}_H(f; x)\right| \le \left|\dot{\Delta}_H(f; x_*)\right| \le \int_0^H \dots \int_0^H w^{-r}(x_* + t_1 + \dots + t_r)\,dt_r \dots dt_1 \le ch^r.$$

■

10. DT_r-Modulus of smoothness

10.1. Spaces C_w^r

For the application of the Ditzian–Totik construction to the investigation of the smoothness of differentiable functions, we need the spaces C_w^r. Recall that

$$w(x) = \sqrt{1-x^2}.$$

We set

$$C_w^0 = C_w = C(I).$$

Definition 10.1. *By* C_w^r, $r \in \mathbb{N}$, *we denote the space of functions* $f \in C^r((-1,1))$ *continuous on* I *and such that*

$$\lim_{x\to 1-} w^r(x) f^{(r)}(x) = 0 \tag{10.1}$$

and

$$\lim_{x\to -1+} w^r(x) f^{(r)}(x) = 0. \tag{10.2}$$

Lemmas 10.1 and 10.2 presented below reveal the close relationship between the space C_w^r and the space $C^r(\mathbb{R})$ of 2π-periodic functions r times continuously differentiable on $\mathbb{R}$.

For every function $f \in C^r(I)$, we set

$$\tilde{f}(t) := f(\cos t). \tag{10.3}$$

Lemma 10.1. *If* $\tilde{f} \in C^r(\mathbb{R})$, *then* $f \in C_w^r$.

Proof. Let $f_1(x) := f(-x)$. Then

$$\tilde{f}_1(t) = f_1(\cos t) = f(-\cos t) = \tilde{f}_1(t-\pi).$$

Therefore, it suffices to prove only relation (10.1). For this purpose, we subtract from the function $\tilde{f}$ its Taylor polynomial

$$\tilde{T}(t) = \sum_{j=0}^{r} \frac{\tilde{f}^{(j)}(0)}{j!} t^j$$

and set

$$\tilde{g}(t) := \tilde{f}(t) - \tilde{T}(t).$$

Since the derivative $\tilde{f}^{(r)}$ is continuous at zero, we have

$$\tilde{g}^{(r)}(t) = \tilde{f}^{(r)}(t) - \tilde{f}^{(r)}(0) \to 0 \quad \text{as} \quad t \to 0 \tag{10.4}$$

and, for each $j = 0, \ldots, r-1$,

$$t^{j-r}\tilde{g}^{(j)}(t) = \frac{1}{(r-j)!}\tilde{g}^{(r)}(\theta) \to 0 \quad \text{as} \quad t \to 0, \tag{10.5}$$

where $|\theta| < |t|$. We set

$$T(x) := \tilde{T}(\arccos x), \quad g(x) := f(x) - T(x), \qquad x \in I.$$

By induction, one can easily verify the identity

$$g^{(r)}(x) \equiv \sum_{j=1}^{r} (\sin t)^{j-2r}\tilde{g}^{(j)}(t)\tilde{T}_{j,r}(t),$$

where $t = \arccos x$, $x \in (-1, 1)$, and $\tilde{T}_{j,r}$ are fixed trigonometric polynomials that do not depend on g. Therefore, by virtue of (10.4) and (10.5), we get

$$\lim_{x \to 1-0} w^r(x) g^{(r)}(x) = \sum_{j=1}^{r} \tilde{T}_{j,r}(0)\Big(\lim_{t \to 0} (\sin t)^{j-r} g^{(j)}(t)\Big) = 0.$$

Since $\tilde{f}$ is an even function, we have $\tilde{f}^{(j)}(0) = 0$ for odd j. Hence,

$$T(x) = \sum_{j=0}^{[r/2]} \frac{1}{(2j)!}\tilde{f}^{(2j)}(0)(\arccos x)^{2j}.$$

Note that $(\arccos x)^2$ is an infinitely differentiable function (unlike $\arccos x$). Hence, the following finite limit exists:

$$\lim_{x\to 1-0} T^{(r)}(x) =: A.$$

Thus,

$$\lim_{x\to 1-0} f^{(r)}(x)w^{(r)}(x) = \lim_{x\to 1-0} g^{(r)}(x)w^{(r)}(x) + A \lim_{x\to 1-0} w^{(r)}(x) = 0.$$

■

Lemma 10.2. *Let r be odd. If $f \in C_w^r$, then $\tilde{f} \in C^r(\mathbb{R})$.*

Proof. Lemma 10.2 follows from the relation

$$\lim_{t\to 0} \tilde{f}^{(r)}(t) = 0. \tag{10.6}$$

Let us prove it. For this purpose, we use the identity

$$\tilde{g}^{(r)}(t) = \sum_{j=1}^{[r/2]} g^{(j)}(t)T_{j,r}(t) + \sum_{j=1+[r/2]}^{r} g^{(j)}(t)(\sin t)^{r-2j}\, T_{j,r}(t), \tag{10.7}$$

where $g \in C^r((-1, 1))$, $\tilde{g}(t) := g(\cos t)$, $x = \cos t \neq \pm 1$, and $T_{j,r}(t)$ are fixed trigonometric polynomials that do not depend on g. We take $\varepsilon > 0$ and choose a point $x_0 \in (0, 1)$ such that

$$\alpha := \sup_{x\in[x_0,1)} w^{(r)}(x)\left|f^{(r)}(x) - f^{(r)}(x_0)\right| < \frac{\varepsilon}{\sum_{j=1}^{r}\|T_{j,r}\|_{\mathbb{R}}}.$$

We subtract from the function f its Taylor polynomial

$$T(x) = \sum_{j=0}^{r} \frac{1}{j!} f^{(j)}(x_0)(x - x_0)^j$$

and set

$$g(x) := f(x) - T(x) = \frac{1}{(r-1)!}\int_{x_0}^{x} (x-u)^{r-1}(f^{(r)}(u) - f^{(r)}(x_0))\,du.$$

Then, for $x \in [x_0, 1)$ and $j = 1, \dots, r-1$, we obtain

$$\left|g^{(j)}(x)\right| \le \frac{\alpha}{(r-1-j)!}\int_{x_0}^{x}\frac{(x-u)^{r-1-j}}{w^r(u)}\,du \le \begin{cases} 1 & \text{if } j < r/2 \\ (w(x))^{r-2j} & \text{if } j > r/2. \end{cases}$$

Therefore, by virtue of (10.6), we get

$$\left|\tilde{g}^{(r)}(t)\right| \le \alpha\sum_{j=1}^{r}\left\|T_{j,r}\right\|_{\mathbb{R}} < \varepsilon, \qquad 0 < |t| < \arccos x_0.$$

Since $\tilde{T}(t) := T(\cos t)$ is an even function, we have $\tilde{T}^{(r)}(0) = 0$. Hence, there exists $\delta_1 > 0$ such that $\left|\tilde{T}^{(r)}(t)\right| < \varepsilon$, $|t| < \delta_1$. Taking $\delta = \delta(\varepsilon) := \min\{\delta_1, \arccos x_0\}$, we get

$$|\tilde{f}^{(r)}(t)| \le |\tilde{g}^{(r)}(t)| + |\tilde{T}^{(r)}(t)| < 2\varepsilon, \quad 0 < |t| < \delta.$$

■

In exactly the same way as Lemmas 10.1 and 10.2, but a little easier, one can prove Lemmas 10.3 and 10.4 presented below; in their proofs, one should use Taylor polynomials of degree $r-1$ instead of Taylor polynomials of degree r.

Lemma 10.3. *If* $\tilde{f} \in W^r(\mathbb{R})$, *then* $f \in B^r$.

Lemma 10.4. *Let* r *be odd. If* $f \in B^r$, *then* $\tilde{f} \in W^r(\mathbb{R})$.

Remark 10.1. In the case where r is even, the assumption that $f \in C_w^r$ ($f \in B^r$), generally speaking, does not imply that $\tilde{f} \in W^r(\mathbb{R})$. For example, the function $f(x) := (x+1)^{r/2}\sqrt[2]{|\ln(x+1)|}$ ($f(x) := (x+1)^{r/2}\ln(x+1)$ belongs to the space C_w^r (B^r), but $\tilde{f} \notin C^r(\mathbb{R})$ ($\tilde{f} \notin W^r(\mathbb{R})$). The following condition is sufficient (and, in a certain sense, necessary) for the inclusion $\tilde{f} \in C^r(\mathbb{R})$ ($\tilde{f} \in W^r(\mathbb{R})$):

$$\int_0^1 t^{-1}\overline{\omega}_{k,r}(t; f^{(r)})\,dt < \infty,$$

where $\overline{\omega}_{k,r}(t; f^{(r)})$ is the function defined below.

10.2. DT_r-modulus of smoothness

We set

$$w_\delta = w_\delta(x) := \sqrt{\left(1 - x - \frac{1}{2}\delta w(x)\right)\left(1 + x - \frac{1}{2}\delta w(x)\right)}. \tag{10.8}$$

Definition 10.2. *Let $k \in \mathbb{N}$ and $r \in \mathbb{N}$. The DT_r-modulus of smoothness of the rth derivative of a function $f \in C_w^r$ is defined as the function*

$$\overline{\omega}_{k,r}(t; f^{(r)}) := \sup_{0 \le h \le t} \sup_{x} \left| w_{kh}^r \dot{\Delta}_{kw(x)}^k (f^{(r)}, x) \right|, \quad t > 0, \tag{10.9}$$

where the inner supremum is taken over all x such that

$$\left[x - \frac{1}{2}khw(x),\ x + \frac{1}{2}khw(x)\right] \subset (-1, 1). \tag{10.10}$$

Remark 10.2. Let $f \in C^r(-1, 1)$. In order that the function $\overline{\omega}_{k,r}(t; f^{(r)})$ defined by relation (10.9) tend to zero as $t \to 0+$, it is necessary and sufficient that $f \in C_w^r$.

Indeed, let $f \in C_w^r$. We take $\varepsilon > 0$ and find $\delta \in (0, 1)$ such that

$$w^r(x) \left| f^{(r)}(x) \right| < 2^{-k}\varepsilon$$

for all $x \in (-1, 1) \setminus (-1 + \delta, 1 - \delta)$. We set

$$\omega(t) := \omega_k(t; f^{(r)}; [-1 + \delta/3, 1 - \delta/3]).$$

Since the function $f^{(r)}$ is continuous on the closed interval $[-1 + \delta/3, 1 - \delta/3]$, we conclude that $\omega(t) \to 0$ as $t \to 0+$. We choose $t^* > 0$ for which $\omega(t^*) < \varepsilon$ and set $t^0 := \min\{t^*, 2\delta/3k\}$. Assume that $0 < h \le t^0$, $x \in (-1, 1)$, and

$$J := \left[x - \frac{1}{2}khw(x),\ x + \frac{1}{2}khw(x)\right] \subset (-1, 1).$$

If $|x| \le 1 - 2\delta/3$, then $J \subset [-1 + \delta/3, 1 - \delta/3]$ and, therefore,

$$\left| w_{kh}^r(x) \dot{\Delta}_{kw(x)}^k (f^{(r)}; x) \right| \le \left| \dot{\Delta}_{kw}^k (f^{(r)}, x) \right| \le \omega(t^*) < \varepsilon. \tag{10.11}$$

If $|x| > 1 - 2\delta/3$, then $J \subset (-1, 1)\backslash(-1+\delta, 1-\delta)$, whence

$$\left|w_{kh}^r(x)\dot{\Delta}_{kw(x)}^k(f^{(r)}; x)\right| \le w_{kh}^r(x)\,2^k\left|f^{(r)}(\theta)\right|$$

$$\le 2^k w^r(\theta)\left|f^{(r)}(\theta)\right| < \varepsilon, \qquad (10.12)$$

where $\theta \in J$. Inequalities (10.11) and (10.12) yield

$$\overline{\omega}_{k,r}(0+; f^{(r)}) = 0. \qquad (10.13)$$

Now assume that relation (10.13) is satisfied. Let us show that $f \in C_w^r$. We take $\varepsilon > 0$ and choose $h \in (0, 1/k)$ for which $\overline{\omega}_{k,r}(h; f^{(r)}) < \varepsilon$. For $x \in (1/2, 1)$, we find $\theta \in (0, 1)$ such that $\theta + khw(\theta)/2 = x$ and note that, in view of (10.10), one has

$$\left|f^{(r)}(x) - \dot{\Delta}_{kw}^k(f^{(r)}; \theta)\right| \le (2^k - 1)\left\|f^{(r)}\right\|_{[-1+(h/2)^2, 1-(h/2)^2]} =: A_h$$

Therefore,

$$|w^r(x)f^{(r)}(x)| \le w^r(x)|\dot{\Delta}_{kw}^k(f^{(r)}; \theta)| + w^r(x)A_h$$

$$= \left(\frac{w(x)}{w_{kh}(\theta)}\right)^r |w_{kh}^r(\theta)\dot{\Delta}_{kw}^k(f^{(r)}; \theta)| + w^r(x)A_h.$$

Since $w(x) < 2w_{kh}(\theta)$, we have

$$\limsup_{x\to 1} w^r(x)\left|f^{(r)}(x)\right| < 2^r\varepsilon,$$

whence $w^r(x)\left|f^{(r)}(x)\right| \to 0$ as $x \to 1$.

By analogy, one can prove that $w^r(x)\left|f^{(r)}(x)\right| \to 0$ as $x \to -1$.

Remark 10.3. Let $f \in C^r(-1, 1)$. By analogy, one can prove that the condition $f \in B^r$ is necessary and sufficient for the validity of the relation

$$\overline{\omega}_{k,r}(0+; f^{(r)}) < \infty. \qquad (10.14)$$

Remark 10.4. For $r = 0$, we obviously have

$$\overline{\omega}_k \equiv \overline{\omega}_{k,0}. \qquad (10.15)$$

In the next subsection, we show that, similarly to the DT-modulus of continuity, the DT_r-modulus of continuity possesses properties analogous to properties of the "ordinary" modulus of continuity ω_k.

In Example 10.1, which is similar to Example 9.1, we show, that, for every k-majorant φ and $r \in \mathbb{N}$, there exists a function $f \in C_w^r$ such that

$$\varphi(t) \le \overline{\omega}_{k,r}(t; f^{(r)}) \le c\varphi(t), \qquad 0 \le t \le 2/k. \tag{10.16}$$

Example 10.1. Suppose that $r \in \mathbb{N}$, $k \in \mathbb{N}$, $\varphi \in \Phi^k$, $\overline{\varphi}(t) := 2\varphi(\sqrt{t})$,

$$F(y) := \int_1^y (y-u)^{k-1} u^{-k-r/2} \overline{\varphi}(u)\, du, \quad y \in (0, 1],$$

and

$$f^{(r)}(x) := F((x+1)/2), \quad x \in (-1, 1].$$

Then relation (10.16) is true.

Proof. Since $f^{(r)}(1) = F(1) = 0$ and

$$y^{r/2} |F(y)| = y^{r/2} \int_y^{\sqrt{y}} (u-y)^{k-1} u^{-k-r/2} \overline{\varphi}(u)\, du$$

$$+ y^{r/2} \int_{\sqrt{y}}^1 (u-y)^{k-1} u^{-k-r/2} \overline{\varphi}(u)\, du$$

$$\le \overline{\varphi}(\sqrt{y})\, y^{r/2} \int_y^\infty u^{-1-r/2}\, du + \overline{\varphi}(\sqrt{y})\, y^{r/4} \int_{\sqrt{y}}^1 u^{-1}\, du < 4\varphi(\sqrt{y}),$$

we have $f \in C_w^r$. Following the notation of Example 9.1 we establish that if $y_0 \ge \overline{h}$, then

$$|\dot{\Delta}^k_{kw(x)}(f^{(r)}, x)| = \overline{h}^k |F^{(k)}(\theta)| = (k-1)!\,\overline{h}^k \theta^{-k-r/2} \overline{\varphi}(\theta)$$

$$\le c y_0^{-r/2} \varphi(h) < 2^{r/2} c\, w_{kh}^{-r}(x) \varphi(h),$$

where $\theta > y_0$, and if $y_0 < \overline{h}$, then

$$\left|\dot{\Delta}^k_{kw(x)}(f^{(r)}, x)\right| = \left|\Delta^k_{\bar{h}}(F; y_0)\right| \le 2^k \int\limits_{y_0}^{(k+1)\bar{h}} u^{-1-r/2}\bar{\varphi}(u)du$$

$$\le 2^{k+1} y_0^{-r/2}\bar{\varphi}((k+1)\bar{h}) < cw_{kh}^{-r}(x)\varphi(h).$$

Thus, the second inequality in (10.16) is proved.

To estimate $\overline{\omega}_{k,r}$ from below, we fix $h \le 2/k$ and choose y from the condition $y_0 = h^2$. As a result, we get

$$\left|\Delta^k_{\bar{h}}(F; h^2)\right| = \bar{h}^k \left|F^{(k)}(\theta)\right| = (k-1)!\bar{h}^k \theta^{-k-r/2}\bar{\varphi}(\theta)$$

$$\ge ch^{-r}\varphi(h) \ge cw_{kh}^{-r}(x)\varphi(h).$$

■

For what follows, we also need the definition of the DT_r-modulus of continuity for a closed interval $J = [a, b] \subset I$, $r \ne 0$. If $[a, b] \subset (-1, 1)$, then we set $C^r_w(J) = C^r(J)$. If $-1 = a < b < 1$, then $C^r_w(J)$ denotes the set of functions $f \in C^r([a, b])$ for which relation (10.2) is true. If $-1 < a < b = 1$, then $C^r_w(J)$ denotes the set of functions $f \in C^r([a, b])$ for which relation (10.1) is true. We set $C^0_w(I) := C(I)$ and $C^0_w(J) := C(J)$.

Definition 10.3. *Let $k \in \mathbb{N}$, $r \in \mathbb{N}$, $J \subset I$, and $f \in C^r_w(J)$. We set*

$$\overline{\omega}_{k,r}(t; f^{(r)}; J) := \sup_{0 \le h \le t} \sup_x \left| w^r_{kh} \dot{\Delta}^k_{hw(x)}(f^{(r)}, x)\right|, \quad t \ge 0, \tag{10.17}$$

where the inner supremum is taken over all x such that

$$\left[x - \frac{1}{2}khw(x),\ x + \frac{1}{2}khw(x)\right] \subset (a, b). \tag{10.18}$$

Note that

$$\overline{\omega}_{k,r}(t; f^{(r)}; I) \equiv \overline{\omega}_{k,r}(t; f^{(r)}). \tag{10.19}$$

For $J = [a, b]$, we set

$$w(a, b) := \sqrt{(1+a)(1-b)}. \tag{10.20}$$

Similarly to (9.27)–(9.30), we obtain the following relations for $f \in C_w^r(J)$ and $J \subset (-1, 1)$:

$$\omega_k(|J|; f^{(r)}; J) \le \frac{1}{w^r(a, b)} \overline{\omega}_{k,r}(/J/; f^{(r)}; J), \tag{10.21}$$

$$E_{k-1}(f^{(r)})_J \le \frac{c}{w^r(a, b)} \overline{\omega}_{k,r}(/J/; f^{(r)}; J), \tag{10.22}$$

$$\left\| f - L_{k-1}(\cdot, f; J) \right\|_J \le \frac{c}{w^r(a, b)} \overline{\omega}_{k,r}(/J/; f^{(r)}; J), \tag{10.23}$$

$$\left| [x_0, \dots, x_k; f^{(r)}] \right| \le c \frac{1}{w^r(a,b)} \overline{\omega}_{k,r}(/J/; f^{(r)}; J) \max_{j=1,\dots,k} \left| x_j - x_{j-1} \right|^{-k}, \tag{10.24}$$

where $a \le x_0 < x_1 < \dots < x_k \le b$.

10.3. Properties of DT_r-modulus of continuity

With regard for Remark 10.2, we have $\overline{\omega}_{k,r}(0+; f^{(r)}; J)$ for $f \in C_w^r$. It is obvious that $\overline{\omega}_{k,r}(t; f^{(r)}; J)$ is a nondecreasing function. We also get

$$\overline{\omega}_{k,r}(nt; f^{(r)}; J) \le cn^k \overline{\omega}_{k,r}(t; f^{(r)}; J), \quad n \in \mathbb{N}, \quad t \ge 0; \tag{10.25}$$

in particular,

$$\overline{\omega}_{k,r}(nt; f^{(r)}) \le cn^k \overline{\omega}_{k,r}(t; f^{(r)}). \tag{10.26}$$

To prove (10.25) we repeat the arguments of Lemma 9.1, replacing (9.30) by (10.24) and, hence, (9.35) by the following relation:

$$\left| \Delta_j \right| \le c_2 \left| J_j \right|^{-k} \overline{\omega}_{k,r}(c_1 h; f^{(r)}; J) w_{knh}^{-r}(x).$$

To establish an analog of inequality (4.25) we prove the following lemma:

Lemma 10.5. *Let* $m = k + r$, $p := 0, \dots, r - 1$, $J \subset I$, $x \in J$, *and* $h \ge 0$ *be such that*

$$\left[x - \frac{1}{2}(m - p)hw(x),\ x + \frac{1}{2}(m - p)hw(x) \right] \subset J$$

and let

$$G_{k,p,r}(x,h) := \begin{cases} \left(w(x)/w_{(m-p)h}(x) \right)^{2p-r} & \text{if} \quad p > r/2\,, \\ \ln\left(w(x)/w_{(m-p)h}(x) \right) & \text{if} \quad p = r/2\,, \\ 1 & \text{if} \quad p < r/2\,. \end{cases} \tag{10.27}$$

If $f \in C_w^r$, *then*

$$|\dot{\Delta}_{kw(x)}^{m-p}(f^{(p)}, x)| \le c\, \frac{h^{r-p}}{w^p(x)}\, G_{k,p,r}(x,h)\overline{\omega}_{k,r}(h; f^{(r)}; J). \tag{10.28}$$

Proof. We set $H := hw(x)$, $x_* = x - (m-p)H/2$, and $\overline{\omega}(h) := \overline{\omega}_{k,r}(h; f^{(r)}; J)$. By virtue of Definition 10.3 and relation (9.31), for every $t \in [x_*, x_* + (r-p)H]$ we have

$$\left|\Delta_H^k(f^{(r)}, x_* + t)\right| \le (1 + x_* + t)^{-r/2}(1 - x_* - t - kH)^{-r/2}\,\overline{\omega}(c_1 h). \tag{10.29}$$

Therefore, relations (9.5), (3.44), and (10.29) yield

$$\begin{aligned} \left|\dot{\Delta}_H^{m-p}(f^{(p)}, x)\right| &= \left|\Delta_H^{m-p}(f^{(p)}, x_*)\right| \\ &= \left| \int_0^H \ldots \int_0^H \Delta_H^k(f^{(r)}, x_* + t)\, dt_{r-p} \ldots dt_1 \right| \\ &\le \overline{\omega}(c_1 h) \int_0^H \ldots \int_0^H \frac{dt_{r-p} \ldots dt_1}{(1 + x_* + t)^{r/2}(1 - x_* - t - kH)^{r/2}} \\ &\le c\overline{\omega}(h)\, h^{r-p} w^{-p}(x) G_{k,p,r}(x,h), \qquad t = t_1 + \ldots + t_{r-p}. \end{aligned}$$

■

Let us verify some corollaries of Lemma 10.5.

Since

$$w^{-p}(x) G_{k,p,r}(x,h) \le w_{(m-p)h}^{-p}(x),$$

relation (10.28) yields

$$\overline{\omega}_{m-p,p}(t; f^{(p)}; J) \le ct^{r-p}\overline{\omega}_{k,r}(t; f^{(r)}; J), \quad t \ge 0. \tag{10.30}$$

As usual, we denote by $L_j = L_j(t; f; J) \equiv L_j(t; f; [a, b])$ the Lagrange polynomial of degree $\le j$ that interpolates a function f at $j+1$ equidistant points $x_i = a + (i/j)(b-a)$, $i = 0, \dots, j$, of a closed interval $J = [a, b]$. Then, assuming that $J \subset (-1, 1)$ and using Lemma 6.2, we obtain

$$\left\| f^{(p)} - L^{(p)}_{m-1}(\cdot; f; J)\right\|_J \le c\left\| f^{(p)} - L_{m-1-p}(\cdot; f^{(p)}; J)\right\|_J. \tag{10.31}$$

Therefore, using (9.29), we get

$$\begin{aligned}\left\| f^{(p)} - L^{(p)}_{m-1}(\cdot; f; J)\right\|_J &\le c\overline{\omega}_{m-p}(h; f^{(p)}; J)\\ &\le c\frac{h^{r-p}}{w^p\left(\frac{a+b}{2}\right)}\overline{\omega}_{k,r}(h; f^{(r)}; J)G^*_{p,r}(J),\end{aligned} \tag{10.32}$$

where

$$h = /J/ = (b-a)/w\left(\frac{a+b}{2}\right)$$

and

$$G^*_{p,r}(J) = \begin{cases} \left(\dfrac{w\left(\frac{a+b}{2}\right)}{\sqrt{(1+a)(1-b)}}\right)^{2p-r} & \text{if } p > r/2, \\ \ln\left(\dfrac{w\left(\frac{a+b}{2}\right)}{\sqrt{(1+a)(1-b)}}\right) & \text{if } p = r/2, \\ 1 & \text{if } p < r/2. \end{cases}$$

Remark 10.5. Inequality (10.32) also holds for $p = r$. If $p < r/2$, then relation (10.32) holds for $J \subset I$ (i.e., not only for $J \subset (-1, 1)$). Note that if $b - a \le \min\{1 + a, 1 - b\}$, then

$$G^*_{p,r}(J) \le c, \quad p = 0, \dots, r. \tag{10.33}$$

10.4. Classes $B^r\overline{H}$ and $B_0^r\overline{H}$

Properties of the DT_r-modulus of continuity and Theorem 4.1 lead to the following natural definition:

Definition 10.4. *For $k \in \mathbb{N}$, $r \in \mathbb{N}$, $\varphi \in \Phi^k$, and $M = \text{const} > 0$, we denote by $MB^r\,\overline{H}[k, \varphi]$ the set of functions $f \in C_w^r$, such that*

$$\overline{\omega}_{k,r}(t; f^{(r)}) \;\le\; M\varphi(t).$$

For $k \in \mathbb{N}$, $\varphi \in \Phi^k$, and $M = \text{const}$ we denote by

$$MB^0\,\overline{H}[k, \varphi] \;\equiv\; M\overline{H}[k, \varphi]$$

the set of functions $f \in C(I)$ such that

$$\overline{\omega}_k(t; f) \;\le\; M\varphi(t).$$

For $r \in \mathbb{N}$ and $M = \text{const} > 0$, we denote by $M\overline{B}^r$ the set of functions $f \in B^r$ such that

$$\left|w^{r/2} f^{(r)}(x)\right| \;\le\; M \quad \text{a.e. on } I.$$

We set

$$B^r\,\overline{H}[k, \varphi] \;:=\; 1B^r\,\overline{H}[k, \varphi], \qquad \overline{H}[k, \varphi] \;:=\; 1\overline{H}[k, \varphi], \qquad \overline{B}^r \;:=\; 1\overline{B}^r.$$

Inequality (10.30) yields

$$B^r\overline{H}_k^{\varphi} \subset cB^p\overline{H}_{k+p}^{\varphi_{p,r}}, \quad p \;=\; 0, \dots, r-1, \tag{10.34}$$

where $\varphi \in \Phi^k$ and $\varphi_p(t) := t^{r-p}\varphi(t)$.

By analogy, one can prove that

$$\overline{B}^r \subset cB^p\overline{H}_p^{\varphi_p^*}, \quad p \;=\; 0, \dots, r-1, \tag{10.35}$$

where $\varphi_p^*(t) = t^{r-p}$.

Conversely, if $f \in C_w^p$ and

$$\omega_{k,p}(t; f^{(p)}) \le t^k, \tag{10.36}$$

then $f \in \overline{B}^{k+p}$, which can be proved by analogy with Lemma 9.3.

For technical reasons, we introduce classes $M B_0^r \overline{H}[k, \varphi]$. The formal difference between the classes $M B_0^r \overline{H}[k, \varphi]$ and $M B^r \overline{H}[k, \varphi]$ is that we replace the factor $w_{kh}^r(x)$ in Definition 10.3 by the factor $w^r(x)$. This replacement plays a significant role near the endpoints ± 1 of the closed interval I, and it is insignificant in the "interior" of I.

Definition 10.5. *For* $k \in \mathbb{N}$, $r \in \mathbb{N}$, $\varphi \in \Phi^k$, *and* $M = \text{const} > 0$, *we denote by* $M B_0^r \overline{H}[k, \varphi]$ *the set of functions* $f \in C^r(I)$ *such that*

$$\left| w^r(x) \dot{\Delta}_{hw(x)}^k (f^{(r)}, x) \right| \le M \varphi(t) \tag{10.37}$$

for all x *and* h *such that*

$$\left[x - \frac{k}{2} hw(x),\ x + \frac{k}{2} hw(x) \right] \subset I.$$

We set

$$B_0^r \overline{H}[k, \varphi] \ := \ 1 B_0^r \overline{H}[k, \varphi].$$

It is obvious that

$$B_0^r \overline{H}[k; \varphi] \subset B^r \overline{H}[k; \varphi].$$

The following statement is true:

Lemma 10.6. *If*

$$\int_0^1 \frac{\varphi(u)}{u^{r+1}} du < \infty, \tag{10.38}$$

then

$$B^r \overline{H}[k; \varphi] \subset c B_0^r \overline{H}[k; \omega], \tag{10.39}$$

where

$$\omega(t) = t^r \int_0^1 u^{-r-1}\varphi(u)du. \tag{10.40}$$

Proof. We fix $x \in (-1, 1)$ and $h > 0$ for which

$$x_* := x - \frac{k}{2}hw(t) > -1, \quad x^* := x + \frac{k}{2}hw(t) < 1;$$

for simplicity, we assume that $x_* \le 1/2$. We set $\delta := 1 + x_*$ and $H := hw(t)$. The statement of the lemma follows from the inequality

$$\left| w^r \Delta_H^k(f^{(r)}, x_*)\right| \le c\omega(h). \tag{10.41}$$

Let us prove this inequality.

If $H \le 2\delta$, then $1 + x = \delta + kH/2 \le (k+1)\delta$, whence

$$w(x) < 2\sqrt{k+1}\sqrt{(1+x_*)(1-x^*)}.$$

Therefore, relation (10.41) follows directly from Definitions 10.3 and 10.5 and the inequality $\varphi(t) \le \omega(t)$. Thus, it suffices to consider the case $H > 2\delta$. Since $H > 2\delta$, we have $1 + x \le (k+1)H/2$, whence

$$w(x) < (k+1)h, \quad H < (k+1)h^2. \tag{10.42}$$

We choose an integer N such that $\delta 2^N < H/2 \le \delta 2^{N+1}$ and set

$$x_l := x_* + \delta 2^l, \quad l = 0, \dots, N; \quad x_{N+1} := x_* + H, \dots, x_{N+k} := x_* + kH,$$

$$\Delta_l = [x_l, \dots, x_{l+k}; f^{(r)}], \quad l = 0, \dots, N.$$

Since

$$2^{-k}\delta 2^l < x_{l+1} - x_l < 4\delta 2^l, \tag{10.43}$$

according to (10.24) we get

$$|\Delta_l| \le c_1\, \varphi\left(\sqrt{\delta 2^l}\right)(\delta 2^l)^{-k-r/2}. \tag{10.44}$$

Using (3.32) and (3.27), we obtain

$$\begin{aligned}\Delta_H^k(f^{(r)}; x_*) &= H^k k![x, x+H, \dots, x+kH] \\ &= H^k k! \sum_{l=0}^{N} (x_{l+k} - x_l)\Delta_l \alpha_{l,k}, \end{aligned} \tag{10.45}$$

where

$$\left|\alpha_{l,k}\right| \le c_2 \frac{(\delta 2^l)^{k-1}}{H^k}. \tag{10.46}$$

Combining (10.43)–(10.46), we get

$$\begin{aligned}\left|\Delta_H^k(f^{(r)}; x)\right| &\le c_3 \sum_{l=0}^{N} (\delta 2^l)^{-r/2} \varphi\left(\sqrt{\delta 2^l}\right) \\ &\le c_4 \sum_{l=0}^{N} \int_{\delta 2^l}^{\delta 2^{l+1}} \frac{\varphi(\sqrt{v})}{v^{1+r/2}} dv \le c_4 \int_{\delta}^{H} \frac{\varphi(\sqrt{v})}{v^{1+r/2}} dv \\ &= 2c_4 \int_{\sqrt{\delta}}^{\sqrt{H}} \frac{\varphi(u)}{u^{1+r}} du \le 2c_4 \int_{\delta}^{\sqrt{H}} \frac{\varphi(u)}{u^{1+r}} du,\end{aligned}$$

which, together with (10.42), yields (10.41) in the case $H > 2\delta$.

■

Remark 10.6. Note that condition (10.38) implies that $H > 2\delta$.

Remark 10.7. We use the classes $B_0^r \overline{H}[k, \varphi]$ because, for $f \in B_0^r \overline{H}[k, \varphi]$, we have an inequality stronger than (10.32), namely,

$$\left\| f^{(p)} - L_{m-1}^{(p)}(\cdot\,; f; J) \right\|_J \le c\left(/J/\right)^{r-p} w^{-p}\left(\frac{a+b}{2}\right) \overline{\omega}_{k,r}\left(/J/; f^{(r)}; J\right), \tag{10.47}$$

where $p = 0, \dots, r$, $J \subset I$, and

$$/J/ = (b-a) w^{-1}\left(\frac{a+b}{2}\right).$$

10.5. Relationship between $\overline{\omega}_{k,r}(f^{(r)};t)$ and $\omega_k(\tilde{f}^{(r)};t)$

Recall [see (10.3)] that, for every function f continuous in I, we have set

$$\tilde{f}(u) := f(\cos u).$$

L. G. Shakh and O. Yu. Dyuzhenkova proved the following two lemmas:

Lemma 10.7. *Let* $k \in \mathbb{N}$ *and* $(r+1) \in \mathbb{N}$. *If* $\tilde{f} \in C^r(\mathbb{R})$, *then*

$$\overline{\omega}_{k,r}(t; f^{(r)}) \le c\omega_k(t; \tilde{f}^{(r)}). \tag{10.48}$$

Lemma 10.8. *If* $f \in C_w^r$, *then the following relations are true:*

$$\omega_k(t; \tilde{f}^{(r)}) \le c(\overline{\omega}_{k,r}(t; f^{(r)}) + t^k \| f \|) \tag{10.49}$$

if $r = 0$ *and* k *is odd, and if* r *and* k *are odd;*

$$\omega_k(t; \tilde{f}^{(r)}) \le c\left(t^k \int_t^2 \frac{\overline{\omega}_{k,r}(u; f^{(r)})}{u^{k+1}} du + t^k \| f \| \right), \quad 0 \le t \le 2, \tag{10.50}$$

if r *is odd and* k *is even;*

$$\omega_k(t; \tilde{f}^{(r)}) \le c\left(\int_0^t \frac{\overline{\omega}_{k,r}(u; f^{(r)})}{u} du + t^k \| f \| \right) \tag{10.51}$$

if r *is even and* k *is odd; and*

$$\omega_k(t; \tilde{f}^{(r)}) \le c\left(\int_0^t \frac{\overline{\omega}_{k,r}(u; f^{(r)})}{u} du + t^k \int_t^2 \frac{\overline{\omega}_{k,r}(u; f^{(r)})}{u^{k+1}} du + t^k \| f \| \right) \tag{10.52}$$

if r *and* k *are even.*

Remark 10.8. It is clear that, for $f \in C([-1, 1])$, we have

$$c_1 \omega_1(t; \tilde{f}) \le \overline{\omega}_1(t; f) \le c_2 \omega_1(t; \tilde{f}). \tag{10.53}$$

Chapter 4
Extension

In 1934, Whitney proved that if the divided differences of order r for a function f defined on a closed set $E \subset \mathbb{R}$ converge on E, then the function f admits an extension to a certain function $\bar{f} \in C^r(\mathbb{R})$ on the entire straight line $\mathbb{R}$, i.e., there exists a function $\bar{f} \in C^r(\mathbb{R})$ that coincides with f at points $x \in E$. According to Lemma 3.8.3, if $\bar{f} \in C^r(\mathbb{R})$, then the divided differences of order r for a function $\bar{f}$ converge on E. Thus, Whitney, in fact, described the traces of the space $C^r(\mathbb{R})$ on an arbitrary closed set $E \subset \mathbb{R}$.

In this chapter, we prove theorems on extension for the classes $W^r H[k, \varphi]$ (including the classes W^r). Then, using these theorems, we describe the traces of the space $W^r H_k^\varphi$ on an arbitrary set $E \subset \mathbb{R}$, omitting the condition of the closedness of the set E. Let us illustrate this by an example. Let E be an arbitrary subset of the straight line $\mathbb{R}$, let φ be a 1-majorant, and let a function f be given on E such that

$$|f(x_1) - f(x_2)| \leq \varphi(|x_1 - x_2)|) \quad \forall \ x_1, x_2 \in E.$$

By continuity, we define the function f on the closure $\overline{E}$ of the set E. On each of the intervals (α_s, β_s) that form the open set $\mathbb{R} \setminus \overline{E}$, we denote by $\bar{f}(x)$ a linear function equal to $f(\alpha_s)$ at the point α_s and to $f(\beta_s)$ at the point β_s. If $\beta_s = \infty$ ($\alpha_s = -\infty$), then we set $\bar{f}(x) = f(\alpha_s)$ for $x > \alpha_s$ ($\bar{f}(x) = f(\beta_s)$ for $x > \beta_s$). For $x \in \overline{E}$, we set $\bar{f}(x) = f(x)$. It is clear that $\bar{f} \in H[1, \varphi, \mathbb{R}]$.

In Section 7 of Chapter 3, the classes $MW^r H[k, \varphi, J]$ and spaces $W^r H_k^\varphi(J)$ were defined for the interval $J = [a, b]$. This definition can easily be carried over to the case where $J = [a, \infty]$, or $J = [-\infty, b]$, or $J = \mathbb{R}$. Namely, we write $f \in MW^r H[k, \varphi, J]$ if $f \in MW^r H[k; \varphi; [a, b]]$ for any interval $[a, b] \subset J$. Recall that $m = k + r$. We also set

$$W^r H_k^\varphi(J) := \bigcup_{M>0} MW^r H[k; \varphi; J].$$

1. Extension from intervals

Let $f \in C^r([a, b])$. Let

$$f_1(x) = f(a) + \sum_{j=1}^{r} \frac{f^{(j)}(a)}{j!}(x-a)^j,$$

$$f_2(x) = f(b) + \sum_{j=1}^{r} \frac{f^{(j)}(b)}{j!}(x-b)^j$$

denote the Taylor polynomials of the function f at the points a and b. We set $\bar{f}(x) := f_1(x)$ if $x < a$, $\bar{f}(x) := f_2(x)$ if $x > b$, and $\bar{f}(x) = f(x)$ if $x \in [a, b]$. It is clear that $\bar{f} \in C^r(\mathbb{R})$.

Let $f \in C([a, b])$. We set $\bar{f}(x) := f(a)$ if $x < a$, $\bar{f}(x) := f(b)$ if $x > b$, and $\bar{f}(x) := f(x)$ if $x \in [a, b]$. It is clear that

$$\omega_1(t; \bar{f}; \mathbb{R}) \le \omega_1(t; f; [a, b]). \tag{1.1}$$

Theorem 1.1 [Dzyadyk (1958), (1958b)]; Frey (1958)]. *Let $f \in C([0, b])$ and $f(0) = 0$. Denote $\bar{f}(x) := -f(-x)$ for $x \in [-b, 0]$ and $\bar{f}(x) := f(x)$ for $x \in [0, b]$. Then*

$$\omega_2\left(t\text{: } \bar{f}; [-b, b]\right) \le 3\omega_2\left(t\text{: } f; [0, b]\right). \tag{1.2}$$

Proof. If

$$0 \le x_0 < x_0 + 2h \le b$$

or

$$-b \le x_0 < x_0 + 2h \le 0,$$

then, obviously,

$$\left|\Delta_h^2(\bar{f}; x_0)\right| \le \omega_2\left(h; f; [0, b]\right).$$

Now let

$$-b \le x_0 < 0, \quad x_0 + h \ge 0, \quad x_0 + 2h \le b.$$

Denoting

$$g(x) := \bar{f}(x) - L(x; f; 0, x_0 + 2h) = \bar{f}(x) - \frac{xf(x_0 + 2h)}{x_0 + 2h},$$

we obtain

$$\left|\Delta_h^2(\bar{f}; x_0)\right| = \left|\Delta_h^2(g; x_0)\right| = \left|2g(x_0 + h) + g(-x_0)\right| \le 3\| g \|_{[0,\, x_0 + 2h]}.$$

It follows from Theorem 3.5.2 that

$$\left|\Delta_h^2(\bar{f}; x_0)\right| \le 3\| g \|_{[0,\, x_0 + 2h]}$$

$$\le 3\omega_2\big(h + x_0 / 2; f; [0, x_0 + 2h]\big) \le 3\omega_2\big(h; f; [0, b]\big).$$

The case where $x_0 + h > 0$ is considered by analogy.

■

Theorem 1.2 [Besov (1963), (1965)]. *For every function* $f \in C([0, b])$, *there exists a function* $\bar{f} \in C\big([-b, b]\big)$ *such that* $\bar{f}(x) = f(x)$ *for* $x \in [0, b]$ *and*

$$\omega_k\big(t; \bar{f}; [-b, b]\big) \le c\omega_k(t; f; [0, b]). \tag{1.3}$$

Proof. The cases $k = 1, 2$ have already been considered [see (1.1) and (1.2)]. Denote

$$\varphi(t) := \omega_k\big(t; f; [0, b]\big), \qquad J_0 := [0, b], \qquad J := [-b, 0).$$

We use the construction proposed by Hestenes (1941), namely, we find k numbers a_i, $i = 0, \dots, k - 1$, from the following system of equations:

$$\begin{cases} \displaystyle\sum_{i=0}^{k-1} a_i = 1, \\ \displaystyle\sum_{i=1}^{k-1} a_i (i / (k-1))^j = (-1)^j, \quad j = 1, \dots, k - 1. \end{cases}$$

The determinant of this system is called the Vandermonde determinant, and, as is well known, it is not equal to zero. Note that, for all $j = 1, \dots, k - 1$, the following identity is true:

$$(-x)^j = \sum_{i=0}^{k-1} a_i(ix/(k-1))^j .$$

Therefore, for any polynomial P_{k-1} of degree $\le k-1$, we get

$$P_{k-1}(x) = \sum_{i=0}^{k-1} a_i P_{k-1}(-ix/(k-1)).$$

We set

$$\bar{f}(x) := \sum_{i=0}^{k-1} f\big(-ix/(k-1)\big)$$

for $x \in J$ and $\bar{f}(x) := f(x)$ for $x \in J_0$. In this case, if $x_0 \in J_0$ and $(x_0 + kh) \in J_0$, then $\left|\Delta_h^k(\bar{f}; x_0)\right| \le \varphi(\hbar)$. If $x_0 \in J$ and $(x_0 + kh) \in J$, then

$$\left|\Delta_h^k(\bar{f}; x_0)\right| \le \sum_{i=0}^{k-1} |a_i|\, \varphi(ih/(k-1)) \le c_1\varphi(h).$$

Finally, if $x_0 \in J$ and $(x_0 + kh) \in J_0$, then we denote

$$h^* := \max\{x_0 + kh, -x_0\},$$

$$L(x) := L\big(x; f; 0; h^*/(k-1), \dots, h^*\big),$$

and

$$g(x) := \bar{f}(x) - L(x).$$

For $x \in \left[0, h^*\right]$, according to the Whitney inequality (3.6.12), we have

$$|g(x)| \le c_2\varphi(h^*) \le c_3\varphi(h),$$

$$|g(-x)| = \left|\sum_{i=0}^{k-1} a_i g(ix/(k-1))\right| \le c_3\varphi(h) \sum_{i=0}^{k-1} |a_i| \le c_4\varphi(h),$$

whence

$$\|g\|_{[-h^*,h^*]} \le c_4\varphi(h).$$

Thus,

$$\left|\Delta_h^k(\bar f; x_0)\right| = \left|\Delta_h^k(g; x_0)\right| \le 2^k c_4\varphi(h).$$

■

Note that Besov [(1963), (1965)] proved a multidimensional analog of Theorem 1.2 (in particular, for integral metrics).

Theorem 1.3. *If* $f \in H[k; \varphi; [0, \infty)]$, *then there exists a function* $\bar f \in cH[k; \varphi; \mathbb{R}]$ *such that* $\bar f(x) = f(x)$ *for* $x \in [0, \infty)$.

Proof. If we denote $J_0 := [0, \infty)$ and $J := [-\infty, 0)$, then it suffices to repeat the proof of Theorem 1.2.

■

The theorems presented below (Theorems 1.2′ and 1.3′) are simple corollaries of Theorems 1.2 and 1.3, respectively.

Theorem 1.2′. *If* $f \in W^r H[k; \varphi; [a, b]]$, *then there exists a function*

$$\bar f \in cW^r H\left[k;\ \varphi;\ \left[a-(b-a),\ b+(b-a)\right]\right],$$

such that $\bar f(x) = f(x)$ *for* $x \in [a, b]$.

Theorem 1.3′. *Let* $J = [a, \infty)$ *or* $J = (-\infty, b]$. *If* $f \in W^r H[k; \varphi; J]$, *then there exists a function* $\bar f(x) \in cW^r H[k; \varphi; \mathbb{R}]$ *such that* $\bar f(x) = f(x)$ *for* $x \in J$.

Using Theorem 1.2′, one can easily show that a function $f \in W^r H[k; \varphi; [a, b]]$ can be extended to the entire straight line. To this end, we need four simple lemmas.

Lemma 1.1. *Let*

$$f \in W^r H[k; \varphi; [\alpha, \beta]], \qquad [a, b] \subset [\alpha, \beta].$$

If

$$\|f\|_{[a,b]} \le (b-a)^r \varphi(b-a),$$

then

$$\left\| f^{(i)} \right\|_{[\alpha,\beta]} \le c(\beta-\alpha)^{k+r-i}(b-a)^{-k}\varphi(b-a), \qquad i = 0, \dots, r. \tag{1.4}$$

Proof. Denote $h := b-a$ and $H := \beta-\alpha$. By using inequalities (3.5.2), (3.6.17), and (3.4.16), we obtain

$$\begin{aligned}\left\| f^{(i)} \right\|_{[\alpha,\beta]} &\le c_1 H^{r-i}\varphi(H) + c_2 H^{-i} \| f \|_{[\alpha,\beta]} \\ &\le c_1 H^{r-i}\varphi(H) + c_3 H^{m-i} h^{-m} \left(\omega_m(h; f; [a,b]) + \| f \|_{[a,b]}\right) \\ &\le (c_1 + 2c_3) H^{m-i} h^{-k} \varphi(h).\end{aligned}$$

■

Definition 8.1. *Let* $k \in \mathbb{N}$, $a < b$,

$$a^* := a + \frac{b-a}{3}, \qquad \textit{and} \qquad b^* := b - \frac{b-a}{3}.$$

The function

$$S(x) := S(x; k; a; b) := \begin{cases} 0, & \textit{if } x < a^*, \\ \dfrac{\int_{a^*}^{x} (u-a^*)^k (b^*-u)^k\, du}{\int_{a^*}^{b^*} (u-a^*)^k (b^*-u)^k\, du} & \textit{if } x \in [a^*, b^*], \\ 1, & \textit{if } x > b^*, \end{cases} \tag{1.5}$$

is called the gluing function.

A multidimensional analog of the lemma presented below can be found in [Burenkov (1976)].

Lemma 1.2. *Let* $f \in W^r H[k; \varphi; [a,b]]$ *and* $G(x) := f(x) S(x; m; a; b)$. *If* $\| f \|_{[a,b]} \le (b-a)^r \varphi(b-a)$, *then* $G \in c W^r H[k; \varphi; [a,b]]$.

Proof. Denote $J := [a,b]$ and $h := b-a$. Let $j = 0, \dots, k$ and $i = 0, \dots, r$. According to (3.4.16), we have

$$\omega_j(t; S^{(i)}, J) \le t^j \left\| S^{(i+j)} \right\|_J \le c_1 t^j h^{-i-j},$$
$$\omega_{m-i}(t; f^{(i)}, J) \le t^{r-i} \omega_k(t; f^{(r)}, J) \le t^{r-i} \varphi(t). \tag{1.6}$$

Taking into account Lemma 1.1 and the conditions of Lemma 1.2, we get $\left\| f^{(i)} \right\|_J \le c_2 h^{r-i} \varphi(h)$. By using the Marchaud inequality (3.5.6), for $i+j \ne m$ we obtain

$$\begin{aligned}
\omega_j(t; f^{(i)}, J) &\le c_3 t^j \int_t^h u^{-j-1} \omega_{m-i}(u; f^{(i)}, J) du + c_4 t^j h^{-j} \left\| f^{(i)} \right\|_J \\
&\le c_3 t^j \int_t^h u^{r-i-j-1} \varphi(u) du + 2c_2 c_4 t^j h^{r-i-j} \varphi(h) \\
&\le c_3 t^j t^{-k} \varphi(t) \int_t^h u^{m-i-j-1} du + 2c_2 c_4 t^{j-k} h^{m-i-j} \varphi(t),
\end{aligned}$$

i.e.,

$$\omega_j(t; f^{(i)}, J) \le c_5 t^{j-k} h^{m-i-j} \varphi(t). \tag{1.7}$$

In the case where $i+j = m$, i.e., for $i = r$ and $j = k$, estimate (1.7) follows directly from the conditions of the lemma. It now follows from (3.4.15), (3.4.16), (1.6), and (1.7) that

$$\begin{aligned}
\omega_k(t; G^{(r)}, J) &\le \sum_{i=0}^{r} \binom{r}{i} \omega_k(t; f^{(i)} S^{(r-i)}, J) \\
&\le \sum_{i=0}^{r} \binom{r}{i} \sum_{j=0}^{k} \binom{k}{j} \omega_j(t; f^{(i)}, J) \omega_{k-j}(t; S^{(r-i)}, J) \\
&\le c_1 c_5 \sum_{i=0}^{r} \sum_{j=0}^{k} \binom{r}{i} \binom{k}{j} t^{j-k} h^{m-i-j} \varphi(t) t^{k-j} h^{i+j-m} \\
&= c_1 c_5 2^r 2^k \varphi(t) = c\varphi(t).
\end{aligned}$$

■

Lemma 1.3. *Let* $h := b - a$ *and* $d \in (0, h]$. *If* $f \in W^r H[k; \varphi; [a-d, b+d]]$ *and* $\|f\|_{[a,b]} \le h^r \varphi(h)$, *then the function* $\bar{f}(x) := f(x)S(x; k; a-d; a)(1 - S(x; k; b; b+d))$ *for* $x \in [a-d, b+d]$, $\bar{f}(x) := 0$ *for* $x \notin [a-d, b+d]$ *possesses the following properties:*

(a) $\bar{f}(x) = f(x)$ *for* $x \in [a, b]$;

(b) $\bar{f} \in c(h/d)^m W^r H[k; \varphi; \mathbb{R}]$, $m = k + r$.

Proof. Property (a) is obvious. Let us prove property (b). For this purpose, we take $x_0 \in \mathbb{R}$ and $\delta > 0$ and show that

$$|\Delta| := \left|\Delta_\delta^k(\bar{f}^{(r)}; x_0)\right| \le c_1 \left(\frac{h}{d}\right)^m \varphi(\delta). \tag{1.8}$$

We consider four cases.

1. Assume that $\delta > d/3$. By virtue of Lemma 1.1, we have

$$\left\|\bar{f}^{(r)}\right\|_{[a-d,b+d]} \le \left\|f^{(r)}\right\|_{[a-d,b+d]} \le c_2 \left\|f^{(r)}\right\|_{[a,b]} \le c_3 \varphi(h).$$

Hence,

$$|\Delta| \le c_3 2^k \varphi(h) \le c_3 2^k \left(\frac{3h}{d}\right)^k \varphi\left(\frac{d}{3}\right) \le c_3 6^k \left(\frac{h}{d}\right)^k \varphi(\delta).$$

2. Assume that $0 < \delta \le d/3$ and $x_0 \notin [a-d, a-d/3] \cup [b, b+2d/3]$. If $x_0 \in (-\infty, a-d) \cup [b+2d/3, \infty]$, then $\Delta = 0$. If $x \in (a-d/3, b)$, then

$$|\Delta| = \left|\Delta_\delta^k(f^{(r)}; x_0)\right| \le \varphi(\delta).$$

3. Assume that $x_0 \in [a-d, a-d/3]$. Denote $J := [a-d, a]$. Note that $\bar{f}(x) = f(x)S(x; k; a-d; a)$ for $x \in J$. By virtue of Lemma 1.1, we have

$$\|f\|_J \le \|f\|_{[a-d,b+d]} \le c_4 \|f\|_{[a,b]} \le c_4 h^r \varphi(h) \le c_4 \left(\frac{h}{d}\right)^m d^r \varphi(d).$$

According to the conditions of Lemma 1.3, $f \in W^r H[k; \varphi; J] \subset (h/d)^m W^r H[k; \varphi; J]$. Hence, by virtue of Lemma 1.2, we get $\bar{f} \in c_5 (h/d)^m W^r H[k; \varphi; J]$, which yields (1.8) in the case under consideration.

4. The case where $0 < \delta \leq d/3$ and $x_0 \in [b, b + 2d/3]$ is analogous to the previous case.

■

The lemma below is a simple corollary of Lemma 1.3.

Lemma 1.4. *Let $h := b - a$ and $d \in (0, h]$. If $f \in W^r H[k; \varphi; [a - d, b + d]]$, $g \in W^r H[k; \varphi; \mathbb{R}]$, and $\|f - g\|_{[a,b]} \leq h^r \varphi(h)$, then the function $\overline{G}(x)$ defined by the formulas*

$$\overline{G}(x) := G(x; f; g)$$

$$:= g(x) + (f(x) - g(x)) S(x; k; a - d; a)(1 - S(x; k; b; b + d)) \tag{1.9}$$

for $x \in [a - d, b + d]$ and $\overline{G}(x) := g(x)$ for $x \notin [a - d, b + d]$ possesses the following properties:

(a) $\overline{G}(x) = \bar{f}(x)$ *for* $x \in [a, b]$;

(b) $\overline{G} \in c(h/d)^m W^r H[k; \varphi; \mathbb{R}]$.

Theorem 1.2″. *Suppose that $f \in W^r H[k; \varphi; [a, b]]$. Then there exists a function $\bar{f} \in c W^r H[k; \varphi; \mathbb{R}]$ such that $\bar{f}(x) = f(x)$ for $x \in [a, b]$.*

Proof. Let L denote the Lagrange polynomial (of degree $\leq m - 1$) that interpolates the function f at the equidistant points

$$x_i = a + \frac{ib}{m-1}, \qquad i = 0, \ldots, m - 1.$$

According the Whitney inequality (3.6.12), we have

$$\|f - L\|_{[a,b]} \leq c_1 \omega_m(b - a; f; [a, b]) \leq c_1 (b - a)^r \varphi(b - a).$$

Moreover, $\omega_k(t; L^{(r)}; \mathbb{R}) = 0$. By virtue of Theorem 1.2′, one can find a function $f_1 \in c_2 W^r H[k; \varphi; [a-(b-a), b+(b-a)]]$ such that $f_1(x) = f(x)$ for $x \in [a, b]$. It remains to use Lemma 1.4 for the case $d = b-a$.

2. Lemma on gluing

First, we prove four auxiliary lemmas. Lemma 2.3 (we call it the lemma on gluing) plays the key role in the proofs presented in the subsequent two sections.

As before, we assume that $k \in \mathbb{N}$, $\varphi \in \Phi^k$, $(r+1) \in \mathbb{N}$, $m = k+r$, $x_0 < x_1 < \ldots < x_m$, $x_{-1} := x_0 - (x_m - x_0)$, and $x_{m+1} := x_m + (x_m - x_0)$.

Definition 2.1. *Let $(p+1) \in \mathbb{N}$ and $q \in \mathbb{N}$. We write $(p, q) \in B_{k,r}$ if $p + r < q \le m$ and $(p, q) \in B^*_{k,r}(x_0, \ldots, x_m)$ if $(p, q) \in B_{k,r}$ and $2(x_q - x_p) \le \min\{x_{q+1} - x_p, x_q - x_{p-1}\}$. We set [cf. (3.6.32)]*

$$\Lambda^*_r(x_0, \ldots, x_m; \varphi) = \max_{(p,q)\in B^*_{k,r}(x_0,\ldots,x_m)} \Lambda_{p,q,r}(x_0, \ldots, x_m; \varphi). \tag{2.1}$$

It is clear that

$$\Lambda^*_r(x_0, \ldots, x_m; \varphi) \le \Lambda_r(x_0, \ldots, x_m; \varphi). \tag{2.2}$$

Note that there exists at least one pair $(p, q) \in B^*_{k,r}(x_0, \ldots, x_m)$, namely, $(0, m) \in B^*_{k,r}(x_0, \ldots, x_m)$.

Definition 2.2. *For all $(p, q) \in B_{k,r}$, we denote*

$$\tilde{\Lambda}_{p,q,r}(x_0, \ldots, x_m; \varphi) := \frac{\int_{(x_q-x_p)}^{d(p,q)} u^{p+r-q-1}\varphi(u)\,du + (x_q - x_p)^{p+r-q}\varphi(x_q - x_p)}{\prod_{i=0}^{p-1}(x_q - x_p)\prod_{i=q+1}^{m}(x_i - x_p)}. \tag{2.3}$$

We set

$$\tilde{\Lambda}_r(x_0, \ldots, x_m; \varphi) = \max_{(p,q)\in B_{k,r}} \tilde{\Lambda}_{p,q,r}(x_0, \ldots, x_m; \varphi). \tag{2.4}$$

It is clear that

$$\Lambda_r(x_0, \ldots, x_m; \varphi) \le \tilde{\Lambda}_r(x_0, \ldots, x_m; \varphi). \tag{2.5}$$

Lemma 2.1. *The following inequality is true:*

$$\tilde{\Lambda}_r(x_0, \dots, x_m; \varphi) \le c\Lambda^*_r(x_0, \dots, x_m; \varphi). \tag{2.6}$$

Proof. If $(p, q) \in B^*_{k,r}(x_0, \dots, x_m)$, then

$$(x_q - x_p)^{p+r-q}\varphi(x_q - x_p) \le l(1 - 2^{-l}) \int_{x_q - x_p}^{d(p,q)} u^{-l-1}\varphi(u)\,du$$

$$\le (k+1) \int_{x_q - x_p}^{d(p,q)} u^{-l-1}\varphi(u)\,du,$$

where $l := q - p - r$. Hence,

$$\tilde{\Lambda}_{p,q,r}(x_0, \dots, x_m; \varphi) \le (k+2)\tilde{\Lambda}_{p,q,r}(x_0, \dots, x_m; \varphi).$$

If $(p, q) \notin B_{k,r}(x_0, \dots, x_m)$, i.e., $d(p, q) < 2(x_q - x_p)$, then, in the case where $d(p, q) = x_{q+1} - x_p$, we have $x_{q+1} - x_i < 2(x_q - x_i)$, $i = 0, \dots, p$, and

$$\int_{x_q - x_p}^{d(p,q)} u^{-l-1}\varphi(u)\,du + (x_q - x_p)^{-l}\varphi(x_q - x_p) \le 2^{l+1}(x_{q+1} - x_p)^l\varphi(x_{q+1} - x_p).$$

Therefore,

$$\tilde{\Lambda}_{p,q,r}(x_0, \dots, x_m; \varphi) \le c_1\tilde{\Lambda}_{p,q+1,r}(x_0, \dots, x_m; \varphi).$$

In the case where $d(p, q) = x_q - x_{p-1}$, we establish by analogy that

$$\tilde{\Lambda}_{p,q,r}(x_0, \dots, x_m; \varphi) \le c_1\tilde{\Lambda}_{p-1,q,r}(x_0, \dots, x_m; \varphi).$$

Thus, the lemma is proved for $c \le (k+2)c_1^{k-1}$.

∎

Inequalities (2.2), (2.5), and (2.6) yield the estimates

$$\Lambda_r(x_0, \dots, x_m; \varphi) \le \tilde{\Lambda}_r(x_0, \dots, x_m; \varphi)$$

$$\le c\Lambda_r^*(x_0, \dots, x_m; \varphi) \le c\Lambda_r(x_0, \dots, x_m; \varphi). \tag{2.7}$$

Lemma 2.2. *Let* $j = 0, \dots, m$. *If* $f \in W^r H[k; \varphi; \mathbb{R}]$, *then the following inequality holds for all* $x \in [x_j, x_j + (x_{j+1} - x_j)/2]$:

$$\left|[x_0, \dots, x_{j-1}, x, x_{j+1}, \dots, x_m; f]\right| \le c\Lambda_r(x_0, \dots, x_m; \varphi). \tag{2.8}$$

Proof. Denote $t_j := x$, $t_s := x_s$ if $s \ne j$, $s = -1, \dots, m$, $t_{m+1} := x_{m+1}$ if $j \ne m$, and $t_{m+1} := x + (x - x_0)$ if $j = m$. Let $(p, q) \in B_{k,r}^*(t_0, \dots, t_m)$. We set $\Lambda := \Lambda_{p,q,r}(x_0, \dots, x_m; \varphi)$, $\Lambda^* := \Lambda_{p,q,r}(t_0, \dots, t_m; \varphi)$, $d := \min\{x_{q+1} - x_p, x_q - x_{p-1}\}$ $(= d(p, q))$, $d^* := \min\{t_{q+1} - t_p, t_q - t_{p-1}\}$ and prove the inequality

$$\Lambda^* \le c_1\Lambda_r(x_0, \dots, x_m; \varphi). \tag{2.9}$$

We first consider the case where $d = x_j - x_p$, i.e., $q = j - 1$ and $x_j - x_p \le x_q - x_{p-1}$. We represent Λ^* in the form

$$\Lambda^* = (x_j - x_p)(x - x_p)^{-1}\Lambda$$

$$+ \prod_{i=0}^{p-1}(x_q - x_i) \prod_{i=q+1}^{m} (x_i - x_p) \int_d^{d^*} (x - x_p)^{-1} u^{p+r-q-1} \varphi(u)du\,.$$

Taking into account that $2(x_q - x_i) \ge (x_j - x_i)$ for all $i = 0, \dots, p - 1$ and

$$d^* = \min\{x - x_p, x_q - x_{p-1}\} \le \min\{x_{j+1} - x_p, x_j - x_{p-1}\},$$

we get

$$\Lambda^* \le \Lambda + 2^p \Lambda_{p,j,r}(x_0, \dots, x_m; \varphi) \le c_1\Lambda_r(x_0, \dots, x_m; \varphi).$$

In the other cases, we have: $2(t_q - t_i) \ge x_q - x_i$ for $i = 0, \dots, p$, $2(t_i - t_q) \ge x_i - x_q$ for $i = q, \dots, m$, and $d^* \le 2d$. Consequently, $\Lambda^* \le c_2\tilde{\Lambda}_{p,q,r}(x_0, \dots, x_m; \varphi)$, which, by virtue of (2.7), implies that $\Lambda^* \le c_1\Lambda_r(x_0, \dots, x_m; \varphi)$. Taking (3.6.36), (2.7), and (2.9) into account, we obtain

$$|[t_0, \dots, t_m; f]| \le c_3 \Lambda_r(x_0, \dots, x_m; \varphi)$$
$$\le c_4 \Lambda_r^*(t_0, \dots, t_m; \varphi) \le c_4 c_1 \Lambda_r(x_0, \dots, x_m; \varphi).$$

■

Remark 2.1. It can be proved by analogy that relation (2.8) also holds on the interval $[x_j - (x_j - x_{j-1})/2, x_j]$, i.e., that relation (2.8) holds for all points $x \in [x_j - (x_j - x_{j-1})/2, x_j + (x_{j+1} - x_j)/2]$.

Lemma 2.3 [Shevchuk (1979), (1980)]. *Suppose that* $(i, j) \in B_{k,r}^*(x_0, \dots, x_m)$ *and* $d := \min\{x_{j+1} - x_j, x_i - x_{i-1}\}$. *If* $f \in W^r H[k; \varphi; \mathbb{R}]$ *and* $f(x_s) = 0$ *for all* $s = 0, \dots, m$, $s \ne j$, *then there exists a function* $\bar{f}$ *such that*

$$\bar{f}(x) = f(x) \quad \text{for} \quad x \in [x_i; x_j], \tag{2.10}$$

$$\bar{f}(x) = 0 \quad \text{for} \quad x \notin (x_i - d; x_j + d), \tag{2.11}$$

$$\bar{f} \in cMW^r H[k; \varphi; \mathbb{R}], \tag{2.12}$$

where

$$M := \frac{\Lambda_r(x_0, \dots, x_m; \varphi)}{\Lambda_{i,j,r}(x_0, \dots, x_m; \varphi)}.$$

Proof. Without loss of generality, we can set $x_i = 0$. Taking Lemma 3.4.4 into account, we can assume, without loss of generality, that

$$\varphi^{(\nu)}(t) \le c_0 t^{-\nu} \varphi(t), \qquad \nu = 1, \dots, m-1. \tag{2.13}$$

We carry out the proof in several steps.

1. First, we prove the inequalities

$$\| f \|_{[0, kx_j]} \le c_1 M x_j^{j-i} \int_{x_j}^{2d} u^{i+r-j-1} \varphi(u)\, du, \tag{2.14}$$

$$\| f \|_{[0, kd]} \le c_1 M d^{j-i} \int_{x_j}^{2d} u^{i+r-j-1} \varphi(u)\, du. \tag{2.15}$$

Indeed, for all $x \in [x_j, x_j + d/2]$, relation (2.9) yields

$$|f(x)| = \left|[x_0, \ldots, x_{j-1}, x, x_{j+1}, \ldots, x_m; f]\right| \prod_{s=0, s\neq j}^{m} |x - x_s|$$

$$\leq c_2 \Lambda_r(x_0, \ldots, x_m; \varphi) \prod_{s=0, s\neq j}^{m} |x - x_s|$$

$$= c_2 M \Lambda_{i,j,r}(x_0, \ldots, x_m; \varphi) \prod_{s=0, s\neq j}^{s=m} |x - x_s| \leq c_3 M x^{j-i} \int_{x_j}^{x_j+d} u^{i+r-j-1} \varphi(u)\, du.$$

This and (3.6.16) imply that

$$\|f\|_{[0, kx_j]} \leq c_4 x_j^{m-1} \int_{x_j/2}^{kx_j} u^{-k} \varphi(u)\, du + c_4 \|f\|_{[x_j, 3x_j/2]}$$

$$\leq c_5 x_j^r \varphi(x_j) + c_4 c_3 M (kx_j)^{j-i} \int_{x_j}^{2d} u^{i+r-j-1} \varphi(u)\, du$$

$$\leq c_1 M x_j^{j-i} \int_{x_j}^{2d} u^{i+r-j-1} \varphi(u)\, du,$$

$$\|f\|_{[0, kd]} \leq c_4 d^{m-1} \int_{d/2}^{kd} u^{-k} \varphi(u)\, du + c_4 \|f\|_{[x_j, x_j + d/2]}$$

$$\leq c_1 M d^{j-i} \int_{x_j}^{2d} u^{i+r-j-1} \varphi(u)\, du.$$

2. Let us prove Lemma 2.3 for $k = 1$. In this case, the pair $(i, j) \in B_{k,r}$ is unique, namely, $(0, r+1) \in B_{1,r}$ (moreover, $(0, r+1) \in B_{1,r}^*(x_0, \ldots, x_{r+1})$). Therefore, $d = x_{r+1}$ and $M = 1$. By virtue of (2.14), we have

$$\|f\|_{[0, kd]} \leq c_1 d^{r+1} \int_{d}^{2d} u^{-2} \varphi(u)\, du \leq c_1 d^r \varphi(d).$$

Lemma 2.3 now follows from Lemma 1.3.

Taking into account that Lemma 2.3 is proved for $k = 1$, we assume in what follows that $k > 1$.

3. Let us choose $m + 1$ points y_s according to the relations $y_s := x_{i+s}$ for $s = 0, \dots, r$ and $y_s := (s-r)x_i$ for $s = r+1, \dots, m$. We denote $\omega_\mu := \omega_\mu(t) := \omega_\mu(t; f^{(r)};$ $[0, kx_j])$ and $a_\mu := [y_0, \dots, y_{r+m}; f]$ and show that

$$\left| a_\mu \right| \le c_6 x_j^{-\mu} \omega_\mu(x_j) \tag{2.16}$$

for all $\mu = 1, \dots, k$. First, let $r = 0$, Then [see (3.3.32)] $a_\mu = x_j^{-\mu}(\mu!)^{-1}\Delta^\mu_{x_j}(f; 0)$, i.e., relation (2.16) is proved. Now let $r \neq 0$. In this case, the pair $(p, q) \in B^*_{k,r}(y_0, \dots, y_{r+\mu})$ is unique, namely, $(0, r + \mu) \in B^*_{k,r}(y_0, \dots, y_{r+\mu})$. Therefore, by virtue of (3.6.36) and (2.9), we get

$$\|f\|_{[0,kd]} \le c_7 \Lambda(y_0, \dots, y_{r+\mu}; \omega_\mu) \le c_8 \Lambda^*_r(y_0, \dots, y_{r+\mu}; \omega_\mu)$$

$$= c_8 \Lambda_{0,r+\mu,r}(y_0, \dots, y_{r+\mu}; \omega_\mu) \le c_6 x_j^{-\mu} \omega_\mu(x_j).$$

Inequality (2.16) is proved.

4. Let us prove the inequality

$$\omega_\mu(x_j) \le c_9 M x_j^\mu \int_{x_j}^{kd} u^{-\mu-1} \varphi(u)\, du =: c_9 M x_j^\mu b_\mu. \tag{2.17}$$

In the case where $\mu = 1, \dots, j - i - r$, inequality (2.17) follows from relations (2.14) and (3.5.2) and the estimate $\omega_\mu(t) \le 2^\mu \left\| f^{(r)} \right\|_{[0,kx_j]}$; for $\mu = j-1-r+1, \dots, k-1$, this inequality follows from relations (2.15) and (3.5.2) and the Marchaud inequality (3.5.6).

It follows from (2.16) and (2.17) that

$$\left| a_\mu \right| \le c_{10} b_\mu M, \qquad \mu = 1, \dots, k-1. \tag{2.18}$$

5. Since, according to the conditions of the lemma, $f(y_s) = 0$ for $s = 0, \dots, r$, we have $[y_0, \dots, y_s; f] = 0$ for $s = 0, \dots, r$. Therefore, the Newton formula for the Lagrange interpolation polynomial $L = L(x) = L(x; f; y_0, \dots, y_{m-1})$ has the form

$$L(x) = \sum_{\mu=1}^{k-1} a_\mu p_\mu(x), \tag{2.19}$$

where

$$p_\mu(x) := \prod_{s=0}^{\mu+r-1} (x - y_s).$$

Let us prove that

$$\|f - L\|_{[0, kx_j]} \le c_{11} x_j^r \varphi(x_j). \tag{2.20}$$

Indeed, if $x \in [kx_j, (k+1/2)x_j]$, then, by virtue of Lemma 2.2, we have

$$\prod_{s=0}^{m-1} |x - y_s|^{-1} |f(x) - L(x)| \equiv \big|[y_0, \ldots, y_{m-1}, x; f]\big|$$

$$\le c_{12} \Lambda_r(y_0, \ldots, y_m; \varphi) \le c_{13} x_j^{-k} \varphi(x_j),$$

i.e., $|f(x) - L(x)| \le c_{14} x_j^r \varphi(x_j)$. Inequality (2.20) now follows from Lemma 3.7.1.

6. Denote

$$g_\mu(x) := \frac{a_\mu}{b_\mu} \int_x^{kd} u^{-\mu-1} \varphi(u)\, du, \qquad G(x) := \sum_{\mu=1}^{k-1} g_\mu(x) p_\mu(x).$$

Let us prove that

$$\|L - G\|_{[x_j, kx_j]} \le c_{15} M x_j^r \varphi(x_j), \tag{2.21}$$

$$\|G\|_{[x_j, kd]} \le c_{16} M d^r \varphi(d), \tag{2.22}$$

$$G \in c_{17} M W^r H[k; \varphi; [x_j, \infty)]. \tag{2.23}$$

Indeed, let $x \in [x_j, kx_j]$. Then

$$|L(x) - G(x)| = \left|\sum_{\mu=1}^{k-1}(a_\mu - g_\mu(x))p_\mu(x)\right|$$

$$\leq \sum_{\mu=1}^{k-1}|a_\mu|b_\mu^{-1}\left|b_\mu - \int_x^{kd} u^{-\mu-1}\varphi(u)\,du\right|(kx_j)^{\mu+r}$$

$$= \sum_{\mu=1}^{k-1}|a_\mu|b_\mu^{-1}\int_{x_j}^{x} u^{-\mu-1}\varphi(u)\,du(kx_j)^{\mu+r}$$

$$\leq c_{10}Mk^m x_j^r\varphi(x) \leq c_{15}Mx_j^r\varphi(x_j).$$

Let $x \in [x_j, kd]$. Then

$$|G(x)| \leq \sum_{\mu=1}^{k-1}|a_\mu|b_\mu^{-1}\int_{x_j}^{kd} u^{-\mu-1}\varphi(u)\,du(kx_j)^{\mu+r} \leq c_{16}Md^r\varphi(d),$$

i.e., estimates (2.21) and (2.22) are proved.

Let $x \geq x_j$. Then, for all $\nu = 1, \dots, m$, we have

$$\left|p_\mu^{(m-\nu)}(x)\right| \leq c_{18}x^{\mu+r-m-\nu} \qquad (p_\mu^{(m)} \equiv 0).$$

Taking (2.13) into account, we get

$$\left|g_\mu^{(\nu)}(x)\right| = |a_\mu|b_\mu^{-1}\left|(x^{-\mu-1}\varphi(x))^{(\nu-1)}\right| \leq c_{19}Mx^{-\mu-\nu}\varphi(x),$$

whence

$$\left|G^{(m)}(x)\right| = \left|\sum_{\mu=1}^{k-1}\sum_{\nu=1}^{m}\binom{m}{\nu}g_\mu^{(\nu)}(x)p_\mu^{(m-\nu)}(x)\right| \leq c_{20}Mx^r\varphi(x), \tag{2.24}$$

$$\left|G^{(r)}(x) - \sum_{\mu=1}^{k-1}g_\mu(x)p_\mu^{(r)}(x)\right| = \left|\sum_{\mu=1}^{k-1}\sum_{\nu=1}^{m}\binom{r}{\nu}g_\mu^{(\nu)}(x)p_\mu^{(m-\nu)}(x)\right| \leq c_{21}M\varphi(x). \tag{2.25}$$

Let us show that relations (2.24) and (2.25) yield (2.23), i.e.,

$$\left|\Delta_h^k(G^{(r)}, x_*)\right| \le c_{17}M\varphi(h) \tag{2.26}$$

for any $h > 0$ and $x_* \ge x_j$. If $x_* \ge h$, then, according to (3.3.34), we have

$$\left|\Delta_h^k(G^{(r)}, x_*)\right| \le h^k \left\|G^{(m)}\right\|_{[x_*, x_*+kh]} \le c_{20}Mh^k x_*^{-k}\varphi(x_*) \le c_{20}M\varphi(h).$$

If $x_* < h$, then, denoting $G_\mu(x) := g_\mu(x)p_\mu^{(r)}(x)$ and using (2.25), we obtain

$$\left|\Delta_h^k(G^{(r)}, x_*) - \sum_{\mu=1}^{k-1}\Delta_h^k(G_\mu, x_*)\right| \le 2^k c_{21}M\varphi(x_* + kh) \le c_{22}M\varphi(h).$$

Thus, it remains to prove the estimate

$$\left|\Delta_h^k(G^{(r)}, x_*)\right| \le c_{23}M\varphi(h), \tag{2.27}$$

where $x_j \le x_* \le h$. For this purpose, we represent the function G_μ in the form

$$G_\mu(x) = \frac{a_\mu}{b_\mu}\left(p_\mu^{(r)}(x)\int_x^{x_*+kh} u^{-\mu-1}\varphi(u)\,du + p_\mu^{(r)}(x)\int_{x_*+kh}^{kd} u^{-\mu-1}\varphi(u)du\right)$$

$$=: \frac{a_\mu}{b_\mu}(\alpha(x)+\beta(x)).$$

Therefore,

$$\left|\Delta_h^k(\alpha; x_*)\right| \le 2^k\|\alpha\|_{[x_*, x_*+kh]} \le 2^k c_{18}\mu^{-1}\varphi(x_* + kh) \le c_{24}\varphi(h),$$

$$\Delta_h^k(\beta; x_*) = \int_{x_*+kh}^{kd} u^{-\mu-1}\varphi(u)\,du\,\Delta_h^k(p_\mu^{(r)}; x_*) = 0,$$

i.e., estimate (2.27) [and, hence, (2.26)] is proved. Relation (2.23) is proved.

7. Taking into account (2.23) and Theorem 1.3′, we find a function $\overline{G}$ such that $\overline{G}(x) = G(x)$ for $x \ge x_j$ and $G \in c_{25}MW^r H[k; \varphi; \mathbb{R}]$. It follows from (2.20) and (2.21) that the following relation holds for $x \in [x_j, kx_j]$:

$$\left|f(x)-\overline{G}(x)\right| = |f(x)-G(x)| \le |f(x)-L(x)| + |L(x)-G(x)| \le c_{26}Mx_j^r\varphi(x_j).$$

Hence,

$$\left\|f-\overline{G}\right\|_{[0,x_j]} = \left\|f-\overline{G}\right\|_{[0,kx_j]} \le c_{27}\left\|f-\overline{G}\right\|_{[x_j,kx_j]} \le c_{27}c_{26}Mx_j^r\varphi(x_j)$$

by virtue of Lemma 2.1.

Denote

$$\bar{f}_1(x) := \overline{G}(x) + (f(x)-\overline{G}(x))S(x;m;-x_j,0)(1-S(x;m;x_j,2x_j)).$$

According to the Lemma 1.4, $\bar{f}_1 \in c_{28}MW^rH[k;\varphi;\mathbb{R}]$, $\bar{f}_1(x) = f(x)$ if $x \in [0, x_j]$, and $\bar{f}_1(x) = \overline{G}(x)$ if $x \overline{\in} [-2x_j/3, 5x_j/3]$. Furthermore, by virtue of Lemma 2.1, we get

$$c_{16}Md^r\varphi(d) \ge \|G\|_{[x_j,kd]} \ge \|G\|_{[3x_j/5,x_j+d]}$$

$$= \left\|\bar{f}_1\right\|_{[3x_j/5,x_j+d]} \ge c_{29}\left\|\bar{f}_1\right\|_{[-d,x_j+d]},$$

and the function $\bar{f}$ required in Lemma 2.3 can be taken in the form

$$\bar{f}(x) := \bar{f}_1(x)S(x;m;-d,-d+H)(1-S(x;m;x_j+d-H,x_j+d)),$$

where

$$H := \frac{x_j+2d}{3}.$$

■

Lemma 2.4. *Let* $(i,j) \in B_{k,r}$, $h := x_j - x_i$, *and* $d := \min\{x_{j+1}-x_j, x_i - x_{i-1}\}$. *If* $f \in W^rH[k;\varphi;\mathbb{R}]$ *and* $f(x_s) = 0$ *for all* $s = 0,\dots,m$, $s \ne j$, *then there exists a function* $\bar{f} = \bar{f}(x)$ *that possesses the following properties:*

(a) $\bar{f}(x) = f(x)$ *for* $x \in [x_i, x_j]$, (2.28)

(b) $\bar{f}(x) = 0$ *for* $x \notin [x_i - d, x_j + d]$, (2.29)

$$(c)\quad \bar{f} \in c\left(1 + (h/d)^{3m}\right) M W^r H[k; \varphi; \mathbb{R}], \tag{2.30}$$

where

$$M := \frac{\Lambda_r(x_0, \ldots, x_m; \varphi)}{\Lambda_{i,j,r}(x_0, \ldots, x_m; \varphi)}.$$

Proof. In the case where $h \le d$, Lemma 2.4 follows from Lemma 2.3. Therefore, we assume that $h > d$. For all $x \in [x_j, x_j + h/d]$, relation (2.9) yields

$$|f(x)| = c_1 M \Lambda_{i,j,r}(x_0, \ldots, x_m; \varphi) \prod_{s=0, s \ne j}^{m} |x - x_s|$$

$$\le c_2 (x - x_i)^{j-i} \int_h^{h+d} u^{i+r-j-1} \varphi(u)\,du \le c_3 h^r \varphi(h) \le c_3 \left(\frac{h}{d}\right)^m d^r \varphi(d)$$

whence, taking Lemma 2.1 into account, we obtain

$$\| f \|_{[x_i, x_j]} \le \| f \|_{[x_i, x_j + d/2]} \le c_4 \left(\frac{h}{d}\right)^m h^m d^{-k} \varphi(d) \le c_4 \left(\frac{h}{d}\right)^{2m} h^r \varphi(h).$$

Lemma 2.4 now follows from Lemma 2.3.

■

3. Interpolation problem

Consider a finite or an infinite collection of isolated points $x_i \in \mathbb{R}$, $x_i < x_{i+1}$, and let a function f be defined at the points x_i. Let H be some class or space of functions defined on $\mathbb{R}$. To solve the interpolation problem is to find a condition under which there exists a function $\bar{f} \in H$ that interpolates the function f at the points x_i, i.e., $\bar{f}(x_i) = f(x_i)$ for all i, and to construct the function $\bar{f}$.

It is easy to see that, in the case $H = C(\mathbb{R})$, the interpolation problem is solvable for any collection of points x_i and any function f given at the points x_i.

De Boor (1976) showed that the corresponding condition for the class $W[r, \mathbb{R}]$ is the following one:

$$\left|[x_i, \ldots, x_{i+r}; f]\right| \le c \qquad \forall i. \tag{3.1}$$

Earlier, Subbotin (1965) established the exact value of the constant c in (3.1) in the case of equidistant points x_i.

If the function f is defined on an arbitrary set $E \subset \mathbb{R}$ (not necessarily consisting of isolated points), then a problem analogous to the interpolation problem is called the trace problem. The trace problem will be considered in subsequent sections of this chapter. In the present section, we study the interpolation problem for the classes $W^r H[k; \varphi; \mathbb{R}]$. The results presented in Subsections 3.1–3.3 can be found in [Shevchuk (1979), (1980)].

As before, we assume that $k \in \mathbb{N}$, $\varphi \in \Phi^k$, $(r+1) \in \mathbb{N}$, and $m := k + r$.

3.1. On the exactness of Theorem 3.6.3

Theorem 3.6.3 states that if $f \in W^r H[k; \varphi; \mathbb{R}]$, then

$$|[x_0, \ldots, x_m; f]| \le c\Lambda_r(x_0, \ldots, x_m; \varphi).$$

The lemma presented below demonstrates the exactness of this theorem.

Lemma 3.1. *For any choice of* $m+1$ *points* $x_0 < x_1 < \ldots < x_m$, *there exists a function* $g \in W^r H[k; \varphi; \mathbb{R}]$ *such that*

$$|[x_0, \ldots, x_m; g]| \ge c\Lambda_r(x_0, \ldots, x_m; \varphi). \tag{3.2}$$

Proof. We prove this lemma in several steps.

1. Let i^* and j^* denote integer numbers such that

$$(i^*, j^*) \in B^*_{k,r}(x_0, \ldots, x_m; \varphi), \quad \Lambda_{i^*, j^*, r}(x_0, \ldots, x_m; \varphi) = \Lambda^*_r(x_0, \ldots, x_m; \varphi).$$

For convenience, we assume that $\min\{x_{j^*+1} - x_{i^*}, x_{j^*} - x_{i^*-1}\} = x_{j^*+1} - x_{i^*}$ and $x_{i^*} = 0$. Also denote

$$h := \frac{x_{i^*+r+1}}{r+1}, \quad s^* = j^* - i^*.$$

2. Consider the case $s^* = m$, i.e., $i^* = 0$ and $j^* = m$, and, in particular, the case $k = 1$. Here, according to (3.3.7) and (3.6.31), we have

$$c_1 x_m^{-k}\varphi(x_m) \le \Lambda_r(x_0, \dots, x_m; \varphi) \le c_2 x_m^{-k}\varphi(x_m).$$

Therefore, as the function indicated in the lemma, we can take the function g defined on $(-\infty, x_m]$ by the equality $g(x) := c_3 x^m x_m^{-k}\varphi(x_m)$ and extended to $\mathbb{R}$ by using Theorem 1.3.

In the remaining part of the proof, we assume that $s^* \ne m$ and, in particular, that $k \ne 1$.

3. In view of Lemma 3.4.4, we can assume, without loss of generality, that

$$\left|\varphi^{(\nu)}(x)\right| \le c_4 x^{-\nu}\varphi(x), \qquad x > 0, \qquad \nu = 1, \dots, m-1. \tag{3.3}$$

We set

$$p(x) := \prod_{i=i^*}^{i^*+r}(x - x_i), \qquad \Phi(x) := \int_h^x u^{-k}\varphi(u)(x-u)^{k-2}\,du,$$
$$g_1(x) := \Phi(x)p(x), \qquad x > 0, \qquad g_1(0) := 0. \tag{3.4}$$

Let us show that

$$g_1 \in c_5 W^r H[k; \varphi; [h, \infty)], \tag{3.5}$$

i.e., let us prove that the following inequality holds for any $\delta > 0$ and $x_* \ge h$:

$$\left|\Delta_\delta^k(g_1^{(r)}; x_*)\right| \le c_5\varphi(\delta). \tag{3.6}$$

Indeed, we have $p^{(r+2)}(x) \equiv 0$, $\left|p^{(r)}(x)\right| \le c_6 x^{r+1-\nu}$ for $\nu = 0, \dots, r+1$, $x > h$, $\Phi^{(k-1)}(x) = (k-2)!\,x^{-k}\varphi(x)$, and, by virtue of (3.3), $\Phi^{(\nu)}(x) \le c_7 x^{-\nu-1}\varphi(x)$ for $\nu = k-1, \dots, m$. Hence, for $x_* \ge \delta$, we get

$$\left|\Delta_\delta^k(g_1^{(r)}; x_*)\right| = \delta^k\left|g_1^{(m)}(\theta)\right| = \delta^k\left|\sum_{\nu=0}^{r+1}\binom{m}{\nu}p^{(\nu)}(\theta)\Phi^{(m-\nu)}(\theta)\right|$$
$$\le c_6 c_7 2^m \delta^k \theta^{-k}\varphi(\theta) \le c_6 c_7 2^m \varphi(\delta),$$

where $\theta \ge x_*$.

If $x_* < \delta$, then we represent the function g_1 in the form

$$g_1(x) = p(x) \int_h^{x_*+k\delta} u^{-k}\varphi(u)(x-u)^{k-2}\,du + p(x) \int_{x_*+k\delta}^{x} u^{-k}\varphi(u)(x-u)^{k-2}\,du$$

$$=: \alpha(x)+\beta(x).$$

For all $x \in [x_*, x_* + k\delta]$ and $\nu = 0, \ldots, r$, taking (3.3) into account, we obtain

$$\left| \frac{d^\nu}{dx^\nu} \int_{x_*+k\delta}^{x} u^{-k}\varphi(u)(x-u)^{k-2}\,du \right| \le c_8 x^{-\nu-1}\varphi(x_* + k\delta) \le c_9 x^{-\nu-1}\varphi(\delta).$$

Therefore,

$$\left|\beta^{(r)}(x)\right| \le c_6 c_9 2^r \varphi(\delta),$$

whence

$$\left|\Delta_\delta^k(\beta^{(r)}; x_*)\right| \le 2^k \left\|\beta^{(r)}\right\|_{[x_*, x_*+k\delta]} \le c_6 c_9 2^m \varphi(\delta).$$

Moreover, $\Delta_\delta^k(\alpha^{(r)}; x_*) \equiv 0$. Inequality (3.6) is proved.

4. Denote $s := j^* - i^* - r$, $y_i := x_{i^*+r+i}$ for $i = 0, \ldots, s+1$, and $y_i = (i-s)y_{s+1}$ for $i = s+2, \ldots, k$. Let us prove that

$$\left|[x_{i^*}, \ldots, x_{i^*+r}, y_1, \ldots, y_k; g_1]\right| \ge c_{10} y_{s+1}^{s-k} \int_{y_s}^{y_{s+1}} v^{-s-1}\varphi(v)\,dv. \tag{3.7}$$

It follows from (3.3.24) that $[x_0, \ldots, x_{i^*+r}, y_1, \ldots, y_k; g_1] = [y_1, \ldots, y_m; \Phi]$. Since $\Phi^{(k-1)}(x) = (k-2)!x^{-k}\varphi(x)$, by virtue of (3.3.19) we get

$$[y_1, \ldots, y_k; \Phi] = (k-2)! \int_0^1 \int_0^{u_1} \ldots \int_0^{u_{k-2}} u^{-k}\varphi(u)\,du_{k-1} \ldots du_1,$$

where $u := y_1 + (y_2 - y_1)u_1 + \ldots + (y_k - y_{k-1})u_{k-1}$.

Since

$$u \le y_s + u_s(k-s)y_{s+1},$$

we have

$$u^{-k}\varphi(u) \ge (y_s + u_s(k-s)y_{s+1})^{-k}\varphi(y_s + u_s(k-s)y_{s+1}).$$

Therefore,

$$[y_1,\dots,y_k;\Phi]$$

$$\ge (k-2)! \int\limits_{1-1/k}^{1} \int\limits_{1-2/k}^{1-1/k} \dots \int\limits_{1-(s-1)/k}^{1-(s-2)/k} \int\limits_{0}^{1-(s-1)/k} \int\limits_{0}^{u_s} \dots \int\limits_{0}^{u_{k-2}} u^{-k}\varphi(u)du_{k-1}\dots du_1$$

$$\ge (k-2)! \int\limits_{1-1/k}^{1} \int\limits_{1-2/k}^{1-1/k} \dots \int\limits_{1-(s-1)/k}^{1-(s-2)/k} \int\limits_{0}^{1-(s-1)/k} \int\limits_{0}^{u_s} \dots \int\limits_{0}^{u_{k-2}} \frac{\varphi(y_s + u_s(k-s)y_{s+1})}{(y_s + u_s(k-s)y_{s+1})^k} du_{k-1}\dots du_1$$

$$= \frac{(k-2)!}{k^s(k-s-1)!} \int\limits_{0}^{1-(s-1)/k} \frac{\varphi(y_s + u_s(k-s)y_{s+1})}{(y_s + u_s(k-s)y_{s+1})^k} u_s^{k-s-1}\, du_s$$

$$= \frac{(k-2)!}{k^s(k-s-1)!} \int\limits_{0}^{y_s+(1-(s-1)/k)(k-s)y_{s+1}} \frac{\varphi(v)}{v^k} u_s^{k-s-1} du_s \frac{(v-y_s)^{k-s-1}}{y_{s+1}^{k-s}(k-s)^{k-s}}\, dv$$

$$\ge c_{10} y_{s+1}^{s-k} \int\limits_{y_s}^{y_{s+1}} v^{-s-1}\varphi(v)dv.$$

Inequality (3.7) is proved.

5. Taking into account relation (3.5) and Theorem 1.3, we can find a function $g_2 \in c_{11} W^r H[k;\varphi;\mathbb{R}]$ such that $g_2(x) = g_1(x)$ for $x \ge h$.

Let i_0 denote one of the numbers $i = i^*, \dots, i^* + r$ for which $x_{i_0+1} - x_{i_0} \ge h$. We set

$$h_0 := x_{i_0} + \frac{h}{2}.$$

The inequality $\| g_1 \|_{[h,2h]} \le c_{12} h^r \varphi(h)$ and Lemma 1.1 yield the following estimate:

$$\| g_2 \|_{[0, x_{i_0} + h/2]} \le \| g_2 \|_{[0,(r+1)h]} \le c_{13} h^r \varphi(h).$$

By using Lemma 1.3, we find a function $g_3 \in c_{14} W^r H[k; \varphi; \mathbb{R}]$ such that $g_3(x) = g_2(x)$ if $x \notin [-h/2, x_{i_0} + h/2]$ and $g_3(x) = 0$ if $x \in [0, x_{i_0} + h/2]$. In particular,

$$g_3(x_i) = g_1(x_i) = 0, \quad i = i^*, \ldots, i^* + r. \tag{3.8}$$

6. Denote $t_i := x_{i^*+i}$ if $i = 0, \ldots, s^*$ and $t_i := (i - s^*) x_{j^*+1}$ if $i = s^* + 1, \ldots, m$. Let us show that

$$\Lambda_{0,s^*,r}(t_0, \ldots, t_m; \varphi) \ge c_{15} \Lambda_r(t_0, \ldots, t_m; \varphi). \tag{3.9}$$

Indeed, assume that there is a pair $(p, q) \in B^*_{k,r}(t_0, \ldots, t_m)$. If $q = r + 1, \ldots, s^* + 1$, then, taking into account that

$$\Lambda_{p+i^*, q+i^*, r}(x_0, \ldots, x_m; \varphi) \le \Lambda_{i^*, j^*, r}(x_0, \ldots, x_m; \varphi),$$

we obtain

$$\frac{\Lambda_{p,q,r}(t_0, \ldots, t_m; \varphi)}{\Lambda_{0,s^*,r}(t_0, \ldots, t_m; \varphi)}$$

$$= \frac{\Lambda_{p+i^*, q+i^*, r}(x_0, \ldots, x_m; \varphi)}{\Lambda_{i^*, j^*, r}(x_0, \ldots, x_m; \varphi)} \prod_{\nu=s^*+2}^{m} \frac{(y_\nu - y_0)}{(y_\nu - y_p)} \prod_{\nu=0}^{i^*-1} \frac{(x_{q+i^*} - x_\nu)}{(x_{j^*} - x_\nu)} \prod_{\nu=s^*+2}^{m} \frac{(x_\nu - x_{p+i^*})}{(x_\nu - y_{i^*})}$$

$$\le 2^{m-s-1} \le 2^{m-2}.$$

Since $(p, q) \in B^*_{k,r}(t_0, \ldots, t_m; \varphi)$, the case $q > s^* + 1$ is possible only if $r = 0$ and $p = q - 1$, but then

$$\Lambda_{p,q,r}(t_0, \ldots, t_m; \varphi) \le c_{16} x_{j^*+1}^{-k} \varphi(x_{j^*+1}) \le c_{17} \Lambda_{0,s^*,r}(t_0, \ldots, t_m; \varphi).$$

Inequality (3.9) is proved.

Denote $g_4(x) := g_3(x) - L(x; g_3; t_0, \dots, t_{s^*-1}, t_{s^*+1}, \dots, t_m)$. It follows from inequality (3.9) and Lemma 2.3 that there exists a function $g_5 \in c_{18} W^r H[k; \varphi; \mathbb{R}]$ such that $g_5(x) = g_4(x)$ if $x \in [t_0, t_{s^*}] \equiv [x_{i^*}, x_{j^*}]$ and $g_5(x) \equiv 0$ if $x \notin [x_{i^*-1}, x_{j^*+1}]$.

7. Let us prove that

$$\big|[x_0, \dots, x_m; g_5]\big| \geq \frac{c_{10}}{2} \Lambda_{i^*, j^*, r}(x_0, \dots, x_m; \varphi). \tag{3.10}$$

Indeed, denote

$$A := \prod_{i=0, i \neq j^*}^{m} \left|x_{j^*} - x_i\right|, \quad B := \prod_{i=0, i \neq j^*}^{m} \left|t_{s^*} - t_i\right|.$$

Then

$$\begin{aligned} [x_0, \dots, x_m; g_5] &= A^{-1} g_5(x_{j^*}) = A^{-1} g_4(x_{j^*}) \\ &= A^{-1} B [t_0, \dots, t_m; g_3] = A^{-1} B [t_0, \dots, t_m; g_1]. \end{aligned}$$

Therefore, by virtue of (3.7), we get

$$\begin{aligned} &\big|[x_0, \dots, x_m; g_5]\big| \\ &\geq \big|[x_0, \dots, x_m; g_5]\big| \geq \left|A^{-1} B\right| c_{10} x_{j^*+1}^{j^*-i^*-m} \int_{x_{j^*}}^{x_{j^*+1}} v^{i^*+r-j^*-1} \varphi(v)\, dv \\ &= c_{10} \left|A^{-1} B\right| \Lambda_{i^*, j^*, r}(x_0, \dots, x_m; \varphi) x_{j^*+1}^{j^*-i^*-m} \prod_{i=0}^{i^*-1} (x_{j^*} - x_i) \prod_{i=j^*+1}^{m} (x_i - x_{i^*}) \\ &= c_{10} \Lambda_{i^*, j^*, r}(x_0, \dots, x_m; \varphi) \prod_{i=j^*+1}^{m} \frac{(x_i - x_{i^*})}{(x_i - x_{j^*})} x_{j^*+1}^{j^*-i^*-m} \prod_{i=s^*+1}^{m} (t_i - t_{s^*}) \\ &\geq \frac{c_{10}}{2} \Lambda_{i^*, j^*, r}(x_0, \dots, x_m; \varphi) \geq c_{19} \Lambda_r(x_0, \dots, x_m; \varphi). \end{aligned}$$

Thus, we can take the function $g(x) := c_{19}^{-1} g_5(x)$ as that indicated in the lemma.

■

3.2. On extension from "minimal" sets

Theorem 3.1. *Let $m+1$ points $x_0 < x_1 < \ldots < x_m$ be given. If a function f defined at the points x_i, $i = 0, \ldots, m$, satisfies the inequality*

$$\left|[x_0, \ldots, x_m; f]\right| \le \Lambda_r(x_0, \ldots, x_m; \varphi),$$

then there exists a function $\bar{f} \in cW^r H[k; \varphi; \mathbb{R}]$ such that $\bar{f}(x) = f(x)$ for $x = x_i$, $i = 0, \ldots, m$.

Proof. Let g be the function from Lemma 3.1. We can take the function $\bar{f}$ indicated in Theorem 3.1 in the form

$$\bar{f}(x) := M(g(x) - L(x; g; x_0, \ldots, x_{m-1})) + L(x; f; x_0, \ldots, x_{m-1}),$$

where

$$M := \frac{[x_0, \ldots, x_m; f]}{[x_0, \ldots, x_m; g]}.$$

Indeed, by virtue of the conditions of Theorem 3.1 and estimate (3.2), we get $|M| \le c_1$, whence $\bar{f} \in c_1 W^r H[k; \varphi; \mathbb{R}]$. If $i = 0, \ldots, m-1$, then, according to the definition of Lagrange polynomials (Definition 3.3.1), we have $g(x_i) - L(x; g; x_0, \ldots, x_{m-1}) = 0$, i.e., $\bar{f}(x_i) = L(x; f; x_0, \ldots, x_{m-1}) = f(x_i)$. If $i = m$, then the equality $\bar{f}(x_m) = f(x_m)$ follows from (3.3.8).

∎

3.3. Extension from an arbitrary set

Using Theorem 3.1 and Lemma 2.3, one can prove Theorem 3.2 presented below. The proof of this theorem is rather cumbersome, and we do not present it here.

Definition 3.1. *Suppose that $k \in \mathbb{N}$, $\varphi \in \Phi^k$, $(r+1) \in \mathbb{N}$, $m = k + r$, and $E \subset \mathbb{R}$ is an arbitrary set. The class $W^r \breve{H}[k; \varphi; E]$ is defined as the set of functions f given on E each of which, in any collection of $m+1$ point $x_0 < x_1 < \ldots < x_m$, $x_i \in E$, satisfies the inequality*

$$\left|[x_0, \ldots, x_m; f]\right| \le \Lambda_r(x_0, \ldots, x_m; \varphi). \tag{3.11}$$

Theorem 3.2. *For every function* $f \in W^r \breve{H}[k;\varphi;E]$, *there exists a function* $\bar{f} \in cW^r H[k;\varphi;\mathbb{R}]$ *such that* $\bar{f}(x) = f(x)$ *for* $x \in E$.

In the case where $r = 0$ and $\varphi(t) = t^{k-1}$, Theorem 3.2 was proved by Jonsson (1980). For $r = 0$ and $k = 2$, it was proved by Brudnyi and Shvartsman (1982), Shvartsman (1982), and Dzyadyk and Shevchuk (1983). In the other cases, this theorem was proved in [Shevchuk (1984), (1984a)]. The idea of its proof (due to Dzyadyk) is based on the fact that Definition 3.1 uses both Marchaud inequality and Whitney inequality. In explicit form, this idea was formulated by Dzyadyk (1975c). Brudnyi and Shvartsman [Brudnyi and Shvartsman (1982), (1985); Shvartsman (1982), (1984)] established a multidimensional analog of Theorem 3.2 for $r = 0$ and $k = 2$. It appears to be of great interest to obtain the corresponding analog for $k + r > 2$.

4. Traces. Extension of complexes

Let $E \subset \mathbb{R}$. The restriction of a function $f\colon \mathbb{R} \to \mathbb{R}$ to E is understood as a function $f_E\colon E \to \mathbb{R}$ whose values coincide with the values of the function f on E, i.e., $f_E(x) = f(x)$ for $x \in E$.

Definition 4.1. *Let* $H(\mathbb{R})$ *be some space of functions* $f\colon \mathbb{R} \to \mathbb{R}$ *and let* $\breve{H}(E)$ *be a space of functions* $f\colon E \to \mathbb{R}$. *We say that the space* $\breve{H}(E)$ *is the trace of the space* $H(\mathbb{R})$ *on* E *and write*

$$\breve{H}(E) = H(\mathbb{R})\,|_E, \tag{4.1}$$

if the following conditions are satisfied:

(a) the restriction f_E *of every function* $f \in H(\mathbb{R})$ *belongs to* $\breve{H}(E)$, *i.e.,* $f_E \subset \breve{H}(E)$;

(b) for every function $f \in \breve{H}(E)$, *there exists a function* $\bar{f} \in H(\mathbb{R})$ *such that* $f = \bar{f}_E$.

Using the results obtained in the previous sections, we describe the traces of various spaces $H(\mathbb{R})$ on E.

Let $\breve{C}^r(E)$, $r \in \mathbb{N}$, denote the space of functions $f\colon E \to \mathbb{R}$ whose rth divided differences converge on E (see Definition 3.8.4).

Theorem 4.1 [Whitney (1934)]. *If E is a closed set, then*

$$C^r(\mathbb{R})\,|_E \;=\; \breve{C}^r(E). \tag{4.2}$$

Let $k \in \mathbb{N}$, $(r+1) \in \mathbb{N}$, and $\varphi \in \Phi^k$. By $W^r\breve{H}_k^{\varphi}(E)$ we denote the space of functions $f\colon E \to \mathbb{R}$ such that the following inequality holds for any collection of points $x_i \in E$, $x_0 < x_1 < \ldots < x_m$, $m = k + r$:

$$\left|[x_0, \ldots, x_m; f]\right| \;\le\; M\Lambda_r(x_0, \ldots, x_m; \varphi),$$

where $\Lambda_r(x_0, \ldots, x_m; \varphi)$ is the rth divided majorant defined by formula (3.6.32) and $M = M(f) = \text{const}$.

In particular, let $\breve{W}^r(E)$, $r \in \mathbb{N}$, denote the space of functions $f\colon E \to \mathbb{R}$ such that the following inequality holds for any collection of different points $x_i \in E$:

$$\left|[x_0, \ldots, x_m; f]\right| \le M, \qquad M = M(f) = \text{const}.$$

Theorems 3.6.4 and 3.2 yield the following statement:

Theorem 4.2. *The space $W^r\breve{H}_k^{\varphi}(E)$ is the trace of the space $W^rH_k^{\varphi}(\mathbb{R})$ on E, i.e.,*

$$W^r\breve{H}_k^{\varphi}(E) \;=\; W^rH_k^{\varphi}(\mathbb{R})\,|_E \quad (\textit{in particular},\;\; \breve{W}^r(E) \;=\; W^r(\mathbb{R})\,|_E). \tag{4.3}$$

Marrien (1966) and then V. Galan (1991) considered the case where not only the values of a function but also its derivatives are given on E. In this connection, we introduce the notion of a complex of functions on a nonincreasing system of sets.

Let $\mathfrak{M}_r := \{E^{(j)}\}_{j=0}^r$ be a nonincreasing system of sets: $E = E^{(0)} \supset E^{(1)} \supset \ldots \supset E^{(r)}$. Assume that, on each set $E^{(j)}$, some function f_j is defined. For $x \in E^{(j)}\backslash E^{(j+1)}$ ($j = 0, \ldots, r$, $E^{(r+1)} := \varnothing$), we set $[f] := (f_0, \ldots, f_j)$ and say that $[f]$ is a complex of functions defined on the system $\mathfrak{M}_r$.

We use the notation of Section 8 of Chapter 3. We write $\bar{x} \in \gamma(\mathfrak{M}_r)$ if $y_s \in E = E^{(0)}$ for all $s = 1, \ldots, q$ and, for every $j = 0, \ldots, r$, the inclusion $y_s \in E^{(j)}\backslash E^{(j+1)}$ yields $p_s \le j$.

The expression

$$[\bar{x}; [f]] \;=\; \sum_{s=1}^{q} \sum_{i=0}^{i=p_s} \frac{f_{p_s-i}(y_s)B_s^{(i)}(y_s)}{(p_s - i)!\,i!}, \tag{4.4}$$

where B_s is defined by (3.8.1), is called the divided difference of a complex $[f]$ with respect to the collection $\bar{x} \in \gamma(\mathfrak{M}_r)$ [cf. (3.8.9)].

Definition 4.2. *Let* $k \in \mathbb{N}$, $(r+1) \in \mathbb{N}$, $m = k + r$, $\varphi \in \Phi^k$, *and* $M = \text{const}$. *The class* $MW\breve{H}[k; \varphi; \mathfrak{M}_r]$ *is defined as the set of complexes* $[f]$ *given on the system* $\mathfrak{M}_r$ *each of which, in any collection* $\bar{x} \in \gamma(\mathfrak{M}_r)$, *satisfies the following estimate:*

$$|[\bar{x}; [f]]| \leq M\Lambda_r(\bar{x}; \varphi), \tag{4.5}$$

where $\Lambda_r(\bar{x}; \varphi)$ *is the divided majorant defined by (3.6.32) [with regard for (3.8.27)]. We set* $W^r\breve{H}[k; \varphi; \mathfrak{M}_r] = 1W^r\breve{H}[k; \varphi; \mathfrak{M}_r]$. *The union*

$$W^r\breve{H}_k^{\varphi}(\mathfrak{M}_r) := \bigcup_{M>0} MW\breve{H}[k; \varphi; \mathfrak{M}_r]$$

is called the space $W^r\breve{H}_k^{\varphi}(\mathfrak{M}_r)$.

The following theorem is true:

Theorem 4.3 [V. Galan (1991)]. *If* $[f] \in W^r\breve{H}[k; \varphi; \mathfrak{M}_r]$, *then there exists a function* $f \in cW^r H[k; \varphi; \mathbb{R}]$ *such that* $f^{(j)}(x) = f_j(x)$ *for all* $j = 0, \ldots, r$ *and* $x \in E^{(j)}$.

Corollary 4.1. *If a complex* $[f]$ *defined on the system* $\mathfrak{M}_{m-1}$ *satisfies, in any collection* $\bar{x} \in \gamma(\mathfrak{M}_{m-1})$, *the inequality* $|[\bar{x}; [f]]| \leq 1$, *then there exists a function* $f \in cW^r[\mathbb{R}]$ *such that* $f^{(j)}(x) = f_j(x)$ *for all* $j = 0, \ldots, m-1$ *and* $x \in E^{(j)}$.

Definition 4.3. *Let* $H(\mathbb{R})$ *be a space of functions* $f: \mathbb{R} \to \mathbb{R}$ *and let* $\breve{H}(\mathfrak{M}_r)$ *be the space of complexes* $[f]$ *defined on* $\mathfrak{M}_r$. *We say that the space* $\breve{H}(\mathfrak{M}_r)$ *is the trace of the space* $H(\mathbb{R})$ *on* $\mathfrak{M}_r$ *and write*

$$\breve{H}(\mathfrak{M}_r) = H(\mathbb{R})\,|_{\mathfrak{M}_r}$$

if the following conditions are satisfied:

(a) the complex $[f]$ *that consists of functions* f_j *each of which is the restriction of the derivative* $f^{(j)}$, $j = 0, \ldots, r$, *to* $E^{(j)}$ *belongs to the space* $\breve{H}(\mathfrak{M}_r)$;

(b) *for every complex* $[f] \in \breve{H}(\mathfrak{M}_r)$, *there exists a function* $\bar{f} \in H(\mathbb{R})$ *such that* $\bar{f}^{(j)}(x) = f_j(x)$ *for all* $j = 0, \dots, r$ *and* $x \in E^{(j)}$.

The theorem below is a corollary of Theorem 4.3 and Lemma 3.8.4.

Theorem 4.4. *The space* $W^r \breve{H}_k^{\varphi}(\mathfrak{M}_r)$ *is the trace of the space* $W^r H_k^{\varphi}(\mathbb{R})$ *on* $\mathfrak{M}_r$, *i.e.,*

$$W^r \breve{H}_k^{\varphi}(\mathfrak{M}_r) = W^r H_k^{\varphi}(\mathbb{R})\,|_E.$$

Chapter 5
Direct theorems on the approximation of periodic functions

The Weierstrass theorems establish the qualitative fact that any function continuous on a segment can be arbitrarily exactly approximated by polynomials. However, the following problems remain open:

1. With what accuracy any given continuous function can be approximated by polynomials of a given degree?

2. What properties of the function determine the possibility of its "good" approximation?

A fairly complete answer to these questions for the case of periodic functions was given by Jackson (see [Jackson (1911), (1912)].

It turns out that the higher the smoothness of a function, the faster the deviation of the approximating polynomial from this function approaches zero.

Any theorem establishing the estimates of deviations (in a certain sense) of a given function (or a class of functions) from polynomials or any other elements onto which this function (or the class of functions) is mapped by a sequence of operators (in the theory of approximation, these operators are, most often, polynomial) is called a direct approximation theorem.

In the present chapter, we analyze the dependence of these estimates on the smoothness of a given function (or a class of functions) and the choice of a sequence of operators.

In Section 2, we establish the direct Jackson theorems. Section 1 contains some necessary preliminary information.

1. Singular integrals and Lebesgue constants

First, we introduce the following definition:

Definition 1.1. *For any* 2π*-periodic polynomial kernel* K_n *of the form*

$$K_n(t) = \frac{1}{2} + \sum_{j=1}^{n} \gamma_j \cos jt \tag{1.1}$$

and any 2π*-periodic function* f *summable on the period* $[-\pi, \pi]$, *an operator*

$$T_{K_n}(t) = T_{K_n}(f; t) := \frac{1}{\pi} \int_{-\pi}^{\pi} f(u) K_n(t-u)\,du = \frac{1}{\pi} \int_{-\pi}^{\pi} f(t-u) K_n(u)\,du, \tag{1.2}$$

is called a singular K_n*-integral for the function* f.

As follows from the definition of the Fourier coefficients of a function f, relations (1.2) and (1.1) imply that if

$$f(t) \sim \frac{a_0}{2} + \sum_{j=1}^{\infty} (a_j \cos jt + b_j \sin jt),$$

then

$$T_{K_n}(t) = \frac{1}{\pi} \int_{-\pi}^{\pi} f(u) K_n(t-u)\,du = \frac{a_0}{2} + \sum_{j=1}^{\infty} \gamma_j (a_j \cos jt + b_j \sin jt) \tag{1.3}$$

and, for any kernel of the form (1.1), the singular integral $T_{K_n}(f; t)$ is a trigonometric polynomial of degree n.

Integrals of the form (1.2) are most often encountered if the kernel K_n is chosen in the form of Dirichlet D_n, Fejér F_n, Jackson J_n, and other kernels. In this case, it is customary to say that the corresponding integrals are the *nth polynomials* or the *nth Dirichlet, Fejér, Jackson, etc. singular integrals.*

In particular, we note that, by using equality (1.3) and the properties of the kernels D_n and F_n [see (2.3.2′) and (2.3.6′)], one can easily show that the nth Dirichlet integral coincides with the nth partial sum of the Fourier series of a given function:

$$T_{D_n}(f; t) = S_n(f; t)$$

and the nth Fejér integral coincides with the arithmetic mean of the sums $S_k(f; t)$:

$$T_{F_n}(f; t) = \frac{1}{n} \sum_{j=0}^{n-1} S_j(f; t).$$

The quantities

$$L_n(K_n) := \frac{1}{\pi}\int_{-\pi}^{\pi}|K_n(u)|du, \tag{1.4}$$

i.e., the norms of operators $T_{\pi^{-1}K_n(f;t)}$ given by relation (1.2), are the Lebesgue constants for the kernels $\frac{1}{\pi}K_n(t)$ (see Section 3 in Chapter 2).

According to relation (2.3.3), for the Dirichlet kernels, these constants do not exceed $\frac{4}{\pi^2}\ln n + 3$. At the same time, for the Fejér, de la Vallée Poussin, Rogosinski, Jackson, and other kernels, according to relations (2.3.7), (2.3.13), (2.3.13′), (2.3.21), (2.3.29), etc., we get

$$L_n(F_n) = 1, \quad L_n(V_n^{2n}) = \frac{1}{3} + \frac{2\sqrt{3}}{\pi}, \quad L_n(V_n^{3n}) = \frac{4}{\pi},$$

$$L_n(R_n) = \frac{2}{\pi}\operatorname{si}\pi - r_n, \quad 0 < r_n < \frac{5}{12}n^{-2}, \quad L_n(J_n) = 1, \quad \text{etc.} \tag{1.5}$$

2. Direct theorems

Let $\tilde{C}$ denote the space of continuous 2π-periodic functions f equipped with the uniform norm

$$\|f\| := \max_{x\in\mathbb{R}}|f(x)|$$

let

$$E_n(f) := \inf_{T_n}\|f - T_n\|$$

be the value of the best uniform approximation of a function $f \in \tilde{C}$ by trigonometric polynomials

$$T_n(t) = \frac{1}{2} + \sum_{k=1}^{n}(a_k\cos kt + b_k\sin kt)$$

of degree $\le n$, let

$$\omega_k(f, t) := \omega_k(t, f, \mathbb{R}) \equiv \omega_k(t, f, [0, (k+1)2\pi])$$

be the modulus of continuity of order k for a function f, and let $\omega(f, t) := \omega_1(f, t)$.

For each $r = 1, 2, \dots$, we denote by $\tilde{C}^r$ the space of r times continuously differentiable functions $f \in \tilde{C}$ and set $\tilde{C}^0 := \tilde{C}$.

In 1911, Jackson obtained a substantial strengthening of the Weierstrass theorems. Actually, for the first time, he proposed procedures of construction of polynomials of sufficiently good approximation for the functions from all spaces $\tilde{C}^0 := \tilde{C}$, $\tilde{C}^1$, $\tilde{C}^2$ and established sufficiently good estimates of approximations by polynomials of a given degree n for each space $\tilde{C}^r$, $r = 0, 1, 2, \dots$. In the periodic case, the final results in this direction (in the sense of accuracy) were obtained by Korneichuk (1970).

Theorem 2.1 (Jackson). *If $f \in \tilde{C}^r$, where r is a nonnegative integer, then the following relation holds for any natural n:*

$$\tilde{E}_n(f) \le \frac{c}{n^r}\,\omega\left(f^{(r)}, \frac{1}{n}\right), \tag{2.1}$$

where c is a constant dependent only on r, and $\omega(f^{(r)}, \cdot)$ is the first modulus of continuity.

Theorem 2.1 readily follows from Lemma 2.1.

Lemma 2.1. *If $f \in \tilde{C}^r$, where r is a nonnegative integer, then, for any natural n, a trigonometric polynomial $T_n(f; r; t) = T_n(t)$ of degree $n' = 2[n/2] \le n$ of the form*

$$T_n(t) = T_n(f; r; t) = \frac{a_0}{2} + \sum_{k=1}^{n'} (1 - \lambda_k^{r+1})(a_k \cos kt + b_k \sin kt), \tag{2.2}$$

where a_k and b_k are the Fourier coefficients of the function f and λ_k are numbers specified by the Jackson kernel

$$J_{[n/2]+1}(t) = \frac{1}{2} + \sum_{k=1}^{2[n/2]} j_k \cos kt = \frac{1}{2} + \sum_{k=1}^{n'} j_k \cos kt \tag{2.3}$$

according to the formula $\lambda_k = 1 - j_k$, approximates the function f with an accuracy characterized by the inequality

$$|f(t) - T_n(t)| \le M_r(f^{(r)}; n)\,\frac{1}{n^r}, \tag{2.4}$$

where

$$M_r(f^{(r)};n) = 12^r \min\left\{12\omega\left(f^{(r)};\frac{1}{n}\right);\ 64\omega_2\left(f^{(r)};\frac{1}{n}\right)\right\} \tag{2.5}$$

and $\omega(f^{(r)};t)$ *and* $\omega_2(f^{(r)};t)$ *are the first and the second moduli of continuity of the function* $f^{(r)}$, *respectively.*

Proof. 1. For $f \in \tilde{C}^0 = \tilde{C}$, i.e., if the function f is simply continuous, we preliminarily need the following properties of the Jackson integral:

If the function f is continuous, then the following inequalities are true for all $\nu \in \mathbb{N}$:

$$\left|f(t) - T_{J_\nu}(t)\right| \le 6\omega\left(f;\frac{1}{\nu}\right), \quad \left|f(t) - T_{J_\nu}(t)\right| \le 16\omega_2\left(f;\frac{1}{\nu}\right). \tag{2.6}$$

Indeed, by using properties (a) and (b) of the Jackson kernels, we obtain

$$\begin{aligned} f(t) - T_{J_\nu}(t) &= \frac{1}{\pi}\int_{-\pi}^{\pi} f(t)J_\nu(u)\,du - \frac{1}{\pi}\int_{-\pi}^{\pi} f(t-u)J_\nu(u)\,du \\ &= \frac{2}{\pi}\int_0^{\pi} f(t)J_\nu(u)\,du - \frac{1}{\pi}\left[\int_{-\pi}^{0} + \int_0^{\pi} f(t-u)J_\nu(u)\,du\right] \\ &= -\frac{1}{\pi}\int_0^{\pi}[f(t-u) - 2f(t) + f(t+u)]J_\nu(u)\,du, \end{aligned}$$

whence, according to the properties of the first moduli of continuity and the properties (a)–(c) of the Jackson kernels, we find

$$\begin{aligned} \left|f(t) - T_{J_\nu}(t)\right| &\le \frac{1}{\pi}\int_0^{\pi}(|f(t+u) - f(t)| + |f(t-u) - f(t)|)J_\nu(u)\,du \\ &\le \frac{2}{\pi}\int_0^{\pi}\omega\left(\nu u\frac{1}{\nu}\right)J_\nu(u)\,du \le \omega\left(\frac{1}{\nu}\right)\frac{2}{\pi}\int_0^{\pi}(\nu u + 1)J_\nu(u)\,du \\ &\le \omega\left(\frac{1}{\nu}\right)\left(2\nu\frac{2.5}{\nu} + 1\right) = 6\omega\left(\frac{1}{\nu}\right) \end{aligned}$$

and, in view of the properties of the second moduli of continuity and the properties (b)–(d) of the Jackson kernels, we obtain

$$\left|f(t) - T_{J_\nu}(t)\right| \le \frac{1}{\pi}\int_0^\pi \omega_2\left(\nu u \frac{1}{\nu}\right) J_\nu(u)du \le \omega_2\left(\frac{1}{\nu}\right)\frac{1}{\pi}\int_0^\pi (\nu u + 1)^2 J_\nu(u)du$$

$$\le \omega_2\left(\frac{1}{\nu}\right)\left(\nu^2\frac{\pi^2}{\nu^2} + 2\nu\frac{2.5}{\nu} + \frac{1}{2}\right) < 16\omega_2\left(\frac{1}{\nu}\right),$$

as required.

Further, by using relations (2.1), (2.2), and (1.3), we get

$$T_n(t) = \frac{a_0}{2} + \sum_{k=1}^{n'}(1-\lambda_k)(a_k\cos kt + b_k \sin kt)$$

$$= \frac{a_0}{2} + \sum_{k=1}^{n'} j_k(a_k\cos kt + b_k\sin kt) = T_{J_{\left[\frac{n+1}{2}\right]}}(f;t). \tag{2.1'}$$

Hence, by virtue of the properties of the Jackson integral (2.6) and (2.6′) and the facts that $\omega(t)\uparrow$, $\omega_2(t)\uparrow$, $\omega(nt) \le n\omega(t)$, and $\omega_2(nt) \le n^2\omega_2(t)$, we obtain

$$|f(t) - T_n(t)| \le |f(t) - T_{J_{[n/2]+1}}(t)| \le 6\omega\left(\frac{1}{[n/2]+1}\right) \le 6\omega\left(\frac{2}{n}\right) \le 12\omega\left(f;\frac{1}{n}\right) \tag{2.7}$$

and, similarly,

$$|f(t) - T_n(t)| \le 16\omega_2\left(\frac{2}{\nu}\right) \le 64\omega_2\left(f;\frac{1}{n}\right). \tag{2.7'}$$

Thus, the theorem is proved for the analyzed case.

Note that if an absolutely continuous function φ with period 2π almost everywhere possesses the derivative $\varphi'(t)$ satisfying the inequality

$$|\varphi'(t)| \le M, \tag{2.8}$$

then, according to Theorem 3.2.1, the following estimate is true for its first modulus of continuity:

$$\omega(\varphi;t) \le Mt. \tag{2.9}$$

Hence, in view of inequality (2.7), this function admits the estimate

$$|\varphi(t) - T_{J_{[n/2]+1}}(\varphi;t)| \le \frac{12}{n}M, \tag{2.10}$$

which is valid for all $n = 1, 2, \ldots$.

2. By induction, we assume that Lemma 2.1 is true for the spaces $\tilde{C}^0, \tilde{C}^1, \ldots, \tilde{C}^r$, $r \geq 0$. Further, suppose that $f \in \tilde{C}^{r+1}$ and, hence, $f' \in \tilde{C}^r$.

Note that, in view of relation (2.1) and the formula

$$f'(t) \sim \sum_{k=1}^{\infty}(a_k \cos kt + b_k \sin kt)',$$

we can write

$$\frac{d}{dt} T_n(f; r; t) = T_n(f'; r; t).$$

Thus, following [Natanson (1952)], by virtue of inequality (2.4), we conclude that

$$|(f(t) - T_n(f; r; t))'| - |f'(t) - T_n(f'; r; t)| \leq M_r(f^{(r+1)}; n)n^{-r},$$

whence, in view of relations (2.10) and (2.5), we get

$$|f(t) - T_n(f; r; t) - T_{J_{[n/2]+1}}[f - T_n(f; r; \cdot); t]|$$

$$\leq \frac{12}{n} M_r(f^{(r+1)}; n)n^{-r} = M_{r+1}(f^{(r+1)}; n)n^{-r-1}$$

and, hence, the required inequality

$$|f(t) - T_n(f; r+1; t)| \leq M_{r+1}(f^{(r+1)}; n)n^{-r-1}$$

because, by virtue of relations (2.1) and (2.1′) and the formula $1 - j_k = \lambda_k$, one can easily see that

$$T_n(f; r; t) + T_{J_{[n/2]=1}}[f - T_n(f; r; \cdot); t]$$

$$= \frac{a_0}{2} + \sum_{k=1}^{n'}(1 - \lambda_k^{r+1})(a_k \cos kt + b_k \sin kt) + \frac{a_0}{2} + \sum_{k=1}^{n'} j_k(a_k \cos kt + b_k \sin kt)$$

$$- \left[\frac{a_0}{2} + \sum_{k=1}^{n'} j_k(1 - \lambda_k^{r+1})(a_k \cos kt + b_k \sin kt)\right]$$

$$= \frac{a_0}{2} + \sum_{k=1}^{n'}(1 - \lambda_k^{r+2})(a_k \cos kt + b_k \sin kt) = T_n(f; r+1; t). \tag{2.11}$$

Thus, the Jackson theorem is proved for the functions from the class $\tilde{C}^{r+1}$ and, therefore, for all classes $\tilde{C}^r$, $r = 0, 1, 2, \ldots$.

■

Let $MW^r\tilde{H}^\alpha$, $0 < \alpha < 1$, and $MW^r\tilde{Z}$ be the classes of functions $f \in \tilde{C}$ such that $\omega(f, t) \leq Mt^\alpha$ and $\omega_2(f, t) \leq t$, respectively, and denote $M\tilde{H}^\alpha := MW^0\tilde{H}^\alpha$ and $\tilde{H}^\alpha := 1\,\tilde{H}^\alpha$.

Corollary 2.1. *If* $f \in MW^r\tilde{H}^\alpha$, $0 < \alpha < 1$, $r = 0, 1, 2, \ldots$, *then,*

$$|f(t) - T_n(f; r; t)| \leq \frac{cM}{n^{r+\alpha}}, \qquad c = c(r) = \text{const}, \tag{2.12}$$

and if $f \in MW^r\tilde{Z}$, *then*

$$|f(t) - T_n(f; r; t)| \leq \frac{cM}{n^{r+1}}, \qquad c = c(r) = \text{const}, \tag{2.12$'$}$$

where $T_n(f; r; t)$ *is the trigonometric polynomial of degree* n *given by relation (2.1).*

As the second corollary of Theorem 2.1, we present the well-known Dini–Lipschitz criterion of uniform convergence for the Fourier series:

Corollary 2.2. *If a continuous* 2π*-periodic function* f *belongs to the Dini–Lipschitz class*

$$\left[\textit{i.e.,} \quad \omega(f; t) = o\left(\frac{1}{|\ln t|}\right) \quad \textit{as} \quad t \to 0 \right],$$

then its Fourier series uniformly converges to f.

Indeed, the validity of this assertion follows from the facts that, first,

$$|f(t) - S_n(f; t)| \leq \left(4 + \frac{4}{\pi^2}\ln n\right) E_n(f) \tag{2.13}$$

by virtue of Theorem 2.3.1 [Lebesgue inequality (2.3.4)] and, second,

$$E_n(t) = o\left(\frac{1}{|\ln t|}\right)$$

in view Theorem 2.1 and the conditions of the corollary.

Remark 2.1. By analogy with the proof of the second part of the Jackson theorem, one can easily establish the following somewhat more general assertion:

Theorem 2.2. *Let* A_n *be a linear polynomial operator given in the space* $\tilde{C}$ *and associating every function* $f \in \tilde{C}$ *with Fourier series of the form*

$$f(t) \sim \frac{a_0}{2} + \sum_{k=1}^{\infty}(a_k \cos kt + b_k \sin kt) \tag{2.14}$$

with a polynomial of degree n *of the form*

$$A_n(f; t) = \frac{a_0}{2} + \sum_{k=1}^{n}(1 - \lambda_k)(a_k \cos kt + b_k \sin kt). \tag{2.15}$$

Further, let

$$\sup_{\varphi \in H} \|\varphi - A_n(\varphi; \cdot)\| := \mathscr{E}_n = \mathscr{E}(A_n) \tag{2.16}$$

so that $|\varphi(t) - A_n(\varphi; t)| \le M\mathscr{E}_n$ *for* $\varphi \in M\tilde{H}^1$.

Then the following inequality is true for any integer $r \ge 0$ *and any function* $f \in \tilde{C}^r$ *with Fourier series of the form (2.14):*

$$\left| f(t) - \frac{a_0}{2} - \sum_{k=1}^{n}(1 - \lambda_k^{r+1})(a_k \cos kt + b_k \sin kt) \right| \le \left\| f^{(r)} - A_n(f^{(r)}; \cdot) \right\| \mathscr{E}_n^r. \tag{2.17}$$

Proof. For $f \in \tilde{C} = \tilde{C}^0$ (i.e., for $r = 0$), inequality (2.17) is obvious. Thus, we assume that it is true for the classes $\tilde{C} = \tilde{C}^0, \tilde{C}^1, \ldots, \tilde{C}^r$, where $r \ge 0$, and prove its validity for $f \in \tilde{C}^{r+1}$. Indeed, by setting

$$\varphi(t) = f(t) - \frac{a_0}{2} - \sum_{k=1}^{n}(1 - \lambda_k^{r+1})(a_k \cos kt + b_k \sin kt),$$

in view of the facts that $f' \in \tilde{C}^r$ and the Fourier series of the function f' is obtained as a result of the term-by-term differentiation of the Fourier series of the function f, by virtue of (2.17), we find

$$|\varphi'(t)| \le \left| f'(t) - \sum_{k=1}^{n}(1-\lambda_k^{r+1})(a_k \cos kt + b_k \sin kt)' \right|$$

$$\le \left\| f^{(r+1)} - A_n(f^{(r+1)};\cdot) \right\| \mathcal{E}_n^r,$$

whence we get $\varphi \in M\tilde{H}^1$, where

$$M = \left\| f^{(r+1)}(t) - A_n(f^{(r+1)}; t) \right\|_C \mathcal{E}_n^r$$

and, therefore, relation (2.16) implies that

$$|\varphi(t) - A_n(\varphi; t)| = \Bigg| f(t) - \frac{a_0}{2} - \sum_{k=1}^{n}(1-\lambda_k^{r+1})(a_k \cos kt + b_k \sin kt)$$

$$- \sum_{k=1}^{n}(1-\lambda_k)\lambda_k^{r+1}(a_k \cos kt + b_k \sin kt)\Bigg|$$

$$= \left| f(t) - \frac{a_0}{2} - \sum_{k=1}^{n}(1-\lambda_k^{r+2})(a_k \cos kt + b_k \sin kt) \right|$$

$$\le M\mathcal{E}_n = \left\| f^{(r+1)} - A_n(f^{(r+1)};\cdot) \right\| \mathcal{E}_n^{r+1}.$$

This means that inequality (2.17) is also true in the case where $f \in \tilde{C}^{r+1}$. Theorem 2.2 is thus proved.

Note that Stechkin generalized Theorem 2.1 to the case of moduli of continuity of order k for all $k \ge 3$.

Theorem 2.3 [Stechkin (1951a)]. *Let a natural k be given. For any natural n and any continuous 2π-periodic function f, one can always find a trigonometric polynomial $T_n(f;\cdot)$ of degree n such that*

$$|f(t) - T_n(f;t)| \le c\omega_k\left(f; \frac{1}{n}\right), \tag{2.18}$$

where c is a constant depending only on k.

Proof. We set

$$T_n(f;t) := -\frac{(-1)^k}{\pi}\int_{-\pi}^{\pi}\sum_{j=1}^{k}(-1)^{k-j}\binom{k}{j}f(t+j\tau)J_{l,n'}(\tau)d\tau$$

$$= \frac{1}{\pi}\int_{-\pi}^{\pi}[f(t)-(-1)^k\Delta_\tau^k(f;t)]J_{l,n'}(\tau)d\tau,$$

where the natural number l is chosen to guarantee the validity of the inequality $k \le 2l-2$ and $n' = \left[\frac{n}{l}\right]+1$ is such that $T_n(f;t)$ is a trigonometric polynomial of degree

$$l(n'-1) = l\left[\frac{n}{l}\right] \le n.$$

Then

$$f(t) - T_n(f;t) = \frac{(-1)^k}{\pi}\int_{-\pi}^{\pi}\Delta_\tau^k f(t)J_{l,n'}(\tau)d\tau$$

and

$$|f(t)-T_n(f;t)| \le \frac{1}{\pi}\int_{-\pi}^{\pi}\omega_k(f;|\tau|)J_{l,n'}(\tau)d\tau.$$

By using the properties of the kth moduli of continuity and equality (2.3.37), we obtain

$$\frac{1}{\pi}\int_{-\pi}^{\pi}\omega_k(f;\tau)J_{l,n'}(\tau)d\tau \le \frac{1}{\pi}\int_{-\pi}^{\pi}(n'|\tau|+1)^k\omega_k\left(f;\frac{1}{n'}\right)J_{l,n}(\tau)d\tau$$

$$\le c_1\omega_k\left(f;\frac{1}{n'}\right) \le c_1\omega_k\left(f;\frac{l}{n}\right) \le c_2\omega_k\left(f;\frac{1}{n}\right)$$

and, hence,

$$|f(t)-T_n(f;t)| \le c\omega_k\left(f;\frac{1}{n}\right).$$

Theorem 2.3 is thus proved.

In conclusion, we present two more theorems on the approximation of continuous functions.

Theorem 2.4 [Stechkin (1951b)]. *If $f \in \tilde{C}$, then, for any $n = 1, 2, \dots,$*

$$|f(t) - R_n(f; t)| \le c\omega_2\left(f; \frac{1}{n}\right), \tag{2.19}$$

where $R_n(f; \cdot)$ are the Rogosinski polynomials for the case $\gamma_n = \dfrac{\pi}{2n}$ and c is an absolute constant.

Proof. Since

$$R_n(f; t) = \frac{1}{\pi}\int_{-\pi}^{\pi} f(t-u)\frac{D_n\left(u - \frac{\pi}{2n}\right) + D_n\left(u + \frac{\pi}{2n}\right)}{2}\,du$$

$$= \frac{1}{2}\left[S_n\left(f; t - \frac{\pi}{2n}\right) + S_n\left(f; t + \frac{\pi}{2n}\right)\right]$$

and, by virtue of the Jackson theorem for the polynomial T_n^* of the best approximation of the function f,

$$\left|f(t) - T_n^*(t)\right| \le 64\omega_2\left(f; \frac{1}{n}\right),$$

we get

$$|f(t) - R_n(f; t)| = \left|f(t) - \frac{1}{2}\left[S_n\left(f; t - \frac{\pi}{2n}\right) + S_n\left(f; t + \frac{\pi}{2n}\right)\right]\right|$$

$$= \left|f(t) - \frac{1}{2}\left[S_n\left(f - T_n^*; t - \frac{\pi}{2n}\right) + S_n\left(f - T_n^*; t + \frac{\pi}{2n}\right)\right]\right.$$

$$\left. - \frac{1}{2}\left[T_n^*\left(t - \frac{\pi}{2n}\right) + T_n^*\left(t + \frac{\pi}{2n}\right)\right]\right|$$

$$\le \left|f(t) - \frac{1}{2}\left[f\left(t - \frac{\pi}{2n}\right) + f\left(t + \frac{\pi}{2n}\right)\right]\right| + \frac{1}{2}\left|R_n(f - T_n^*; t)\right|$$

$$+ \frac{1}{2}\left|f\left(t - \frac{\pi}{2n}\right) - T_n^*\left(t - \frac{\pi}{2n}\right)\right| + \frac{1}{2}\left|f\left(t + \frac{\pi}{2n}\right) - T_n^*\left(t + \frac{\pi}{2n}\right)\right|$$

$$\leq c_1\omega_2\left(f; \frac{1}{n}\right) + \frac{1}{2\pi}\int_{-\pi}^{\pi} |R_n(t)|dt\, 64\omega_2\left(f; \frac{1}{n}\right) + 64\omega_2\left(f; \frac{1}{n}\right)$$

$$\leq c\omega_2\left(f; \frac{1}{n}\right).$$

Theorem 2.4 is proved.

Theorem 2.5. *If $f \in \tilde{C}$, then, for any $n = 1, 2, \ldots$ and $n' > n$ such that $n' - n \geq \varepsilon n$,*

$$\left|f(t) - V_n^{n'}(f; t)\right| \leq \left(1 + \frac{2}{\varepsilon}\right) E_n(f), \tag{2.20}$$

$$\left|f(t) - V_n^{2n}(f; t)\right| \leq \left(\frac{4}{3} + \frac{2\sqrt{3}}{\pi}\right) E_n(f) < \frac{5}{2} E_n(f), \tag{2.20'}$$

and

$$\left|f(t) - V_n^{3n}(f; t)\right| \leq \left(\frac{4}{\pi} + 1\right) E_n(f), \tag{20''}$$

where $V_n^{n'}(f; t)$ are the de la Vallée Poussin polynomials formed by using the kernels of the form (2.3.10).

By virtue of relations (2.3.11), (2.3.13), and (2.3.13′), this theorem follows from inequality (2.3.10″).

Chapter 6
Inverse theorems on the approximation of periodic functions

1. Theorems on evaluation of the absolute value of the derivative of a polynomial

1. The problem of evaluation of the absolute value of the derivative of a polynomial via its values plays an important role in numerous fields of mathematical analysis. In what follows, we present some results of evaluation of the absolute value of the derivative for trigonometric and algebraic polynomials given on the real axis.

Theorem 1.1 (Bernstein inequality [Bernstein (1912)]).[1] *If the absolute value of a trigonometric polynomial* T_n *of degree* n *which takes real values on the real axis is bounded by a number* M, *then its derivative is bounded by the number* nM, *i.e.,*

$$\max_t |T_n(t)| \leq M \Rightarrow \max_t |T_n'(t)| \leq nM. \tag{1.1}$$

This inequality is exact in a sense that the number n *on the right-hand side of (1.1) cannot be replaced by any other smaller number.*

Proof. The proof is based on the comparison of the plot of the polynomial T_n with the plot of an auxiliary function $T_n^*(t) = M\cos(nt+\alpha)$ and the evaluation of the number of zeros of the difference $T_n - T_n^*$.[2]

Assume the opposite, i.e., that the inequality $|T_n'(t_0)| > nM$ holds at a certain point t_0 instead of the Bernstein inequality (1.1) and, at the same time, $|T_n(t_0)| \leq M$. In this case, for any $c \in (-\infty, \infty)$, this inequality is also true for each polynomial $\pm T_n(t-c)$ at the point $t_0 + c$. In view of this fact, we can assume, without loss of generality, that the polynomial $T_n(t)$ and the point t_0 satisfy the conditions

$$T_n(t_0) = M\cos nt_0, \quad t_0 \in \left(-\frac{\pi}{n}, 0\right), \quad T_n'(t_0) > nM. \tag{1.2}$$

Relations (1.2) imply that

$$T_n'(t_0) - \frac{d}{dt}[M\cos nt]_{n=t_0} > 0.$$

Therefore, in view of the fact that $T_n(t) \le M$, we conclude that the plot of the function T_n crosses the increasing branch of the cosine curve $M\cos nt$ which passes through the point $(t_0, M\cos nt_0)$ at at least three points. Moreover, it crosses each of the remaining $2n-1$ branches at least once (counting the double zeros). Therefore, the polynomial $T_n(t) - M\cos nt$ must have at least $2n+2$ zeros in the period $(-\pi, \pi]$ but this is impossible because $T_n(t) \not\equiv M\cos nt$ by virtue of inequality (1.2).

The fact that the constant n on the right-hand side of inequality (1.1) cannot be made smaller is established by analyzing an example of the polynomial $T_n(t) = \cos nt$.

This proves Theorem 1.1.

■

Remark 1.1 (van der Corput and Schaake)**.** The proof presented above shows that we arrive at a contradiction not only in the case where inequality (1.2) is satisfied but also if the following weaker inequality is true:

$$T_n'(t_0) > \frac{d}{dt}[M\cos nt]_{t_0} = Mn|\sin nt_0| = n\sqrt{M^2 - T_n^2(t_0)}\,.$$

Therefore, inequality (1.1) can be replaced by a more exact inequality

$$\max_t |T_n(t)| \le M \Rightarrow |T_n'(t)| \le n\sqrt{M^2 - T_n^2(t)}\,. \tag{1.1'}$$

Later, the Bernstein inequality (1.1) was somewhat strengthened in a different direction:

Theorem 1.1′ [Stechkin (1958)]**.** *The inequalities*

$$\max_t \left|T_n^{(r)}(t)\right| \le \left(\frac{n}{2\sin\dfrac{nh}{2}}\right)^r \max_t \left|\Delta_h^r(T_n;t)\right| \tag{1.$\tilde{1}$}$$

hold for the derivatives of any trigonometric polynomial of degree n *with any natural* r *and any* $h \in \left(0, \dfrac{2\pi}{n}\right)$ *and, in particular,*

$$\left\|T_n^{(r)}\right\| \le \left(\frac{n}{2}\right)^r \left\|\Delta_{\pi/n}^r(T_n;\cdot)\right\|.^{\dagger} \tag{1.$\tilde{1}'$}$$

† For the case $h = \pi/n$, this inequality was simultaneously established by Nikol'skii [Nikol'skii (1948)].

Proof. First, we consider the case $r = 1$ and choose a point t_0 and a number L such that the following equalities are true:

$$\|T_n'\| = |T_n'(t_0)| = L. \tag{1.3}$$

For the sake of definiteness, we assume that $T_n'(t_0) = L$. By using the same argument as in the proof of Theorem 1.1, we conclude that the inequality

$$T_n'(t_0 + \tau) \geq L \cos n\tau \tag{1.4}$$

holds for all $\tau \in \left(-\frac{\pi}{n}, \frac{\pi}{n}\right)$ [in the opposite case, the difference $T_n'(t_0+\tau) - L\cos n\tau$ would be a trigonometric polynomial of degree n not identically equal to zero with at least $2n + 1$ zeros in the period: a double zero at the point $\tau = 0$ and $2n - 1$ more zeros].

Integrating inequality (1.4) from $-h/2$ to $h/2$, we get

$$T_n\left(t_0 + \frac{h}{2}\right) - T_n\left(t_0 - \frac{h}{2}\right) \geq \frac{2L}{n}\sin\frac{nh}{n}.$$

By virtue of relation (1.3), this yields

$$\|T_n'\| = L \leq \frac{n}{2\sin\frac{nh}{2}}\dot{\Delta}_h T_n(t_0) \leq \frac{n}{2\sin\frac{nh}{2}}\|\Delta_h T_n, \cdot\|.$$

We proceed by induction. Assume that inequality $(1.\tilde{1})$ is true for all derivatives up to the rth degree $(r \geq 1)$, inclusive. As a result, in the next step, we get

$$\left\|\Delta_h^{r+1} T_n, \cdot\right\| = \left\|\Delta_h^r[T_n(\cdot + h) - T_n(\cdot)]\right\| \geq \left(\frac{2\sin\frac{nh}{2}}{n}\right)^r \left\|T_n^{(r)}(\cdot + h) - T_n^{(r)}(\cdot)\right\|$$

$$= \left(\frac{2\sin\frac{nh}{2}}{n}\right)^r \|\Delta_h T_n, \cdot\| \geq \left(\frac{2\sin\frac{nh}{2}}{n}\right)^{r+1} \|T_n^{(r+1)}\|.$$

This means that Theorem 1.1′ remains true in the case where the $(r + 1)$th derivative is involved.

This completes the proof of Theorem 1.1′.

■

Note that, since $\left\|\Delta_{\pi/n}^r T_n\right\| \leq 2^r \|T_n\|$, inequality $(1.\tilde{1}')$ immediately yields (1.1).

Corollary 1.1 (Bernstein inequality). *If an algebraic polynomial P_n of degree n satisfies the inequality*

$$|P_n(x)| \le M \tag{1.5}$$

on the segment $[-1, 1]$, *then its derivative* $P_n'(x)$ *satisfies the inequalities*[3]

$$|P_n'(x)| \le \frac{n\sqrt{M^2 - P_n^2(x)}}{\sqrt{1-x^2}} \le \frac{nM}{\sqrt{1-x^2}} \tag{1.5'}$$

at any point $x \in (-1, 1)$.

Indeed, for a given algebraic polynomial P_n, we construct the trigonometric polynomial $T_n(t) := P_n(\cos t)$. It is clear that the degree of the polynomial T_n also does not exceed n. Thus, by virtue of inequality (1.1′), we get

$$|T_n'(t)| \le |P_n'(\cos t)||\sin t| \le n\sqrt{M^2 - P_n^2(\cos t)}.$$

Further, we set $\cos t = x$ and arrive at inequality (1.5′).

In exactly the same way as in the proof of inequality (1.1), we establish the following assertion:

Theorem 1.2 (Chebyshev). *If an algebraic polynomial P_n satisfies the inequality*

$$|P_n(x)| \le M, \qquad x \in [-1, 1], \tag{1.6}$$

on the segment $[-1, 1]$, *then, for all* $x \notin [-1, 1]$,

$$|P_n(x)| \le MT_n(x) = M\cosh n \operatorname{arccosh}|x|, \qquad x \notin [-1, 1]. \tag{1.6'}$$

In order to check the validity of this inequality, it suffices to take into account the fact that, by virtue of condition (1.6), the polynomial P_n crosses each of the n branches of the Chebyshev polynomial MT_n at least once on the segment $[-1, 1]$ and that the number of intersections of this sort becomes greater than n if inequality (1.6′) does hold at at least one point $x_0 \notin [-1, 1]$.

Note that the right-hand side of inequalities (1.5′) behaves well at points distant from the ends of the segment $[-1, 1]$ but this is not so as $|x| \to 1$. In this case, the so-called Markov inequality seems to be more useful.

Theorem 1.3 [A. A. Markov (1884)]. *If an algebraic polynomial P_n of degree n is bounded in the absolute value on the segment $[-1, 1]$ by a number M, then its derivative $P_n'(x)$ is bounded by the number Mn^2 for all $x \in [-1, 1]$, i.e.,*

$$\max_{x\in[-1,1]} |P_n(x)| \le M \Rightarrow |P_n'(x)| \le Mn^2, \qquad x \in [-1, 1]. \tag{1.7}$$

Proof. If $x \in \left[-\cos\frac{\pi}{n},\ \cos\frac{\pi}{n}\right]$, then inequality (1.5) implies that

$$|P_n'(x)| \le \frac{nM}{\sqrt{1-x^2}} \le \frac{nM}{\sin\frac{\pi}{n}} < n^2 M .$$

In order to check the validity of inequality (1.7) for

$$x \in \left[-1,\ -\cos\frac{\pi}{n}\right) \cup \left(\cos\frac{\pi}{n},\ 1\right],$$

we assume (by contradiction) that the inequality $P_n'(x_0) > Mn^2$ holds, e.g., at a point

$$x_0 \in \left(\cos\frac{\pi}{n},\ 1\right].$$

In this case, we compare the plot of the polynomial P_n with the plot of the polynomial $MT_n(x-c) = M\cos n \arccos(x-c)$, where c is chosen so that the rightmost branch of the Chebyshev polynomial $MT_n(x-c)$ passes through the point $(x_0, P_n(x_0))$. Then one can easily show that the plot of the polynomial P_n passes through the rightmost (infinite) branch of the polynomial $MT_n(x-c)$ at at least two points (even at three points if $c>0$), and for $c \le 0$ it crosses each of the remaining $n-1$ branches at at least one point; for $c>0$, it crosses $n-2$ branches of the remaining ones. Therefore, the polynomial $P_n(x) - MT_n(x-c)$ of degree n must have at least $n+1$ zeros, which is impossible.

Theorem 1.3 is thus proved.

■

Note that if the conditions of this theorem are satisfied, then as a result of its successive application, we find

$$\left|P_n^{(j)}(x)\right| \le Mn^2(n-1)^2 \ldots (n-j+1)^2, \qquad j = 1, 2, \ldots, n,$$

It turns out that, for $j > 1$, the quantity appearing on the right-hand side can be made smaller. The best (unimprovable) estimate was established by V. A. Markov:

Theorem 1.3′ [V. A. Markov (1892)]. *The estimate*

$$\left|P_n^{(j)}(x)\right| \leq MT_n^{(j)}(1), \tag{1.7′}$$

where $T_n(x) = \cos n \arccos x$, *is true under the conditions of Theorem 1.3.*

In particular, for $j = n$, *this estimate gives* $\left|P_n^{(n)}(x)\right| \leq M2^{n-1}n!$.

Theorem 1.4 [Dzyadyk (1971a)]. *Let* D *be a convex domain and let* $\overset{*}{P}_n$ *be an algebraic polynomial of degree* n *all zeros of which are located in* D. *If, for all* z *located on the boundary* ∂D *of the set* D, *a polynomial* P_n *of degree* n *satisfies the inequality*

$$|P_n(z)| < \left|\overset{*}{P}_n(z)\right|, \qquad z \in \partial D, \tag{1.8}$$

then, for all $z \in \partial D$, *the absolute value of its derivative satisfies the inequality*

$$|P_n'(z)| < \left|\overset{*}{P}_n'(z)\right|, \qquad z \in \partial D.^{\dagger} \tag{1.9}$$

Proof (see [Dzyadyk (1971a)] and [Meiman (1952)]). Assume the contrary, i.e., that the inequality

$$|P_n'(z_0)| > \left|\overset{*}{P}_n'(z_0)\right| \tag{1.10}$$

contradicting (1.9) is true at a certain point $z_0 \in \partial D$. Without loss of generality, we can assume that

(a) the origin of coordinates $O \in D$;

(b) the point z_0 lies on the positive side of the OX-axis;

(c) the vector l with origin at the point z_0 and end at a point $z_0 + i$ is located outside $\operatorname{int} D$.

These conditions are satisfied as a result of transformations of the form $\tilde{z} = ze^{i\alpha} + b$ (rotation and translation) that do not violate the conditions of Theorem 1.4. In addition,

† In the case where D is a circle, this theorem was proved by Bernstein [Bernstein (1930)] by a different method.

since, parallel with P_n, the polynomial $e^{i\gamma}P_n$ also satisfies the conditions of Theorem 1.4 for any $\gamma \in [-1, 1]$, we can assume that $\arg P_n'(z_0) = 0$.

If we now move along the vector l from the point z_0 to a sufficiently close point z' and take into account both the Taylor formula and relation (1.10), then we get

$$\Delta := [\overset{*}{P}_n(z') - P_n(z')] - [\overset{*}{P}_n(z_0) - P_n(z_0)]$$

$$= (z' - z_0)[\overset{*}{P}{}_n'(z_0) - P_n'(z_0)] + O(|z' - z_0|^2), \tag{1.11}$$

$$\operatorname{Arg}\Delta = \arg(z' - z_0) + \pi + \arg[P_n'(z_0) - \overset{*}{P}{}_n'(z_0)] + o(1)$$

$$= \frac{\pi}{2} + \pi + r + o(1) + 2k\pi = \frac{3}{2}\pi + r + o(1) + 2k\pi, \quad k = 0, \pm 1, \ldots, \tag{1.12}$$

where $|r| < \dfrac{\pi}{2}$.

This means that, as a result of the transition along l from the point z_0 to the point z', the argument of the polynomial $\pi_n(z) := \overset{*}{P}_n(z) - P_n(z)$ decreases. But this is impossible because, by virtue of the Rouche theorem, all zeros of the polynomial π_n belong to the domain D and, hence, in view of its convexity, the argument of the polynomial must increase as we move along l.

Thus, we arrive at a contradiction with inequality (1.10). Finally, we note that the sign of equality in (1.9) is also impossible because, in this case, the polynomial $\tilde{P}_n(z) := (1+\varepsilon)P_n(z)$ would satisfy inequalities (1.8) and (1.9) for sufficiently small $\varepsilon > 0$.

Theorem 1.4 is proved.

■

Remark 1.1. If the requirement of convexity of the domain is not satisfied, then the assertion of Theorem 1.4 is, generally speaking, not true. To prove this, we consider a nonconvex domain D_0 bounded by a lemniscate

$$|z^2 - 1| = 1 + \frac{1}{25} \tag{1.13}$$

and the following polynomials:

$$\overset{*}{P}_2(z) = z^2 - 1 \quad \text{and} \quad P_2(z) = \frac{z}{2}.$$

It is easy to see that both zeros ± 1 of the polynomial $\overset{*}{P}_2$ are located inside the domain D_0 and the following inequality is true for all $\xi \in \partial D_0$:

$$|P_2(z)| < \left|\overset{*}{P}_2(\xi)\right|.$$

At the same time, at the point $\xi_0 = \frac{i}{5} \in \partial D_0$, we have the opposite inequality

$$\left|\overset{*}{P}_2(\xi_0)\right| = \frac{1}{2} > \left|\overset{*}{P}{}_2'(\xi_0)\right| = \frac{2}{5}.$$

Remark 1.2. It is easy to see that the assertion of Theorem 1.4 is true not only for polynomials but also for all functions $\overset{*}{f}$ and f analytic in D, continuous on $\overline{D}$, and satisfying the conditions

$$\arg[\overset{*}{f}(z) - e^{i\alpha} f(z)]\uparrow \quad \text{for all} \quad \alpha \in [0, 2\pi] \tag{1.14}$$

and

$$|f(z)| < |\overset{*}{f}(z)| \quad \text{for all} \quad z \in \partial D. \tag{1.15}$$

Note that Theorem 1.4 yields the following assertion:

Theorem 1.5 [Bernstein (1926)]. *If an algebraic polynomial P_n of degree n satisfies the inequality*

$$|P_n| \leq M \tag{1.16}$$

on the circle $|z| \leq 1$, then its derivative satisfies the inequality

$$\left|P_n'(z)\right| \leq nM \tag{1.17}$$

at all points of this circle.

Indeed, the circle $K := \{z : |z| \leq 1\}$ is a convex domain and, for any $z \in \partial K$ and $\varepsilon > 0$, we have $|P_n(z)| < \overset{*}{P}_n(\varepsilon, z)$, where $\overset{*}{P}_n(\varepsilon, z) := (1+\varepsilon)Mz^n$ and all its n zeros lie in $\operatorname{int} K$. Therefore,

$$\left|P_n'(z)\right| < \left|\overset{*}{P}_n'(\varepsilon, z)\right| = (1+\varepsilon)nM$$

for any $z \in \partial K$. In view of the arbitrariness of ε, this yields inequality (1.17).

2. **Theorem 1.6** (on the estimation of intermediate derivatives).[4] *If, for any natural k, the function f is absolutely continuous on a segment $[a, b]$ together with all its derivatives up to the order $k-1$, inclusive, and, moreover,*

$$\max_{x\in[a,b]} \left|f(x)\right| := M_0 < \infty \quad \text{and} \quad \operatorname{essup}_{x\in[a,b]} \left|f^{(k)}(x)\right| := M_k < \infty, \tag{1.18}$$

then its derivatives $f^{(j)}$, $j = 1, \ldots, k-1$, satisfy the inequalities

$$\max_{x\in[a,b]} \left|f^{(j)}(x)\right| \le A_{kj} M_0 \tau^j \quad \text{and} \quad \tau := \max\left\{\frac{2}{b-a}; \left(\frac{M_k}{M_0}\right)^{1/k}\right\}, \tag{1.19}$$

where

$$A_{kj} = \text{const} \le T_k^{(j)}(1)\left(1 + \frac{1}{k!}\right) + \frac{1}{(k-j)!}.$$

Proof. If $M_k = 0$, then the function f is the polynomial P_{k-1} of degree $k-1$. Hence, by setting $x = \dfrac{a+b}{2} + \dfrac{b-a}{2}t$ and

$$\hat{P}_{k-1}(t) := P_{k-1}\left(\frac{a+b}{2} + \frac{b-a}{2}t\right), \qquad t \in [-1, 1],$$

we conclude that, according to Theorem 1.3′ (Markov), the inequalities

$$\left|\hat{P}_{k-1}^{(j)}(t)\right| \le M_0 T_k^{(j)}(1)$$

hold for all $t \in [-1, 1]$ and $j = 1, \ldots, k-1$ and, hence, the inequalities

$$\left|P_{k-1}^{(j)}(x)\right| \le M_0 \left(\frac{2}{b-a}\right)^j T_k^{(j)}(1) \tag{1.19$'$}$$

hold for all $x \in [a, b]$ and $j = 1, \ldots, k-1$. This proves that inequality (1.18) is true in the analyzed case with constant $T_k^{(j)}(1)$.

For $M_k > 0$ and any point $x_0 \in [a+h, b-h]$, where

$$h = \frac{1}{\tau} = \min\left\{\frac{b-a}{2};\ \left(\frac{M_0}{M_k}\right)^{1/k}\right\},$$

we can represent the function f by using the Taylor formula with remainder r_k as follows:

$$f(x) = P_{k-1}(x; x_0) + r_k(x; x_0), \tag{1.20}$$

where

$$P_{k-1}(x; x_0) = \sum_{j=0}^{k-1} \frac{f^{(j)}(x_0)}{j!}(x - x_0)^j,$$

$$r_k(x; x_0) = \frac{1}{(k-1)!}\int_{x_0}^{x}(x-t)^{k-1} f^{(k)}(t)\,dt. \tag{1.20′}$$

In view of relations (1.20′) and (1.18), equality (1.20) implies that the following inequality holds on the segment $[x_0 - h, x_0 + h]$ for the polynomial $P_{k-1}(x; x_0)$:

$$\left|P_{k-1}(x; x_0)\right| \le |f(x)| + \frac{1}{(k-1)!}\left|\int_{x_0}^{x}(x-t)^{k-1} f^{(k)}(t)\,dt\right|$$

$$\le M_0 + \frac{M_k}{k!}h^k \le M_0\left(1 + \frac{1}{k!}\right).$$

According to (1.19′), this yields, in particular that, for any $x_0 \in [a+h, b-h]$ and $x \in [x_0 - h, x_0 + h]$, we have

$$\left|P_{k-1}^{(j)}(x; x_0)\right| \le \tilde{A}_{kj} M_0 \left(\frac{2}{2h}\right)^j = \tilde{A}_{kj} M_0 \tau^j, \tag{1.21}$$

$$\left|f^{(j)}(x_0)\right| = \left|P_{k-1}^{(j)}(x; x_0)\right| \le \tilde{A}_{kj} M_0 \tau^j, \tag{1.21′}$$

where

$$\tilde{A}_{kj} = T_k^{(j)}(1)\left(1 + \frac{1}{k!}\right), \qquad j = 1, 2, \ldots, k-1.$$

In order to prove a similar inequality for all $x \in [a, b]$, we demonstrate its validity, e.g., for $x \in [b-h, b]$. In this case, by setting $x_0 = b - h$ and taking into account inequality (1.21), we obtain

$$\left|f^{(j)}(x)\right| = \left|P_{k-1}^{(j)}(x; x_0) + \frac{1}{(k-j-1)!}\int_{x_0}^{x}(x-t)^{k-j-1} f^{(k)}(t)\,dt\right|$$

$$\leq \left|P_{k-1}^{(j)}(x; x_0)\right| + \frac{M_k}{(k-j)!}(x-x_0)^{k-j}$$

$$\leq \tilde{A}_{kj} M_0 \tau^j + \frac{M_k}{(k-j)!}\frac{1}{\tau^{k-j}} \leq \left[T_k^{(j)}(1)\left(1+\frac{1}{k!}\right) + \frac{1}{(k-j)!}\right] M_0 \tau^j .$$

Note that, at the points x located far from a and b, this inequality can be significantly improved. Thus, e.g., on the segment $[-1, 1]$, for $x_0 = 0$ and any $x \in [-h, h]$, we get

$$\left|P_{k-1}(x; 0)\right| \leq M_0\left(1+\frac{1}{k!}\right)$$

and, hence,

$$\left|f'(0)\right| = \left|P_{k-1}'(0, 0)\right| \leq M_0\left(1+\frac{1}{k!}\right)(k-1)\tau$$

by virtue of (1.20′) and the Bernstein inequality (1.5′).

Theorem 1.7 (Kolmogorov inequality [Kolmogorov (1938)]). *Assume that* $f\colon R^1 \to R^1 \in W^r$, *i.e., that the function* f *is bounded in* R^1 *and possesses, for some natural* $r \geq 2$, *absolutely continuous derivatives up to the* $(r-1)$*th order, inclusive; the* r*th derivative of the function* f *is bounded for almost all* $x \in R^1$. *Moreover, let*

$$B_0 = B_0(f) := \sup_{x \in R^1}|f(x)| \quad \textit{and} \quad B_r = B_r(f) := \operatorname*{ess\,sup}_{R^1}\left|f^{(r)}(x)\right|.$$

Then, for any natural $j \in (0, r)$, *the derivative* $f^{(j)}$ *is also bounded and satisfies the inequality*

$$B_j = B_j(f) := \sup_{x \in R^1}\left|f^{(j)}(x)\right| \leq \frac{M_{r-j}}{M_r^{1-j/r}} B_0^{1-j/r} B_r^{j/r}, \tag{1.22}$$

where M_ν, $\nu = 1, 2, \ldots$, *are constants given by the formulas*

$$M_\nu := \frac{4}{\pi}\sum_{k=0}^{\infty}\frac{(-1)^{k(\nu+1)}}{(2k+1)^{\nu+1}}. \tag{1.23}$$

Inequality (1.22) is exact in a sense that the constant factor $\dfrac{M_{r-j}}{M_r^{1-j/r}}$ *cannot be replaced by a smaller quantity.*[4]

Proof. The unimprovability of inequality (1.22) follows from the fact that, for the function

$$\mathring{f}(x) = \frac{4}{\pi}\sum_{k=0}^{\infty}\frac{\sin\left[(2k+1)t - \frac{r\pi}{2}\right]}{(2k+1)^{r+1}},$$

we get

$$\mathring{f}^{(j)}(x) = \frac{4}{\pi}\sum_{k=0}^{\infty}\frac{\sin\left[(2k+1)t - \frac{r-j}{2}\pi\right]}{(2k+1)^{r+1-j}},$$

$$\mathring{f}^{(r)}(x) \sim \frac{4}{\pi}\sum_{k=0}^{\infty}\frac{\sin(2k+1)t}{2k+1} \sim \operatorname{sgn}\sin t$$

and, according to relation (1.23),

$$B_0 = \left\|\frac{4}{\pi}\sum_{k=0}^{\infty}\frac{\sin\left[(2k+1)\cdot - \frac{r\pi}{2}\right]}{(2k+1)^{r+1}}\right\| = M_r, \quad B_j = M_{r-j}, \quad B_r = M_0 = 1.$$

Thus, it is easy to see that inequality (1.22) turns into the equality for the function $\mathring{f}$.

We now prove inequality (1.22). Since the function $f_c(x) := C_1 f(C_2 x)$ belongs to the class W^r together with the function f for any $C_j = \text{const}$, $j = 1, 2$, and satisfies the equality $B_j(f_c) = |C_1|B_j|C_2|^j$, the validity of inequality (1.22) for the function f yields its validity for the function f_c, and vice versa. For this reason, from the very beginning, we choose constants C_ν for which the following equalities are true:

$$B_0(f_c) = B_0(\overset{\circ}{f}) = M_r, \quad B_r(f_c) = B_r(\overset{\circ}{f}) = M_0 = 1,$$

and, for any $\varepsilon > 0$, one can find points x_ε such that

$$f_c^{(j)}(x_\varepsilon) = \left|f_c^{(j)}(x_\varepsilon)\right| > B_j(f_c) - \varepsilon.$$

We now set $f_c(x) = f(x)$. First, we establish the validity of inequality (1.22) under the assumption that the function f is periodic with period $2\pi\nu$, where ν is an arbitrary natural number, for $j = 1$. By contradiction, we assume that inequality (1.22) is not true and compare the function $f(x-c) := \hat{f}(x)$, where c is a constant, with the function $\overset{\circ}{f}(x)$ for which this inequality definitely holds. This yields

$$B_1(\hat{f}) > B_1(\overset{\circ}{f}) = \sup_{x\in[0,2\pi]} \left|\overset{\circ}{f}{}'(x)\right|.$$

Hence, by the proper choice of the constant c, one can find at least one point $\mathring{x}$ such that

$$\hat{f}(\mathring{x}) = \overset{\circ}{f}(\mathring{x}) \quad \text{and} \quad \hat{f}'(\mathring{x}) > \overset{\circ}{f}{}'(\mathring{x}) > 0.$$

In this case, since $\left|\hat{f}(x)\right| \le \left\|\overset{\circ}{f}\right\|$, the plot of the function $\hat{f}$ in a certain half interval $[a, a + 2\nu\pi)$ containing the point $\mathring{x}$ and 2ν branches of the function $\overset{\circ}{f}$ must cross the increasing branch of the function $\overset{\circ}{f}$ that passes through the point $(\mathring{x}, \overset{\circ}{f}(\mathring{x}))$ at at least three points, and each of the remaining $2\nu - 1$ branches of the function $\overset{\circ}{f}$ is crossed at least once. Hence, the difference $\overset{\circ}{f} - \hat{f}$ must have at least $2\nu + 2$ zeros in the interval $[a, a + 2\nu\pi)$. For sufficiently small $\varepsilon > 0$, this assertion is also true for the difference $\overset{\circ}{f} - f_\varepsilon$, where $f_\varepsilon(x) := (1-\varepsilon)\hat{f}(x)$. Thus, in view of of the fact that the function $\hat{f} - f_\varepsilon$ is continuous together with its derivatives up to the $(r-1)$th order, inclusive, and periodic with period $2\nu\pi$, we conclude that the difference $\overset{\circ}{f}{}^{(r-1)} - f_\varepsilon^{(r-1)}$ must have at least $2\nu + 2$ zeros in the interval $[a, a + 2\nu\pi]$ and, hence, at least two zeros ξ_1 and ξ_2 in at least one half interval $I_j := [a + j\pi, a + j\pi + \pi)$. Thus, in view of the fact that the function $\overset{\circ}{f}{}^{(r-1)}$ is linear in I_j and $\left|\overset{\circ}{f}{}^{(r)}(x)\right| = 1$ for any $x \in \operatorname{int} I_j$, there exists at least one segment $[\xi_1', \xi_2'] \subset (\xi_1, \xi_2)$, $\xi_2' > \xi_1'$, such that

$$\left|f_\varepsilon^{(r)}(\xi)\right| = (1-\varepsilon)\left|f^{(r)}(\xi)\right| \geq 1 \Rightarrow \left|f^{(r)}(\xi)\right| > 1 = B_r(f), \quad \xi \in [\xi_1', \xi_2'],$$

at all points ξ of this segment, which is impossible. This contradiction proves that the inequality $B_1(f) \leq B_1(\mathring{f})$ is indeed true. After this, in exactly the same way, we can prove that

$$B_2(f) \leq B_2(\mathring{f}), \ldots, B_{r-1}(f) \leq B_{r-1}(\mathring{f}).$$

We now prove the theorem for an arbitrary function from W^r. To do this, we note that, for any $\delta > 0$, there exist a natural number $\nu = \nu(\delta)$ and a function $\upsilon \in W^{r+1}$ such that $\upsilon(x) = 1$ for $|x| \leq 1$, $\upsilon(x) = 0$ for $|x| \geq \pi\nu$, $|\upsilon(x)| \leq 1$ for $x \in (-\infty, \infty)$, and $\left|\upsilon^{(j)}(x)\right| \leq \delta$ for $x \in (-\infty, \infty)$ and $j = 1, 2, \ldots, r$. Thus, in view of the fact that, according to Theorem 1.6, $B_j \leq 3T_r^{(j)}(1)$ in the entire line, we find $|\upsilon(x)f(x)| \leq B_0$ and, hence,

$$\left|[\upsilon(x)f(x)]^{(r)}\right| = \left|\sum_{i=0}^{r}\binom{r}{i}\upsilon^{(i)}(x)f^{(r-i)}(x)\right| \leq B_r + \delta\sum_{i=1}^{r}\binom{r}{i}B_{r-i} = B_r + \varepsilon,$$

where $\varepsilon = \varepsilon(\delta)$ as $\delta \to 0$. Therefore, if the function υf is periodically extended onto the entire line with period $2\nu\pi$, then, by using the relation $\upsilon(x)f(x) = 1$ for $x \in [-1, 1]$ and the periodic case of Theorem 1.7 studied above, we arrive at the inequality

$$\left|f^{(j)}(0)\right| = \left|[f(x)\upsilon(x)]_{x=0}^{(j)}\right| \leq \frac{M_{r-j}}{M_r^{1-j/r}} B_0^{1-j/r}(B_r + \varepsilon)^{j/r}, \quad j = 1, 2, \ldots, r-1. \tag{1.24}$$

Since the numbers $\delta < 0$ and, hence, $\varepsilon = \varepsilon(\delta) > 0$ can be made arbitrarily small, for any $x_0 \in (-\infty, \infty)$ and any function $f(x_0 + \cdot) \in W^r$, we get $\left|f^{(j)}(x_0)\right|$ on the left-hand side of inequality (1.24) instead of $\left|f^{(j)}(0)\right|$. Thus, by passing to the limit as $\delta \to 0$, we establish the validity of inequality (1.22). This completes the proof of Theorem 1.7.

■

2. On the estimates of errors of application of the method of grids to the Chebyshev theory of approximation of functions

In a natural way, the results established above lead us to the analysis of the following problem:

In Sections 5 and 6 of Chapter 1, we studied the algorithms of approximate construction of the polynomial P_n^* of the best approximation for a given function f on a certain set X. Since, in the course of numerical calculations, it is, as a rule, impossible to take into account the values of the function f at all points $x \in X$, the original problem is almost always replaced by a problem in which the values of the analyzed function are taken into account not in the entire set X but at finitely many points x_k, $k = 1, 2, \ldots, N$, of this set (or, in other words, the values of f are taken into account on a grid). This enables us to find (with any desired degree of accuracy) the polynomial $P_n^0(x) = P_n^0(f; x; \{x_i\}_1^N)$ of the best uniform approximation of the function f only on the system of points x_k but not the required polynomial P_n^*.

Thus, it necessary to analyze the error $\left\| f - P_n^0 \right\|_x := E_n^{(0)}(f)$ of approximation of the function $f(x)$ on X by the polynomial P_n^0 and compare it with the error $E_n(f)$ of approximation of this function by the polynomial P_n^*.

In the present section, we study the posed problem for a special choice of points x_1, $x_2, \ldots, x_N$ in the following three cases:

1) f is periodic and $X = [0, 2\pi]$;

2) f is given on $X = [a, b]$;

3) X is the set of points of the unit circle and, generally speaking, f is a complex-valued function.

To do this, we need the following lemma:

Lemma 2.1 [Dzyadyk (1978)]. *If, for some* $t_0 \in (-\infty, +\infty)$ *and* $\Delta \in [0, \xi)$, *where*

$$\xi := \min\left\{\frac{\pi}{2N}, \pi\frac{N-n}{2nN}\right\} \quad \text{and} \quad N > n,$$

a trigonometric polynomial T_n *of degree* n *satisfies* $2N$ *inequalities of the form*

$$\left|T_n(\tilde{t}_k)\right| \le M, \qquad k = 0, 1, \ldots, 2N-1, \tag{2.1}$$

where $\tilde{t}_k = t_0 + \dfrac{k\pi}{N} + \delta_y$ *and* $|\delta_k| \le \Delta$, *then the following inequality is also true:*

$$\max_t \left|T_n(t)\right| = \|T_n\| \le \frac{M}{\cos n\left(\dfrac{\pi}{2N} + \Delta\right)}. \tag{2.2}$$

Proof. Without loss of generality, we can assume that the polynomial T_n satisfies the strict inequalities at the points $\tilde{t}_k$, i.e.,

$$\left|T_n(\tilde{t}_k)\right| < M, \qquad k = 0, 1, \ldots, 2N-1. \tag{2.1'}$$

The validity of this assumption follows from the fact that if, for any $\varepsilon > 0$, the strict inequalities $\left|T_n(\tilde{t}_k)\right| < M + \varepsilon$ yield the inequality

$$\| T_n \| \le \frac{M + \varepsilon}{\cos n\left(\dfrac{\pi}{2N} + \Delta\right)},$$

then inequality (2.2) is also true in view of the arbitrariness of $\varepsilon > 0$.

To prove inequality (2.2), we assume (by contradiction) that inequality (2.1′) is satisfied and, at the same time, the inequality

$$\left|T_n(t^*)\right| > \frac{M}{\cos n\left(\dfrac{\pi}{2N} + \Delta\right)} \tag{2.2'}$$

holds at a point t^*. Without loss of generality, we can set $t_0 = -\dfrac{\pi}{2N}$ and $\left|T_n(t^*)\right| = T_n(t^*)$, $t^* \in (\tilde{t}_0, \tilde{t}_1)$. Moreover, let

$$\alpha(t) := \frac{M}{\cos n\left(\dfrac{\pi}{2N} + \Delta\right)} \cos nt \quad \text{and} \quad \varphi(t) := \alpha(t) - T_n(t). \tag{2.3}$$

Further, for each $\nu = 1, 2, \ldots, 2n-1$, let $\tilde{\xi}_\nu$ be a point of the set $\{\tilde{t}_k\}_{k=1}^{2N-1}$ from the interval

$$I_\nu := \left(\frac{\nu\pi}{n} - \frac{\pi}{2N} - \Delta, \frac{\nu\pi}{n} + \frac{\pi}{2N} + \Delta\right).$$

Note that

(a) $\tilde{t}_0 < t^* < \tilde{t}_1 < \tilde{\xi}_1 < \tilde{\xi}_2 < \ldots < \; < \tilde{\xi}_{2n-1} < \tilde{t}_0 + 2\pi$;

(b) the inequality

$$(-1)^\nu \alpha(t) > M \;\Rightarrow\; (-1)^\nu \alpha(\tilde{\xi}_\nu) > M, \quad \nu = 1, 2, \ldots, 2n-1,$$

holds for all $t \in I_\nu$;

(c) $\left|T_n(\tilde{\xi}_\nu)\right| < M$;

(d) $\tilde{t}_0, \tilde{t}_1 \in I_0 := \left(-\frac{\pi}{2N} - \Delta, \frac{\pi}{2N} + \Delta\right)$.

Thus, $\alpha(t) > M$ for all $t \in I_0$ and we conclude that $\varphi(t)$ satisfies the inequalities

$$(-1)^\nu \varphi(\tilde{\xi}_\nu) = (-1)^\nu [\alpha(\tilde{\xi}_\nu) - T_n(\tilde{\xi}_\nu)] > (-1)^\nu \alpha(\tilde{\xi}_\nu) - M > 0$$

and, hence, in view of (2.1) and (2.2′), the inequalities

$$\varphi(\tilde{t}_0) > 0, \quad \varphi(t^*) < 0, \quad \varphi(\tilde{t}_1) > 0, \quad \varphi(\tilde{\xi}_1) < 0,$$

$$\varphi(\xi_2) > 0, \ldots, \varphi(\tilde{\xi}_{2^{n-1}}) < 0, \quad \text{and} \quad \varphi(\tilde{t}_0 + 2\pi) > 0.$$

In view of relations (2.3), this means that the trigonometric polynomial $\varphi(t)$ of degree n has at least $2n+2$ zeros in the period $[\tilde{t}_0, \tilde{t}_0 + 2\pi)$, which is impossible.
Lemma 2.1 is thus proved.

Remark 2.1. Inequality (2.2) is unimprovable in a sense that if N is a multiple of n, i.e., $N = \nu_0 n$, where ν_0 is a natural number, $\delta_{j\nu_0} = -\Delta$, $\delta_{j\nu_0+1} = +\Delta$ and, therefore,

$$\tilde{t}_{j\nu_0} = \frac{j\pi}{n} - \frac{\pi}{2N} - \Delta \quad \text{and} \quad \tilde{t}_{j\nu_0+1} = \frac{j\pi}{n} + \frac{\pi}{2N} + \Delta,$$

then inequality (2.2) turns into the equality.

Remark 2.2. For $\Delta = 0$, inequality (2.2) was proved in a different way by Bernstein [Bernstein (1931)].

In what follows, for the sake of simplicity, we present the proofs of Theorems 2.1–2.3 for the case $\Delta = 0$.

Theorem 2.1.[5] *If, for some 2π-periodic function f, a polynomial T_n^* of degree n is the polynomial of its best uniform approximation on $[0, 2\pi]$ and T_n^0 is the polynomial of its best uniform approximation at $2N$ points*

$$t_0 + \frac{k\pi}{N}, \quad k = 0, 1, \dots, 2N-1, \quad N > n, \quad t_0 \in (-\infty, +\infty),$$

then the following inequalities hold:

$$E_n^{(0)}(f) = \left\| f - T_n^0 \right\| \le \left(\frac{2}{\cos \frac{n\pi}{2N}} + 1 \right) \left\| f - T_n^* \right\| \tag{2.4}$$

or

$$E_n^0(t) \le \left(\frac{2}{\cos \frac{n\pi}{2N}} + 1 \right) E_n(f)$$

and, for all $\nu = n+1, \dots, N-1$,

$$E_n^0(f) \le \frac{1}{\cos \frac{\nu\pi}{2N}} E_n(f) + \left(1 + \frac{3}{\cos \frac{\nu\pi}{2N}} \right) E_\nu(f). \tag{2.4'}$$

Proof. Indeed, according to the definition of the polynomial T_n^0, we have

$$\max_k \left| f\left(t_0 + \frac{k\pi}{N}\right) - T_n^0\left(t_0 + \frac{k\pi}{N}\right) \right|$$
$$\le \max_k \left| f\left(t_0 + \frac{k\pi}{N}\right) - T_n^*\left(t_0 + \frac{k\pi}{N}\right) \right| \le E_n(f).$$

Hence, for all $k = 0, 1, \dots, 2N-1$,

$$\left| T_n^0\left(t_0 + \frac{k\pi}{N}\right) - T_n^*\left(t_0 + \frac{k\pi}{N}\right) \right| \le 2E_n(f)$$

and, therefore, according to Lemma 2.1,

$$\left\| T_n^0 - T_n^* \right\| \le \frac{2E_n(f)}{\cos \frac{n\pi}{2N}}. \tag{2.5}$$

This enables us to conclude that

$$\left\| f - T_n^0 \right\| \le \left\| f - T_n^* \right\| + \left\| T_n^* - T_n^0 \right\| \le \left(\frac{2}{\cos \frac{n\pi}{2N}} + 1 \right) E_n(f).$$

Inequality (2.4) is thus proved.

Finally, in view of the fact that, for all $k = 0, 1, \dots, 2N-1$,

$$\left| T_\nu^0\left(t_0 + \frac{k\pi}{N}\right) - T_n^0\left(t_0 + \frac{k\pi}{N}\right) \right| \le E_\nu(f) + E_n(f)$$

by using inequality (2.4) and Lemma 2.1, we obtain

$$\begin{aligned} \left\| f - T_n^0 \right\| &\le \left\| f - T_\nu^0 \right\| + \left\| T_\nu^0 - T_n^0 \right\| \\ &\le \left(\frac{2}{\cos \frac{\nu\pi}{2N}} + 1 \right) E_\nu(f) + \frac{E_\nu(f) + E_n(f)}{\cos \frac{\nu\pi}{2N}}. \end{aligned}$$

This yields inequality (2.4′). The proof of Theorem 2.1 is completed.

Theorem 2.2. *If, for a function f given on $[-1, 1]$, P_n^* is the algebraic polynomial of its best uniform approximation on $[-1, 1]$ and P_n^0 is the algebraic polynomial of its best uniform approximation at N points*

$$x_k = \cos\left(\frac{\pi}{2N} + \frac{k\pi}{N}\right), \qquad k = 0, 1, \dots, N-1, \qquad N > n,$$

then

$$\left\| f - P_n^0 \right\|_{[-1,1]} \le \left(\frac{2}{\cos \frac{n\pi}{2N}} + 1 \right) \left\| f - P_n^* \right\| \tag{2.6}$$

and

$$\left\| f - P_n^0 \right\|_{[-1,1]} \le \frac{E_n(f)}{\cos \frac{\nu\pi}{2N}} + \left(1 + \frac{3}{\cos \frac{\nu\pi}{2N}} \right) E_\nu(f), \quad \nu = n+1, \dots, N-1, \tag{2.6′}$$

Proof. We set $\hat{f}(t) = f(\cos t)$ and, as in the proof of Theorem 2.1, denote by T_n^* and T_ν^0, $\nu = n+1, \dots, N-1$, the polynomials of the best uniform approximation for the constructed 2π-periodic even function $\hat{f}$ on the entire axis and at the N points

$$\frac{\pi}{2N} + \frac{k\pi}{N}, \qquad k = 0, 1, \dots, N-1,$$

respectively.

In this case, according to Theorem 2.1, we get

$$\left\| T_n^0 - T_n^* \right\| \le \frac{2E_n(\hat{f})}{\cos\dfrac{n\pi}{2N}},$$

$$\left\| f(\cos\cdot) - T_n^0 \right\| \le \left(\frac{2}{\cos\dfrac{n\pi}{2N}} + 1 \right) E_n(\hat{f}), \tag{2.7}$$

$$\left\| T_n^0 - T_\nu^0 \right\| \le \frac{1}{\cos\dfrac{\nu\pi}{2N}} [E_n(\hat{f}) + E_\nu(\hat{f})], \qquad \nu = n+1, \ldots, N-1. \tag{2.7'}$$

Since the function $\hat{f}$ is even and the points $\dfrac{k\pi}{N}$ are located symmetrically about the origin, both polynomials T_n^* and T_n^0 are also even. Therefore, the functions $P_n^*(x) := T_n^*[\arccos x]$ and $P_n^0(x) := T_n^0(\arccos x)$ are algebraic polynomials of degree n. Moreover, by virtue the Chebyshev theorem, they are the polynomials of the best uniform approximation for the function $f(x) = \hat{f}(\arccos x)$ on the segment $[-1, 1]$ and at the points x_k, respectively. In view of relation (2.7), the following inequalities hold at the indicated points (after the change of variables $\cos t = x$):

$$\left\| P_n^0 - P_n^* \right\| \le \frac{2E_n(f)}{\cos\dfrac{n\pi}{2N}},$$

$$\left\| f - P_n^0 \right\|_{[-1,1]} \le \left(\frac{2}{\cos\dfrac{n\pi}{2N}} + 1 \right) E_n(f), \tag{2.8}$$

$$\left\| P_n^0 - P_\nu^0 \right\|_{[-1,1]} \le \frac{1}{\cos\dfrac{\nu\pi}{2N}} [E_n(f) + E_\nu(f)].$$

These inequalities immediately yield the assertion of Theorem 2.2.

Theorem 2.3. *If, for a function f given on the unit circle $|z| = 1$, $P_n^*(z)$ and $P_n^0(z)$ are the polynomials of its best uniform approximation on $|z| = 1$ and at the points $z_k = e^{ik\pi/N}$, $k = 0, 1, 2, \ldots, 2N-1$, $N > n$, respectively, then*

$$\max_{|z|=1} \left| f(z) - P_n^0(z) \right| \le \left(\frac{2}{\sqrt{\cos\dfrac{\pi n}{2N}}} + 1 \right) E_n(f) \tag{2.9}$$

and

$$\max_{|z|=1} \left| f(z) - P_n^0(z) \right| \le \frac{1}{\sqrt{\cos \frac{\nu\pi}{2N}}} E_n(f) + \left(1 + \frac{3}{\sqrt{\cos \frac{\nu\pi}{2N}}} \right) E_\nu(f), \tag{2.9'}$$

$$\nu = n+1, \dots, N-1.$$

Proof. Indeed, it is clear that, for any polynomial of the form

$$P_n(z) = \sum_{k=0}^{n} c_k z^k ,$$

the function

$$\left| P_n(e^{it}) \right|^2 = \sum_{j=0}^{n} \sum_{k=0}^{n} c_j \bar{c}_k \, e^{i(j-k)t}$$

is a real trigonometric polynomial of degree n. Now let $T_n(t)$ be a trigonometric polynomial of degree n of the form

$$T_n(t) := \left| P_n^0(e^{it}) - P_n^*(e^{it}) \right|^2. \tag{2.10}$$

Thus, in exactly the same way as in the proof of Theorem 2.1, we obtain

$$\left| P_n^0(z_k) - P_n^*(z_k) \right| \le 2E_n(f),$$

$$T_n\left(\frac{k\pi}{N}\right) = \left| P_n^0(e^{ik\pi/N}) - P_n^*(e^{ik\pi/N}) \right|^2 \le 4E_n^2(f),$$

$$\left| P_n^0(e^{it}) - P_n^*(e^{it}) \right|^2 \le \| T_n \| \le \frac{4E_n^2(f)}{\cos \frac{n\pi}{2N}}.$$

Finally, by virtue of the maximum-modulus principle, this yields

$$\left| P_n^0(z) - P_n^*(z) \right| \le \max_t \left| P_n^0(e^{it}) - P_n^*(e^{it}) \right| \le \frac{2E_n(f)}{\sqrt{\cos \frac{n\pi}{2N}}}, \tag{2.11a}$$

$$\max_t \left| f(z) - P_n^0(z) \right| \le \left(\frac{2}{\sqrt{\cos \frac{n\pi}{2N}}} + 1 \right) E_n(f). \tag{2.11b}$$

The proof of inequality (2.9) is similar to the proof of inequality (2.4′).
Theorem 2.3 is thus proved.

3. Inverse theorems

In the theory of approximation of functions, any assertion is called *inverse theorem* if it establishes the degree of smoothness of a function (or a class of functions) depending on the rate of vanishing of the difference between this function (or the class of functions) and its approximations.

The notion of inverse theorem was introduced by Bernstein (1912). He obtained the first important results in this direction. Later, his results in the periodic case were supplemented by de la Vallée Poussin (1919), Zygmund (1945), and other researchers. In this field, the following inverse theorem is of principal importance[6]:

Theorem 3.1 ([Bernstein (1912); de la Vallée Poussin (1919); Stechkin (1951)]). *Assume that a function* ω *satisfies the following four conditions for* $r = 0$:

(i) ω *is a continuous function;*

(ii) ω *is a monotonically increasing function;*

(iii) $\omega(0) = 0,$

(iv) for any $t > 0,$

$$\omega(2t) \le C\omega(t), \qquad C = \text{const.} \tag{3.1}$$

Moreover, suppose that this function satisfies the following fifth condition for $r \ge 1$:

(v) $$\int_0^1 \frac{\omega(t)}{t}\, dt < \infty. \tag{3.2}$$

If, for some integer $r \ge 0$ *and a* 2π*-periodic function* f*, there exists a sequence of* *trigonometric polynomials* T_n *of degree* n *approximating the function* f *and such that the inequalities*

$$\left|f(t) - T_n(t)\right| \leq \frac{A}{n^r}\,\omega\left(\frac{1}{n}\right) \tag{3.3}$$

are true for $n = 1, 2, \ldots,$ *then* $f \in \tilde{C}^r$, *and, for a fixed number* $k = 1, 2, \ldots,$ *the* k*th modulus of continuity* $\omega_k(f^{(r)}; t)$ *of the derivative* $f^{(r)}$ *of the function* f *satisfies the inequality*

$$\omega_k(f^{(r)}; t) \leq \begin{cases} A_1 A t^k \displaystyle\int_t^1 \frac{\omega(u)}{u^{k+1}}\,du, \quad A_1 = \text{const}, & \text{for} \quad r = 0, \\ A_1 A \left[t^k \displaystyle\int_t^1 \frac{\omega(u)}{u^{k+1}}\,du + \int_0^t \frac{\omega(u)}{u}\,du \right] & \text{for} \quad r \geq 1. \end{cases} \tag{3.4}$$

Proof. 1. First, we consider the case $r = 0$. For any $h > 0$ and any natural N, we have

$$\Delta_h^k f(t) = \Delta_h^k T_1(t) + \sum_{j=1}^{N} \Delta_h^k [T_{2^j}(t) - T_{2^{j-1}}(t)] + \Delta_h^k [f(t) - T_{2^N}(t)]. \tag{3.5}$$

Further, we choose a number N such that

$$\frac{1}{2^{N+1}} < h \leq \frac{1}{2^N}. \tag{3.6}$$

For $r = 0$, inequality (3.3) implies that

$$\begin{aligned} \left|T_{2^j}(t) - T_{2^{j-1}}(t)\right| &\leq \left|T_{2^j}(t) - f(t)\right| + \left|f(t) - T_{2^{j-1}}(t)\right| \\ &\leq 2A\omega\left(\frac{1}{2^{j-1}}\right) \leq 2AC\omega\left(\frac{1}{2^j}\right) \end{aligned} \tag{3.7}$$

and, hence, in view of the monotonicity of the function ω, that

$$2^{kj}\,\omega\left(\frac{1}{2^j}\right) = \frac{2^{k+1}}{2^j}\,\frac{\omega\left(\frac{1}{2^j}\right)}{\frac{1}{2^{(k+1)(j-1)}}} \leq 2^{k+1} \int_{1/2^j}^{1/2^{j-1}} \frac{\omega(u)}{u^{k+1}}\,du. \tag{3.8}$$

Therefore, taking into account relations (3.5)–(3.8), the inequality for the absolute value of the derivative of a trigonometric polynomial, and inequality (3.3) for $r = 0$, we obtain

$$\left|\Delta_h^k f(t)\right| \le O(h^k) + 2CAh^k \sum_{j=1}^{N} 2^{jk}\omega\left(\frac{1}{2^j}\right) + 2^k A\omega\left(\frac{1}{2^N}\right)$$

$$\le O(h^k) + 2^{k+2}CAh^k \sum_{j=1}^{N} \int_{1/2^j}^{1/2^{j-1}} \frac{\omega(u)}{u^{k+1}}du + 2^k A\omega(2h)$$

$$\le O(h^k) + 2^{k+2}CAh^k \int_h^1 \frac{\omega(u)}{u^{k+1}}du + 2^k CA\omega(h) \le A_1 Ah^k \int_h^1 \frac{\omega(u)}{u^{k+1}}du, \qquad (3.9)$$

where A_1 is a constant. This yields the validity of Theorem 3.1 for $r = 0$.

2. Consider the case $r \ge 1$.

We represent the function f in the form of a series

$$f(t) = T_1(t) + \sum_{j=1}^{\infty}[T_{2^j}(t) - T_{2^j-1}(t)]. \qquad (3.10)$$

Since, by virtue of inequality (3.3),

$$\left|T_{2^k}(t) - T_{2^{k-1}}(t)\right| \le 2\frac{A}{2^{(k-1)r}}\,\omega\left(\frac{1}{2^{k-1}}\right),$$

the Bernstein inequality implies that

$$\left|T_{2^k}^{(r)}(t) - T_{2^{k-1}}^{(r)}(t)\right| \le 2^{r+1}A\omega\left(\frac{1}{2^{k-1}}\right) \le 2^{r+1}CA\left(\omega\left(\frac{1}{2^k}\right)\right).$$

By using this inequality, the fact that, for each $j = 1, 2, \ldots$, the following inequality holds in view of the fact that $\omega\uparrow$:

$$\int_{1/2^j}^{1/2^{j-1}} \frac{\omega(u)}{u}du \ge \ln 2\,\omega(2^{-j}),$$

and relation (3.10), we readily conclude that $f \in \tilde{C}^r$.

In this case, for any $j = 1, 2, \ldots,$ we get

$$\left|f^{(r)}(t) - T_{2^j}^{(r)}(t)\right| \leq \sum_{k=j}^{\infty}\left|T_{2^k}^{(r)}(t) - T_{2^{k-1}}^{(r)}(t)\right|$$

$$\leq 2^{r+1}AC\sum_{k=j}^{\infty}\omega(2^{-k}) \leq 2^{r+2}AC\int_0^{2^{-j}}\frac{\omega(u)}{u}du = \tilde{A}\Omega(2^{-j}), \tag{3.11}$$

where $\tilde{A} = 2^{r+2}AC$ and

$$\Omega(t) := \int_0^t \frac{\omega(u)}{u}\,du. \tag{3.12}$$

By virtue of inequality (3.2), the function Ω exists for all $t \in [0, 1]$. Moreover, this function satisfies the conditions:

(a) Ω is a continuous function;

(b) Ω is a monotonically increasing function;

(c) $\Omega(0) = 0$;

(d) for any $t > 0$,

$$\Omega(2t) = \int_0^{2t}\frac{\omega(u)}{u}\,du = \int_0^t\frac{\omega(2u)}{u}\,du \leq C\int_0^t\frac{\omega(u)}{u}\,du \leq C\Omega(t).$$

Therefore, in view of relation (3.11) and case 1 $(r = 0)$, we get

$$\omega_k(f^{(r)}; t) \leq A_1\tilde{A}t^k\int_t^1\frac{\Omega(u)}{u^{k+1}}\,du = \frac{A_1\tilde{A}t^k}{k}\left[\Omega(u)\frac{1}{u^k}\bigg|_1^t + \int_t^1\frac{\omega(u)}{u^{k+1}}\,du\right]$$

$$\leq A_1\tilde{A}\left[t^k\int_t^1\frac{\omega(u)}{u^{k+1}}\,du + \int_0^t\frac{\omega(u)}{u}\,du\right].$$

The proof of Theorem 3.1 is completed.

■

This theorem yields the following corollaries:

Corollary 3.1. *If* ω *is a function of the* k*th-modulus-of-continuity type, then the following inequality is true:*

$$\omega_{k+1}(f^{(r)};t) \le A_1\,\omega(t), \quad A_1 = \text{const}, \tag{3.13}$$

under the conditions of Theorem 3.1 for $r = 0$ *and with the following additional condition for* $r \ge 1$:

$$\int_0^t \frac{\omega(u)}{u}\,du \le A_2\,\omega(t).$$

Indeed, according to the first inequality in (3.4) for $r = 0$, we get

$$\omega_{k+1}(f;t) \le A_1 A t^{k+1} \int_t^1 \frac{\omega(u)}{u^{k+2}}\,du$$

$$\le A_1 A t^{k+1} \int_t^1 \left(2\frac{u}{t}\right) \frac{\omega(t)}{u^{k+2}}\,du \le \tilde{A}_1\omega(t), \quad \tilde{A}_1 = \text{const}.$$

Corollary 3.2. *If, under the conditions of Theorem 3.1,* $\omega(t) = t^{\alpha}$, *where* $0 < \alpha < 1$, *then* $f \in \tilde{C}^r$ *and, for any integer* $r \ge 0$,

$$\omega(f^{(r)};t) \le A t^{\alpha}, \tag{3.14}$$

i.e., $f \in W^r \tilde{H}^{\alpha}$.

Corollary 3.3. *If, under the conditions of Theorem 3.1,* $\omega(t) = t$, *then* $f \in \tilde{C}^r$ *and, for any integer* $r \ge 0$,

$$\omega_2(f^{(r)};t) \le A t, \tag{3.15}$$

i.e., $f \in W^r \tilde{Z}$.

4. On the constructive characteristics of periodic functions of the Hölder and Zygmund classes

1. In the case where the inverse theorems for a class of functions Φ completely supplement the direct theorems in a sense that the collection of direct and inverse theorems establishes the conditions that are both necessary and sufficient for functions to belong to the class Φ, we say that the constructive characteristic is obtained for the class Φ.

The Jackson theorem applied to periodic functions from the spaces $W^r\tilde{H}^\alpha$ $(0 < \alpha < 1)$ and $W^r\tilde{Z}$ and Corollaries 3.2 and 3.3 of the inverse Theorem 3.1 yield the following two theorems on the constructive characteristics of functions from the Hölder spaces $W^r\tilde{H}^\alpha$ $(0 < \alpha < 1)$ and Zygmund spaces $W^r\tilde{Z}$, respectively.

Theorem 4.1. *In order that a function f belong to the space $W^r\tilde{H}^\alpha$ for integer $r \geq 0$ and $\alpha \in (0, 1)$, it is necessary and sufficient that, for any natural n, one can find a trigonometric polynomial T_n of degree n such that the inequality*

$$|f(t) - T_n(t)| \leq \frac{A}{n^{r+\alpha}}, \tag{4.1}$$

where A is a constant independent of n, holds for all $t \in [0, 2\pi]$ or, equivalently, it is necessary and sufficient that the best approximations $E_n(f)$ of the function f satisfy the conditions

$$\tilde{E}_n(f) \leq \frac{A}{n^{r+\alpha}}, \tag{4.1$'$}$$

where $r \geq 0$ is an integer number.

Theorem 4.2. *In order that a function f belong to the Zygmund space $W^r\tilde{Z}$ for a nonnegative integer r, it is necessary and sufficient that, for any natural n, one can find a trigonometric polynomial T_n of degree n such that the inequality*

$$|f(t) - T_n(t)| \leq \frac{A}{n^{r+1}}, \tag{4.2}$$

where A is a constant independent of n, holds for all $t \in [0, 2\pi]$ or, equivalently, it is necessary and sufficient that the best approximations $E_n(f)$ of the function f satisfy the inequalities

$$\tilde{E}_n(f) \leq \frac{A}{n^{r+1}}, \tag{4.2$'$}$$

where $r \geq 0$ is an integer number.

2. We draw the attention of the reader to the fact that the constructive characteristic of functions from the Hölder classes $W^r H^{\alpha}$, where r is a nonnegative integer, was obtained solely for the case where $0 < \alpha < 1$. For $\alpha = 1$, the required characteristic was obtained for somewhat broader classes $W^r Z$ but not for the classes $W^r H^1$.

Trigub (1965) (see also [Znamenskii (1950)]) indicated that, in order to get the characteristic of functions f from the spaces $W^r \tilde{H}^1$ (and $W^r \tilde{H}^{\omega}$) in terms of the theory of approximation of functions, one can use somewhat "spoiled" polynomials, instead of "good" polynomials traditionally used for the approximation of functions from these spaces whose behavior is so good that they approximate the functions from both classes ($W^r \tilde{H}^1$ and $W^r \tilde{Z} \supset W^r \tilde{H}^1$) with the same accuracy (equal to n^{-r-1}) for any fixed nonnegative integer $r \geq 0$. The indicated "spoiled" polynomials $\tau_n(t) = \tau_n(f; t)$ of degree n should be such that $|f(t) - \tau_n(t)| = O(n^{-r-1})$ for any function $f \in W^r \tilde{H}^1$ but $|f(t) - \tau_n(t)|\, n^{r+1} \to \infty$ for any function $f \in W^r \tilde{Z} \setminus W^r \tilde{H}^1$.

We first present the results from [Dzyadyk (1975)] which are somewhat less exact than the results obtained by Trigub (1965) but more general and be extended, unlike the results from [Trigub (1965)] to the case of approximation of nonperiodic functions.

Theorem 4.3. *In order that a* 2π*-periodic function* f *belong to the space* $\tilde{H}_k^{\omega}$ *for a* k*-majorant* ω, *it is necessary and sufficient that there exist a sequence of trigonometric polynomials* U_n *of degrees* $n = 1, 2, \ldots$ *with the following properties:*

$$|f(t) - U_n(t)| \leq A\omega\left(\frac{1}{n}\right) \tag{4.3}$$

and

$$\left|f(t) - U_n^*(t)\right| \leq A\omega\left(\frac{1}{n}\right), \tag{4.4}$$

where A *is a constant independent of* n *and*

$$U_n^*(t) := U_n(t) - \Delta_{\pi/n}^k(U_n; t). \tag{4.5}$$

Proof. *Necessity.* If $f \in \tilde{H}_k^{\omega}$, then, according to Theorem 5.2.3, for any $n = 1, 2, \ldots$, there exists a polynomial $U_n(t) = U_n(f; t)$ of degree n such that

$$|f(t) - U_n(t)| \leq A\omega_k\left(f; \frac{1}{n}\right) < A\omega\left(\frac{1}{n}\right)$$

and, hence, by virtue of (4.5),

$$\left|f(t)-U_n^*(t)\right| \le |f(t)-U_n(t)| + |\Delta_{\pi/n}^k(f-U_n;t)| + |-\Delta_{\pi/n}^k(f;t)|$$

$$\le A\omega\left(\frac{1}{n}\right)[1+2^k] + \omega_k\left(f;\frac{1}{n}\right) \le A_1\omega\left(\frac{1}{n}\right),$$

$$A_1 = \text{const.}$$

Sufficiency. If, for the function f, there exist polynomials U_n and U_n^* connected by relation (4.5) and satisfying inequalities (4.3) and (4.4), then, in view of equality (4.5), we get

$$\left|\Delta_{\pi/n}^k(U_n;t)\right| = \left|U_n - U_n^*\right| \le 2A\omega\left(\frac{1}{n}\right).$$

Thus, by using the Nikol'skii–Stechkin inequality $(1.\tilde{1}')$ for any M and setting $n = \left[\frac{1}{h}\right]$, we get

$$\left\|\Delta_h^k f\right\| \le \left\|\Delta_h^k(f-U_n)\right\| + \left\|\Delta_h^k U_n\right\| \le 2^k A\omega\left(\frac{1}{n}\right) + h^k\left\|U_n^{(k)}\right\|$$

$$\le 2^k A\omega\left(\frac{1}{n}\right) + h^k\left(\frac{n}{2}\right)^k \left\|\Delta_{\pi/n}^k(U_n,\cdot)\right\|$$

$$\le 2^k A\omega\left(\frac{1}{n}\right) + \left(\frac{1}{2}\right)^k 2A\omega\left(\frac{1}{n}\right) = A_1\omega\left(\frac{1}{n}\right), \quad A_1 = \text{const.}$$

Theorem 4.3 is thus proved.

Theorem 4.3′. *In order that a 2π-periodic function f belong to the space $\tilde{H}_k^\omega$ for a k-majorant ω, it is necessary and sufficient that the inequalities*

$$\left|f(t)-U_n^*(t)\right| \le A\omega\left(\frac{1}{n}\right), \quad A = \text{const}, \tag{4.6}$$

be true for all natural n with polynomials $U_n^*(t)$ introduced by the formula

$$U_n^*(t) := U_n(t) - \Delta_{\pi/n}^k(U_n;t),$$

where $U_n(t) = U_n(f;\, t)$ *are trigonometric polynomials (operators) of degree* n *approximating any continuous* 2π*-periodic function* f *so that the following inequalities are satisfied:*

$$|f(t) - U_n(f;\, t)| \le A_1 \omega_{k+1}\left(f;\, \frac{1}{n}\right), \qquad A_1 = \text{const}, \tag{4.7}$$

Proof. *Necessity.* If $f \in \tilde{H}_k^{\omega}$, then, according to Theorem 5.2.3, for any $n = 1, 2, \ldots$, there exists a polynomial $U_n(t) = U_n(f;\, t)$ of degree n such that

$$|f(t) - U_n(t)| \le A_1 \omega_{k+1}\left(f;\, \frac{1}{n}\right).$$

Since

$$\omega_{k+1}\left(f;\, \frac{1}{n}\right) \le 2\omega_k\left(f;\, \frac{1}{n}\right) \le 2\omega\left(\frac{1}{n}\right)$$

(in view of the fact that $f \in \tilde{H}_k^{\omega}$), we proceed in exactly the same way as in the final part of the proof of necessity in Theorem 4.3.

Sufficiency. If, for a given modulus of continuity ω of order k, there exists a sequence of trigonometric polynomials U_n^* of the form (4.5) such that

$$\left|f(t) - U_n^*(f;\, t)\right| \le A\omega\left(\frac{1}{n}\right), \tag{4.8}$$

then, according to Corollary 3.1, $\omega_{k+1}(f;\, t) \le A_1 \omega(t)$ and, hence,

$$|f(t) - U_n(f;\, t)| \le A_2 \omega_{k+1}\left(f;\, \frac{1}{n}\right) \le A_3 \omega\left(\frac{1}{n}\right) \tag{4.9}$$

by virtue of (4.7). In view of Theorem 4.3 and inequalities (4.8) and (4.9), we conclude that $f(t) \in H_k$.

This completes the proof of Theorem 4.3′.

Note that a function f belongs to $W^r \tilde{H}^1$, where $r \ge 0$ is an integer, if and only if (see [Marchaud (1927)])

$$\omega_{r+1}(f;\, t) \le A\, t^{r+1}, \qquad A = \text{const}. \tag{4.10}$$

As a consequence of Theorem 4.3′, one can obtain the following result [Dzyadyk (1975)]:

Theorem 4.4. *In order that a* 2π*-periodic function* f *belong to the space* $W^r\tilde{H}^1$ *for a nonnegative integer* r, *it is necessary and sufficient that the following inequalities hold for all* $n = 1, 2, \dots$:

$$\left|f(t) - U_n^*(f; t)\right| \le An^{-r-1}, \quad A = \text{const}, \tag{4.11}$$

where $U_n^*(f; \cdot)$ *are polynomials determined as indicated in Theorem 4.3′ for* $k = r + 1$.

3. In conclusion, we present a result taken from [Trigub (1965)] and valid, unlike Theorem 4.3, not only for the entire sequence of numbers $n = 1, 2, \dots$ but also for each fixed number n separately.

Theorem 4.5. *For any* 2π*-periodic continuous function* f *and any natural numbers* k *and* n, *the polynomials* $\tau_{k,n}$ *constructed by using kernels (2.3.27), i.e.,*

$$\begin{aligned}\tau_{k,n}(f; t) &= D_n(t) + \frac{1}{2^k}\sum_{\nu=0}^{k}\binom{k}{\nu}(-1)^{k-\nu+1}D_n\left(t + \frac{\nu\pi}{n}\right)\\ &= D_n(t) - \frac{1}{2^k}\Delta_{\pi/n}^k(D_n; t),\end{aligned} \tag{4.12}$$

where D_n *is the Dirichlet kernel, by the formula*

$$\tau_{k,n}(f) = \tau_{k,n}(f; t) = \frac{1}{\pi}\int_{-\pi}^{\pi} f(u)\tau_{k,n}(t-u)\,du = S_n(f; t) - \frac{1}{2^k}\Delta_{\pi/n}^k(S_n; t), \tag{4.13}$$

where $S_n(f; t)$ *are partial sums of the Fourier series for the function* f, *possess the following property:*

There exist two positive constants $c_1 = c_1(k)$ *and* $c_2 = c_2(k)$ *independent of* n *such that the inequalities*

$$c_1\omega_k\left(f; \frac{\pi}{n}\right) \le \left\|f - \tau_{k,n}(f)\right\| \le c_2\omega_k\left(f; \frac{\pi}{n}\right) \tag{4.14}$$

hold uniformly for all $n = 1, 2, \dots$.

Proof. By T_n^* we denote the polynomial of degree n of the best uniform approximation of the function f. By setting $\|f - T_n^*\| = E_n$ and taking into account relation (4.13) and the fact that $S_n(T^*; t) = T_n^*(t)$, we obtain

$$f - \tau_{k,n}(f) = f - T_n^* + \frac{1}{2^k} \Delta_{\pi/n}^k T_n^* - \tau_{k,n}(f - T_n^*)$$

$$= f - T_n^* - \tau_{k,n}(f - T_n^*) - \frac{1}{2^k} \Delta_{\pi/n}^k (f - T_n^*) + 2^{-k} \Delta_{\pi/n}^k f. \quad (4.15)$$

According to relation (2.3.27′), we have $\|\tau_{k,n}\| \le 2k\pi$. In view of equality (4.13), this yields

$$\left\|f - T_n^* - \tau_{k,n}(f - T_n^*) - 2^{-k} \Delta_{\pi/n}^k (f - T_n^*)\right\|$$

$$\le E_n + 2kE_n + 2^{-k} 2^k E_n = (2k+2)E_n.$$

On the one hand, by using this inequality, identity (4.15), and the generalized Jackson theorem (Theorem 5.2.3), we obtain

$$\|f - \tau_{k,n}(f)\| \le (2k+2)E_n + 2^{-k}\omega_k\left(f; \frac{\pi}{n}\right) \le [(2k+2)A_1 + 2^{-k}]\omega_k\left(f; \frac{\pi}{n}\right). \quad (4.16)$$

On the other hand, we can write

$$2^{-k}\left\|\Delta_{\pi/n}^k f\right\| \le \|f - \tau_{k,n}(f)\| + (2k+2)E_n \le (2k+3)\|f - \tau_{k,n}(f)\|,$$

whence, in view of the Nikol'skii–Stechkin inequality (1.$\tilde{1}$′), we conclude that

$$\omega_k\left(f; \frac{\pi}{n}\right) = \sup_{0 \le h \le \pi/n} \left\|\Delta_{\pi/n}^k f\right\| \le \sup_{h \in [0,\pi/n]} \left(\left\|\Delta_{\pi/n}^k T_n^*\right\| + \left\|\Delta_{\pi/n}^k (f - T_n^*)\right\|\right)$$

$$\le \sup_{h \in [0,\pi/n]} \left(t^k \left\|T_n^{*(k)}\right\| + 2^k \tilde{E}_n\right) \le \left(\frac{\pi}{n}\right)^k \left(\frac{n}{2}\right)^k \left\|\Delta_{\pi/n}^k T_n^*\right\| + 2^k \tilde{E}_n$$

$$\le \left(\frac{\pi}{2}\right)^k \left(\left\|\Delta_{\pi/n}^k f\right\| + \left\|\Delta_{\pi/n}^k (T_n^* - f)\right\|\right) + 2^k \tilde{E}_n$$

$$\le \left(\frac{\pi}{2}\right)^k \left\|\Delta_{\pi/n}^k f\right\| + (2^k + 2^k)\tilde{E}_n$$

$$\leq \pi^k(2k+3)\|f - \tau_{k,n}(f)\| + (2^k + 2^k)E_n \leq A\|f - \tau_{k,n}(f)\|, \qquad (4.17)$$

where $A = \pi^k(2k+3) + 2^{k+1} = \text{const}$.

Inequalities (4.16) and (4.17) yield the assertion of Theorem 4.5.

Remarks to Chapter 6

1. S. N. Bernstein established inequality (1.1) in the form presented in our book for the case where the trigonometric polynomial T_n is even or odd. For trigonometric polynomials of the general form, he established a somewhat less exact inequality: $|T_n'(t)| \leq 2nM$. However, it turned out that the factor 2 can be removed fairly easily. This was done by M. Riesz (1914) and F. Riesz (1914) and, somewhat later, by Bernstein himself and other researchers.

2. The idea of the proof apparently appeared, for the first time, in the monograph by de la Vallée Poussin [de la Vallée Poussin (1919)].

3. The Bernstein inequality [Bernstein (1912)] $|P_n'(t)| \leq \dfrac{nM}{\sqrt{1-t^2}}$ follows from the inequalities established for $|P_n'(t)|$ by Markov (1884). However, the proof proposed by Markov was quite complicated and the results were not presented in the compact form convenient for subsequent applications.

4. The problem of the estimation of the norms of intermediate derivatives was studied by numerous researchers. Here, we mention only the most significant results. The first simplest problem of this sort was solved by Hadamard (1914):

If $f \in W^2(-\infty, \infty)$, then

$$\|f'(x)\| \leq \sqrt{2\|f(x)\|\|f''(x)\|}, \quad \text{where} \quad \|\varphi(x)\| := \operatorname*{ess\,sup}_{-\infty<x<\infty} |\varphi(x)|.$$

Later, Kolmogorov (1938) established the general result (Theorem 1.7).

For a finite segment, inequalities of the form (2.19) (with somewhat coarser constants) were deduced for the first time by Gorny (1938) and Cartan (see, e.g., [Mandelbrojt (1955)]).

Later, similar problems were studied in the metric of $L(-\infty, \infty)$ (see, e.g., [Taikov (1968)]), on the semiaxis (see, e.g., [Matorin (1955)] and [Tikhomirov (1955)]), in the case of several intermediate derivatives (see, e.g., [Rodov (1956)] and [Dzyadyk and Dubovik (1974), (1975)], in R^2 on the space C [Konovalov (1977)], etc.

5. This and all subsequent theorems in Section 2 were proved by Dzyadyk in a somewhat weaker form [Dzyadyk (1971a)].

Note that the results similar to these theorems (but somewhat less exact) can readily be established by using Theorems 2.1 and 2.2 for the case of approximation of continuous functions in R^n (periodic and nonperiodic functions given on a parallelepiped $[a_1, b_1] \times \ldots \times [a_n, b_n]$).

6. Bernstein (1913) posed, for the first time, the problem of inverse theorems and demonstrated, for periodic functions, that if $\tilde{E}_n(f) \leq \frac{1}{n^{r+\alpha}}$, where $r \geq 0$ is an integer and $0 < \alpha \leq 1$, then $f \in W^r \tilde{H}^{\alpha-\varepsilon}$ for arbitrarily small $\varepsilon > 0$. Under the same assumptions, de la Vallée Poussin (1919) proved that if $\alpha \neq 1$, then $f \in W^r \tilde{H}^{\alpha}$. Zygmund (1945) introduced the spaces $W^r \tilde{H}_2^{\omega}$ and established, for $\omega(t) = t$, the inverse theorems for the classes $W^r \tilde{Z}$. The inverse theorems for the spaces $\tilde{H}^{\omega}$ in the uniform metric were first proved by Salem [(1935), (1940)]. For the spaces $W^r \tilde{H}_k^{\omega}$, the inverse theorems in the metric L^p with $1 \leq p < \infty$ were obtained by A. Timan and M. Timan (see [A. Timan (1950)]) and, in the uniform metric, by Stechkin (1951a).

Later, it was shown (see [Lozinskii (1952)] and [Bari and Stechkin (1956a)]) that the inequalities

$$E_n(f) \leq A_3 \omega\left(\frac{1}{n}\right), \quad n = 1, 2, \ldots, \quad A_3 = \text{const}, \tag{A}$$

where ω is a function of the modulus-of-continuity type, imply that the kth modulus of continuity for a function f satisfies the condition

$$\omega_k(f; \delta) \leq A_4 \omega(\delta) \tag{B}$$

and, vice versa, it is clear that (A) follows from (B) if and only if there exists at least one constant $A > 1$ such that

$$\limsup_{\delta \to 0} \frac{\omega(A\delta)}{\omega(\delta)} < A^k.$$

Chapter 7
Approximation by polynomials

1. Introduction

1.1. Let f be a 2π-periodic function and let $\tilde{E}_n(f)$ be the value of its best uniform approximation by trigonometric polynomials of degree n. Recall that Jackson [(1912), (1930)] proved the direct theorem according to which one has

$$\tilde{E}_n(f) < c\omega_1\left(f; \frac{1}{n}\right). \tag{1.1}$$

The direct Jackson theorems [Jackson (1912), (1930)] and the inverse Bernstein [Bernstein (1912), (1930)] and de la Vallée Poussin [de la Vallée Poussin (1919)] theorems give a constructive characteristic of the Hölder spaces $\tilde{H}^\alpha$ for $\alpha \in (0; 1)$, namely

$$\tilde{E}_n(f) = O\left(\frac{1}{n^\alpha}\right) \Leftrightarrow \omega_1(f; t) = O(t^\alpha), \tag{1.2}$$

and a constructive characteristic of the spaces $W^r\tilde{H}^\alpha$ for $\alpha \in (0; 1)$:

$$\tilde{E}_n(f) = O\left(\frac{1}{n^{r+\alpha}}\right) \Leftrightarrow \omega_1(f^{(r)}, t) = O(t^\alpha). \tag{1.3}$$

The interesting case where $\tilde{E}_n(f) = O\left(\frac{1}{n}\right)$ does not fit these formulations because they essentially use the fact that $\alpha \neq 0$ and $\alpha \neq 1$. It is known that

$$\omega_1(f, t) = O(t) \Rightarrow \tilde{E}_n(f) = O\left(\frac{1}{n}\right). \tag{1.4}$$

However, the converse statement is not true. Zygmund [(1945), (1959)] noted for the first time that this case also admits a formulation in terms of equivalence if, instead of first-order moduli of continuity, one uses a second-order modulus of continuity. He proved that the relations

$$\tilde{E}_n(f) = O\left(\frac{1}{n}\right) \quad \text{and} \quad \omega_2(f; t) = O(t) \tag{1.5}$$

are equivalent, and the same is also true for the relations

$$\tilde{E}_n(f) = O\left(\frac{1}{n^r}\right) \quad \text{and} \quad \omega_1(f^{(r-1)}, t) = O(t). \tag{1.6}$$

For the spaces $W^r\tilde{H}_k^{\varphi}$ of periodic functions, a constructive characteristic was obtained by Zygmund [(1924), (1945)] (for $k=2$ and $\omega_2(f;t)\le t$), Akhiezer (for $k=2$; see the first edition of the book [Akhiezer (1965)]), and Stechkin [(1949), (1951), (1951a)] (for $k\ge 3$).

1.2. In the nonperiodic case, i.e., for $f\in C([a,b])$, the estimate

$$E_n(f) \le c\omega_1\left(\frac{b-a}{n}; f; [a,b]\right) \tag{1.7}$$

can easily be derived from (1.1) by using the change of variables $2x = (b-a)\cos t + b + a$. However, relations (1.2), (1.3), (1.5), and (1.6) are not true in this case (V. Motornyi; see also Theorem 3.7.2). S. Nikol'skii, Dzyadyk, and Timan showed that, in the nonperiodic case, a constructive characteristic is realized not in terms of $E_n(f)$ but in terms of a pointwise estimate for the deviation of the function f from the approximating polynomial.

First, S. Nikol'skii (1946a) established that if

$$|f(x') - f(x'')| \le |x' - x''|, \qquad x', x'' \in [-1, 1] =: I,$$

then one can construct a sequence $\{P_n\}$ of polynomials P_n of degree n for f such that

$$|f(x) - P_n(x)| \le \frac{\pi}{2}\left(\frac{\sqrt{1-x^2}}{n} + O\left(\frac{\ln}{n^2}\right)\right), \qquad x\in I. \tag{1.8}$$

Later, Timan [(1951), (1960)] proved that if $f\in W^rH[1;\varphi;I]$, then there exists a sequence $\{P_n\}$ of polynomials P_n such that

$$|f(x) - P_n(x)| \le c\left(\frac{1}{n^2} + \frac{\sqrt{1-x^2}}{n}\right)^r \varphi\left(\frac{1}{n^2} + \frac{\sqrt{1-x^2}}{n}\right), \qquad x\in I. \tag{1.9}$$

Denote

$$\rho_n(x) := \frac{1}{n^2} + \frac{\sqrt{1-x^2}}{n}, \qquad \rho_0(x) :\equiv 1. \tag{1.10}$$

In 1956, Dzyadyk (1956) obtained the following inequality for the modulus of the derivative of an algebraic polynomial:

$$\left\| \rho_n^{s+j} P_n^{(j)} \right\|_I \leq j! M \left\| \rho_n^s P_n \right\|_I, \tag{1.11}$$

where $s \in R$, $M = M(s)$, and $j \in N$. Using this inequality, he established the following inverse theorem: Let $(r+1) \in \mathbb{N}$ and $\alpha \in (0; 1)$. If there exists a sequence $\{P_n\}$ of polynomials P_n such that

$$|f(x) - P_n(x)| \leq \rho_n^{r+\alpha}(x), \qquad x \in I, \tag{1.12}$$

then $f \in W^r H^\alpha(I)$. Thus, as a result, a constructive characteristic was found for the Hölder spaces $W^r H^\alpha([a, b])$, $0 < \alpha < 1$.

The direct theorem for the class $W^r H[2; \varphi; [a, b]]$, in particular, for the Zygmund class $W^r Z[[a, b]]$, was proved by Dzyadyk [(1958), (1958b)] and Freud (1959); for the class $W^r H[k; \varphi; [a, b]]$, $k > 2$, this theorem was proved by Brudnyi [(1963), (1968)].

The inverse theorem was proved by Dzyadyk [(1958), (1958b)] for the Zygmund space $W^r Z[[a, b]]$ and by Timan (1957) ($k = 1$), Lebed' [(1957), (1975)], and Brudnyi (1959) for the spaces $W^r H_k^\varphi([a, b])$.

1.3. We formulate the indicated results on the approximation of functions by polynomials on a segment in the form of three theorems, calling them the classical theorems of approximation without restrictions. Recall that c stands everywhere for constants that may depend only on k and r.

Theorem 1.1 (direct theorem). *Suppose that $k \in \mathbb{N}$, $(r+1) \in \mathbb{N}$, and $m = k + r$. If $f \in C^r(I)$, then, for each natural $n \geq m - 1$, there exists an algebraic polynomial P_n of degree n for which*

$$|f(x) - P_n(x)| \leq c\rho_n^r(x)\omega_k(\rho_n(x); f^{(r)}; I), \qquad x \in I. \tag{1.13}$$

Theorem 1.2 (inverse theorem). *Suppose that $k \in \mathbb{N}$, $(r+1) \in \mathbb{N}$, $\varphi \in \Phi^k$, and $m = k + r$. If, for a function f defined on I, for each $n \geq m - 1$, there exists an algebraic polynomial P_n of degree n such that*

$$|f(x) - P_n(x)| \leq c\rho_n^r(x)\varphi(\rho_n(x)), \qquad x \in I, \tag{1.14}$$

then

$$\omega_k(t; f^{(r)}; I) \le c\left(\int_0^t ru^{-1}\varphi(u)du + t^k\int_t^1 u^{-k-1}\varphi(u)du\right), \qquad 0 \le t \le \frac{1}{2}. \qquad (1.15)$$

We say that a function $\varphi \in \Phi$ satisfies the Zygmund–Stechkin condition and write $\varphi \in S(k, r)$ if

$$\int_0^t ru^{-1}\varphi(u)du + t^k\int_t^1 u^{-k-1}\varphi(u)du = O(\varphi(t)). \qquad (1.16)$$

Theorem 1.3 (constructive characteristic). *Suppose that* $k \in N$, $(r+1) \in N$, $m = k + r$, *and* $\varphi \in S(k, r)$. *A function* f *belongs to* $W^r H_k^{\varphi}(I)$ *if and only if, for each* $n \ge m-1$, *there exists an algebraic polynomial* P_n *of degree* n *such that*

$$|f(x) - P_n(x)| \le M\rho_n^r(x)\varphi(\rho_n(x)), \qquad (1.17)$$

where $M =$ const *is independent of* n *and* x.

1.4. In this chapter, we use the following notation: $I := [-1; 1]$, $\omega_k(t; f) := \omega_k(t; f; I)$, $W^r H[k; \varphi] := W^r H[k; \varphi; I]$, and $W^r H_k^{\varphi} := W^r H_k^{\varphi}(I)$.

In what follows, we also write ρ instead of $\rho_n(x)$, i.e.,

$$\rho = \rho_n(x) = \frac{1}{n^2} + \frac{\sqrt{1-x^2}}{n}.$$

Note that if $x^2 \ne y^2$, then $\rho \ne \rho_n(y)$, and if $n_1 \ne n$, then $\rho \ne \rho_{n_1}(x)$. We also denote $\alpha := \arccos y$ and $\beta := \arccos x$, $x, y \in I$. In particular,

$$\rho = \frac{1}{n^2} + \frac{1}{n}\sin\beta,$$

$$\rho_n(y) = \frac{1}{n^2} + \frac{1}{n}\sin\alpha.$$

In this chapter, we often use the following obvious estimate:

$$\rho_n^2(y) < 4\rho(|x-y| + \rho), \qquad x \in I, \quad y \in I. \qquad (1.18)$$

In particular, estimate (1.18) readily yields

$$2(|x-y|+\rho) > |x-y| + \rho_n(y) > \frac{1}{2}(|x-y|+\rho), \qquad x \in I, \quad y \in I. \tag{1.19}$$

2. Inequality for the modulus of the derivative of an algebraic polynomial. Inverse theorem

2.1. The Dzyadyk inequality (1.11) is a generalization of the Markov inequality

$$\| P_n' \| \le n^2 \| P_n \| \tag{2.1}$$

and the Bernstein inequality [Bernstein (1912)]

$$\sqrt{1-x^2}\,|P_n'(x)| \le \| P_n \|, \qquad x \in I. \tag{2.2}$$

In turn, Lebed' [(1957), (1975)] and Brudnyi (1959) generalized inequality (1.11), namely, they proved that if $s \in \mathbb{R}$ and $\varphi \in \Phi^k$, then

$$\left\| \rho_n^{s+1} \varphi^{-1}(\rho_n) P_n' \right\| \le M \left\| \rho_n^s \varphi^{-1}(\rho_n) P_n \right\|, \qquad M = M(s,k) = \text{const.} \tag{2.3}$$

In particular,

$$\left\| \rho_n \varphi^{-1}(\rho_n) P_n' \right\| \le c \left\| \varphi^{-1}(\rho_n) P_n \right\|. \tag{2.4}$$

We prove inequalities (2.3) and (2.4) in Subsections 2.2–2.4. For this purpose, we use Dzyadyk's arguments based on the maximum-modulus principle.

2.2. Let

$$z = \Psi(w) := \frac{1}{2}\left(w + \frac{1}{w} \right)$$

be a Zhukovskii function that maps the exterior of the unit circle $|w| = 1$ conformally and univalently onto the exterior of I, i.e., onto the set $\mathbb{C}\backslash I$, where $\mathbb{C}$ is the complex plane. Let $w = \Phi(z)$ denote the inverse mapping. Note that $|\Phi(z)| = 1$ for $z \in I$.

The image of the circle $|w| = 1 + 1/n$ under the mapping $z = \Psi(w)$ is called the nth level line of the segment I and is denoted by Γ_n, i.e.,

$$\Gamma_n = \left\{z \in \mathbb{C} : |\Phi(z)| = 1 + \frac{1}{n}\right\}.$$

Note that the level line Γ_n is an ellipse. Let $d_n(x)$ denote the distance from a point $x \in I$ to Γ_n. Since, for $x \in I$, one has

$$\left|\Psi\left(\frac{n+1}{n}e^{it}\right) - x\right|^2 = \left(\frac{n+1/2}{n(n+1)}\right)^2 (1-x^2) + \left(\cos t - x\,\frac{n^2+n+1/2}{n^2+n}\right)^2,$$

where $t \in [0, 2\pi]$, we conclude that

$$d_n^2(x) = \min_{t\in[0,2\pi]} \left|\Psi\left(\frac{n+1}{n}e^{it}\right) - x\right|^2$$

$$= \left(\frac{n+1/2}{n(n+1)}\right)^2 (1-x^2) + \left(\max\left\{0; |x|\frac{n^2+n+1/2}{n^2+n} - 1\right\}\right)^2.$$

Therefore,

$$d_n(x) = \begin{cases} \dfrac{n+1/2}{n^2+n}\sqrt{1-x^2} & \text{if } |x| \le \dfrac{n^2+n}{n^2+n+1/2}, \\[2ex] 1-|x|+\dfrac{1}{2n^2+2n} & \text{if } \dfrac{n^2+n}{n^2+n+1/2} \le |x| \le 1, \end{cases}$$

whence

$$\frac{1}{6}\rho_n(x) \le d_n(x) \le \rho_n(x), \qquad n \ne 1. \tag{2.5}$$

Lemma 2.1. *Let $(l+1) \in N$, $n \in N$, $x_0 \in N$, $d = d_n(x_0)$, and let $\gamma = \{z : |z - x_0| = d\}$ be a circle of radius d centered at the point x_0. If*

$$|P_n(x)| \le (|x - x_0| + d)^l, \qquad x \in I, \tag{2.6}$$

then the following inequality holds for all $z \in \gamma$:

$$|P_n(z)| \le Md^l, \qquad M = M(l) = \text{const.} \tag{2.7}$$

Proof. Since $d < 1$, at least one of the two points $x_0 + d$ and $x_0 - d$ belongs to I. Assume, for definiteness, that $(x_0 + d) \in I$. The function

$$z = \Psi_1(v) = \frac{d}{4}\left(v + \frac{1}{v}\right) + x_0 + \frac{d}{2}$$

maps the exterior of the circle $|v| = 1$ conformally and univalently onto the exterior of the segment $[x_0, x_0 + d]$. Let $v = \Phi_1(z)$ denote the mapping inverse to $\Psi(v)$. Note that, for all $z \in \mathbb{C}$, one has

$$\frac{d}{4}|\Phi_1(z)| \le |z - x_0| + d \le 2d|\Phi_1(z)|. \tag{2.8}$$

Indeed,

$$\frac{d}{4}|\Phi_1(z)| \equiv \frac{d}{4}|v| \le \frac{d}{4}\left|v + \frac{1}{v} + 2\right| + \frac{3d}{4}$$

$$= |\Psi_1(v) - x_0| + \frac{3d}{4} = |z - x_0| + \frac{3d}{4}$$

$$< |z - x_0| + d = |\Psi_1(v) - x_0| + d = \frac{d}{4}\left|v + \frac{1}{v} + 2\right| + d$$

$$\le \frac{d}{4}|v| + \frac{3d}{4} + d \le 2d|v| \equiv 2d|\Phi_1(z)|.$$

We introduce the function $F(z) := \Phi^{l-n-1}(z)\Phi_1^{-l}(z)P_n(z)$ analytic in $\mathbb{C}\backslash I$. Since

$$\lim_{z\to\infty} \frac{\Phi(z)}{z} = \lim_{w\to\infty} \frac{w}{\Psi(w)} = 2, \qquad \lim_{z\to\infty} \frac{\Phi_1(z)}{z} = \lim_{v\to\infty} \frac{v}{\Psi_1(v)} = \frac{4}{d},$$

we have $F(\infty) = 0$. By virtue of the maximum-modulus principle, taking (2.8) into account, we obtain the following relation for all $z \in \mathbb{C}$:

$$|F(z)| \le \max_{x\in I} |F(x)| = \max_{x\in I} \frac{|P_n(x)|}{|\Phi_1(x)|^l} \le \max_{x\in I} \frac{(|x - x_0| + d)^l}{|\Phi_1(x)|^l} \le (2d)^l.$$

In particular, $|F(z)| \le (2d)^l$ for $z \in \gamma$. Therefore, taking (2.8) into account, we obtain the following relation for $z \in \gamma$:

$$|P_n(z)| \le (2d)^l |\Phi(z)|^{n+1-l} |\Phi_1(z)|^l \le (2d)^l |\Phi(z)|^{n+1-l} 8^l.$$

It remains to note that γ lies inside the level line Γ_n, and, hence, the preimage of the circle γ lies in the ring $1 \le |w| = 1 + 1/n$, so that $1 \le |\Phi(z)| = 1 + 1/n$, i.e.,

$$|\Phi(z)|^{n+1-l} \le \left(1 + \frac{1}{n}\right)^{n+1} < 4.$$

■

Corollary 2.1. *Under the conditions of Lemma 2.1, the following inequality is true:*

$$\left|P_n^{(j)}(x_0)\right| \le j! M d^{l-j}. \tag{2.9}$$

Indeed, using the Cauchy integral formula and relation (2.7), we get

$$\left|P_n^{(j)}(x_0)\right| = \frac{j!}{2\pi}\left|\int_\gamma \frac{P_n(z)}{(z-x_0)^{j+1}}\,dz\right| \le \frac{j!}{2\pi}\frac{Md^l}{d^{j+1}}\operatorname{mes}\gamma = j! M d^{l-j}.$$

Lemma 2.2. *Suppose that* $(l+1) \in \mathbb{N}$, $n \in \mathbb{N}$, *and* $x_0 \in I$. *If*

$$|P_n(x)| \le (|x - x_0| + \rho_n(x))^l, \qquad x \in I, \tag{2.10}$$

then

$$|P_n'(x_0)| \le M\rho_n^{l-1}(x_0), \qquad M = M(l) = \text{const}. \tag{2.11}$$

Lemma 2.2 is a corollary of inequalities (2.5) and (2.9).

2.3. We prove inequality (2.4). Let $|P_n(x)| \le \varphi(\rho)$, $x \in I$. We take a point $x_0 \in I$. By virtue of (1.9), we get

$$\begin{aligned}\varphi(\rho) &\le \varphi(2|x-x_0| + 2\rho_n(x_0)) \\ &\le 2^k(|x-x_0| + \rho_n(x_0))^k \rho_n^{-k}(x_0)\,\varphi(\rho_n(x_0)) \\ &=: A_0(|x-x_0| + \rho_n(x_0))^k.\end{aligned}$$

Therefore, by virtue of (2.11), we obtain

$$|P_n'(x_0)| \le A_0 M(k)\rho_n^{k-1}(x_0) = M(k)\rho_n^{-1}(x_0)\varphi(\rho_n(x_0)),$$

which proves inequality (2.4).

2.4. We prove inequality (2.3). For $s \le 0$, inequality (2.3) is a trivial corollary of (2.4) because $t^{-s}\varphi(t) \in \Phi^{k*}$, where $k^* = k - [s]$ and $[s]$ is the integer part of s.

Let us prove inequality (2.3) for $s > 0$. Denote $s^* = [s] + 1$. Let $P_n(x) \le \rho^{-s}\varphi(\rho)$, $x \in I$. We take a point $x_0 \in I$. By virtue of (1.19), we get

$$\varphi(\rho) \le 2^k(|x - x_0| + \rho_n(x_0))^k \rho_n^{-k}(x_0)\,\varphi(\rho_n(x_0));$$

according to (1.18) and (1.19), we have

$$\rho \ge (4(|x - x_0| + \rho))^{-1}\rho_n^2(x_0) \ge (8(|x - x_0| + \rho_n(x_0))^{-1}\rho_n^2(x_0)).$$

Hence,

$$\begin{aligned}|P_n(x)| &\le (16)^k\rho_n^{-s}(x_0)\rho_n^{-k-s}(x_0)(|x - x_0| + \rho_n(x_0))^{k+s}\varphi(\rho)\\ &\le (16)^k\rho_n^{-s}(x_0)\rho_n^{-k-s^*}(x_0)(|x - x_0| + \rho_n(x_0))^{k+s^*}\varphi(\rho_n(x_0))\\ &=: B_0(|x - x_0| + \rho_n(x_0))^{k+s^*}.\end{aligned}$$

Therefore, by virtue of (2.11), we obtain

$$|P_n'(x_0)| \le B_0 M(s^*)\rho_n^{k+s^*-1}(x_0) = (16)^k M(s^*)\rho_n^{-s-1}(x_0)\varphi(\rho_n(x_0)).$$

■

2.5. Let us prove Theorem 1.2 for $r = 0$.

Denote $\bar{\rho}_n(x) := \rho_{2^n}(x)$. Note that $\bar{\rho}_{n+1}(x) \le \bar{\rho}_n(x) \le 4\bar{\rho}_{n+1}(x)$, $x \in I$. Assume, for simplicity, that $k \ne 1$. We choose an integer n_0 from the condition $2^{n_0} \le k - 1 < 2^{n_0+1}$. We fix a point $x_0 \in I$ and a number $h \in (0, k^{-3}]$ so that $(x_0 + kh) \in I$ and choose an integer n_* from the condition $\bar{\rho}_{n_*+1}(x_0) \le kh \le \bar{\rho}_{n_*}(x_0)$. For $n \le n_* + 1$ and $x \in [x_0, x_0 + kh]$, using (1.18), we obtain the estimates $c_1\rho_n(x_0) \le \bar{\rho}_n(x) \le c_2\rho_n(x_0)$. Note that $n_* \ge n_0$ because $kh \le k^{-2}$. We set

$$Q_{n_0+1}(x) := P_{2^{n_0+1}}(x) - P_{k-1}(x)$$

and

$$Q_n(x) := P_{2^n}(x) - P_{2^{n-1}}(x)$$

for $n > n_0 + 1$. We expand the function f into the Bernstein telescopic sum, i.e., we represent it in the form

$$f(x) = f(x) - P_{2^{n_*+1}}(x) + \sum_{n=n_0+1}^{n_*+1} Q_n(x) + P_{k-1}(x).$$

Since

$$\left|\Delta_h^k\left(f - P_{2^{n_*+1}}; x_0\right)\right| \le 2^k \left\|f - P_{2^{n_*+1}}\right\|_{[x_0, x_0+kh]}$$

$$\le 2^k \varphi(c_0 \overline{\rho}_{n_*+1}(x_0)) \le c_3 \varphi(h), \qquad \Delta_h^k(P_{k-1}; x_0) = 0,$$

the problem reduces to the estimation of the kth difference $\Delta_h^k(Q_n; x_0)$. It follows from condition (1.14) of Theorem 1.2 that, for $n > n_0 + 1$, one has

$$|Q_n(x)| \le \left|P_{2^n}(x) - f(x)\right| + \left|f(x) - P_{2^{n-1}}(x)\right|$$

$$\le \varphi(\overline{\rho}_n(x)) + \varphi(\overline{\rho}_{n-1}(x)) \le 2\varphi(\overline{\rho}_{n-1}(x)), \qquad x \in I,$$

$$\left|Q_{n_0+1}(x)\right| \le 2\varphi(\rho_{k-1}(x)), \qquad x \in I,$$

whence

$$|Q_n(x)| \le 2\varphi(4\overline{\rho}_n(x)) \le 2 \cdot 4^k \varphi(\overline{\rho}_n(x)), \qquad x \in I, \quad n \ge n_0 + 1.$$

Using inequality (2.4), we get

$$\left|Q_n^{(k)}(x)\right| \le c_4 \overline{\rho}_n^{\,-k}(x) \varphi(\overline{\rho}_n(x)), \qquad x \in I, \quad n \ge n_0 + 1.$$

Therefore, for all $x \in [x_0, x_0 + kh]$ and $n = n_0 + 1, \dots, n_* + 1$, we have

$$
\begin{aligned}
\left|Q_n^{(k)}(x)\right| &\le c_4\overline{\rho}_n^{\,-k}(x)\varphi(\overline{\rho}_n(x)) \le c_4\overline{\rho}_n^{\,-k}(x)\varphi(c_2\overline{\rho}_n(x)) \\
&\le c_5\overline{\rho}_n^{\,-k}(x_0)\,\varphi(\rho_n(x_0)) \\
&= c_5 \int\limits_{\overline{\rho}_n(x_0)}^{\overline{\rho}_{n-1}(x_0)} \overline{\rho}_n^{\,-k}(x_0)\,\varphi(\overline{\rho}_n(x_0))(\overline{\rho}_{n-1}(x_0)-\rho_n(x))^{-1}du \\
&\le c_5 \int\limits_{\overline{\rho}_n(x_0)}^{\overline{\rho}_{n-1}(x_0)} 4^k\overline{\rho}_{n-1}^{\,-k}(x_0)\,\varphi(u)2\overline{\rho}_{n-1}^{\,-k}(x_0)du \\
&\le c_6 \int\limits_{\overline{\rho}_n(x_0)}^{\overline{\rho}_{n-1}(x_0)} u^{-k-1}\varphi(u)du\,.
\end{aligned}
$$

Hence,

$$
\begin{aligned}
\left|\Delta_h^k(f;x_0)\right| &\le c_3\varphi(h) + c_6h^k \sum_{n=n_0+1}^{n_*+1} \int\limits_{\overline{\rho}_n(x_0)}^{\overline{\rho}_{n-1}(x_0)} u^{-k-1}\varphi(u)\,du \\
&= c_3\varphi(h) + c_6h^k \int\limits_{\overline{\rho}_{n_*+1}(x_0)}^{\overline{\rho}_{n_0}(x_0)} u^{-k-1}\varphi(u)\,du \\
&\le c_3\varphi(h) + c_6h^k \int\limits_{kh/4}^{2} u^{-k-1}\varphi(u)\,du \le c_7h^k \int\limits_h^1 u^{-k-1}\varphi(u)\,du\,.
\end{aligned}
$$

Thus, Theorem 1.2 is proved for $r = 0$.

2.6. Let us prove Theorem 1.2 for $r \neq 0$.

Denote $\overline{\varphi}(t) := t^r\varphi(t)$. According to the result proved above, we have

$$
\omega_m(t;f) \le c_1t^m \int\limits_t^1 u^{-m-1}\overline{\varphi}(u)\,du\,.
$$

Therefore, by virtue of inequality (3.5.16), we get

$$\omega_k(t; f^{(r)}) \le c_2 \int_0^t u^{-r-1}\omega_m(u, f)du \le c_1c_2 \int_0^t u^{k-1}\left(\int_u^1 v_1^{-m-1}\overline{\varphi}(v)dv\right)du$$

$$= k^{-1}c_1c_2\left(\int_0^t u^{-1}\varphi(u)du + t^k\int_t^1 u^{-k-1}\varphi(u)du\right).$$

2.7. Lemma 2.3. *Suppose that* $k \in N$, $(r+1) \in N$, $m = k + r$, $h > 0$, $(x_0 + kh) \in I$, $n_0 \in N$, $(n_0+1)^{-2} < h \le n_0^{-2}$, *and* $\overline{\varepsilon} = \{\varepsilon_n\}$ *is a nonincreasing sequence of positive numbers* ε_n, $(n+1) \in \mathbb{N}$. *If, for all* n, *one has*

$$E_n(f) \le \varepsilon_n, \tag{2.12}$$

then

$$\left|\Delta_h^k(f^{(r)}; x_0)\right| \le c\sum_{j=n_0+1}^{\infty} rj^{2r-1}\varepsilon_j + cn_0^{-2k}\sum_{j=1}^{n_0} j^{2(r+k)-1}\varepsilon_j. \tag{2.13}$$

Let us prove Lemma 2.3 for $r \ne 0$. Denote $\overline{\varepsilon}_n := \overline{\varepsilon}_{2^n}$. Let $P_{2^n} = P_{2^n}(x)$ be a polynomial of degree 2^n for which $\left\|f - P_{2^n}\right\| \le 2\overline{\varepsilon}_n$ and let $Q_n = Q_n(x) := P_{2^n}(x) - P_{2^{n-1}}(x)$. We choose an integer n_* from the condition $2^{n_*} < n_0 \le 2^{n_*+1}$ and represent f in the form

$$f(x) = P_1(x) + \sum_{n=1}^{n_*} Q_n(x) + \sum_{n=n_*+1}^{\infty} Q_n(x) := i_1(x) + i_2(x) + i_3(x).$$

Taking into account that $\left\|Q_{n^n}\right\| \le 4\overline{\varepsilon}_{n-1}$ and using the Markov inequality (2.1), we establish that

$$\left\|Q_h^{(r)}\right\| \le 4\cdot 2^{2nr}\overline{\varepsilon}_{n-1} \quad \text{and} \quad \left\|Q_h^{(m)}\right\| \le 4\cdot 2^{2nm}\overline{\varepsilon}_{n-1},$$

whence

$$\left|\Delta_h^k(i_2^{(r)}; x_0)\right| \le h^k\left\|i_2^{(m)}\right\| \le 4h^k\sum_{n=1}^{n_*} 4^{nm}\overline{\varepsilon}_{n-1},$$

$$\left|\Delta_h^k(i_3^{(r)}; x_0)\right| \le 2^k\left\|i_3^{(m)}\right\| \le 4\cdot 2^k\sum_{n=n_*+1}^{\infty} 4^{rn}\overline{\varepsilon}_{n-1}.$$

Taking into account that

$$4^{nl} \le \sum_{j=2n+1}^{2^{n+1}} j^{2l-1}, \qquad l \in \mathbb{N},$$

we get

$$\begin{aligned}
\Delta_h^k(f^{(r)}; x_0) &\le 4h^k \sum_{n=1}^{n_*} 4^{nm}\bar{\varepsilon}_{n-1} + 4\cdot 2^k \sum_{n=n_*+1}^{\infty} 4^{rn}\bar{\varepsilon}_{n-1} \\
&< 4h^k 4^m \bar{\varepsilon}_0 + 4^{2m+1}h^k \sum_{n=2}^{n_*} \sum_{j=2^{n-2}+1}^{2^{n-1}} j^{2m-1}\bar{\varepsilon}_{n-1} \\
&\quad + 4^{2r-1}2^k \sum_{n=n_*+1}^{\infty} \sum_{j=2^{n-2}+1}^{2^{n-1}} j^{2r-1}\bar{\varepsilon}_{n-1} \\
&\le 4^{m+1}h^k\varepsilon_1 + 4^{2m+1}h^k \sum_{n=2}^{n_*} \sum_{j=2^{n-2}+1}^{2^{n-1}} j^{2m-1}\varepsilon_j \\
&\quad + 4^{2r-1}2^k \sum_{n=n_*+1}^{\infty} \sum_{j=2^{n-2}+1}^{2^{n-1}} j^{2r-1}\varepsilon_j \\
&= 4^{m+1}h^k\varepsilon_1 + 4^{2m+1}h^k \sum_{j=1}^{2^{n_*-1}} j^{2m-1}\varepsilon_j + 4^{2r-1}2^k \sum_{j=2^{n_*}}^{\infty} j^{2r-1}\varepsilon_j \\
&\le c \sum_{j=n_0+1}^{\infty} rj^{2r-1}\varepsilon_j + cn_0^{-2k} \sum_{j=1}^{n_0} j^{2m-1}\varepsilon_j .
\end{aligned}$$

Thus, Lemma 2.3 is proved for $r \neq 0$.

For $r = 0$, estimate (2.13) is proved in the same way as Theorem 1.2 for $r = 0$; moreover, the proof becomes even simpler because, by analogy with the arguments presented above, one should use the Markov inequality (2.1) instead of inequality (2.3).

■

Note that, in the case where $r = 0$ and $k = 1$, one must add the term $cn_0^{-2k}\varepsilon_0$ to (2.13).

Lemma 2.3 means that relation (3.7.24) yields (3.7.25).

Lemma 2.3 also yields the inverse theorem for DT_r-moduli of continuity (see [Ditzian and Totik (1987)] for $r = 0$).

Theorem 2.1. *Let* $k \in \mathbb{N}$, $(r+1) \in \mathbb{N}$, $\varphi \in \Phi$, *and* $f \in C(I)$. *If*

$$E_n(f) \le \frac{1}{n^r}\varphi\left(\frac{1}{n}\right), \qquad n \ge k+r-1,$$

then

$$\overline{\omega}_{k,r}(t; f^{(r)}; I) \le c\int_0^t \frac{r\varphi(u)}{u}\,du + ct^k\int_t^1 \frac{\varphi(u)}{u^{k+1}}\,du, \qquad 0 \le t \le \frac{1}{2}. \tag{2.14}$$

3. Polynomial kernels. Direct theorem

3.1. In this section, we study Dzyadyk-type polynomial kernels (blendings) of the form

$$D(y, x) = \sum_{\nu=0}^{n} \alpha_\nu(y)x^\nu.$$

It turns out that the algebraic polynomials

$$P_n(x) = \int_{-1}^{1} f(y)D(y, x)dy$$

are a fairly universal tool for the approximation of functions $f \in C(I)$, $I := [-1; 1]$.

Recall (see Section 1) that we write ρ instead of $\rho_n(x)$, $\alpha := \arccos y$, and $\beta := \arccos x$, $x, y \in I$. By C (in contrast to c) we denote positive numbers whose values may depend only on fixed natural numbers l and r and on fixed nonnegative integer numbers p, q, and s.

3.2. Recall some information on trigonometric kernels. The trigonometric polynomial

$$F_n(t) = \frac{\sin^2 nt/2}{2n\sin^2 t/2} = \frac{1}{2} + \sum_{\nu=1}^{n-1}\left(1 - \frac{\nu}{n}\right)\cos \nu t \tag{3.1}$$

is called a Fejér kernel.

The function

$$J_{n,r}(t) = \frac{1}{\gamma_{n,r}} \left(\frac{\sin nt/2}{\sin t/2} \right)^{2(r+1)}, \tag{3.2}$$

where

$$\gamma_{n,r} = \int_{-\pi}^{\pi} \left(\frac{\sin nt/2}{\sin t/2} \right)^{2(r+1)} dt,$$

is called a Jackson-type kernel.

The following estimates are true:

$$c^{-1} n^{2r+1} < \gamma_{n,r} < c n^{2r+1}, \tag{3.3}$$

$$\int_0^{\pi} J_{n,r}(t) t^j \, dt \le c n^{-j}, \quad j = 0, \dots, 2r \tag{3.4}$$

It follows from relation (3.1) that $J_{n,r}(t)$ is a trigonometric polynomial of degree $(r+1)(n-1)$, i.e.,

$$J_{n,r}(t) = \frac{1}{2\pi} + \frac{1}{\pi} \sum_{\nu=1}^{(r+1)(n-1)} j_{\nu,n,r} \cos \nu t, \tag{3.5}$$

where

$$j_{\nu,n,r} = \frac{1}{\pi} \int_{-\pi}^{\pi} j_{\nu,n,r} \cos^2 \nu t \, dt = \int_{-\pi}^{\pi} J_{n,r}(t) \cos \nu t \, dt.$$

3.3. Definition 3.1. *The function*

$$D_{l,n,r}(y, x) = \frac{1}{(l-1)!} \frac{\partial^l}{\partial x^l} (x-y)^{l-1} \int_{\beta-\alpha}^{\beta+\alpha} J_{n,r}(t) dt, \tag{3.6}$$

where $l, n, r \in \mathbb{N}$, $x, y \in I$, $\alpha := \arccos y$, *and* $\beta := \arccos x$, *is called a Dzyadyk-type polynomial kernel.*

It follows from representation (3.5) that the kernel $D_{l,n,r}(y, x)$ is a polynomial of degree $(r+1)(n-1)-1$ in the variable x, and $D_{l,1,r}(y, x) \equiv 0$.

3.4. Lemma 3.1. *The following inequality is true:*

$$\left|\frac{\partial^p}{\partial x^p}\left[\frac{1}{\sin\beta}(J_{n,r}(\alpha+\beta)-J_{n,r}(\alpha-\beta))\right]\right| \le \frac{C\rho_n^{r-p-1}(x)}{(|x-y|+\rho_n(x))^r}. \tag{3.7}$$

We divide the proof of the lemma into three parts.

1. We show that the Fejér kernel (3.1) satisfies the estimate

$$|F_n(\alpha+\beta)-F_n(\alpha-\beta)| \le 2\rho^{-1}\sin\beta. \tag{3.8}$$

Indeed, taking (3.3) into account, for $\sin\beta < 1/n$ we obtain

$$|F_n(\alpha+\beta)-F_n(\alpha-\beta)| = 2\left|\sum_{\nu=1}^{n-1}\left(1-\frac{\nu}{n}\right)\sin\nu\alpha\sin\nu\beta\right|$$

$$\le 2\sum_{\nu=1}^{n-1}\left(1-\frac{\nu}{n}\right)\nu\sin\nu\beta < n^2\sin\beta < 2\rho^{-1}\sin\beta.$$

For $\sin\beta \ge 1/n$, we have $|F_n(\alpha+\beta)-F_n(\alpha-\beta)| \le n \le 2\rho^{-1}\sin\beta$. Estimate (3.8) is proved.

2. Let us prove inequality (3.7) for $p=0$. To this end, we represent the left-hand side of (3.7) in the form

$$\frac{1}{\sin\beta}(J_{n,r}(\alpha+\beta)-J_{n,r}(\alpha-\beta)) = \frac{1}{\sin\beta}\gamma_{n,r}^{-1}2^{r+1}n^{2r+1}(F_n(\alpha+\beta)-F_n(\alpha-\beta))A,$$

where

$$A := n^{-r}\sum_{\nu=0}^{r}F_n^{\nu}(\alpha+\beta)F_n^{r-\nu}(\alpha-\beta). \tag{3.9}$$

It follows from (3.2) and (3.8) that

$$\frac{1}{\sin\beta}\gamma_{n,r}^{-1}2^{r+1}n^{2r+1}|F_n(\alpha+\beta)-F_n(\alpha-\beta)| \le c\rho^{-1}.$$

Therefore, it remains to estimate A. If $|x-y| \leq \rho$, then

$$A \leq n^{-r}(r+1)(2n)^{-r}n^{2r} \leq (r+1)2^{-r} \leq (r+1)\rho^r(|x-y|+\rho)^{-r}.$$

If $|x-y| > \rho$, then, bringing the sum in (3.9) to the common denominator $2^{-r}n^r(x-y)^{2r}$ and using the inequalities

$$\sin^2\left(\frac{\alpha-\beta}{2}\right) \leq \sin^2\left(\frac{\alpha+\beta}{2}\right) \leq 2\sin\left(\frac{\alpha+\beta}{2}\right)\sin\left(\frac{\alpha-\beta}{2}\right) \equiv |x-y| + 1 - x^2,$$

we get

$$A \leq (r+1)2^r\left(\sin\left(\frac{\alpha-\beta}{2}\right)\right)^{2r}(n(x-y))^{-2r}$$

$$\leq (r+1)2^r(|x-y|+1-x^2)^r(n^2(x-y)^2)^{-r} \leq (r+1)2^r|x-y|^{-r}\rho^r.$$

The lemma is proved for $p = 0$.

3. Let $p \in N$. Denote

$$D(y,x) := D_{1,n,r}(y,x) = \frac{1}{\sin\beta}(J_{n,r}(\alpha-\beta) - J_{n,r}(\alpha+\beta)),$$

$$D^{(p)}(y,x) := \frac{\partial^p}{\partial x^p}D(y,x).$$

By induction, assume that (3.7) is true for the number $p-1$, which is equivalent to the system of two inequalities

$$\left|D^{(p-1)}(y,x)\right| \leq C\rho^{-p}, \quad \left|(x-y)^r D^{(p-1)}(y,x)\right| \leq C\rho^{r-p}.$$

By virtue of (1.11), we obtain the following inequalities:

(i) $$\left|D^{(p1)}(y,x)\right| \leq C\rho^{-p-1},$$

(ii) $$\left|\frac{\partial}{\partial x}((x-y)^r D^{(p-1)}(y,x))\right| \leq C\rho^{r-p-1}.$$

Using inequality (ii) and the induction hypothesis, we get

$$\left|(x-y)^r D^{(p)}(y,x)\right| = \left|\frac{\partial}{\partial x}((x-y)^r D^{(p-1)}(y,x)) - r(x-y)^{r-1}D^{(p-1)}(y,x)\right|$$

$$\leq C\rho^{r-p-1} + C|x-y|^{-1}\rho^{r-p} \leq C\rho^{r-p-1}(1+|x-y|^{-1}\rho).$$

Inequality (3.7) follows from the estimate obtained and estimate (i).

■

3.5. Denote

$$\varphi_{l,n,r}(x,y) := (x-y)^{l-1}\int_{\beta-\alpha}^{\beta+\alpha} J_{n,r}(t)\,dt. \tag{3.10}$$

We define numbers $a_{l,p,q}$ as follows: if $p > l+q$, then $a_{l,p,q} = 0$; if $p = l+q$, then $a_{l,p,q} = (-1)^q(l-1)!q!$; if $l \leq p < l+q$, then $a_{l,p,q} = 0$; and if $p < l$, then

$$a_{l,p,q} = \binom{l-1}{p}\frac{p!}{l+q-p}.$$

We also introduce the following notation: $\bar{l} = l$ if l is even and $\bar{l} = l+1$ if l is odd.

Lemma 3.2. *The function*

$$A_{l,n,p,q,r}(x) := \int_{-1}^{1}(x-y)^q\frac{\partial^p}{\partial x^p}\varphi_{l,n,r}(x,y)\,dy - a_{l,p,q}(x+1)^{l+q-p} \tag{3.11}$$

is an algebraic polynomial of degree $\leq l+q-p$ *(if* $l+q-p<0$*, then* $A_{l,n,p,q,r} \equiv 0$*). Furthermore,*

$$\left\|A_{l,n,p,q,r}\right\| \leq Cn^{-l^*}, \tag{3.12}$$

where $l^* = \min\{2r+1, l^{**}\}$ *and* $l^{**} = \max\{\bar{l}, \overline{l+q-p}\}$.

Proof. The lemma can easily be proved by induction on q with the use of the recurrence relation

$$A_{l,n,p,q+1,r}(x) = A_{l+1,n,p,q,r}(x) - pA_{l,n,p-1,q,r}(x).$$

Therefore, only the cases $p = 0$ and $q = 0$ must be verified.

1. Let $p = 0$. Integrating (3.11) by parts, we obtain

$$A_{l+q}(x) := A_{l,n,0,q,r}(x)$$

$$= -(l+q)^{-1} \int_0^{\pi} (\cos\beta - \cos\alpha)^{l+q} (J_{n,r}(\alpha+\beta) - J_{n,r}(\alpha-\beta))\, d\alpha .$$

Taking into account the orthogonality of the system $\{\cos \nu\alpha\}$ on $[0, \pi]$ and the identity

$$J_{n,r}(\alpha+\beta) - J_{n,r}(\alpha-\beta) = \frac{2}{\pi} \sum_{\nu=1}^{(r+1)(n-1)} j_{\nu,n,r} \cos\nu\beta \cos\nu\alpha + \frac{1}{\pi},$$

which follows from (3.5), we conclude that $A_{l+q}(x)$ is an algebraic polynomial of degree $\le l+q$. Furthermore, representing $A_{l+q}(x)$ in the form

$$A_{l+q}(x) = -2^{l+q}(l+q)^{-1} \int_{-\pi}^{\pi} \left(\sin\frac{t}{2}\right)^{l+q} \left(\sin\left(\beta + \frac{t}{2}\right)\right)^{l+q} J_{n,r}\, dt$$

and using (3.4), we get

$$2^{-l-q}(l+q)\| A_{l+q} \| \le \int_{-\pi}^{\pi} \left(\sin\frac{t}{2}\right)^{\overline{l+q}} J_{n,r}\, dt < Cn^{-\overline{(l+q)}}$$

for $\overline{l+q} < 2r+1$ and

$$\| A_{l+q} \|_I \le 2^{l+q+1}\pi(l+q)^{-1}\gamma_{n,r}^{-1} < Cn^{-2r-1}$$

for $\overline{l+q} > 2r+1$.

2. By virtue of the identity $A_{l,n,p,0,r}(x) = A_l^{(p)}(x)$, the case $q = 0$ follows from the case $p = 0$ considered above.

■

3.6. Relation (3.5), Lemmas 3.1 and 3.2, and the representation

$$\frac{\partial^p}{\partial x^p} D_{l,n,r}(y,x) = \sum_{\nu=1}^{l}(l-j)!\binom{l+p}{\nu-1}(x-y)^{l-\nu}\frac{\partial^{p+l-\nu}}{\partial x^{p+l-\nu}} D_{1,n,r}(y,x)$$

yield the following statement:

Theorem 3.1 [Shevchuk (1989), (1992)]. *The polynomial kernel*

$$D_{l,n,r}(y,x) = \frac{1}{(l-1)!}\frac{\partial^l}{\partial x^l}(x-y)^{l-1}\int_{\beta-\alpha}^{\beta+\alpha} J_{n,r}(t)\,dt$$

is an algebraic polynomial of degree $(r+1)(n-1)-1$ in the variable x. Furthermore, this kernel satisfies the inequalities

$$\left|\frac{\partial^p}{\partial x^p} D_{l,n,r}(y,x)\right| \le C\rho_n^{r-p-1}(x)(|x-y|+\rho_n(x))^{-r}, \tag{3.13}$$

$$\left|\frac{1}{p!}\int_{-1}^{1}(y-x)^q\frac{\partial^p}{\partial x^p} D_{l,n,r}(y,x)\,dy - \delta_{q,p}\right| \le Cn^{-\min\{2r+1,l\}}, \tag{3.14}$$

where $\delta_{q,p}$ is the Kronecker symbol. Moreover, the integral in (3.14) is an algebraic polynomial of degree $\le q-p$ (for $q-p<0$, it is identically equal to zero).

3.7. The idea of the representation of the Dzyadyk kernel in the form (3.6) is based on the application of the function

$$\varphi_{n,r}(x,y) := \varphi_{1,n,r}(x,y) = \int_{\beta-\alpha}^{\beta+\alpha} J_{n,r}(t)\,dt$$

used by DeVore and Yu (1985). We set $(x-y)_+^0 := 1$ if $x \ge y$ and $(x-y)_+^0 := 0$ if $x<y$.

Theorem 3.2 [DeVore and Yu (1985)]. *For all $x \in I$ and $y \in I$, one has*

$$\left|\varphi_{n,r}(x,y) - (x-y)_+^0\right| \le cn^{-2r}$$

if $|x-y| > 1$, and

$$\left|\varphi_{n,r}(x,y)-(x-y)_+^0\right| \le c\big(\rho_n(x)+\rho_n(y)\big)^{2r}\big(|x-y|+\rho_n(x)+\rho_n(y)\big)^{-2r}$$

if $|x-y| \le 1.$

Proof. Denote $A := \left|\varphi_{n,r}(x,y)-(x-y)_+^0\right|$, $\varphi(x,y) := \varphi_{n,r}(x,y)$, and $J(t) := J_{n,r}(t)$. We take into account the estimate $|\beta-\alpha| \le |\beta+\alpha| \le 2\pi - |\beta-\alpha|$ and the evenness and periodicity of the Jackson-type kernel $J(t)$. First, let $x<y$, i.e., $\beta>\alpha$. Then

$$A = \varphi(x,y) = \int_{\beta-\alpha}^{\beta+\alpha} J(t)\,dt \le \int_{\beta-\alpha}^{\pi} J(t)\,dt.$$

Let $x>y$, i.e., $\beta<\alpha$. Then

$$A = 1-\varphi(x,y) = \int_{-\pi}^{\pi} J(t)\,dt - \int_{\beta-\alpha}^{\beta+\alpha} J(t)\,dt$$

$$= \int_{-\pi}^{\beta-\alpha} J(t)\,dt - \int_{\beta+\alpha}^{\pi} J(t)\,dt \le 2\int_{\alpha-\beta}^{\pi} J(t)\,dt,$$

i.e., for all $x \ne y$, we have

$$A \le 2\int_{|\alpha-\beta|}^{\pi} J(t)\,dt \le 2\int_{|\alpha-\beta|}^{\pi}\left(\frac{t}{|\alpha-\beta|}\right)^{2r} J(t)\,dt.$$

Therefore, by virtue of (3.4), we get

$$A \le 2|\alpha-\beta|^{-2r}\int_0^{\pi} t^2 J(t)\,dt \le c_1(n|\alpha-\beta|)^{-2r}.$$

If $|x-y|>1$, then $|\beta-\alpha|>1$ and, consequently, $A<c_1 n^{-2r}$. If $1 \ge |x-y| \ge (\sin\alpha+\sin\beta)/n$, then $1 \le n|\tan(\alpha-\beta)/2| < n|\alpha-\beta|$, whence

$$\frac{1}{2n|\alpha-\beta|} \le \frac{1}{n|\tan((\alpha-\beta)/2)|+1}$$

$$= \frac{(\sin\alpha+\sin\beta)/n}{|x-y|+(\sin\alpha+\sin\beta)/n} < \frac{\rho_n(y)+\rho}{|x-y|+\rho_n(y)+\rho}.$$

Finally, if $|x-y| < (\sin\alpha + \sin\beta)/n$, then

$$A \le 1 < 2^{2r}(\rho_n(y)+\rho)^{2r}(|x-y|+\rho_n(y)+\rho)^{-2r}.$$

Theorem 3.2 and inequalities (1.18) and (1.19) yield the following statement:

Corollary 3.1. *The following estimates are true:*

$$\left|\varphi_{n,r}(x,y) - (x-y)_+^0\right| \le c\rho_n^r(y)(|x-y|+\rho_n(y))^{-r},$$
$$\left|(x-y)_+^{l-1} - \varphi_{l,n,r}(x,y)\right| \le C|x-y|^{l-1}(|x-y|+\rho)^{-r}\rho^r. \tag{3.15}$$

The last estimate immediately yields the direct theorem (Theorem 1.1) for $f \in W^r[I]$. Indeed, denoting

$$P_n(x) = \sum_{\nu=0}^{r-1}(\nu!)^{-1}f^{(\nu)}(-1)(x+1)^\nu + ((r-1)!)^{-1}\int_{-1}^{1} f^{(r)}(y)\varphi_{r,n,r+1}(x,y)\,dy,$$

we obtain

$$|f(x)-P_n(x)| = ((r-1)!)^{-1}\left|\int_{-1}^{1} f^{(r)}(y)[(x-y)_+^{r-1} - \varphi_{r,n,r+1}(x,y)]dy\right|$$
$$\le c((r-1)!)^{-1}\int_{-1}^{1}|x-y|^{r-1}(|x-y|+\rho)^{-r-1}\rho^{r+1}dy \le c\rho^r.$$

For the approximation of a continuous (not necessarily differentiable) function, it is convenient to use the kernels $D_{l,n,r}$.

3.8. Prior to the proof of the direct theorem (Theorem 1.1), we prove Lemma 3.3.

For the segment $J := [a,b]$, function $f \in C(J)$, and number $m \in N$, we denote by $L_{m-1}(x;f;J)$ the Lagrange polynomial that interpolates the function at m equidistant points of J, including its endpoints, i.e., $L_0(x;f;J) = f(a)$ and $L_{m-1}(x;f;J) = L(x;f;a,a+(b-a)/(m-1),\dots,b)$ for $m \ne 1$.

Lemma 3.3. *Suppose that* $l \in \mathbb{N}$, $m \in \mathbb{N}$, $l > 2m$, $n \in \mathbb{N}$, $n \ne 1$, $(p+1) \in N$, $\varphi \in \Phi^m$, $x_0 \in I$, $\rho_0 := \rho_n(x_0)$, $J := [x_0 - \rho_0, x_0 + \rho_0] \cap I$, $\delta > 0$, $f \in C(I)$, *and*

$$\mathcal{D}_{l,m,n}(x;f) := L_{m-1}(x;f;I) + \int_{-1}^{1}(f(y) - L_{m-1}(y;f;I))D_{2l+1,n,l}(y,x)\,dy$$

is a polynomial of degree $<(l+1)(n-1)$. *If* $f \in H[m,\varphi]$, *then*

$$\left|\mathcal{D}^{(p)}_{l,m,n}(x_0;f) - L^{(p)}_{m-1}(x_0;f;J)\right|$$

$$\le C\rho_0^{-p}\left(\omega_m(\rho_0;f;[x_0-\delta,x_0+\delta]\cap I) + \left(\frac{\rho_0}{\delta}\right)^{l-2m}\varphi(\rho_0)\right). \tag{3.16}$$

Proof. Without loss of generality, we assume that $\rho_0 \le \delta \le 2$. Denote

$$D^{(p)}(y,x) := \frac{\partial^p}{\partial x^p} D_{2l+1,n,l}(y,x), \qquad g(x) := f(x) - L_{m-1}(x;f;I),$$

and $L(x) := L_{m-1}(x;g;J)$ and note that

$$\mathcal{D}^{(p)}_{l,m,n}(x;f) - L^{(p)}_{m-1}(x;f;J) = \int_{-1}^{1}(g(y) - L(y))D^{(p)}(y,x)\,dy$$

$$+ \int_{-1}^{1} L(y)D^{(p)}(y,x)\,dy - L^{(p)}(x)$$

$$=: i_1(x) + i_2(x) - L^{(p)}(x),$$

$$\Delta^m_h(f;x) \equiv \Delta^m_h(g,x), \qquad x,(x+mh) \in I. \tag{3.17}$$

1. Let us estimate $i_1(x_0)$. Denote

$$J_\delta := [x_0-\delta, x_0+\delta]\cap I \quad \text{and} \quad \omega(x) := \omega_m(t;f;J_\delta) \equiv \omega_m(t;g;J_\delta).$$

Using (3.6.15), we obtain

$$|g(x) - L(x)| = c_1\rho_0^{-m}(|x-x_0|+\rho_0)^m\omega(\rho_0), \qquad x \in J_\delta,$$

$$|g(x) - L(x)| = c_1\rho_0^{-m}(|x-x_0|+\rho_0)^m\varphi(\rho_0), \qquad x \in I.$$

Furthermore, according to (3.13), we have $\left|D^{(p)}(y, x_0)\right| \le C_2\rho_0^{l-p-1}(|y-x_0|+\rho_0)^{-l}$. Decomposing the integral $i_1(x_0)$ into two integrals (over I_δ and $I \setminus I_\delta$), we get

$$\left|\int_{J_\delta} (g(y)-L(y))D^{(p)}(y, x_0)\,dy\right|$$

$$\le c_1 C_2 \rho_0^{l-p-m-1}\omega(\rho_0)\int_{J_\delta}(|y-x_0|+\rho_0)^{m-l}\,dy$$

$$\le c_1 C_2 \rho_0^{l-p-m-1}\omega(\rho_0)\int_{-\infty}^{\infty}(|y-x_0|+\rho_0)^{m-l}\,dy \le C_3\rho_0^{-p}\omega(\rho_0),$$

$$\left|\int_{I\setminus J_\delta} (g(y)-L(y))D^{(p)}(y, x)\,dy\right|$$

$$\le 2c_1 C_2 \rho_0^{l-p-m-1}\varphi(\rho_0)\int_{\delta}^{\infty}(|y-x_0|+\rho_0)^{m-l}\,dy$$

$$< C_4\rho_0^{-p}\varphi(\rho_0)\left(\frac{\rho_0}{\delta}\right)^{l-m-1} \le C_4\rho_0^{-p}\varphi(\rho_0)\left(\frac{\rho_0}{\delta}\right)^{l-2m}.$$

2. Let us estimate the difference $i_2(x_0) - L^{(p)}(x_0)$. We expand the polynomial $L(x)$ into a Taylor series:

$$L(x) = \sum_{q=0}^{m-1}\frac{(x-x_0)^q}{q!}L^{(q)}(x_0).$$

Then

$$i_2(x_0) - L^{(p)}(x_0) = \sum_{q=0}^{m-1}\left(\frac{1}{q!}\int_{-1}^{1}(y-x_0)^q D^{(p)}(y, x_0)\,dy - \delta_{q,p}\right)L^{(q)}(x_0)$$

$$=: \sum_{q=0}^{m-1} A_q L^{(q)}(x_0).$$

By virtue of the Whitney inequality (see (3.6.12)), we have $\|g\|_I \le c_5\varphi(1)$. Therefore, using (3.3.48), we get $\left|L^{(q)}(x_0)\right| \le c_6\rho_0^{-q}\varphi(1)$. Furthermore, by virtue of (3.14), we obtain $\left|A_q\right| \le c_7 n^{-2l-1}$. Hence,

$$\begin{aligned}\left|i_2(x_0) - L^{(p)}(x_0)\right| &\le \sum_{q=0}^{m-1} c_6 c_7 n^{-2l-1}\rho_0^{-q}\varphi(1)\\ &\le mc_6C_7 n^{-2l-1}\rho_0^{-m}\varphi(1) < C_8\rho_0^{l-2m}\varphi(\rho_0)\\ &\le C_8 2^{l-2m+p}\rho_0^{-p}\left(\frac{\rho_0}{\delta}\right)^{l-2m+p}\varphi(\rho_0)\\ &= C_9\rho_0^{-p}\left(\frac{\rho_0}{\delta}\right)^{l-2m-p}\varphi(\rho_0) = C_9\rho_0^{-p}\left(\frac{\rho_0}{\delta}\right)^{l-2m}\varphi(\rho_0).\end{aligned}$$

Thus,

$$\begin{aligned}\left|\mathcal{D}^{(p)}_{l,m,n}(x_0; f) - L^{(p)}_{m-1}(x_0; f; J)\right| &\le \left|i_1(x_0)\right| + \left|i_2(x_0) - L^{(p)}(x_0)\right|\\ &\le (c_3 + c_8)\rho_0^{-p}\omega(\rho_0) + (c_4 + c_9)\rho_0^{-p}\left(\frac{\rho_0}{\delta}\right)^{l-2m}\varphi(\rho_0)\\ &\le C\rho_0^{-p}\left(\omega(\rho_0) + \left(\frac{\rho_0}{\delta}\right)^{l-2m}\varphi(\rho_0)\right).\end{aligned}$$

3.9. We now prove the direct theorem (Theorem 1.1). More exactly, we prove Theorem 3.3, which somewhat generalizes both Theorem 1.1 and the Trigub theorem [Trigub (1962)] on the joint approximation of a function and its derivatives and on an estimate for the modulus of the derivative of approximating polynomials.

Theorem 3.3. *If $f \in W^r H[k, \varphi]$, then, for each $n \ge k + r - 1$, there exists an algebraic polynomial P_n of degree $\le n$ such that, for all $x \in I$, the following inequalities are true:*

$$\left|f^{(p)}(x) - P_n^{(p)}(x)\right| \le c\rho_n^{r-p}(x)\varphi(\rho_n(x)), \quad p = 0, \ldots, r, \tag{3.18}$$

$$\left|P_n^{(p)}(x)\right| \le \frac{c}{\rho_n^p(x)}\left(\rho_n^r(x)\varphi(\rho_n(x)) + (k + r - p)\|f\|_{[x-\rho_n(x), x+\rho_n(x)]\cap I}\right), \tag{3.19}$$

$$p = 0, \ldots, m.$$

Proof. First, assume that $m-1 \le n \le 2m+1$. Then $\rho \ge c_1$ and we can take $P_n(x) := L_{m-1}(x; f; I)$. Estimate (3.18) follows from the Whitney inequality (see (3.6.12)), and estimate (3.19) follows from (3.3.48).

Let $n > 2m+1$. We set $n_1 := 1 + [(n+1)/(2m+2)]$, where $[\ ]$ denotes the integer part of a number. We also introduce the polynomial $P_n(x) := \mathcal{D}_{2m+1,m,n_1}(x; f)$ of degree $\le (2m+2)(n_1-1)-1 \le n$. Taking $\delta = 2$ and $x_0 \in I$, denoting $\rho_0 := \rho_{n_1}(x_0)$ and $J := [x_0-\rho_0, x_0+\rho_0] \cap I$, and using Lemma 3.3, we get

$$\left|P_n^{(p)}(x_0) - L_{m-1}^{(p)}(x_0; f; J)\right| \le c_2\rho_0^{-p}\left(\omega_m(\rho_0; f; I) + \left(\frac{\rho_0}{\rho_0+2}\right)\omega_m(\rho_0; f; I)\right)$$

$$\le 2c_2\rho_0^{-p}\omega_m(\rho_0; f; I) \le 2c_2\rho_0^{r-p}\varphi(\rho_0).$$

Furthermore, by virtue of Lemma 4.3, we obtain

$$\left|f^{(p)}(x) - L_{m-1}^{(p)}(x_0; f; J)\right| \le c_3\rho_0^{-p}\omega_{m-p}(\rho_0; f^{(p)}; I) \le c_3\rho_0^{r-p}\varphi(\rho_0).$$

By virtue of (1.48), we have

$$\left|L_{m-1}^{(p)}(x_0; f; J)\right| \le c_4(m-p)\rho_0^{-p}\|f\|_J.$$

Therefore,

$$\left|f^{(p)}(x_0) - P_n^{(p)}(x_0)\right| \le \left|f^{(p)}(x_0) - L_{m-1}^{(p)}(x_0; f; J)\right| + \left|L_{m-1}^{(p)}(x_0; f; J) - P_n^{(p)}(x)\right|$$

$$\le (c_3 + 2c_2)\rho_0^{r-p}\varphi(\rho_0),$$

$$\left|P_n^{(p)}(x_0)\right| \le \left|P_n^{(p)}(x_0) - L_{m-1}^{(p)}(x_0; f; J)\right| + \left|L_{m-1}^{(p)}(x_0; f; J)\right|$$

$$\le 2c_2\rho_0^{r-p}\varphi(\rho_0) + c_4(m-p)\rho_0^{-p}\|f\|_J$$

$$\le 2c_2\rho_0^{r-p}\varphi(\rho_0) + c_4(m-p)\rho_0^{-p}\|f\|_J.$$

It remains to note that $n_1 > (n+1)/(2m+2)$, which implies that $\rho_0 < c_5\rho_n(x_0)$, i.e., $\rho_0^{r-p}\varphi(\rho_0) \le c_5^m\rho_n^{r-p}(x_0)\varphi(\rho_n(x_0))$.

∎

Remark 3.1. If $p \geq m$, then $L_{m-1}^{(p)}(x_0; f; J) \equiv 0$ and, hence,

$$\left|P_n^{(p)}(x_0)\right| \leq \left|P_n^{(p)}(x_0) - L_{m-1}^{(p)}(x_0; f; J)\right| = \left|\mathcal{D}_{2m+1,n_1}^{(p)}(x_0; f) - L_{m-1}^{(p)}(x_0; f; J)\right|$$

$$\leq C\rho_0^{-p}\varphi(\rho_0) < C\rho_n^{-p}(x_0)\varphi(\rho_n(x_0)),$$

where $C = C(k, m, p, r)$, i.e.,

$$\left|P_n^{(p)}(x)\right| \leq C\rho_n^{-p}(x)\varphi(\rho_n(x)), \quad x \in I, \quad p = m, \ldots, n. \tag{3.20}$$

Remark 3.2. For $n < m-1$, Theorem 3.3 is, generally speaking, not true. Indeed, let us take an arbitrary $A > 0$ and consider the Chebyshev polynomial

$$f(x) = A\cos(m-1)\arccos x.$$

Since $\omega_k(t; f^{(r)}; I) \equiv 0$, we have $f \in W^r H[k, \varphi]$ for any $\varphi \in \Phi^k$. At the same time, it is well known that, for any polynomial $P_n(x)$ of degree $n < m-1$, there exists a point $x_0 \in I$ at which $|f(x_0) - P_n(x_0)| \geq A$. On the other hand, we have

$$\rho_n^r(x_0)\varphi(\rho_n(x_0)) \leq 2^m\varphi(2),$$

i.e.,

$$|f(x_0) - P_n(x_0)| \geq A2^{-m}\varphi^{-1}(2)\rho_n^r(x_0)\varphi(\rho_n(x_0)).$$

3.10. The theorem on a constructive characteristic (Theorem 1.3) is a direct corollary of Theorems 1.1 and 1.2.

It follows from Theorem 3.7.2 that if the condition $\varphi \in S(k, r)$ is not satisfied, then Theorem 1.3 is, generally speaking, not true. In other words, the condition $\varphi \in S(k, r)$ is necessary and sufficient for the constructive characterization of functions from the space $W^r H_k^\varphi(I)$.

Note that the necessity of the sufficient condition $\varphi \in S(k, r)$ for the constructive characterization of the approximation of periodic functions of the space $W^r H_k^\varphi(R)$ by periodic polynomials was established by Bari and Stechkin [Stechkin (1949), (1951a); Bari and Stechkin (1956)] and Lozinskii (1952).

Lemmas 3.3 and 3.10.5 also yield the direct theorem for DT_r-moduli of continuity (see [Ditzian and Totik (1987)] for $r = 0$).

Theorem 3.4. *Let* $k \in \mathbb{N}$, $\varphi \in \Phi^k$, $(r+1) \in \mathbb{N}$, *and* $m = k + r$. *If* $f \in B^r H[k, \varphi]$, *then, for each* $n \geq m - 1$, *there exists an algebraic polynomial* P_n *of degree* $\leq n$, *such that*

$$\| f - P_n \| \leq \frac{c}{n^r} \varphi\left(\frac{1}{n}\right),$$

$$\left\| (f^{(p)} - P_n^{(p)}) \rho^p \right\| \leq \frac{c}{n^r} \varphi\left(\frac{1}{n}\right), \quad p < \frac{r}{2},$$

$$\left\| (f^{(p)} - P_n^{(p)}) \rho^p \right\|_{\left[-1+\frac{1}{n^2}, 1-\frac{1}{n^2}\right]} \leq \frac{c}{n^r} \varphi\left(\frac{1}{n}\right), \quad p \leq r,$$

and, for all $x \in I$ *and* $p = 0, \ldots, m$,

$$\left| P_n^{(p)}(x) \right| \leq c\left(\varphi\left(\frac{1}{n}\right) + (m-p)\| f \|_{[x-\rho, x+\rho]}\right) \rho^{r-p}.$$

Corollary 3.2. *If* $f \in B^r$, *then*

$$E_n(f) \leq \frac{c}{n^r} \left\| w^r f^{(r)} \right\|,$$

where $w(x) = \sqrt{1 - x^2}$.

3.11. Let us find an estimate for the K-functional.

Lemma 3.4. *If* $f \in H[k; \varphi; I]$, *then, for every natural* $n \geq k - 1$, *there exists an algebraic polynomial* P_n *of degree* $\leq n$ *such that*

$$\| f - P_n \|_I \leq c\varphi\left(\frac{1}{n}\right), \tag{3.21}$$

$$\left\| P_n^{(k)} \right\| \leq cn^k \varphi\left(\frac{1}{n}\right). \tag{3.22}$$

Proof. Using Theorem 4.1.2′, we extend the function f from $I = [-1, 1]$ to the segment $[-2, 2]$, i.e., we construct a function $\bar{f}$ such that $\omega_k(t; \bar{f}; [-2, 2]) \leq c_1 \varphi(t)$ and $\bar{f}(x) = f(x)$ for $x \in I$. Denote $f_*(x) := \bar{f}(2x)$, $x \in I$. It is obvious that $\omega_k(t; f_*; I) \equiv \omega_k(2t; \bar{f}; [-2, 2]) \leq 2^k c_1 \varphi(t)$. By virtue of Theorem 3.3, there exists a polynomial $Q_n(x)$ for which

$$|f_*(x) - Q_n(x)| \le c_2\varphi(\rho), \qquad \left|Q_n^{(k)}(x)\right| \le c_3\rho^{-k}\varphi(\rho), \qquad x \in I. \tag{3.23}$$

We set $P_n(x) := Q_n(x/2)$ and note that, for $x \in \left[-\frac{1}{2}, \frac{1}{2}\right]$,

$$\frac{1}{2n} \le \frac{1}{n^2} + \frac{\sqrt{3}}{2n} = \rho_n\left(\frac{1}{2}\right) \le \rho_n\left(\frac{x}{2}\right) \le \rho_n(0) = \frac{1}{n^2} + \frac{1}{n} \le \frac{2}{n}.$$

Therefore, for $x \in I$, relation (3.23) yields

$$|f(x) - P_n(x)| = \left|f_*\left(\frac{x}{2}\right) - Q_n\left(\frac{x}{2}\right)\right| \le c_2\varphi\left(\rho_n\left(\frac{x}{2}\right)\right) \le c_2\varphi\left(\frac{2}{n}\right) \le 2^k c_2\varphi\left(\frac{1}{n}\right),$$

$$\left|P_n^{(k)}(x)\right| = 2^{-k}\left|Q_n^{(k)}\left(\frac{x}{2}\right)\right| \le 2^{-k}c_3\rho_n^{-k}\left(\frac{x}{2}\right)\varphi\left(\rho_n\left(\frac{x}{2}\right)\right)$$

$$\le 2^{-k}c_3\rho_n^{-k}\left(\frac{1}{2}\right)\varphi\left(\rho_n\left(\frac{1}{2}\right)\right) \le c_3 n^k\varphi\left(\frac{1}{2n}\right) \le c_3 n^k\varphi\left(\frac{1}{n}\right).$$

■

Lemma 3.5. *If $f \in H[k; \varphi; I]$, then*

$$K_k(t^k; f; I) \le c\varphi(t), \qquad t \in \left[0, \frac{1}{k}\right]. \tag{3.24}$$

Proof. Let $n := [1/t]$ be the integer part of $1/t$ and let $P_n(x)$ be the polynomial from Lemma 3.4. Taking (3.21) and (3.22) into account, we get

$$K_k(t^k; f; I) \le \|f - P_n\|_I + n^{-k}\left\|P_n^{(k)}(x)\right\|_I \le c_1\varphi\left(\frac{1}{n}\right) \le c\varphi(t).$$

■

4. On the application of the method of decomposition of unity to approximation of functions

4.1. In this section, we follow the arguments of Kopotun, Leviatan, and Shevchuk (2005).

The method of decomposition of unity was applied to the proof of direct theorems by Freud (1959), Brudnyi (1968), Dzyadyk and Konovalov (1973), and others. In the last cited paper, Dzyadyk proposed to use this method with polynomials

$$\int_{-1}^{x} \left(\frac{T_n(u)}{u - \tilde{x}} \right)^l du,$$

where $\tilde{x}$ is a zero of the Chebyshev polynomial T_n.

In Subsection 4.2, we prove Lemma 4.1 for these polynomials. In Subsection 4.3, we describe the idea of the method. In Subsection 4.4, we give a new proof of Corollary 2.1.

4.2. In what follows,

$$\rho = \rho_n(x), \qquad m = k + r,$$

and constants c may depend only on k, r, and m.

We fix $n \in \mathbb{N}$, $n > 2$. Let $T_n(x) = \cos n \arccos x$ be the Chebyshev polynomial, let $x_j = \cos \frac{j\pi}{n}$, $j = 0, \dots, n$, be extremum points of T_n on I, let $\tilde{x}_j = \cos\left(j - \frac{1}{2}\right)\frac{\pi}{n}$, $j = 1, \dots, n$, be zeros of T_n, let $I_j := [x_j, x_{j-1}]$, and let $|I_j| = x_{j-1} - x_j$, $j = 1, \dots, n$, be the lengths of the intervals I_j. One can easily verify the following inequalities for each $j = 1, \dots, n$:

$$\rho < |I_j| < 5\rho, \qquad x \in I_j, \tag{4.1}$$

$$|I_{j\pm1}| < 3|I_j|, \tag{4.2}$$

$$|I_j| < 4|x_j - \tilde{x}_j|, \qquad |I_j| < 4|x_{j-1} - \tilde{x}_j|, \tag{4.3}$$

$$1 < n / I_j / < \pi. \tag{4.4}$$

For each $j = 1, \dots, n$, we consider the following algebraic polynomials of degree $n - 1$:

$$t_j(x) := \frac{T_n(x)}{x - \tilde{x}_j} |I_j|, \quad x \neq x_j, \qquad \text{and} \qquad t_j(\tilde{x}_j) := T_n'(\tilde{x}_j)|I_j|.$$

It is clear that

$$|t_j(x)| \le \frac{|I_j|}{|x - \tilde{x}_j|}, \qquad x \in I \setminus I_j. \tag{4.5}$$

Lemma 4.1. *For each $j = 1, \dots, n$, the following relation is true:*

$$\frac{4}{3} < |t_j(x)| < 4, \qquad x \in I_j. \tag{4.6}$$

Moreover, the constants in (4.5) are exact and cannot be improved.

Proof. First, we note that $|t_{n-j+1}(-x)| = |t_j(x)|$ and $\{-x \mid x \in I_{n-j+1}\} = I_j$. Hence, without loss of generality, one can assume that $j \le [(n+1)/2]$. Note that t_j is a polynomial of degree $n-1$ having exactly $n-1$ real zeros $\tilde{x}_i$, $1 \le i \le n$, $i \ne j$. Therefore, by virtue of the Rolle theorem, t_j' has exactly $n-2$ different zeros. In particular, t_j' has a unique zero in $[\tilde{x}_{j+1}, \tilde{x}_{j-1}] \supset I_j$ if $j \ge 2$, and so $|t_j(x)| \ge \min\{|t_j(x_j)|, |t_j(x_{j-1})|\}$ for $x \in I_j$ and $j \ge 1$. Hence, for $x \in I_j$, we get

$$|t_j(x)| \ge |t_j(x_j)| = \frac{|I_j|}{\tilde{x}_j - x_j} \ge \frac{|I_1|}{\tilde{x}_1 - x_1} = \frac{4\cos^2(\pi/4n)}{4\cos^2(\pi/4n) - 1} > \frac{4}{3},$$

which is the lower estimate in (4.6).

Denoting $\tau := \arccos x$ (hence, the inclusion $x \in I_j$ implies that $(j-1)\pi/n \le \tau \le j\pi/n$) and $\tau_j := \left(j - \frac{1}{2}\right)\frac{\pi}{n}$, for $x \in I_j$ we get

$$\begin{aligned} |t_j(x)| &= |I_j| \left| \frac{\sin n(\tau - \tau_j)/2}{\sin(\tau - \tau_j)/2} \, \frac{\sin n(\tau + \tau_j)/2}{\sin(\tau + \tau_j)/2} \right| \le \frac{n|I_j|}{\sin(\tau + \tau_j)/2} \\ &\le \frac{n|I_j|}{\sin((j-1)\pi/n + \tau_j)/2} \\ &= \frac{2n\sin(\pi/2n)\sin(j-1/2)\pi/n}{\sin(j-3/4)\pi/n} \le \pi\,\frac{j-1/2}{j-3/4}. \end{aligned}$$

Therefore, if $j \ne 1$, then $\|t_j\|_{I_j} \le 6\pi/5 < 4$.

Finally, if $j = 1$, then, since t_1 is positive and strictly increasing on $[\tilde{x}_2, 1]$, we have

$$0 < t_1(x) \le t_1(1) = \frac{|I_j|}{1 - \tilde{x}_1} = 4\cos^2\left(\frac{\pi}{4n}\right) < 4, \qquad x \in I_1.$$

Inequalities (4.6) are thus verified. The constants in (4.6) are exact because

$$t_1(1) \;=\; 4\cos^2\left(\frac{\pi}{4n}\right) \to 4 \quad \text{and} \quad t_1(1) \;=\; \frac{4\cos^2\left(\frac{\pi}{4n}\right)}{4\cos^2\left(\frac{\pi}{4n}\right) - 1} \to \frac{4}{3} \quad \text{as} \quad n \to \infty.$$

■

4.3. Decomposition of unity

For all $j = 1, \dots, n,$ we set

$$\chi_j(x) \;:=\; \begin{cases} 1 & \text{if} \quad x \ge x_j, \\ 0 & \text{otherwise,} \end{cases}$$

and

$$Q_j(x) \;:=\; \frac{\int_{-1}^{x} t_j^l(u)\,du}{\int_{-1}^{1} t_j^l(u)\,du},$$

where l is a positive even number. Then relation (4.6) yields

$$\int_{-1}^{1} t_j^l(u)\,du \;\ge\; \int_{I_j} t_j^l(u)\,du \;>\; \left(\frac{4}{3}\right)^l |I_j| \;>\; |I_j|.$$

Therefore, if $x \in [-1, x_j),$ then it follows from (4.5) that

$$\begin{aligned} \left|Q_j(x) - \chi_j(x)\right| \;&=\; Q_j(x) \;<\; |I_j|^{l-1} \int_{-1}^{x} \frac{du}{(\tilde{x}_j - u)^l} \\ &<\; \frac{|I_j|^{l-1}}{(l-1)(\tilde{x}_j - x)^{l-1}} \;<\; \left|x - \tilde{x}_j\right|^{1-l} |I_j|^{l-1}. \end{aligned} \tag{4.7}$$

Similarly, if $x \in [x_{j-1}, 1],$ then

$$\left|Q_j(x) - \chi_j(x)\right| = \frac{\int_x^1 t_j^l(u)du}{\int_{-1}^1 t_j^l(u)du} < \left|I_j\right|^{l-1} \int_{-1}^{x} \frac{du}{(u - \tilde{x}_j)^l} < \left|x - \tilde{x}_j\right|^{1-l} \left|I_j\right|^{l-1}. \quad (4.8)$$

We now set $Q_0 :\equiv 0$,

$$R_j := Q_j - Q_{j-1}, \quad j = 1, \ldots, n-1, \quad \text{and} \quad R_n := 1 - Q_{n-1}.$$

Lemma 4.2. *The following relations are true:*

$$\sum_{j=1}^{n} R_j \equiv 1 \quad (4.9)$$

and, for each $j = 1, \ldots, n$,

$$\left|R_j(x)\right| \le C(l) \left(\frac{\left|I_j\right|}{\left|x - x_j\right| + \left|I_j\right|} \right)^{l-1}, \quad x \in I. \quad (4.10)$$

Proof. Identity (4.9) is obvious. Therefore, we verify (4.10) for $j \ne 1, n$ (for $j = 1$ and $j = n$, arguments are similar). If $x \notin I_j \cup I_{j-1}$, then $\chi_j(x) = \chi_{j-1}(x)$, whence

$$R_j(x) = (Q_j(x) - \chi_j(x)) - (Q_{j-1}(x) - \chi_{j-1}(x)).$$

Consequently, relations (4.7) and (4.5) yield

$$R_j(x) = \left|x - \tilde{x}_j\right|^{1-l} \left|I_j\right|^{l-1} + \left|x - \tilde{x}_{j-1}\right|^{1-l} \left|I_{j-1}\right|^{l-1} \le C(l) \left(\frac{\left|I_j\right|}{\left|x - x_j\right| + \left|I_j\right|} \right)^{l-1},$$

where we have used (4.2) and (4.3) to obtain the last estimate. If $x \in I_j \cup I_{j-1}$, then

$$\left|R_j(x)\right| \le \left\|Q_j\right\| + \left\|Q_{j-1}\right\| = 2,$$

which yields (4.10) with $C(l) < 4^l$.

■

Now let a function $f \in C(I)$ and a natural number m be given. For each $j = 1, \ldots, n$, we denote by l_j the Lagrange polynomial of degree $\le m - 1$ that interpolates f at m

equidistant points of I_j, including the endpoints of I_j. By virtue of (3.6.15), for all $x \in I_j$ we have

$$\left|f(x) - l_j(x)\right| \le c\left(\frac{|x - x_j| + |I_j|}{|I_j|}\right)^m \omega_m(|I_j|, f)$$

$$\le c\left(\frac{|x - x_j| + |I_j|}{|I_j|^m \rho^m}\right)^{2m} \omega_m(\rho, f). \qquad (4.11)$$

We set

$$P_n(x) := P_n(x, f, m, l) := \sum_{j=1}^{n} l_j(x) R_j(x)$$

and prove, that, for a properly chosen l, P_n is the required polynomial in Theorem 1.1. The following statement is true:

Lemma 4.3. *If $f \in W^r H[k, \varphi]$, then*

$$|P_n(x) - f(x)| \le c\rho^r \varphi(\rho).$$

Proof. We set $\varphi^*(t) := t^r \varphi(t)$. Then $\omega_m(t, f) \le \varphi^*(t)$. We take $l = 3m + 4$. Identity (4.9) yields

$$f - P_n = \sum_{j=1}^{n} (f - l_j) R_j.$$

Hence, relations (4.10) and (4.11) imply, for $x \in I$, that

$$|f(x) - P_n(x)| \le c\,\frac{\varphi^*(\rho)}{\rho^m} \sum_{j=1}^{n} \frac{|I_j|^{l-1-m}}{(|x - x_j| + |I_j|)^{l-1-2m}} =: c\varphi^*(\rho)\sigma.$$

Using (1.18), (1.19), and (4.1), we get

$$\frac{|I_j|^{l-1-m}}{(|x - x_j| + |I_j|)^{l-1-2m}} \le \frac{c\rho^{m+1}|I_j|^{l-3m-3}}{(|x - x_j| + \rho)^{l-3m-2}} = \frac{c\rho^{m+1}|I_j|}{(|x - x_j| + \rho)^2}.$$

Therefore,

$$\sigma \le c\rho \sum_{j=1}^{n} \frac{|I_j|}{\left(|x-x_j|+\rho\right)^2} \le c\rho \int_{-1}^{1} \frac{du}{\left(|x-x_j|+\rho\right)^2} \le 2c\rho \int_{0}^{\infty} \frac{dv}{(v+\rho)^2} = 2c.$$

■

Note that, for a properly chosen l, P_n is the required polynomial in Theorems 3.3 and 3.4 as well.

4.4. We now show that Lemma 4.1 also yields Corollary 2.1.

Lemma 4.4. *Let $x \in \mathbb{N}$ and $x_0 \in I$. If a polynomial P_n of degree $\le n$ satisfies the relation*

$$|P_n(x)| \le \left(|x-x_0|+\rho_n(x_0)\right)^r, \quad x \in I,$$

then

$$\left|P_n^{(r)}(x_0)\right| \le c.$$

Proof. First, let $r = 1$. By j we denote an index such that $x_0 \in I_j$. We introduce the following polynomial of degree $< 2n$:

$$Q_n(x) := P_n(x)\frac{t_j(x)}{t_j(x_0)}.$$

Estimates (4.6), (4.5), (4.2), and (4.3) yield

$$|Q_n(x)| \le (|x-x_0|+\rho_n(x_0))\frac{3}{4}\left|t_j(x)\right| \le 6|I_j|, \quad x \in I.$$

Therefore, it follows from the Markov and Bernstein inequalities that

$$|Q_n'(x)| \le \frac{12|I_j|}{\rho_{2n}(x)} < \frac{48|I_j|}{\rho}, \quad x \in I.$$

Hence,

$$|P_n'(x_0)| \le \left| Q_n'(x_0) - P_n(x_0) \frac{t_j'(x)}{t_j(x_0)} \right| \le 48 \frac{|I_j|}{\rho_n(x_0)} + \rho_n(x_0) \frac{3}{4} \frac{2\| t_j \|}{\rho_n(x_0)}$$

$$< 48 \cdot 5 + 6 < 250.$$

For $r \ge 1$, Lemma 4.4 is proved by induction with the use of the same arguments with Q_n replaced by $Q_{n,r} = (t_j / t_j(x_0))^r P_n$.

■

The following corollary is often useful:

Corollary 4.1. *If, for some* $l \ge 0$, *one has*

$$P_n(x) \le 1 + x^l n^{2l}, \qquad x \in [0, 1],$$

then the following inequality holds for each $r \in \mathbb{N}$:

$$\left| P_n^{(r)}(0) \right| < C(l, r) n^{2r}.$$

5. Extreme functions

Let us return to the spaces $W^r H_k^\varphi := W^r H_k^\varphi(I)$, $k \in \mathbb{N}$, $\varphi \in \Phi^k$, $(r+1) \in \mathbb{N}$ (see Definition 3.7.2).

Definition 5.1. *A function* f *is called extreme in the space* $W^r H_k^\varphi$ *if*

$$\omega_k(t; f^{(r)}; [-1, 1]) =: \omega_k(t; f^{(r)}) \sim \varphi(t) \tag{5.1}$$

and

$$E_n(f) \sim n^{-2r} \varphi\left(\frac{1}{n^2}\right). \tag{5.2}$$

Theorem 1.1 (direct theorem) gives the following estimate for $f \in W^r H_k^\varphi$:

$$E_n(f) \le c n^{-r} \varphi\left(\frac{1}{n}\right).$$

Therefore, in terms of uniform approximations, the approximation properties of extreme functions are substantially better than in general in the space $W^r H_k^\varphi$. Moreover, if $\varphi \in S(k, r)$ (see (1.16)), then, in these terms, the properties of extreme functions are the best in order among the functions satisfying relation (5.1) because, in this case,

$$\limsup_{n\to\infty} \frac{n^{2r} E_n(f)}{\varphi\left(\frac{1}{n^2}\right)} > 0. \tag{5.3}$$

Inequality (5.3) is a simple corollary of Lemma 5.1, which is proved by using the arguments of Bari and Stechkin (1956).

Lemma 5.1. *Suppose that* $\varphi \in S(k, r)$, *i.e.,*

$$\int_0^t r u^{-1} \varphi(u)\,du + t^k \int_t^1 u^{-k-1} \varphi(u)\,du \le A\varphi(t), \tag{5.4}$$

$0 < t \le 1/2$, $A = \text{const} > 0$, $f \in C^r(I)$, *and* $\omega_k(t; f^{(r)}) \sim \varphi(t)$. *If*

$$E_n(f) \le n^{-2r} \varphi\left(\frac{1}{n^2}\right), \qquad n \in \mathbb{N}, \tag{5.5}$$

then

$$E_n(f) \sim n^{-2r} \varphi\left(\frac{1}{n^2}\right). \tag{5.6}$$

Proof. By virtue of Lemma 3.1 and the conditions of Lemma 5.1, there exists a function $\alpha = \alpha(t)$ that does not decrease on $[0, 1]$ and possesses the following properties:

$$\alpha\left(\frac{1}{n^2}\right) = E_n(f), \qquad \alpha(t) \le t^r \varphi(t),$$

$$\varphi(t) \le A_1 \left(\int_0^t r u^{-r-1} \alpha(u)\,du + t^k \int_t^1 u^{-m-1} \alpha(u)\,du \right),$$

$$t \in \left(0, \frac{1}{2}\right], \qquad A_1 = \text{const} > 0, \qquad m = k + r. \tag{5.7}$$

We fix an arbitrary point $t \in (0, 1/2]$ and denote $a^m := \alpha(t)/t^r\varphi(t)$. Let us prove that

$$\ln\left(\frac{1}{a}\right) \le A_2 := 4AA_1(1+A). \tag{5.8}$$

Indeed, let $t_* := t\sqrt{a}$. According to condition (5.4), we have

$$A\varphi(t_*) \ge \int_0^{t_*} ru^{-1}\varphi(u)\,du \ge \int_{at}^{t_*} ru^{-1}\varphi(u)\,du \ge r\varphi(at)\ln\left(\frac{1}{\sqrt{a}}\right),$$

$$A\varphi(t_*) \ge t_*^k \int_{t_*}^{t} ru^{-k-1}\varphi(u)\,du \ge \left(\frac{t_*}{t}\right)^k \varphi(t)\ln\left(\frac{1}{\sqrt{a}}\right) \ge a^{k/2}\varphi(t)\ln\left(\frac{1}{\sqrt{a}}\right),$$

whence

$$\begin{aligned}
&\int_0^{t_*} ru^{-r-1}\alpha(u)\,du + t_*\int_{t_*}^{1} u^{-m-1}\alpha(u)\,du \\
&\le \int_0^{at} ru^{-1}\varphi(u)\,du + \int_{at}^{t_*} ru^{-r-1}a^m t^r\varphi(t)\,du \\
&\qquad + \int_{t_*}^{t} u^{-m-1}a^m t^r\varphi(t)\,du + t_*^k\int_t^1 u^{-k-1}\varphi(u)\,du \\
&\le Ar\varphi(at) + a^k\varphi(t) + \left(\frac{t}{t_*}\right)^r a^m\varphi(t) + A\left(\frac{t_*}{t}\right)^k\varphi(t) \\
&\le 2A\ln^{-1}\left(\frac{1}{a}\right)\varphi(t_*)(A + a^{k/2} + a^{m/2} + A) \\
&\le 4A\ln^{-1}\left(\frac{1}{a}\right)\varphi(t^*)(1+A).
\end{aligned}$$

The required relation (5.8) now follows from (5.7). Relation (5.8) means that

$$\frac{n^{2r}E_n(f)}{\varphi\left(\frac{1}{n^2}\right)} \equiv \frac{n^{2r}\alpha\left(\frac{1}{n^2}\right)}{\varphi\left(\frac{1}{n^2}\right)} \ge A_3 := e^{-mA_2}, \qquad n \in \mathbb{N}, \qquad n \ne 1.$$

■

In order to establish a theorem on the existence of extreme functions (Theorem 5.1), we first give Definition 5.2 and prove Lemma 5.2.

Definition 5.2. *We set*

$$F_{k,r,\varphi}(y) := \int_0^y \frac{(y-u)^{r-1} F(u; \varphi_{m+2}; k)\,du}{(r-1)!},$$

where $m := k + r$, φ_{m+2} *is defined by (3.4.41), and* $F(u; \varphi_{m+2}; k)$ *is defined by (3.4.27), and*

$$F^h_{k,r,\varphi}(y) := \int_0^y \frac{(y-u)^{r-1}(F(u; \varphi_{m+2}; k) - F(4mh^2; \varphi_{m+2}; k))\,du}{(r-1)!};$$

$$f_{k,r,\varphi}(x) := F_{k,r,\varphi}\left(\frac{x+1}{2}\right), \quad x \in I. \tag{5.9}$$

Theorem 3.4.2 and Lemma 3.4.3 yield

$$\varphi(t) \le \omega_k(t; F^{(r)}_{k,r}; [0,1]) \le c\varphi(t),$$

whence

$$2^{-k}\varphi(t) \le \omega_k(t; f^{(r)}_{k,r,\varphi}) \le c\varphi(t). \tag{5.10}$$

Lemma 5.2. *The following estimate is true:*

$$\overline{\omega}_{2m}(t; f) \le ct^{2r}\varphi(t^2), \quad t \ge 0, \quad m = k + r. \tag{5.11}$$

Proof. Taking $x \in I$, $h \ge 0$, $(x + 2m\rho(h, x)) \in I$, and $y := (x+1)/2$ and denoting

$$\rho_* := h\sqrt{y(1-y)} + \frac{h^2}{2},$$

we get

$$\Delta^{2m}_\rho(f_{k,r,\varphi}; x) = \Delta^{2m}_\rho(F_{k,r,\varphi}; y).$$

Hence, if $y \ge h^2/2$, then

$$\left|\Delta_{\rho_*}^{2m}(F_{k,r,\varphi};\,y)\right| = \rho_*^{2m}\left|F_{k,r,\varphi}^{(2m)}(\theta)\right| = \rho_*^{2m}\left|(\theta^{1-k}\varphi'(\theta))^{(m)}\right|$$

$$\le c_1\rho_*^{2m}y^{-m-k}\varphi(y) \le c_2h^{2r}\varphi(h^2), \quad \theta\in[y,\,y+2m\rho_*],$$

and if $y < h^2/2$, then

$$\left|\Delta_{\rho_*}^{2m}(F_{k,r,\varphi};\,y)\right| = \left|\Delta_{\rho_*}^{2m}(F_{k,r,\varphi}^{h};\,y)\right| \le 2^{2m}\left\|F_{k,r,\varphi}^{h}\right\|_{[0,4mh^2]} \le c_3h^{2r}\varphi(h^2).$$

■

It follows from Lemma 5.2 and Theorem 3.4 that

$$E_n(f_{k,r,\varphi}) \le cn^{-2r}\varphi\left(\frac{1}{n^2}\right), \quad n\in\mathbb{N}. \tag{5.12}$$

Relations (5.10) and (5.12) and Lemma 5.1 yield the following theorem:

Theorem 5.1 [Shevchuk (1986), (1989)]. *If* $\varphi\in S(k,r)$, *then* $f_{k,r,\varphi}(x)$ *is an extreme function in the space* $W^rH_k^{\varphi}$, *i.e., an extreme function exists in every space* $W^rH_k^{\varphi}$ *with* $\varphi\in S(k,r)$.

Theorem 5.1 generalizes the following well-known theorem:

Theorem 5.2 [Bernstein (1952), Ibragimov (1946)]. *Let* $(r+1)\in\mathbb{N}$ *and* $0<\alpha<1$. *The function* $f(x) = (1+x)^{r+\alpha}$ *is an extreme function in the space* W^rH^{α}, *and* $f(x) = (1+x)^{r+1}\ln(1+x)$ *is an extreme function in the space* W^rZ.

In connection with Theorems 5.1 and 5.2, note the following example proposed by Brudnyi (1963): A continuous function $f_{\alpha,\beta}\colon [0,1]\to\mathbb{R}$ defined on $(0,1]$ by the relation $f_{\alpha,\beta}(x) = x^{\alpha}\sin(x^{-\beta})$, where $\alpha,\beta>0$, has the modulus of continuity $\omega_k(t; f_{\alpha,\beta}; [-1,1]) \sim t^{\alpha/(\beta+1)}$ for $k>\alpha/(\beta+1)$, whereas $E_n(f)_{[0,1]} \sim n^{-2\alpha/(2\beta+1)}$.

6. Shape-preserving approximation

6.1. The problem of the best approximation of functions by polynomials with monotonicity condition was investigated for the first time by Chebyshev [(1948), p. 41]. He constructed a monotone polynomial $P_n(x) = x^n + a_1x^{n-1} + \ldots + a_n$ that least deviates from 0 on I.

Definition 6.1. *Let $q \in N$. By Δ^q we denote the set of functions $f \in C(I)$ such that $\Delta_h^q(f; x) \geq 0$ for all $x \in I$ and $h > 0$, $(x + qh) \in I$.*

Note that Δ^1 is the set of functions nondecreasing on I, and Δ^2 is the set of functions convex on I.

Definition 6.2. *The number*

$$E_n^{(q)}(f) := \inf_{p_n \in \Delta^q} \| f - p_n \| \tag{6.1}$$

is called the value of the best uniform shape-preserving approximation of a function $f \in \Delta^q$ by algebraic polynomials p_n of degree $\leq n$. The number $E_n^{(1)}(f)$ is called the value of the best uniform monotone approximation of the function f. The number $\Delta_n^{(2)}(f)$ is called the value of the best uniform convex approximation of the function f.

An analog of the Weierstrass theorem is true for $E_n^{(q)}(f)$, namely, if $f \in \Delta^q$, then $E_n^{(q)}(f) \to 0$ as $n \to \infty$ (see [G. Lorentz (1953)]).

Indeed, let

$$B_n(x; f) := 2^{-n} \sum_{j=0}^{n} \binom{n}{j} f\left(\frac{n-2j}{n}\right)(1+x)^{n-j}(1-x)^j$$

denote the Bernstein polynomial. It is well known that $\| f - B_n \| \to 0$ as $n \to \infty$. At the same time, we have

$$B_n^{(q)}(x; f)$$

$$= 2^{-n} \frac{n!}{(n-q)!} \sum_{j=0}^{n-q} \binom{n-q}{j} \Delta_{2j/n}^{q}\left(f; \frac{n-2(j+q)}{n}\right)(1+x)^{n-j-q}(1-x)^j \geq 0$$

for $x \in I$ if $f \in \Delta^q$.

G. Lorentz and Zeller (1969) constructed a function $f \in \Delta^q$ such that

$$\limsup_{n \to \infty} \frac{E_n^{(q)}(f)}{E_n(f)} = \infty.$$

Therefore, the problem of the estimation of $E_n^{(q)}(f)$ does not reduce to the problem of the estimation of $E_n(f)$. Nevertheless, G. Lorentz and Zeller (1968) proved the following inequality for $f \in \Delta^1$:

$$E_n^{(1)}(f) \le c\omega_1\left(\frac{1}{n}; f\right), \qquad n \in \mathbb{N}. \tag{6.2}$$

Moreover, if $f \in \Delta^q$, then

$$E_n^{(q)}(f) \le c\omega_2\left(\frac{1}{n}; f\right), \qquad n \in \mathbb{N}, \tag{6.3}$$

which, in particular, implies that

$$E_n^{(q)}(f) \le cn^{-1}\omega_1\left(\frac{1}{n}; f'\right), \qquad n \in \mathbb{N}, \tag{6.4}$$

provided that $f \in \Delta^q \cap C^1(I)$. Inequalities (6.3) and (6.4) were established by G. Lorentz (1972) [(6.4) for $q = 1$], DeVore (1976) [(6.3) for $q = 1$], Beatson (1978) [(6.4) for $q \in \mathbb{N}$], and Shvedov [(1979), (1980), (1981)] [(6.3) for $q \in \mathbb{N}$]. Note that Shvedov also proved (6.3) for integral metrics [Shvedov (1979), (1980), (1981)] and obtained a multi-dimensional analog of (6.2) in [Shvedov (1981a)].

Later in this section, we prove Theorems 6.2 and 6.3 (due to Beatson, DeVore, Yu, and Leviatan), which strengthen relations (6.2)–(6.4) for $q = 2$.

6.2. The theorem presented below shows that $E_n^{(q)}(f)$ cannot be estimated from above in terms of the modulus of continuity of arbitrary order; for $q = 1$, estimate (6.3) cannot be obtained with a modulus of continuity of order higher than 2; for $q = 2$, estimates (6.3) and (6.4) cannot be obtained with a modulus of continuity of order higher than 3 or 2, respectively.

Theorem 6.1 [Shvedov (1980), (1981)]. *Let $A > 0$, $q \in \mathbb{N}$, $k \in \mathbb{N}$, $(r + 1) \in \mathbb{N}$, and $m = k + r$. If $r < q < m - 1$, then, for each $n \ge m - 1$, $n \in \mathbb{N}$, there exists a function $f \in \Delta^q \cap C^q$ such that*

$$E_n^{(q)}(f) \ge A(m-1)^{-r}\omega_k\left(\frac{1}{m-1}; f^{(r)}\right) \ge An^{-r}\omega_k\left(\frac{1}{n}; f^{(r)}\right). \tag{6.5}$$

Proof. We choose a number $b \in (0, 1)$ from the condition

$$(q+1-r)!\left(n^{-2q}b^{r-q} - \frac{b^r}{(q+1)!}\right) = A(m-1)^{-r}2^k.$$

We set $x_0 := b - 1$,

$$P(x) := \frac{(x-x_0)^{q+1}}{(q+1)!},$$

and

$$f(x) := \frac{(x-x_0)_+^{q+1}}{(q+1)!},$$

i.e., $f(x) = P(x)$ if $x \geq x_0$ and $f(x) \equiv 0$ if $x < x_0$. Note that

$$\left\| f^{(j)} - P^{(j)} \right\| = \left| P^{(j)}(-1) \right| = \frac{b^{q+1-j}}{(q+1-j)!} \qquad \text{for} \quad j = 0, \dots, q.$$

Let $Q_n = Q_n(x)$ be a polynomial of degree $\leq n$ such that $Q_n \in \Delta^q$, i.e., $Q_n^{(q)}(x) \geq 0$ for $x \in I$. Taking into account that $Q_n^{(q)}(-1) \geq 0$ and using the Markov inequality (2.1), we get

$$\begin{aligned} b &= -P^{(j)}(-1) \leq Q_n^{(q)}(-1) - P^{(q)}(-1) \leq \left\| Q_n^{(q)} - P^{(q)} \right\| \\ &\leq n^{2q}\| Q_n - P \| \leq n^{2q}\| Q_n - f \| + n^{2q}\| f - P \| \\ &= n^{2q}\| f - Q_n \| + \frac{n^{2q}b^{q+1}}{(q+1)!}, \end{aligned}$$

whence

$$\| f - Q_n \| \geq n^{-2q}b - \frac{b^{q+1}}{(q+1)!} = A(m-1)^{-r}2^k \left\| f^{(r)} - P^{(r)} \right\|.$$

On the other hand, by virtue of (2.22), we have

$$\omega_k\left(\frac{1}{m-1}; f^{(r)}\right) \equiv \omega_k\left(\frac{1}{m-1}; f^{(r)} - P^{(r)}\right) \leq 2^k \left\| f^{(r)} - P^{(r)} \right\|,$$

i.e.,

$$\|f - Q_n\| \geq A(m-1)^{-r}\omega_k\left(\frac{1}{m-1}; f^{(r)}\right).$$

■

Note that, in [Shvedov (1981)], Theorem 6.1 was proved in the case $r = 0$ for all integral metrics of L_p, $0 < p \leq \infty$. The applicability of Shvedov's arguments to the case $r \neq 0$ was noticed by S. Manya.

Corollary 6.1. *Suppose that* $B > 0$, $q \in \mathbb{N}$, $k \in \mathbb{N}$, $(r+1) \in \mathbb{N}$, *and* $m = k + r$. *If* $r < q < m - 1$, *then, for each* $n \geq m - 1$, $n \in \mathbb{N}$, *there exists a function* $f \in \Delta^q \cap C^q$ *such that, for any polynomial* $Q_n \in \Delta^q$ *of degree* $\leq n$, *the following estimates are true:*

$$\left\|\rho_n^{-r}\varphi^{-1}(\rho_n)(f - Q_n)\right\| \geq \left\|\rho_{m-1}^{-r}\varphi^{-1}(\rho_{m-1})(f - Q_n)\right\| \geq B.$$

6.3. Theorem 6.2 [DeVore and Yu (1985)]. *If a function* f *does not decrease on* I *and* $f \in H[2; \varphi]$, *then, for each* $n \in \mathbb{N}$, *there exists an algebraic polynomial* P_n *of degree* $\leq n$ *nondecreasing on* I *and such that*

$$|f(x) - P_n(x)| \leq c\varphi(\rho_n(x)), \quad x \in I. \tag{6.6}$$

The proof of this theorem is given in Subsections 6.4–6.8.

6.4. For all $j = 0, \ldots, n$, we set $x_j = \cos j\pi/n$,

$$\chi_j := \begin{cases} 1 & \text{if } x > x_j, \\ 0 & \text{if } x \leq x_j, \end{cases}$$

and

$$\Phi_j(x) := (x - x_j)_+ = \int_{-1}^{x} \chi_j(u)\,du.$$

For $j = 1, \ldots, n$, we denote $I_j := [x_j, x_{j-1}]$ and $h_j = x_{j-1} - x_j$ and note that if $x \in I_j$, then

$$\rho < h_j < 5\rho. \tag{6.7}$$

Let L denote a continuous broken line with nodes at the points x_j, $j = 0, \dots, n$. By virtue of (3.9), for $x \in I_j$ we have

$$|f(x) - L(x)| \le \varphi\left(\frac{h_j}{2}\right) < \left(\frac{5}{2}\right)^2 \varphi(\rho). \tag{6.8}$$

It is easy to verify that the broken line L can be represented in the form

$$L(x) = \sum_{j=1}^{n-1} [x_{j+1}, x_j, x_{j-1}; f](x_{j-1} - x_{j+1})\Phi_j(x) + [x_n, x_{n-1}; f](x+1) + f(-1), \tag{6.9}$$

or, which is the same,

$$L(x) = f(-1) + \sum_{j=1}^{n} [x_j, x_{j-1}; f](\Phi_j(x) - \Phi_{j-1}(x)). \tag{6.10}$$

Thus, the problem is reduced to the approximation of the functions Φ_j.

6.5. We introduce the following notation: $J(t) = J_{n,6}(t)$ is a Jackson-type kernel (see Section 3), $\alpha = \arccos y$, $\beta = \arccos x$, and

$$\varphi(x, y) := \varphi_{n,r}(x, y) = \int_{\beta-\alpha}^{\beta+\alpha} J(t)\,dt, \qquad r = 6.$$

By virtue of (3.5), $\varphi_{n,r}(x, y)$ is a polynomial of degree $(r+1)(n-1)$ in x.

It is obvious that $\varphi(x, 1) \equiv 0$, $\varphi(x, -1) \equiv 1$, and, for every fixed x, the function $\varphi(x, y)$ decreases with respect to y. Therefore, the equation

$$1 - x_j = \int_{-1}^{1} \varphi(x, y_j)\,dx \tag{6.11}$$

has a unique solution $y_j \in I$, and, furthermore, $-1 = y_n < y_{n-1} < \dots < y_1 < y_0 = 1$. Let $j^* = 1, \dots, n$ be a number for which $y_j \in [x_{j^*}, x_{j^*-1})$, $j = 1, \dots, n$, and let $0^* := 0$. Denote

$$Q_j(x) := \int_{-1}^{x} \varphi(u, x_j)du, \qquad j = 0, \ldots, n.$$

In particular, $Q_0(x) \equiv 0$ and $Q_n(x) = x + 1$. Using (6.11) and the monotonicity of φ with respect to y, we obtain the estimates

$$Q_{j^*}(1) \geq 1 - x_j \geq Q_{j^*-1}(1), \qquad j = 1, \ldots, n.$$

Therefore, for each $j = 1, \ldots, n$, there exists a number $a_j \in (0, 1)$ such that

$$a_j Q_{j^*}(1) + (1 - a_j) Q_{j^*-1}(1) = 1 - x_j.$$

We set $R_0(x) :\equiv 0$ and $R_j(x) := a_j Q_{j^*}(1) + (1 - a_j) Q_{j^*-1}(1)$, $j = 1, \ldots, n$, and note that

$$R_j(1) = 1 - x_j. \tag{6.12}$$

We have

$$R_j'(x) \geq R_{j-1}'(x), \qquad j = 1, \ldots, n, \qquad x \in I. \tag{6.13}$$

Indeed, since $y_j < y_{j-1}$, we get $j^* \geq (j-1)^*$. Therefore, if $j^* > (j-1)^*$, then

$$\begin{aligned} R_j' - R_{j-1}' &= a_j Q_{j^*}' + (1 - a_j) Q_{j^*-1}' - a_{j-1} Q_{(j-1)^*}' - (1 - a_{j-1}) Q_{(j-1)^*-1}' \\ &= a_j (Q_{j^*}' - Q_{(j-1)^*}') + (1 - a_j)(Q_{j^*-1}' - Q_{(j-1)^*}') \\ &\quad + (1 - a_{j-1})(Q_{(j-1)^*}' - Q_{(j-1)^*-1}') > 0. \end{aligned}$$

If $j^* = (j-1)^*$, then, obviously, $a_j > a_{j-1}$.

6.6. Let us prove the inequality

$$\left| R_j(x) - \Phi_j(x) \right| \leq c\rho_n^6(x_j)(|x - x_j| + \rho_n(x_j))^{1-6}. \tag{6.14}$$

Using (6.11) and (3.15), we obtain

$$|x_j - y_j| = \left| \int_{-1}^{1} ((u - y_j)_+^0 - \varphi(u, y_j))\,du \right|$$

$$\leq c_1 \int_{-1}^{1} \rho_n^r(y_j)(|u - y_j| + \rho_n(y_j))^{-r}\,du < c_1\rho_n(y_j).$$

Therefore, by virtue of (1.18), we have

$$\rho_n(y_j) \leq c_2\rho_n(x_j)$$

and, with regard for (6.7),

$$\left|x_j - x_{j^*}\right| \leq |x_j - y_j| + h_{j^*} < (c_1 + 5)\rho_n(y_j) < c_3\rho_n(x_j),$$

$$\left|x_j - x_{j^*-1}\right| < c_5\rho_n(x_{j^*}), \qquad c_4\rho_n(x_{j^*}) < \rho_n(x_j) < c_5\rho_n(x_{j^*}),$$

$$c_4\rho_n(x_{j^*-1}) < \rho_n(x_j) < c_5\rho_n(x_{j^*-1}). \tag{6.15}$$

Inequality (3.15) means that

$$\begin{aligned}
\left|R_j'(x) - \chi_j(x)\right| &\equiv a_j(\varphi(x, x_{j^*}) - \chi_{j^*}(x)) \\
&\quad + (1 - a_j)(\varphi(x, x_{j^*-1}) - \chi_{j^*-1}(x)) + a_j(\chi_{j^*}(x) - \chi_j(x)) \\
&\quad + (1 - a_j)(\chi_{j^*-1}(x) - \chi_j(x)) \\
&\leq c_1\rho_n^6(x_{j^*})(|x - x_{j^*}| + \rho_n(x_{j^*}))^{-6} \\
&\quad + c_1\rho_n^6(x_{j^*-1})(|x - x_{j^*-1}| + \rho_n(x_{j^*-1})) \\
&\quad + \left|\chi_j(x) - \chi_{j^*}(x)\right| + \left|\chi_j(x) - \chi_{j^*-1}(x)\right|.
\end{aligned}$$

Hence, by virtue of (6.15), we obtain

$$\left|R_j'(x) - \chi_j(x)\right| \leq c_6\rho_n^6(x_j)(|x - x_j| + \rho_n(x_j))^{-6}.$$

If $x < x_j$, then

$$|R_j(x) - \Phi_j(x)| = \left| \int_{-1}^{x} (R_j'(u) - \chi_j(u))du \right| \le c_6 \rho_n^6(x_j) \int_{-\infty}^{x} (|u - x_j| + \rho_n(x_j))^{-6} du$$

$$\le c_6 \rho_n^6(x_j)(|x - x_j| + \rho_n(x_j))^{-5}.$$

If $x > x_j$, then, using (6.12), establish, by analogy, that

$$|R_j(x) - \Phi_j(x)| = \left| \int_{x}^{1} (R_j'(u) - \chi_j(u))du \right| \le c_6 \rho_n^6(x_j)(|x - x_j| + \rho_n(x_j))^{-5}.$$

Inequality (6.14) is proved.

It follows from (6.14), (1.18), and (1.19) that

$$|R_j(x) - \Phi_j(x)| = \frac{c_6 \rho_n^6(x_j) \rho^2}{(|x - x_j| + \rho_n(x_j))^5} \le \frac{c_7 \rho_n^6(x_j) \rho^2}{(|x - x_j| + \rho)^3}. \tag{6.16}$$

6.7. Denote $A_j := [x_{j+1}, x_j, x_{j-1}; f](x_{j-1} - x_{j+1})$ and

$$P_n(x) := f(-1) + [x_n, x_{n-1}; f](x+1) + \sum_{j=1}^{n-1} A_j R_j(x).$$

Let us prove (6.6). Indeed, with regard for (6.7), it follows from (3.5.8) that $|A_j| \le c_8 h_j \rho_n^{-2}(x_j) \varphi(\rho_n(x_j))$, which, by virtue of (1.18), yields

$$|A_j| \le c_8 h_j \rho_n^{-2}(x_j) \varphi\left(2\sqrt{\rho(|x - x_j| + \rho)}\right) \le \frac{2c_8 h_j (|x - x_j| + \rho)\varphi(\rho)}{\rho_n^2(x_j)\rho}. \tag{6.17}$$

Combining (6.6), (6.16), and (6.17), we get

$$|f(x) - P_n(x)| \le |f(x) - L(x)| + |L(x) - P_n(x)|$$

$$\le \left(\frac{5}{2}\right)^2 \varphi(\rho) + \sum_{j=1}^{n-1} |A_j| |R_j(x) - \Phi_j(x)|$$

$$\le \left(\frac{5}{2}\right)^2 \varphi(\rho) + 2c_6 c_8 \rho \varphi(\rho) \sum_{j=1}^{n-1} h_j (|x - x_j| + \rho)^{-2} < c_9 \varphi(\rho).$$

6.8. The monotonicity of the polynomial $P_n(x)$ follows from the representation

$$P_n'(x) = \sum_{j=1}^{n}[x_j, x_{j-1}; f](R_j'(x) - R_{j-1}'(x)), \tag{6.10'}$$

where $[x_j, x_{j-1}; f] \geq 0$ by the condition of the theorem, and $R_j'(x) - R_{j-1}'(x) \geq 0$ by virtue of (6.13).

■

6.9. Remark 6.1. Relation (6.6) yields the estimate

$$|f(x) - P_n(x)| \leq c\omega_1(\rho_n(x); f), \tag{6.18}$$

which was established by Beatson (see [DeVore and Yu (1985)]).

Remark 6.2. DeVore and Yu (1985) proved the estimate

$$|f(x) - P_n(x)| < c\varphi\left(\frac{\sqrt{1-x^2}}{n}\right), \tag{6.19}$$

which is more exact than (6.6).

Indeed, for $x \notin I_1 \cup I_n$ estimate (6.19) follows from (6.6) because, in this case,

$$\rho_n(x) < \frac{3\sqrt{1-x^2}}{n}.$$

If $x \in I_n$, then

$$1 + x \leq \frac{\pi}{\sqrt{2}} \frac{\sqrt{1-x^2}}{n}$$

and

$$|R_j(x) - \Phi_j(x)| = \left|\int_{-1}^{x}(R_j'(u) - \chi_j(u))du\right| \leq c_6(x+1)\rho_n^6(x_j)\left(|x - x_j| + \rho_n(x_j)\right)^{-6}$$

$$\leq c_{10}\frac{(x+1)\rho\rho_n^2(x_j)}{(|x - x_j| + \rho)^3} \leq c_{10}\frac{\sqrt{1-x^2}}{n}\frac{\rho\rho_n^2(x_j)}{(|x - x_j| + \rho)^3},$$

whence

$$|L(x) - P_n(x)| < c_{12}\varphi\left(\frac{\sqrt{1-x^2}}{n}\right).$$

Furthermore, $L(x) \equiv L(x; f; -1, x_{n-1})$ for $x \in I_n$. Therefore, relation (3.5.11) yields

$$|f(x) - L(x)| < c_{13}\varphi\left(\frac{\sqrt{1-x^2}}{n}\right).$$

The case $x \in I_1$ is considered by analogy with the case $x \in I_n$.

For approximation without restrictions, the corresponding improvement of an estimate of the form (6.6) to an estimate of the form (6.19) was made by Telyakovskii (1966) (for $k = 1$) and DeVore (1976) (for $k = 2$) (see also [Dahlhaus (1989)]).

6.10. Theorem 6.3 [Leviatan (1986)]. *If a function f is convex on I and $x \in H[2;\varphi]$, then, for each $n \in \mathbb{N}$, there exists an algebraic polynomial P_n of degree $\le n$ convex on I and such that*

$$|f(x) - P_n(x)| < c\varphi(\rho_n(x)), \quad x \in I. \tag{6.20}$$

Theorem 6.3 can be proved by analogy with Theorem 6.2. One should only establish (instead of the reasoning presented in Subsection 6.8) that P_n is convex on I, i.e., that $P_n''(x) \ge 0$. According to the conditions of the theorem, the function f is convex. Hence, $[x_{j+1}, x_j, x_{j-1}; f] \ge 0$, i.e., $A_j \ge 0$. In view of the representation

$$P_n''(x) = \sum_{j=1}^{n-1} A_j R_j''(x) = \sum_{j=1}^{n-1} A_j\left(a_j Q_{j^*}''(x) + (1-a_j)Q_{j^*-1}''(x)\right),$$

it is sufficient to establish that $Q_j''(x) \ge 0$, $x \in (-1, 1)$. Indeed, we have $Q_0''(x) \equiv 0$, $Q_n''(x) \equiv 0$, and, for $j = 1, \dots, n-1$,

$$Q_j''(x) = \varphi'(x, x_j) = -\frac{1}{\sin\beta}\frac{d}{d\beta}\int_{\beta - j\pi/n}^{\beta + j\pi/n} J(u)\,du = \frac{1}{\sin\beta}\left(J\left(\beta + \frac{j\pi}{n}\right) - J\left(\beta - \frac{j\pi}{n}\right)\right)$$

$$= \frac{\sin^{14}((n\beta + j\pi)/2)}{\gamma_{n,6}\sin\beta}\left(\sin^{-14}\left(\frac{\beta - j\pi/n}{2}\right) - \sin^{-14}\left(\frac{\beta + j\pi/n}{2}\right)\right).$$

It remains to use the inequality

$$\sin\left(\frac{\alpha+\beta}{2}\right) \geq \left|\sin\left(\frac{\alpha-\beta}{2}\right)\right|.$$

■

Leviatan (1986) also proved the following estimates:

$$E_n^{(1)}(f) \leq c\overline{\omega}_1\left(\frac{1}{n}, f\right) \tag{6.21}$$

and

$$E_n^{(2)}(f) \leq c\overline{\omega}_2\left(\frac{1}{n}, f\right), \qquad n \in \mathbb{N}. \tag{6.22}$$

6.11. Now let $f \in C^r(I)$. DeVore (1977a) proved that if $f \in \Delta^1 \cap C^r(I)$, then

$$E_n^{(1)}(f) \leq cn^{-r}\left(\frac{1}{n}; f^{(r)}; I\right), \qquad n = r, r+1, \ldots . \tag{6.23}$$

The following general theorem is true:

Theorem 6.4 [Shevchuk (1989), (1992)]. *Suppose that $r \in \mathbb{N}$, $k \in \mathbb{N}$, and $m = k + r$. If a function $f = f(x)$ does not decrease on I and $f \in W^r H[k; \varphi]$, then, for each $n \in \mathbb{N}$, $n \geq m - 1$, there exists an algebraic polynomial $P_n = P_n(x)$ of degree $\leq n$ nondecreasing on I and such that*

$$|f(x) - P_n(x)| < c\rho_n^r(x)\varphi(\rho_n(x)), \qquad x \in I. \tag{6.24}$$

Unfortunately, in contrast to Theorems 6.2 and 6.3, estimates (6.23) and (6.24) are proved by using nonlinear methods.

Theorems 6.1, 6.2, and 6.4 give a complete answer to the question of whether the direct theorem (Theorem 1.1) on approximation without restrictions remains true for the comonotone approximation; namely, it is true in the cases $r \in N$, $k \in N$ and $r = 0$, $k = 1, 2$ and it is not true in the case $r = 0$, $k = 3, 4, \ldots$.

A modification of the proof of Theorem 6.4 enables one to establish the following result:

Theorem 6.5 (S. Manya). *Suppose that $r \in N$, $r \neq 1$, $k \in N$, and $m = k + r$. If a function f is convex on I and $f \in W^r H[k; \varphi]$, then, for each $n \in \mathbb{N}$, $n \geq m - 1$,* *there exists an algebraic polynomial P_n of degree $\le n$ convex on I and satisfying inequality (6.24).*

Inequality (6.21) yields the following estimate for $f \in \Delta^1 \cap B^r$ and $r = 1, 2$:

$$E_n^{(1)}(f) \le \frac{c}{n^r} \left\| \varphi^r f^{(r)} \right\|, \qquad n \in \mathbb{N}. \tag{6.25}$$

For $r > 2$, this estimate was proved by Dzyubenko, Listopad, and Shevchuk (1993).

Similarly, relation (6.22) yields the following estimate for $f \in \Delta^2 \cap B^r$ and $r = 1, 2$:

$$E_n^{(2)}(f) \le \frac{c}{n^r} \left\| \varphi^r f^{(r)} \right\|, \qquad n \in \mathbb{N}. \tag{6.26}$$

For $r > 4$ and $r = 3$, this estimate was proved by Kopotun (1992).

What was unexpected was the fact that estimate (6.26) is not true for $r = 4$.

Theorem 6.6 [Kopotun (1992)]. *For every $n \in \mathbb{N}$ and $M = \text{const} > 0$, there is a function $f \in \Delta^2 \cap B^4$ such that*

$$E_n^{(2)}(f) > M \left\| \varphi^4 f^{(\mathrm{IV})} \right\|.$$

Proof. Let

$$g(x) := (x+1)^2 \ln \frac{1}{x+1}, \qquad g(-1) := 0.$$

We take a positive $b < 1$, set $y_1 := -1 + b$, and represent g in the form

$$g(x) := T(x) + \frac{1}{3!} \int_{y_1}^{x} (x-t)^3 g^{\mathrm{IV}}(t)\, dt,$$

where T is the Taylor polynomial of g at the point y_1. Let us show that the function

$$f(x) := \frac{2}{3!} \int_{y_1}^{x} (x-t)^3 \frac{dt}{(1+t)^2} = \frac{1}{3!} \int_{y_1}^{x} (x-t)^3 g^{\mathrm{IV}}(t)\, dt$$

with suitable b is the required one.

It is clear that $f \in \Delta^{(2)}$ and, for $x \in (-1, 1]$,

$$\varphi^4(x) f^{(IV)}(x) = (1-x^2)^2 \frac{2}{(1+x)^2} = 2(1-x)^2 < 8.$$

Therefore, it remains to prove that, for every convex polynomial P_n, we have

$$\| f - P_n \| > \frac{1}{8} M .$$

Assume the contrary, namely, let

$$\| f - P_n \| \le \frac{1}{8} M .$$

Then, using Markov inequality, we get

$$\begin{aligned} -3 + 2\ln\frac{1}{b} &= f''(y_1) = T''(y_1) \le T''(y_1) + P_n''(y_1) \le \| T'' + P_n'' \| \\ &\le n^2 \| T + P_n \| = n^2 \| P_n - f + g \| \le n^2 (\| P_n - f \| + \| g \|) \\ &\le \frac{n^2 M}{8} + 4\ln 2, \end{aligned}$$

which is impossible for sufficiently small b.

■

For more recent results on the uniform polynomial shape-preserving approximation, see the papers of the authors cited above and the papers by Bondarenko, Gilewicz, Hu, Nissim, Pleshakov, Popov, Shatalina, Yushchenko, Zhou, etc.

7. On rational approximation

7.1. Let us show that if functions of some class H^α, $0 < \alpha \le 1$, are approximated not by algebraic polynomials but by rational polynomials of degree n of the form (see Section 2.4)

$$R_n(x) = \frac{P_n(x)}{Q_n(x)}, \tag{7.1}$$

then, in the sense of the order of smallness, we do not reach an accuracy better than in the case of approximation of functions of this class by algebraic polynomials of degree n (it will be shown in Subsection 7.2 that, for individual functions, this statement is, generally speaking, not true).

To verify this, we consider, on the segment $[-1, 1]$, the even $2/n$-periodic function f_0 defined on the half-period by the equality

$$f_0(x) = x^\alpha - \frac{1}{2}\left(\frac{1}{n}\right)^\alpha, \quad x \in \left[0, \frac{1}{n}\right]. \tag{7.2}$$

This function obviously belongs to the class H^α; furthermore, on the segment $[-1, 1]$, it has $2n+1$ extrema with equal absolute values at the points $x_k = \frac{k}{n}$, $k = 0, \pm 1, \ldots, \pm n$. Since the polynomials P_n and Q_n in (7.1) can be assumed to be irreducible, by virtue of the fact that the polynomials R_n must approximate the function f well on $[-1, 1]$ we can assume in what follows without loss of generality that the polynomial Q_n does not take the zero value on $[-1, 1]$ and, moreover, $Q_n(x) > 0$ for all $x \in [-1, 1]$. In this case, the assumption that, for a certain polynomial

$$R_n^0(x) = \frac{P_n^0(x)}{Q_n^0(x)},$$

one has

$$\left\| f_0 - R_n^0 \right\| = \left\| f_0 - \frac{P_n^0}{Q_0^n} \right\| < \frac{1}{2}\left(\frac{1}{n}\right)^\alpha$$

implies that the numerator P_n^0 of the polynomial R_n^0 must satisfy the following condition at all points x_k:

$$\operatorname{sign} P_n^0(x_k) = \operatorname{sign} f_0(x_k) = (-1)^{k-1}, \quad k = 0, \pm 1, \ldots, \pm n.$$

This, in turn, implies that the algebraic polynomial P_n^0 of degree n must have at least $2n > n+1$ roots on the segment $[-1, 1]$, which is impossible. It follows from the contradiction obtained that

$$\sup_{f \in H^\alpha} \inf_{R_n} \| f - R_n \| \geq \inf_{R_n} \| f_0 - R_n \| \geq \frac{1}{2n^\alpha}. \tag{7.3}$$

On the other hand, if a function f belongs to H^α, then the function $\bar{f}(t) = f(\cos t)$ a fortiori belongs to H^α, and, therefore, one can find an algebraic polynomial P_n of degree n such that

$$|f(x) - P_n(x)| \le \frac{A}{n^\alpha}. \tag{7.4}$$

The required statement now follows from (7.3) and (7.4). Also note that analogous arguments are also true in the case of approximation of functions from the spaces $W^r H^\alpha$.

7.2. It follows from the previous subsection that, in the case of approximation of "rapidly oscillating" functions from the spaces $W^r H^\alpha$, rational polynomials of degree n do not have any advantages over algebraic polynomials. However, it turns out that functions f sufficiently smooth everywhere except one or several points may sometimes be approximated by rational polynomials much better than by algebraic polynomials.

Indeed, in Section 2.4, we have established the Newman theorem (Theorem 2.4.1), according to which, for the function $|x|$, one can find a rational polynomial R_n of degree n such that (see (2.4.10))

$$\||x| - R_n\| \le 3e^{-\sqrt{n}}. \tag{7.5}$$

Let us show that the following relation holds for any algebraic polynomial P_n of degree $n \ge 6$:

$$\||x| - P_n\| \ge \frac{1}{80}\frac{1}{n}. \tag{7.6}$$

The problem of approximation of the function $|x|$ was thoroughly studied by Bernstein (1912), who showed that the following, much stronger, statement is true: The limit $\lim\limits_{n\to\infty} nE_n(|x|)$ exists and is approximately equal to 0.282.

We present the proof of inequality (7.6) given by Dzyadyk (1966). For another simple proof of this inequality proposed by de la Vallée Poussin, see [Natanson (1949), p. 215].

First, note that if a certain algebraic polynomial $P_n(x)$ of degree $n \ge 6$ satisfies the condition

$$\max_{x\in[-1,1]} |P_n(x)| = |P_n(x_0)| = M, \qquad x_0 \in \left[-\frac{1}{4}, \frac{1}{4}\right], \tag{7.7}$$

then

$$\min_{x \in \bar{u}(x_0, 1/2n)} |P_n(x)| \geq \frac{M}{3}, \qquad u\left(x_0, \frac{1}{2n}\right) = \left(x_0 - \frac{1}{2n}, x_0 + \frac{1}{2n}\right). \tag{7.8}$$

Indeed, if inequality (7.8) is not true, then one can find a point

$$\zeta \in \bar{u}\left(x_0, \frac{1}{2n}\right) \subset \left[-\frac{1}{3}, \frac{1}{3}\right]$$

such that

$$|P_n'(\zeta)| \geq \frac{M - M/3}{1/2n} = \frac{4}{3} Mn > \frac{Mn}{\sqrt{1-\zeta^2}},$$

and, hence, we arrive at a contradiction with the Bernstein inequality for the modulus of the derivative of an algebraic polynomial.

By contradiction, assume that inequality (7.6) does not hold for some polynomial. In this case, one can obviously find an even polynomial $P_n(x)$ that satisfies the conditions

$$P_n(0) = 0, \qquad \| |x| - P_n(x) \| < \frac{1}{40n}. \tag{7.9}$$

This implies that the following inequality holds for all $x \in [-1, 1]$:

$$\left| \frac{P_n(x)}{x} \right| \leq 9. \tag{7.10}$$

Indeed, assuming, by contradiction, that

$$\max_{x \in [-1,1]} \left| \frac{P_n(x)}{x} \right| = \left| \frac{P_n(x_0)}{x_0} \right| = M > 9$$

and taking into account that, by virtue of (7.9), one has $|P_n(x)| \leq 2$, we conclude that $x_0 \in \left[-\frac{1}{4}, \frac{1}{4}\right]$ and, hence, according to (7.8),

$$\min_{x \in \bar{u}(x_0, 1/2n)} \left| \frac{P_n(x)}{x} \right| > \frac{M}{3} > 3, \qquad |P_n(x)| > 3|x|, \qquad x \in \bar{u}\left(x_0, \frac{1}{2n}\right),$$

$$\max_{x \in \bar{u}(x_0, 1/2n)} |P_n(x) - |x|| > \max_{x \in \bar{u}(x_0, 1/2n)} |x| \geq \frac{1}{n},$$

i.e., we arrive at a contradiction with (7.9), which proves inequality (7.10).

By virtue of inequality (7.10) and the Bernstein inequality, the following relations hold for all $x \in \left[0, \frac{1}{4}\right]$:

$$\left|\left(\frac{P_n(x)}{x}\right)'\right| \le \frac{9n}{\sqrt{1-x^2}} < 10n,$$

$$\left|\frac{P_n(x)}{x}\right| = \left|\int_0^x \left(\frac{P_n(t)}{t}\right)' dt\right| < 10nx, \qquad P_n(x) \le 10nx^2.$$

Hence,

$$\max \left||x| - P_n(x)\right| \ge \frac{1}{20n} - 10n\left(\frac{1}{20n}\right)^2 = \frac{1}{40}\frac{1}{n},$$

which contradicts (7.9). The required statement is proved.

References

Ahlberg, J. H., Nilson, E. N., and Walsh, J. L.

(1967) *The Theory of Splines and Their Applications,* Academic Press, New York (1967).

Ahlfors, L. V.

(1966) *Lectures on Quasiconformal Mappings,* Nostrand, Toronto (1966).

Akhiezer, N. I.

(1928) "Über einige Funktionen die in gegebenen Intervallen am wenigsten von Null abweichen," *Izv. Kaz. Fiz.-Mat. Obshch.,* **3**, 1–69 (1928).

(1965) *Lectures on Approximation Theory,* Nauka, Moscow (1965) [in Russian].

Akhiezer, N. I., and Krein, M. G.

(1937) "On the best approximation of differentiable periodic functions by trigonometric sums," *Dokl. Akad. Nauk SSSR,* **15**, 107–111 (1937).

Alexits, G.

(1952) "Sur l'ordre de grandeur de l'approximation d'une fonction périodique par les sommes de Fejér," *Math. Hung.,* **3**, 29–42 (1952).

Alibekov, G. A., and Kurbanov, A. A.

(1978) "Refinement of one S. N. Mergelyan's theorem on the approximation of functions of a complex variable in domains with smooth boundary," in: *Investigations in the Theory of Approximation of Functions and Their Applications,* Institute of Mathematics, Ukrainian Academy of Sciences, Kiev (1978), pp. 3–15 [in Russian].

Al'pert, S. Ya.

(1955) "On uniform approximation of functions of a complex variable in a closed domain," *Izv. Akad. Nauk SSSR, Ser. Mat.,* **19**, No. 3, 423–444 (1955).

(1959) "On asymptotic values of the best approximation of analytic functions in a complex domain," *Usp. Mat. Nauk,* **16**, No. 1 (85), 131–134 (1959).

Andrashko, M. I.

(1964) "Inequalities for a derivative of an arbitrary polynomial in the metric of L_p $(p \geq 1)$ in a domain with corners," *Ukr. Mat. Zh.,* **16**, 431–444 (1964).

Andrievskii, V. V.

(1981) "Geometric properties of Dzyadyk domains," *Ukr. Mat. Zh.,* **33**, No. 6, 723–727 (1981).

(1985) "Geometric structure of domains and direct theorems of the constructive theory of functions," *Mat. Sb.,* **126**, No. 1, 41–58 (1985).

Andrievskii, V. V., and German, S. P.

(1982) "Approximation of functions defined by smoothnesses of the kth order in domains with piecewise-quasiconformal boundary," *Ukr. Mat. Zh.,* **34**, No. 3, 273–278 (1982).

Antonyuk, P. E.

(1968) "Inverse theorems on uniform approximation of functions continuous on closed sets with corners," *Dokl. Akad. Nauk Ukr. SSR, Ser. A,* No. 9, 771– 773 (1968).

(1971) "On uniform approximation of functions continuous on closed sets with corners," *Dokl. Akad. Nauk Ukr. SSR, Ser. A,* No. 6, 487–489 (1971).

Babenko, K. I.

(1958) "On the best approximations of one class of analytic functions," *Izv. Akad. Nauk SSSR, Ser. Mat.,* **22**, No. 5, 631–640 (1958).

(1985) "On some problems of approximation theory and numerical analysis," *Usp. Mat. Nauk,* **40**, No. 1, 3–28 (1985).

Bari, N. K.

(1955) "On the best approximation of two conjugate functions by trigonometric polynomials," *Izv. Akad. Nauk SSSR, Ser. Mat.,* **19**, No. 5, 285–302 (1955).

(1961) *Trigonometric Series,* Fizmatgiz, Moscow (1961) [in Russian].

Bari, N. K., and Stechkin, S. B.

(1956) "Best approximation and differential properties of two conjugate functions," *Tr. Mosk. Mat. Obshch.,* **5**, 483–522 (1956).

Beatson, R. K.

(1978) "The degree of monotone approximation," *Pacific J. Math.,* **74**, No. 1, 5–14 (1978).

Belinskii, P. P.

(1954) *Quasiconformal Mappings,* Author's Abstract of the Candidate-Degree Thesis, Lvov University (1954) [in Russian].

Belyi, V. I.

(1965) "Problems of approximation of functions of certain classes in a complex domain. I," *Ukr. Mat. Zh.,* **17**, No. 1, 3–17 (1965).

(1969) "On approximation of quasismooth functions of complex variables," *Izv. Akad. Nauk Arm. SSR, Ser. Mat,* **5**, No. 4, 356–393 (1969).

(1977) "Conformal mappings and approximation of analytic functions in domains with quasiconformal boundary," *Mat. Sb.,* **102 (144)**, No. 3, 331–361 (1977).

(1977a) "Asymptotic behavior of a conformal mapping near the boundary and conditions for a domain to belong to a class of sets (B)," *Dokl. Akad. Nauk Ukr. SSR, Ser. A,* No.4, 291–294 (1977).

(1979) "Approximation of functions of the classes $A^{(r)}(G)$ in finite domains with quasiconformal boundary," in: *Theory of Functions and Mappings,* Naukova Dumka, Kiev (1979), pp. 37–62 [in Russian].

Belyi, V. I., and Miklyukov, V. M.

(1974) "Some properties of conformal and quasiconformal mappings and direct theorems of constructive theory of functions," *Izv. Akad. Nauk SSSR, Ser. Mat.,* **38**, No. 6, 1343–1361 (1974).

Bergh, J., and Löfström, J.

(1976) *Interpolation Spaces. An Introduction,* Springer, Berlin (1976).

Bernstein, S. N.

(1912) "On the best approximation of continuous functions by polynomials of given degree" (1912), in: S. N. Bernstein, *Collected Works,* Vol. 1, Academy of Sciences of the USSR, Moscow (1952), pp. 11–104 [in Russian].

(1926) *Leçons sur les Propriétés Extrémales et la Meilleure Approximation des Fonctions Analytiques d'une Variable Réelle,* Gauthier-Villars, Paris (1926).

(1928) "Absolutely monotone functions" (1928), in: S. N. Bernstein, *Collected Works,* Vol. 1, Academy of Sciences of the USSR, Moscow (1952), pp. 370–425 [in Russian].

(1930) "On estimates for derivatives of polynomials" (1930), in: S. N. Bernstein, *Collected Works,* Vol. 1, Academy of Sciences of the USSR, Moscow (1952), pp. 497–499 [in Russian].

(1931) "On the restriction of values of a polynomial $P_n(x)$ of degree n on the entire segment on the basis of its values at $n+1$ points" (1931), in: S. N. Bernstein, *Collected Works*, Vol. 2, Academy of Sciences of the USSR, Moscow (1954), pp. 107–126 [in Russian].

(1931a) "On one class of interpolation formulas" (1931), in: S. N. Bernstein, *Collected Works*, Vol. 2, Academy of Sciences of the USSR, Moscow (1954), pp. 146–154 [in Russian].

(1935) "On some extremal properties of successive integrals" (1935), in: S. N. Bernstein, *Collected Works*, Vol. 2, Academy of Sciences of the USSR, Moscow (1954), pp. 107–172 [in Russian].

(1937) *Extremal Properties of Polynomials and the Best Approximation of Continuous Functions of One Real Variable,* Part 1, ONTI, Moscow (1937) [in Russian].

(1938) "On the best approximation of $|x-c|^p$," *Dokl. Akad. Nauk SSSR,* **18**, 379–384 (1938).

Besov, O. V.

(1962) "On the extension of a function with preservation of properties of the integral modulus of smoothness of the second order," *Mat. Sb.,* **58**, No. 2, 673–684 (1962).

(1963) "Extension of functions with preservation of differential-difference properties in L_p," *Dokl. Akad. Nauk SSSR,* **150**, No. 3, 963–966 (1963).

(1965) "Extension of functions outside the limits of the domain with preservation of the differential-difference properties in L_p," *Mat. Sb.,* **66**, No. 1, 80–96 (1965).

Besov, O. V., Il'in, V. P., and Nikol'skii, S. M.

(1975) *Integral Representations of Functions and Imbedding Theorems,* Nauka, Moscow (1975) [in Russian].

Besov, O. V., and Stechkin, S. B.

(1975) "Description of moduli of continuity in L_p," *Tr. Mat. Inst. Steklov.,* **134**, 23–25 (1975).

Bijvoets, L., Hogeveen, W., and Korevaar, J.

(1980) *Inverse Approximation Theorems of Lebedev and Tamrasov,* Preprint **1**, No. 6, University of Amsterdam (1980).

Binev, P.

(1985) "$O(n)$ bounds of Whitney constants," *C. R. Acad. Bul. Sci.,* **38**, No. 10, 1303–1305 (1985).

Binev, P. G., and Ivanov, K. G.

(1985) "On a representation of mixed finite differences," *SERDICA Bul. Math. Publ.*, **11**, 259–268 (1985).

Bohr, H.

(1935) "Un théorème général sur l'intégration d'un polynôme trigonométrique," *Comptes Rendus*, **200**, 1276–1277 (1935).

Boikov, I. V.

(1987) "Algorithms for calculation of integrals with optimal accuracy," in: *Optimal Calculation Methods and Their Application*, Penza Polytechnic Institute, Penza (1987), pp. 4–22 [in Russian].

Borel, E.

(1905) *Leçons sur les Fonctions de Variables Réelles*, Gauthier-Villars, Paris (1905).

Borodin, V. A.

(1974) "Local moduli of continuity and properties of harmonic functions in domains with corners," in: *Theory of Approximation of Functions and Its Applications*, Institute of Mathematics, Ukrainian Academy of Sciences, Kiev (1974), pp. 4–14 [in Russian].

Brudnyi, Yu. A.

(1959) "Approximation by entire functions in the exterior of a segment and a half axis," *Dokl. Akad. Nauk SSSR*, **124**, No. 4, 739–742 (1959).

(1963) "Generalization of one theorem of A. F. Timan," *Dokl. Akad. Nauk SSSR*, **188**, No. 6, 1237–1240 (1963).

(1964) "On one theorem of the theory of local approximation," *Uch. Zap. Kazan. Univ.*, **124**, Book 6, 43–49 (1964).

(1965) "On one method for approximation of bounded functions defined on a segment," in: *Investigation of the Contemporary Problems in the Constructive Theory of Functions*, Baku (1965), pp. 40–45 [in Russian].

(1968) "Approximation of functions by algebraic polynomials," *Izv. Akad. Nauk SSSR, Ser. Mat.*, **32**, No. 4, 780–787 (1968).

(1970) "Approximation of functions of n variables by quasipolynomials," *Izv. Akad. Nauk SSSR, Ser. Mat.*, **34**, No. 6, 564–583 (1970).

(1983) "Whitney constants," in: *Abstracts of the Conference on Contemporary Problems in the Theory of Functions* (1983), p. 24 [in Russian].

Brudnyi, Yu. A., and Gopengauz, I. E.

(1960) "On the measure of the set of points of normal deviation," *Izv. Akad. Nauk SSSR, Ser. Mat.*, **24**, No. 1, 129–144 (1960).

Brudnyi, Yu. A., Krein, S. G., and Semenov, E. M.

(1987) *Interpolation of Linear Operators,* VINITI Series in Mathematical Analysis, Vol. 24, VINITI, Moscow (1987) [in Russian].

Brudnyi, Yu. A., and Shvartsman, P. A.

(1982) "Description of the trace of a function from a generalized Lipschitz space on an arbitrary compact set," in: *Investigations in the Theory of Functions of Many Real Variables,* Yaroslavl University, Yaroslavl (1982), pp. 16– 24 [in Russian].

(1985) "Linear operator of extension for a space of smooth functions defined on a closed subset of R^n," *Dokl. Akad. Nauk SSSR,* **280**, No. 2, 268–272 (1985).

Brui, I. N.

(1975) *Linear Summation Methods for Series in Faber Polynomials,* Candidate-Degree Thesis, Minsk Pedagogic Institute (1975) [in Russian].

Bugaets, P. T.

(1951) "Approximation of continuous periodic functions of two variables that satisfy the Lipschitz condition by interpolation trigonometric polynomials," *Dokl. Akad. Nauk SSSR,* **79**, No. 3, 381–384 (1951).

(1951a) "Asymptotic estimate for the remainder in the case of approximation of functions of two variables by Fourier sums," *Dokl. Akad. Nauk SSSR,* **79**, No. 4, 557–560 (1951).

Bugaets, V. P., and Martynyuk, V. T.

(1974) "Exact constants of approximation of continuous periodic functions by Jackson integrals," *Ukr. Mat. Zh.,* **26**, No. 4, 435–444 (1974).

Burenkov, V. I.

(1976) "On the extension of functions with preservation of a seminorm," *Dokl. Akad. Nauk SSSR,* **228**, No. 4, 779–782 (1976).

Burkill, H.

(1952) "Cesàro–Perron absolute periodic functions," *Proc. London Math. Soc.,* **3**, 160–174 (1952).

Bushanskii, A. V.

(1974) "On the problem of the best uniform approximation by entire functions on classes of functions represented in the form of sources on the real axis," *Izv. Severo-Kavkazsk. Nauchn. Tsentra Vyssh. Shk., Ser. Estestv. Nauk,* No. 4, 54–60 (1974).

(1974a) "On the best uniform approximation by entire functions for some classes of functions represented in the form of sources on the real axis," in: *Theory of Approximation of Functions and Its Applications,* Institute of Mathematics, Ukrainian Academy of Sciences, Kiev (1974), pp. 28–40 [in Russian].

Butzer, P. L.

(1956) "Sur la théorie des demi-groupes et classes de saturation de certaines intégrales singulières," *Comptes Rendus Acad. Sci.,* **243**, No. 20, 1473–1475 (1956).

Butzer, P. L., and Nessel, R. J.

(1971) *Fourier Analysis and Approximation,* Birkhäuser, Basel (1971).

Cartan, H.

(1940) "Sur les classes de fonctions définies par des inégalités portant sur leurs dérivées successives," *Actualités Scient. Industr.,* No. 867 (1940).

Chebyshev, P. L.

(1854) "Theory of mechanisms known as parallelograms" (1854), in: P. L. Chebyshev, *Complete Works,* Vol. 2, Academy of Sciences of the USSR, Moscow (1948), pp. 23–51 [in Russian].

(1859) "Problems of the least quantities related to approximate representation of functions" (1859), in: P. L. Chebyshev, *Complete Works,* Vol. 2, Academy of Sciences of the USSR, Moscow (1948), pp. 151–235 [in Russian].

(1859a) "Sur les questions des minima" (1859), in: P. L. Chebyshev, *Complete Works,* Vol. 2, Academy of Sciences of the USSR, Moscow (1948), pp. 146–250 [in Russian].

(1948) *Complete Works,* Vol. 3, Academy of Sciences of the USSR, Moscow (1948) [in Russian].

Cheney, E. W.

(1966) *Introduction to Approximation Theory,* McGraw-Hill, New York (1966).

Chirka, E. M.

(1975) "Analytic representation of *CR*-functions," *Mat. Sb.,* **98**, No. 4, 591–623 (1975).

Ciesielski, Z.

(1979) *Lectures on Spline Functions,* Gdansk University, Gdansk (1979) [in Polish].

Curtiss, J. H.

(1947) "A note on the degree of polynomial approximation," *Bull. Amer. Math. Soc.,* **42**, 29–30 (1947).

Dahlhaus, R.

(1989) "Pointwise approximation by algebraic polynomials," *J. Approxim. Theory,* **57**, No. 3, 274–277 (1989).

Dahmen, W., DeVore, R. A., and Scherer, K.

(1980) "Multi-dimensional spline approximation," *SIAM J. Numer. Anal.,* **17**, 308–402 (1980).

Daugavet, I. K.

(1963) "On one property of completely continuous operators in the space *C*," *Usp. Mat. Nauk,* **18**, No. 5, 157–158 (1963).

Davis, P. J.

(1965) *Interpolation and Approximation,* Blaisdell, New York (1965).

De Boor, C.

(1976) "Splines as linear combinations of *B*-splines," in: G. G. Lorentz, C. K. Chui, and L. L. Schumaker (editors), *Approximation Theory, II,* Academic Press, New York (1976), pp. 1–47.

Denisenko, P. N.

(1974) "On the numerical solution of ordinary differential equations according to the Dzyadyk's scheme," *Ukr. Mat. Zh.,* **26**, No. 3, 377–383 (1974).

DeVore, R. A.

(1976) "Degree of approximation," in: G. G. Lorentz, C. K. Chui, and L. L. Schumaker (editors), *Approximation Theory, II,* Academic Press, New York (1976), pp. 117–162.

(1977) "Pointwise approximation by splines," in: *Theory of Approximation of Functions,* Nauka, Moscow (1977), pp. 132–141 [in Russian].

(1977a) "Monotone approximation by polynomials," *SIAM J. Math. Anal.,* **8**, No. 5, 905–921 (1977).

DeVore, R. A., and Yu., X. M.

(1985) "Pointwise estimates for monotone polynomial approximation," *Constr. Approxim.,* **1**, No. 4, 323–331 (1985).

Dickmeis, W., and Nessel, R. J.

(1982) *Quantitative Prinzipien gleichmäßiger Beschränktheit und Schärfe von Fehlerabschätzungen,* Westdt. Verlag, Opladen (1982).

Dickmeis, W., Nessel, R. J., and Wickeren, E. van

(1984) "A general approach to counterexamples in numerical analysis," *Numer. Math.,* **43**, 249–263 (1984).

Dieudonné, J.

(1960) *Foundations of Modern Analysis,* Academic Press, New York (1960).

Ditzian, Z., and Tolik, V.

(1987) *Moduli of Smoothness,* Springer, New York (1987).

Dolzhenko, E. P.

(1965) "Smoothness of analytic functions at boundary points of a domain," *Izv. Akad. Nauk SSSR, Ser. Mat.,* **29**, No. 5, 1069–1084 (1965).

Dolzhenko, E. P., and Sevast'yanov, E. A.

(1978) "On the dependence of properties of functions on the rate of their approximation by polynomials," *Izv. Akad. Nauk SSSR, Ser. Mat.,* **42**, No. 2, 272–304 (1978).

Dunford, N., and Schwartz, J. T.

(1958) *Linear Operators. Part I: General Theory,* Interscience, New York (1958).

Dyn'kin, E. M.

(1974) "On the uniform approximation by polynomials in a complex domain," in: *Investigation of Linear Operators in the Theory of Functions,* Vol. 5, Nauka, Leningrad (1974), pp. 164–165 [in Russian].

Dzhrbashyan, M. M.

(1966) *Integral Transformations and Representations of Functions in a Complex Domain,* Nauka, Moscow (1966) [in Russian].

Dzyadyk, V. K.

(1951) "On the best approximation of periodic functions with singularities in the mean," *Dokl. Akad. Nauk SSSR,* **77**, No. 6, 949–952 (1951).

(1953) "On the best approximation on the class of periodic functions with bounded sth derivative $(0 < s < 1)$," *Izv. Akad. Nauk SSSR, Ser. Mat.,* **17**, No. 2, 135–162 (1953).

(1953a) "On the best approximation of periodic functions with singularities in the mean," *Nauk. Zap. Luts'k. Pedagog. Inst., Ser. Fiz.-Mat.,* **1**, 51–65 (1953).

(1955) "On specific features of fractional differentiation and integration of functions," *Nauk. Zap. Luts'k. Pedagog. Inst., Ser. Fiz.-Mat.,* **2**, 19–28 (1955).

(1956) "On the constructive characteristic of functions satisfying the condition $\mathrm{Lip}\,\alpha$ $(0 < \alpha < 1)$ on a finite segment of the real axis," *Izv. Akad. Nauk SSSR, Ser. Mat.,* **20**, No. 2, 623–642 (1956).

(1956a) "On the extension of functions that satisfy the Lipschitz condition in the metric of L_p," *Mat. Sb.,* **40 (82)**, No. 2, 239–242 (1956).

(1958) "On the approximation of functions by ordinary polynomials on a finite segment of the real axis," *Izv. Akad. Nauk SSSR, Ser. Mat.,* **22**, No. 3, 337–354 (1958).

(1958a) "On one new proof of the Weierstrass theorem on the approximation of continuous functions by polynomials," in: *Proceedings of the Conference of Lecturers in Physics and Mathematics of Pedagogic Institutes of Ukr. SSR,* Ukrainian Ministry of Education, Kiev State Pedagogic Institute, Kiev (1958) [in Ukrainian].

(1958b) "Further improvement of the Jackson theorem on approximation of continuous functions by ordinary polynomials," *Dokl. Akad. Nauk SSSR,* **121**, No. 3, 403–406 (1958).

(1959) "On S. M. Nikol'skii's problem in a complex domain," *Izv. Akad. Nauk SSSR, Ser. Mat.,* **23**, No. 5, 697–736 (1959).

(1959a) "On the best approximation on classes of periodic functions defined by kernels that are integrals of absolutely monotone functions," *Izv. Akad. Nauk SSSR, Ser. Mat.,* **23**, No. 6, 933–950 (1959).

(1959b) On the best trigonometric approximation of some functions in the metric of *L*," *Dokl. Akad. Nauk SSSR,* **129**, No. 1, 19–22 (1959).

(1961) "On the problem of the best approximation of absolutely monotone and some other functions in the metric of L using trigonometric polynomials," *Izv. Akad. Nauk SSSR, Ser. Mat.,* **25**, No. 2, 173–238 (1961).

(1962) "On the problem of approximation of continuous functions in closed domains with corners and on S. M. Nikol'skii's problem. I," *Izv. Akad. Nauk SSSR, Ser. Mat.,* **26**, No. 6, 796–824 (1962).

(1963) "Inverse theorems of the theory of approximation of functions in complex domains," *Ukr. Mat. Zh.,* **15**, No. 4, 365–375 (1963).

(1963a) "On the theory of approximation of analytic functions continuous in closed domains and on S. M. Nikol'skii's problem. II," *Izv. Akad. Nauk SSSR, Ser. Mat.,* **27**, No. 5, 1135–1164 (1963).

(1965) "On the approximation of analytic functions in domains with smooth and piecewise-smooth boundary," in: *Proceedings of the Third Summer Mathematical School,* Naukova Dumka, Kiev (1965), pp. 29–82 [in Russian].

(1966) "Approximation of functions of the classes H^{ω} by polynomials on closed domains," in: *Investigation of Contemporary Problems of Constructive Theory of Functions,* Baku (1966), pp. 273–283 [in Russian]; "On the constructive characteristic of functions of Hölder classes," in: *Contemporary Problems of Constructive Theory of Analytic Functions,* Nauka, Moscow (1966), pp. 137–141 [in Russian].

(1966a) "On a new method for approximation of functions," *Ukr. Mat. Zh.,* **18**, No. 1, 36–47 (1966).

(1966b) "On approximation of functions by linear positive operators and singular integrals," *Mat. Sb.,* **70 (112)**, No. 4, 508–517 (1966).

(1967) "On analytic and harmonic transformations of functions and approximation of harmonic functions," *Ukr. Mat. Zh.,* **19**, No. 5, 33–57 (1967).

(1968) "On the constructive characteristic of functions of Hölder classes on closed sets with piecewise-smooth boundary admitting zero angles," *Ukr. Mat. Zh.,* **20**, No. 5, 603–619 (1968).

(1969) "Investigations in the theory of approximation of analytic functions carried out at the Institute of Mathematics of the Ukrainian Academy of Sciences," *Ukr. Mat. Zh.,* **21**, No. 2, 173–192 (1969).

(1970) "On the application of linear methods to the approximation of solutions of ordinary differential equations and Hammerstein integral equations by polynomials," *Izv. Akad. Nauk SSSR, Ser. Mat.,* **34**, No. 4, 827–848 (1970).

(1971) "On one method for the construction of normal solutions in Tikhonov's sense for systems of linear equations," *Ukr. Mat. Zh.,* **23**, No. 2, 235–239 (1971).

(1971a) "On error estimates for the grid method in the theory of Chebyshev approximation of functions," *Dokl. Akad. Nauk Ukr. SSR, Ser. A,* No. 6, 500– 504 (1971).

(1972) "On the application of generalized Faber polynomials to the approximation of Cauchy-type integrals and functions of the classes A^r in domains with smooth and piecewise-smooth boundary," *Ukr. Mat. Zh.,* **24**, No. 1, 3–19 (1972).

(1973) "On one extremal problem," *Dokl. Akad. Nauk Ukr. SSR, Ser. A,* No. 4, 299–300 (1973).

(1973a) "On the efficient construction of polynomials that realize an approximation of functions e^x, $\sin x$, etc., close to the best one," *Ukr. Mat. Zh.,* **25**, No. 4, 435–453 (1973).

(1974) "Interpolation method for the approximation of solutions of linear differential equations by algebraic polynomials," *Izv. Akad. Nauk SSSR, Ser. Mat.,* **33**, No. 4, 937–967 (1974).

(1974a) "On conditions for the convergence of Dirichlet series in closed polygons," *Mat. Sb.,* **95 (137)**, No. 4 (12), 475–493 (1974).

(1974b) "On the best approximation on classes of periodic functions defined by integrals of a linear combination of absolutely monotone kernels," *Mat. Zametki,* **16**, No. 5, 691–701 (1974).

(1974c) "On the approximation of functions on sets that consist of finitely many points," in: *Theory of Approximation of Functions and Its Applications,* Institute of Mathematics, Ukrainian Academy of Sciences, Kiev (1974), pp. 69–80 [in Russian].

(1975) "Approximation characteristic of the Lipschitz classes W^rH^1, $r = 0, 1, \ldots$," *Analysis Math.,* No. 1, 19–30 (1975).

(1975a) "On the least upper bounds for the best approximations on certain classes of continuous functions defined on the entire real axis," *Dokl. Akad. Nauk Ukr. SSR, Ser. A,* No. 7, 589–593 (1975).

(1975b) "Application of methods of the theory of quasiconformal mappings to the proof of direct theorems on approximation of functions," *Dokl. Akad. Nauk SSSR,* **225**, No. 3, 491–494 (1975).

(1975c) "On the theory of approximation of functions on closed sets of the complex plane," *Tr. Steklov Mat. Inst.,* **134**, 63–114 (1975).

(1976) "On the application of linear operators to the approximate solution of ordinary differential equations," in: *Problems of the Theory of Approximation of Functions and Its Applications,* Institute of Mathematics, Ukrainian Academy of Sciences, Kiev (1976), pp. 61–96 [in Russian].

(1977) "On the approximation of functions of complex variables on arcs," *Ukr. Mat. Zh.,* **29**, No. 2, 254–258 (1977).

(1977a) "On a problem of Chebyshev and Markov," *Analysis Math.,* No. 3 (3), 174–175 (1977).

(1977b) "On the constructive theory of functions on closed sets of the complex plane," in: *Proceedings of the International Conference on the Theory of Approximation of Functions,* Nauka, Moscow (1977), pp. 157–172 [in Russian].

(1978) "On the behavior of the set of extremal polynomials on the basis in the T-system in the presence of constraints," in: *Investigations in the Theory of Functions and Its Applications,* Institute of Mathematics, Ukrainian Academy of Sciences, Kiev (1978), pp. 56–60.

(1979) "On an error estimate for the polynomial approximation of solutions of ordinary differential equations," *Ukr. Mat. Zh.,* **31**, No. 1, 83–89 (1979).

(1988) *Approximation Methods for the Solution of Differential and Integral Equations,* Naukova Dumka, Kiev (1988) [in Russian].

Dzyadyk, V. K., and Alibekov, G. A.

(1968) "On the uniform approximation of functions of complex variables on closed sets with corners," *Mat. Sb.,* **75 (117)**, No. 4, 502–557 (1968).

Dzyadyk, V. K., and Dubovik, V. A.

(1974) "On the Kolmogorov problem of dependences between upper bounds of derivatives of real functions defined on the entire axis," *Ukr. Mat. Zh.,* **26**, No. 3, 300–317 (1974).

(1975) "On the Kolmogorov problem of dependences between upper bounds of derivatives of real functions defined on the entire axis. II," *Ukr. Mat. Zh.,* **27**, No. 3, 291–299 (1975).

Dzyadyk, V. K., and Galan, D. M.

(1965) "On the approximation of analytic functions in domains with smooth boundary," *Ukr. Mat. Zh.,* **17**, No. 1, 26–38 (1965).

Dzyadyk, V. K., Gavrilyuk, V. T., and Stepanets, A. I.

(1969) "On the approximation of functions of Hölder classes by Rogosinski polynomials," *Dokl. Akad. Nauk Ukr. SSR, Ser. A,* No. 3, 203–206 (1969).

(1970) "On the least upper bound for approximations on classes of differentiable periodic functions using Rogosinski polynomials," *Ukr. Mat. Zh.,* **22**, No. 2, 481–493 (1970).

Dzyadyk, V. K., and Konovalov, V. N.

(1973) "A method for the partition of unity in domains with piecewise-smooth boundary into a sum of algebraic polynomials of two variables possessing some properties of kernels," *Ukr. Mat. Zh.,* **25**, No. 2, 179–192 (1973).

Dzyadyk, V. K., and Krutigolova, E. K.

(1973) "On the representation of analytic functions by Dirichlet series on the boundary of the domain of convergence," *Mat. Zametki,* **14**, No. 6, 769–780 (1973).

Dzyadyk, V. K., and Shevchuk, I. A.

(1972) "On the limit values of a Cauchy-type integral for functions of Zygmund classes," *Ukr. Mat. Zh.,* **24**, No. 5, 601–617 (1972).

(1983) "Extension of functions that are traces of functions with given second modulus of continuity on an arbitrary set of a straight line," *Izv. Akad. Nauk SSSR, Ser. Mat.,* **47**, No. 2, 248–267 (1983).

Dzyadyk, V. K., and Shvai, A. I.

(1967) "On the constructive characteristic of functions of Hölder classes in domains with piecewise-smooth boundary and arbitrary positive external angles," *Dokl. Akad. Nauk Ukr. SSR, Ser. A,* **11**, 1012–1015 (1967).

(1971) "On the approximation of functions of Hölder classes on closed sets with acute external angles," in: *Metric Problems in the Theory of Functions and Mappings,* Issue 2, Naukova Dumka, Kiev (1971), pp. 74–164 [in Russian].

Dzyadyk, V. K., and Stepanets, A. I.

(1971) "On the sequence of zeros of the integral sine," in: *Metric Problems in the Theory of Functions and Mappings,* Issue 2, Naukova Dumka, Kiev (1971), pp. 64–73 [in Russian].

(1972) "Asymptotic equalities for least upper bounds of approximations of functions of Hölder classes by using Rogosinski polynomials," *Ukr. Mat. Zh.,* **24**, No. 4, 476–487 (1972).

Dzyubenko, G. A., Listopad, V. V., and Shevchuk, I. A.

(1993) "Uniform estimates for a monotone polynomial approximation," *Ukr. Mat. Zh.*, **45**, No. 1, 38–43 (1993).

Efimov, A. V.

(1959) "On approximation of periodic functions by de la Vallée Poussin sums. I," *Izv. Akad. Nauk SSSR, Ser. Mat.*, **23**, No. 5, 737–770 (1959).

(1960) "On linear summation methods for Fourier series," *Izv. Akad. Nauk SSSR, Ser. Mat.*, **24**, No. 5, 743–756 (1960).

(1960a) "Approximation of continuous periodic functions by Fourier sums," *Izv. Akad. Nauk SSSR, Ser. Mat.*, **24**, No. 2, 243–296 (1960).

(1960b) "On approximation of periodic functions by de la Vallée Poussin sums. II," *Izv. Akad. Nauk SSSR, Ser. Mat.*, **24**, No. 3, 431–468 (1960).

(1961) "Linear methods for approximation of some classes of continuous periodic functions," *Mat. Sb.*, **54 (96 : 1)**, 51–90 (1961).

Elliot, M. H.

(1951) "On approximation to functions satisfying a generalized continuity condition," *Trans. Amer. Math. Soc.*, **71**, No. 1, 1–23 (1951).

Erdős, P., and Turan, P.

(1961) "An extremal problem in the theory of interpolation," *Acta Math. Acad. Sci. Hung.*, **12**, Fas. 1–2, 221–234 (1961).

Faber, G.

(1903) "Über polynomische Entwicklungen," *Math. Ann.*, **57**, 389–408 (1903).

(1907) "Über polynomische Entwicklungen II," *Math. Ann.*, **64**, 116–135 (1907).

Favard, J.

(1936) "Application de la formule sommatoire d'Euler à la démonstration de quelques propriétés extrémales des intégrale des fonctions périodiques ou presquepériodiques," *Mat. Tidsskr. B*, **4**, 81–94 (1936).

(1937) "Sur les meilleurs procédés d'approximation de certaines classes de fonctions par des polynômes trigonométriques," *Bull. Sci. Math.*, **61**, 209–224, 243–256 (1937).

(1949) "Sur l'approximation dans les espaces vectoriels," *Ann. Math.*, **24**, No. 4, 259–291 (1949).

(1957) "Sur la saturation des procédés de sommation," *J. Math. Pures Appl.*, **36**, No. 9, 359–372 (1957).

Fejér, L.

(1904) "Untersuchungen über Fouriersche Reihen," *Math. Ann.,* **58**, 501–569 (1904).

Freud, G.

(1959) *Über die Approximation reeller stetigen Funktionen durch gewöhnliche Polynome,* Math. Ann., **137**, No. 1, 17–25 (1959).

Freud, G., and Popov, V.

(1972) "On approximation by spline functions," in: *Proceedings of the Conference on the Constructive Theory of Functions (Budapest, 1969),* Acad. Kiado, Budapest (1972), pp. 163–172.

Frey, T.

(1958) "A legjobb polinom approximáció lokalizálásázoe, II, MTA, III," *Oszt. Közl.,* VIII/1, 89–112 (1958).

Fuksman, A. L.

(1965) "Structural characteristic of functions for which $E_n(f; -1, 1) \leq Mn^{-(k+\alpha)}$," *Usp. Mat. Nauk,* **20**, No. 4, 187–190 (1965).

Gaier, D.

(1980) *Vorlesungen über Approximation im Komplexen,* Birkhäuser, Basel (1980).

Galan, D. M.

(1973) "On the problem of classification of continuous functions," *Dopov. Akad. Nauk Ukr. RSR,* No. 1, 8–11 (1973).

(1975) "On one hypothesis of V. K. Dzyadyk," *Ukr. Mat. Zh.,* **27**, No. 5, 579–589 (1975).

Galan, V. D.

(1991) *Smooth Functions and Estimates for Derivatives,* Author's Abstract of the Candidate-Degree Thesis, Kiev (1991) [in Russian].

Galan, V. D., and Galan, D. M.

(1987) "On the degree of smoothness and moduli of continuity in the classification of continuous functions," in: *Tr. Mat. Inst. Akad. Nauk SSSR,* **180**, No. 3, 30–50 (1987).

Garkavi, A. L.

(1964) "On a criterion for an element of the best approximation," *Sib. Mat. Zh.,* **5**, No. 2, 472–476 (1964).

Gavrilyuk, V. T.

(1973) "Approximation of continuous periodic functions of one and two variables by Rogosinski polynomials of the interpolation type," *Ukr. Mat. Zh.,* **25**, No. 5, 637–648 (1973).

(1974) "On the approximation of continuous periodic functions of many variables by trigonometric polynomials," *Dokl. Akad. Nauk Ukr. SSR, Ser. A,* No. 11, 966–969 (1974).

Gavrilyuk, V. T., and Stepanets, A. I.

(1973) "Approximation of differentiable functions by Rogosinski polynomials," *Ukr. Mat. Zh.,* **25**, No. 1, 3–13 (1973).

(1973a) "On the least upper bounds of the deviations of Bernstein sums from functions of Hölder classes," *Ukr. Mat. Zh.,* **25**, No. 2, 158–169 (1973).

Gehring, F. W., Heyman, W. K., and Hinkkanen, A.

(1982) "Analytic functions, satisfying Hölder conditions on the boundary," *J. Approxim. Theory,* **35**, No. 3, 243–249 (1982).

Geit, V. É.

(1972) "Imbedding theorems for some classes of periodic functions," *Izv. Vyssh. Uchebn. Zaved., Ser. Mat.,* No. 4, 67–77 (1972).

Gel'fond, A. O.

(1969) *Calculus of Finite Differences,* Fizmatgiz, Moscow (1969) [in Russian].

Goluzin, G. M.

(1966) *Geometric Theory of Functions of Complex Variables,* Nauka, Moscow (1966) [in Russian].

Gonchar, A. A., and Mergelyan, S. N.

(1970) "Theory of approximation of functions of complex variables," in: *History of Domestic Mathematics,* Vol. 4, Book 1, Naukova Dumka, Kiev (1970), pp. 112–152 [in Russian].

Goncharov, L. V.

(1954) *Theory of Interpolation and Approximation of Functions,* GTTL, Moscow (1954) [in Russian].

Gorny, A.

(1939) "Contribution à l'etude des fonctions dérivables d'une variable réelle," *Acta Math.,* **76**, 317–358 (1939).

Grebenyuk, D. G.

(1960) *Polynomials of Best Approximation with Linearly Related Coefficients,* Academy of Sciences of Uzb. SSR, Tashkent (1960) [in Russian].

Guseinov, E. G.

(1979) "Imbedding theorems for classes of continuous functions," *Izv. Akad. Nauk Azerb. SSR, Ser. Mat.,* No. 2, 57–60 (1979).

Guseinov, E. G., and Il'yasov, N. A.

(1977) "Differential and smoothness properties of continuous functions," *Mat. Zametki,* **22**, No. 6, 785–794 (1977).

Haar, A.

(1918) "Die Minkowskische Geometrie und die Annäherung an stetige Funktionen," *Math. Ann.,* **78**, 293–311 (1918).

Hadamard, J.

(1914) "Sur le module maximum d'une fonction et de ses dérivées," *C. R. Séanses Soc. Math. Fr.,* **42** (1914).

Hardy, G. H., and Littlewood, J. E.

(1928) "Some properties of fractional integrals. I," *Math. Zeitschr.,* **27**, No. 1, 565–606 (1928).

(1932) "Some properties of fractional integrals. II," *Math. Zeitschr.,* **34**, No. 1, 403–439 (1932).

Harshiladze, F. I.

(1955) "Gibbs phenomenon in the summation of Fourier series by the Bernstein–Rogosinski method," *Dokl. Akad. Nauk SSSR,* **101**, No. 3, 425–428 (1955).

(1958) "Classes of saturation for some summation processes," *Dokl. Akad. Nauk SSSR,* **122**, No. 3, 352–355 (1958).

Hasson, M.

(1982) "The sharpness of Timan's theorem on differentiable functions," *J. Approxim. Theory,* **35**, No. 3, 264–274 (1982).

Helly, E.

(1923) "Über Mengen konvexer Körper mit gemeinschaftlichen Punkten," *Jahresber. Dtsch. Math.-Ver.,* **32**, 175–176 (1923).

Hermite, G.

(1878) “Sur la formule d’interpolation de Lagrange,” *J. Reine Angew. Math.*, **84**, 70–79 (1878).

Hestenes, M.

(1941) “Extension of the range of differentiable functions,” *Duke Math. J.*, **8**, 183–192 (1941).

Heuser, P.

(1939) “Zur Theorie der Faberschen Polynomreihen,” *Dtsch. Math.*, **4**, 451–454 (1939).

Ibragimov, I. I.

(1946) “On the asymptotic value of the best approximation of functions having a real singular point,” *Izv. Akad. Nauk SSSR, Ser. Mat.*, **10**, No. 5, 429–460 (1946).

(1971) *Methods for Interpolation of Functions and Their Applications,* Nauka, Moscow (1971) [in Russian].

Ivanov, K., and Takev, M. O.

(1985) “$O(n \ln n)$ bounds of Whitney constants,” *C. R. Acad. Bul. Sci.*, **38**, No. 9, 1129–1131 (1985).

Ivanov, V. K.

(1951) “Minimax problem for a system of linear functions,” *Mat. Sb.*, **28 (70)**, No. 3, 685–706 (1951).

(1952) “On uniform approximation of continuous functions,” *Mat. Sb.*, **30 (72)**, No. 3, 543–558 (1952).

Ivanov, V. V.

(1958) “Some properties of singular integrals,” *Izv. Akad. Nauk SSSR, Ser. Mat.*, **4**, No. 5, 89–95 (1958).

Jackson, D.

(1911) *Über die Genauigkeit der Annäherung stetiger Funktionen durch ganze rationale Funktionen gegebenen Grades und trigonometrische Summen gegebener Ordnung,* Dissertation, Göttingen (1911).

(1912) “On approximation by trigonometric sums and polynomials,” *Trans. Amer. Math. Soc.*, **13**, 491–515 (1912).

(1924) “A general class of problems in approximation,” *Amer. J. Math.*, **46**, 215– 234 (1924).

(1930) *The Theory of Approximation,* American Mathematical Society, New York (1930).

(1941) *Fourier Series and Orthogonal Polynomials,* Carus Mathematical Monographs, No. 6, New York (1941).

Jonsson, A.

(1980) "The trace of the Zygmund class Λ to closed sets and interpolating polynomials," *Univ. Umea, Dept. Math.,* No. 7 (1980).

Jonsson, A., and Wallin, H.

(1977) "The trace on closed sets of functions with Second difference of order $O(h)$," *Univ. Umea, Dept. Math.,* No. 7 (1977).

Karlin, S.

(1963) "Representation theorems for positive functions," *J. Math. Mech.,* **12**, No. 4, 599–618 (1963).

Karlin, S., and Studden, W. J.

(1966) *Tchebycheff Systems: With Applications in Analysis and Statistics,* Interscience, New York (1966).

Keldysh, M. V.

(1945) "Determination of functions of complex variables by series of polynomials in closed domains," *Mat. Sb.,* **16 (58)**, 249–258 (1945).

Kirchberger, P.

(1903) "Über Tchebycheffsche Annäherungsmethoden," *Math. Ann.,* **57**, 509–540 (1903).

Kolesnik, L. I.

(1969) "Inversion of a Kellogg-type theorem," *Ukr. Mat. Zh.,* **21**, No. 1, 104–108 (1969).

Kolmogorov, A. N.

(1923) "Sur l'ordre de grandeur des coefficients de la série de Fourier–Lebesgue," *B. A. P.,* 83–86 (1923).

(1935) "Zur Grössenordnung des Restgliedes Fourierscher Reihen differenzierbarer Funktionen," *Ann. Math.,* **36**, No. 2, 521–526 (1935).

(1938) "Une généralisation de l'inégalité de M. J. Hadamard entre les bornes supérieures des dérivées successives d'une fonction," *C. R. Acad. Sci.,* **207**, 764–765 (1938).

(1938a) "On inequalities between upper bounds of successive derivatives of an arbitrary function on an infinite interval," *Uch. Zap. Moscow Univ., Ser. Mat.,* Issue 30, 3–16 (1938).

(1948) "A remarks on Chebyshev polynomials least deviating from a given function," *Usp. Mat. Nauk,* **3**, Issue 1, 216–221 (1948).

Kolyada, V. I.

(1975) "On imbedding into the classes $\varphi(L)$," *Izv. Akad. Nauk SSSR, Ser. Mat.,* **39**, No. 2, 418–437 (1975).

(1988) "On relations between moduli of continuity in different metrics," *Tr. Mat. Inst. Akad. Nauk SSSR,* **181**, 117–136 (1988).

(1989) "Rearrangement of functions and imbedding theorems," *Usp. Mat. Nauk,* **44**, No. 5 (269), 61–95 (1989).

Konovalov, V. N.

(1972) "Constructive characteristic of the classes $W^r H^\alpha$ of functions of many real variables," *Dokl. Akad. Nauk Ukr. SSR, Ser. A,* No. 10, 882–884 (1972).

(1973) *On Constructive Characteristics of Some Classes of Functions of Many Real Variables,* Candidate-Degree Thesis, Kiev (1973) [in Russian].

(1978) "Exact inequalities for norms of functions, third partial derivatives, second mixed derivatives, and oblique derivatives," *Mat. Zametki,* **23**, No. 1, 67–78 (1978).

(1984) *Description of Traces of Some Classes of Functions of Many Variables,* Preprint No. 84.21, Institute of Mathematics, Ukrainian Academy of Sciences, Kiev (1984) [in Russian].

Konyagin, S. V.

(1983) "On the moduli of continuity of functions," in: *Abstracts of the All-Union School on the Theory of Approximation of Functions Dedicated to the Centenary of Acad. N. N. Luzin (September 10–19, 1983),* Kemerovo (1983), p. 59 [in Russian].

Kopotun, K. A.

(1992) "Uniform estimates for the coconvex approximation of functions by polynomials," *Mat. Zametki,* **51**, No. 3, 24–33 (1992).

Kopotun, K. A., Leviatan, D., and Shevchuk, I. A.

(2005) "Convex polynomial approximation in the uniform norm: Conclusion," *Canad. J. Math.,* **57**, 1224–1248 (2005).

Korkin, A. I., and Zolotarev, E. I.

(1873) "Sur un certain minimum" (1873), in: E. I. Zolotarev, *Complete Works,* Issue 2, Academy of Sciences of the USSR, Leningrad (1932), pp. 138–154.

Korneichuk, N. P.

(1959) "On the approximation of periodic functions satisfying the Lipschitz condition by Bernstein–Rogosinski sums," *Dokl. Akad. Nauk SSSR,* **125**, No. 4, 258–261 (1959).

(1959a) *Some Problems of Approximation of Periodic Functions by Trigonometric Polynomials,* Candidate-Degree Thesis, Dnepropetrovsk (1959) [in Russian].

(1961) "On an estimate for approximations of functions of the class H^{α} by trigonometric polynomials," in: *Investigation of Contemporary Problems of Constructive Theory of Functions,* Moscow (1961), pp. 148–154 [in Russian].

(1962) "Exact constant in the Jackson theorem on the best uniform approximation of continuous periodic functions," *Dokl. Akad. Nauk SSSR,* **145**, No. 3, 514–515 (1962).

(1971) "Extremal values of functionals and the best approximation on classes of periodic functions," *Izv. Akad. Nauk SSSR, Ser. Mat.,* **35**, No. 1, 93–124 (1971).

(1976) *Extremal Problems of Approximation Theory,* Nauka, Moscow (1976) [in Russian].

Korovkin, P. P.

(1959) *Linear Operators and Approximation Theory,* Nauka, Moscow (1959) [in Russian].

Koval'chuk, R. N.

(1965) "Direct theorems on approximation of analytic functions of many complex variables in polycylindric domains," *Dokl. Akad. Nauk Ukr. SSR, Ser. A,* No. 2, 170–174 (1965).

(1965a) "On one generalization of the Kellogg theorem," *Ukr. Mat. Zh.,* **17**, No. 4, 104–108 (1965).

Krasnosel'skii, M. A., and Lifshits, E. A.

(1968) "On the problem of convergence of a sequence of positive operators in linear topological spaces," *Usp. Mat. Nauk,* **23**, No. 2, 213–214 (1968).

Krein, M. G.

(1959) "*L*-problem in an abstract linear normed space," in: *On Some Problems in the Theory of Moments,* DNTVU (1959), pp. 171–199 [in Russian].

Krein, M. G., and Nudel'man, A. A.

(1973) *Problem of Markov Moments and Extremal Problems,* Nauka, Moscow (1973) [in Russian].

Krochuk, V. V.

(1969) "On the polynomial approximation of solutions of ordinary differential equations," in: *Proceedings of the Seminar on Differential and Integral Equations,* Issue 1, Institute of Mathematics, Ukrainian Academy of Sciences, Kiev (1969), pp. 254–261 [in Russian].

(1970) *On the Polynomial Approximation of Solutions of Differential and Integral Equations,* Candidate-Degree Thesis, Kiev (1970) [in Russian].

Kryakin, Yu. V.

(1989) "On Whitney constants," *Mat. Zametki,* **46**, No. 2, 155–157 (1989).

Krylov, V. I.

(1959) *Approximate Calculation of Integrals,* Fizmatgiz, Moscow (1959) [in Russian].

Kudryavtsev, L. D.

(1963) "Interpolation and approximation of functions," in: *Elements of the Theory of Functions,* Fizmatgiz, Moscow (1963), pp. 106–198 [in Russian].

Laurent, P.-J.

(1972) *Approximation et Optimisation,* Hermann, Paris (1972).

Lavrent'ev, M. A.

(1934) "On the theory of conformal mappings," *Tr. Steklov. Fiz.-Mat. Inst.,* 159– 246 (1934).

(1935) "Sur une classe de représentations continues," *Mat. Sb.,* **42**, 407–424 (1935).

(1936) "On some boundary-value problems in the theory of univalent functions," *Mat. Sb.,* **1**, No. 6, 815–844 (1936).

Lebed', G. K.

(1957) "Inequalities for polynomials and their derivatives," *Dokl. Akad. Nauk SSSR,* **117**, No. 4, 570–572 (1957).

(1975) "Some inequalities for trigonometric algebraic polynomials and their derivatives," *Tr. Mat. Inst. Akad. Nauk SSSR,* **134**, 142–160 (1975).

Lebedev, N. A.

(1966) "On inverse theorems on uniform approximation," *Dokl. Akad. Nauk SSSR,* **171**, No. 4, 788–790 (1966).

Lebedev, N. A., and Shirokov, N. A.

(1971) "On the uniform approximation of functions on closed sets that have finitely many corner points with nonzero exterior angles," *Izv. Akad. Nauk Arm. SSR,* **6**, No. 4, 311–341 (1971).

Lebedev, N. A., and Tamrazov, P. M.

(1970) "Inverse theorems on approximation on regular compact sets of the complex plane," *Izv. Akad. Nauk SSSR, Ser. Mat.,* **34**, No. 6, 1340–1390 (1970).

Lebesgue, H.

(1898) "Sur l'approximation des fonctions," *Bull. Sci. Math., Ser. 2,* **22**, 278–287 (1898).

(1909) "Sur les intégrales singulières," *Ann. Toulouse,* **1**, 25–117 (1909).

Leviatan, D.

(1986) "Pointwise estimates for corner polynomial approximation," *Proc. Amer. Math. Soc.,* **98**, No. 3, 471–474 (1986).

Lichtenstein, L.

(1909) "Zur Theorie der linearen partiellen Differentialgleichungen zweiter Ordnung des elliptischen Typus," *Math. Ann.,* **67**, 559–575 (1909).

Lorentz, G. G.

(1953) *Bernstein Polynomials,* Toronto University, Toronto (1953).

(1966) *Approximation of Functions,* Holt, Rinehart, and Watson, New York (1966).

(1972) "Monotone approximation," in: O. Shisha (editor), *Inequalities,* Vol. 3, Academic Press, New York (1972), pp. 201–215.

Lorentz, G. G., and Zeller, K. L.

(1968) "Degree of approximation by monotone polynomials. I," *J. Approxim. Theory,* **1**, No. 4, 501–504 (1968).

(1969) "Degree of approximation by monotone polynomials. II," *J. Approxim. Theory,* **2**, No. 3, 265–269 (1969).

Lorentz, R. A.

(1971) "Uniqueness of best approximation by monotone polynomials," *J. Approxim. Theory,* **4**, 401–418 (1971).

Lozinskii, S. M.

(1944) "On convergence and summability of Fourier series and interpolation processes," *Mat. Sb.,* **14 (56)**, Issue 3, 175–268 (1944).

(1952) "Inversion of the Jackson theorem," *Dokl. Akad. Nauk SSSR,* **83**, No. 5, 645–657 (1952).

Lukacs, F.

(1918) "Verschärfung des ersten Mittelwertsatzes der Integralrechnung für rationale Polynome," *Math. Zeitschr.,* **2**, 295–305 (1918).

Mairhuber, J.

(1956) "On Haar's theorems concerning Chebysheff approximation problems having unique solutions," *PAMS,* **7**, No. 4, 609–615 (1956).

Mandelbrojt, S.

(1952) *Séries Adhérentes. Régularisation des Suites. Applications,* Gauthier-Villars, Paris (1952).

Marchaud, A.

(1927) "Sur les dérivées et sur les différences des fonctions de variables réelles," *J. Math. Pures Appl.,* **6**, 337–425 (1927).

Markov, A. A.

(1884) *On Some Applications of Algebraic Continued Fractions,* Thesis, St.Petersburg (1884).

(1889) "On one Mendeleev Problem" (1889), in: A. A. Markov, *Selected Works,* GITTL, Moscow (1948), pp. 51–75 [in Russian].

(1898) "On the limit values of integrals in connection with interpolation," *Zap. Akad. Nauk, Ser. VIII,* **IV** (1898).

(1906) "Lectures on functions least deviating from zero" (1906), in: A. A. Markov, *Selected Works*, GITTL, Moscow (1948), pp. 244–291 [in Russian].

(1948) *Selected Works,* GITTL, Moscow (1948) [in Russian].

Markov, V. A.

(1892) *On the Functions Least Deviating from Zero on a Given Segment,* St.Petersburg (1892).

Markushevich, A. I.

(1936) "Sur la représentation conforme des domaines à frontières variables," *Mat. Sb.,* **1 (43)**, No. 6, 863–864 (1936).

(1967) *Theory of Analytic Functions,* Nauka, Moscow (1967–1968) [in Russian].

Matorin, A. P.

(1955) "On inequalities between the maximum values of the moduli of functions and their derivatives on a half line," *Ukr. Mat. Zh.,* **7**, No. 3, 262–266 (1955).

Meiman, N. N.

(1952) "Differential inequalities and some problems of distribution of zeros of entire and single-valued analytic functions," *Usp. Mat. Nauk,* **7**, No. 3, 3–62 (1952).

Meinardus, G.

(1964) *Approximation von Funktionen und ihre numerische Behandlung,* Springer, Berlin (1964).

Mergelyan, S. N.

(1951) "On some problems in the constructive theory of functions," *Tr. Mat. Inst. Akad. Nauk SSSR,* **37**, 1–92 (1951).

(1952) "Uniform approximation of functions of complex variables," *Usp. Mat. Nauk,* **7**, No. 2 (48), 31–122 (1952).

Merrien, J.

(1966) "Prolongateurs de fonctions différentiables d'une variable réelle," *J. Math. Pures Appl.,* **45**, No. 3, 291–309 (1966).

Natanson, I. P.

(1949) *Constructive Theory of Functions,* GITTL, Moscow (1949) [in Russian].

(1952) "On the approximation of multiply differentiable periodic functions by using singular integrals," Dokl. Akad. Nauk SSSR, **82**, No. 2, 337–339 (1952).

(1974) *Theory of Functions of Real Variables,* Nauka, Moscow (1974) [in Russian].

Nessel, R. J., and Wickeren, E. van

(1984) "Some negative results in connection with Marchaud-type inequalities," in: *International Series of Numerical Mathematics,* **71**, Birkhäuser, Basel (1984), pp. 221–238.

Newman, D. J.

(1964) "Rational approximation to $|x|$," *Mich. Math. J.,* **11**, No. 1, 11–14 (1964).

(1974) "Jackson's theorem on complex arcs," *J. Approxim. Theory,* **10**, 206–217 (1974).

Nikolaev, V. F.

(1948) "On the problem of approximation of continuous functions by polynomials," *Dokl. Akad. Nauk SSSR,* **61**, No. 2, 201–204 (1948).

Nikol'skii, S. M.

(1940) "On the asymptotic behavior of a remainder in the case of approximation of functions satisfying the Lipschitz condition by Fejér sums," *Izv. Akad. Nauk SSSR, Ser. Mat.,* **4**, No. 6, 501–508 (1940).

(1940a) "On some methods of approximation by trigonometric sums," *Izv. Akad. Nauk SSSR, Ser. Mat.,* **4**, No. 6, 509–520 (1940).

(1941) "Asymptotic estimate for a remainder in the case of approximation by Fourier sums," *Dokl. Akad. Nauk SSSR,* **32**, No. 6, 386–389 (1941).

(1945) "Approximation of periodic functions by trigonometric polynomials," *Tr. Mat. Inst. Steklov.,* **15**, 1–58 (1945).

(1946) "Approximation of functions by trigonometric polynomials in the mean," *Izv. Akad. Nauk SSSR, Ser. Mat.,* **10**, No. 3, 207–256 (1946).

(1946a) "On the best approximation of functions satisfying the Lipschitz condition by polynomials," *Izv. Akad. Nauk SSSR, Ser. Mat.,* **10**, No. 4, 295–317 (1946).

(1946b) "Fourier series with given modulus of continuity," *Dokl. Akad. Nauk SSSR,* **52**, No. 3, 191–194 (1946).

(1947) *Theory of Approximation of Functions. Part. 1. Functional Analysis,* Dnepropetrovsk University, Dnepropetrovsk (1947) [in Russian].

(1947a) "On the best polynomial approximation of functions with singularities of the form $|a-x|^s$ in the mean," *Dokl. Akad. Nauk SSSR,* **55**, No. 3, 195– 198 (1947).

(1948) "On linear methods for summation of Fourier series," *Izv. Akad. Nauk SSSR, Ser. Mat.,* **12**, No. 3, 259–278 (1948).

(1948a) "Generalization of one Bernstein inequality," *Dokl. Akad. Nauk SSSR,* **60**, No. 9, 1507–1510 (1948).

(1958) "Approximation of functions of many variables by polynomials," in: *Proceedings of the Third All-Union Mathematical Congress (Moscow, 1956),* Vol. 4, Academy of Sciences of the USSR, Moscow (1958), pp. 226–231 [in Russian].

(1966) "On one method for covering a domain and an inequality for polynomials of many variables," *Mathematica (Cluj),* **8**, No. 2, 345–356 (1966).

(1969) *Approximation of Functions of Many Variables and Imbedding Theorems,* Nauka, Moscow (1969) [in Russian].

(1974) *Quadrature Formulas,* Nauka, Moscow (1974) [in Russian].

Nikol'skii, V. N.

(1963) "The best approximations of elements of convex sets in linear normed spaces," *Uch. Zap. Kalininsk. Pedagog. Inst.,* **29**, 85–119 (1963).

Nörlund, N. E.

(1926) *Differenzenrechnung,* Springer, Berlin (1926).

Osgood, W. F., and Taylor, E. H.

(1913) "Conformal transformations on the boundaries of their regions of definition," *Trans. Amer. Math. Soc.,* **14**, 277–298 (1913).

Oskolkov, K. I.

(1976) "Uniform modulus of continuity of summable functions on sets of positive measure," *Dokl. Akad. Nauk SSSR,* **229**, No. 2, 304–306 (1976).

Peetre, J. A.

(1968) *A Theory of Interpolation of Normed Spaces,* Instituto de Matemática Pura e Aplicada, Rio de Janeiro, Brasil (1968).

Pinkevich, V. T.

(1940) "On the order of the remainder of the Fourier series of functions differentiable in the Weyl sense," *Izv. Akad. Nauk SSSR,* **4**, No. 6, 521–528 (1940).

Podlipenko, Yu. K.

(1977) *Application of the Approximation Method to the Solution of Cauchy and Goursat Problems for Differential Equations,* Candidate-Degree Thesis, Kiev (1977) [in Russian].

Pokalo, A. K.

(1962) "On one class of linear methods of summation," *Vests. Akad. Nauk Belorus. SSR, Ser. Fiz.-Tekhn. Navuk,* No. 1, 24–27 (1962).

Polyakov, R. V.

(1969) *On the Approximation of Continuous Functions Defined on Bounded Closed Sets,* Candidate-Degree Thesis, Kiev (1969) [in Russian].

(1972) "Approximation of continuous (analytic interior) functions in domains with a smooth boundary," *Ukr. Mat. Zh.,* **24**, No. 1, 57–68 (1972).

Popoviciu, T.

(1934) "Sur quelques propriétés des fonctions d'une ou de deux variables réelles," *Mathematica,* **8**, 1–85 (1934).

(1934a) "Sur l'approximation des fonctions convexes d'ordre supérieur," *Mathematica,* **10**, 49–54 (1934).

Potapov, M. K.

(1981) "Polynomial approximation on a finite interval of the real axis," in: *Constructive Theory of Functions,* Varna (1981), pp. 134–138 [in Russian].

Privalov, A. A.

(1990) *Theory of Interpolation of Functions,* Saratov University, Saratov (1990) [in Russian].

Prypik, A. S.

(1974) "On the application of generalized Faber polynomials to the approximation of functions analytic on closed sets," in: *Theory of Approximation of Functions and Its Applications,* Institute of Mathematics, Ukrainian Academy of Sciences, Kiev (1974), pp. 114–118 [in Russian].

(1976) "On the inverse theorems of functions on regular compact sets," *Ukr. Mat. Zh.,* **28**, No. 1, 43–52 (1976).

Radoslavova, T. V.

(1979) "Decrease orders of the moduli of continuity," *Analysis Math.,* **5**, No. 3, 219–234 (1979).

Remez, E. Ya.

(1935) *On Methods for the Best (in the Chebyshev Sense) Approximate Representation of Functions,* Ukrainian Academy of Sciences, Kiev (1935) [in Ukrainian].

(1953) "Some problems of Chebyshev approximation in the complex plane," *Ukr. Mat. Zh.,* **5**, No. 1, 3–49 (1953).

(1957) "On Chebyshev approximation in the complex plane," *Dokl. Akad. Nauk SSSR,* **77**, No. 6, 965–968 (1957).

(1957) *General Computational Methods of Chebyshev Approximation,* Ukrainian Academy of Sciences, Kiev (1957) [in Russian].

(1969) *Foundations of Numerical Methods of Chebyshev Approximation,* Naukova Dumka, Kiev (1969) [in Russian].

Rice, J. R.

(1964) *The Approximation of Functions, Vol. I: Linear Theory,* Addison-Wesley, Reading, MA (1964).

(1969) *The Approximation of Functions, Vol. II: Nonlinear Theory,* Addison-Wesley, Reading, MA (1969).

Riesz, F.

(1914) "Sur les polynômes trigonométriques," *Comptes R.,* **158**, No. 23, 1657– 1661 (1914).

(1925) "Über die Randwerte einer analytischen Funktion," *Math. Zeitschr.,* **18**, 87–95 (1925).

Riesz, M.

(1914) "Eine trigonometrische Interpolationsformel und einige Ungleichungen für Polynome," *Jahresber. Dtsch. Math.–Ver.,* **23**, 354–368 (1914).

Rodov, A. M.

(1946) "Dependences between upper bounds of derivatives of functions of real variables," *Izv. Akad. Nauk SSSR, Ser. Mat.,* **10**, No. 2, 257–270 (1946).

Rogosinski, W.

(1925) "Über die Abschnitte trigonometrischer Reihen," *Math. Annal.,* **95**, 110– 135 (1925).

Rubel, L. A., Shields, A. L., and Taylor, B. A.

(1975) "Mergelyan sets and modulus of continuity of analytic functions," *J. Approxim. Theory,* **15**, No. 1, 23–40 (1975).

Rudin, W.

(1964) *Principles of Mathematical Analysis,* McGraw-Hill, New York (1964).

Runge, C.

(1885) "Zur Theorie der eindeutigen analytischen Funktionen," *Acta Math.,* **6**, 229–244 (1885).

Salem, R.

(1935) "Sur certaines fonctions continues et les propriétés de leur séries de Fourier," *Comptes R.,* **201**, No. 17, 703–705 (1935).

(1940) "Essais sur les séries trigonométriques," *Actualités Sci. Industr.,* No. 862, 1–85 (1940).

Sendov, B.

(1982) "On the constants of H. Whitney," *C. R. Acad. Bul. Sci.,* **35**, No. 4, 431–434 (1982).

(1985) "The constants of H. Whitney are bounded," *C. R. Acad. Bul. Sci.,* **38**, No. 10, 1299–1302 (1985).

(1986) "On one H. Whitney theorem," *Dokl. Akad. Nauk SSSR,* **291**, No. 6, 1296–1300 (1986).

Sendov, B., and Popov, V.

(1988) *The Averaged Moduli of Smoothness,* Mir, Moscow (1988) [Russian translation].

(1989) *The Averaged Moduli of Smoothness,* Wiley, New York (1989).

Sendov, B., and Takev, M.

(1986) "The theorem of Whitney for integral norm," *C. R. Acad. Bul. Sci.,* **39**, No. 10, 35–38 (1986).

Sewell, W. E.

(1937) "Degree of approximation by polynomials in complex domain. Generalized derivatives and approximation by polynomials," *Trans. Amer. Math. Soc.,* **41**, No. 1, 84–121 (1937).

(1939) "Jackson summation of the Faber development," *Bull. Amer. Math. Soc.,* **45**, 187–189 (1939).

(1942) *Degree of Approximation by Polynomials in the Complex Domain,* Princeton University Press, Princeton (1942).

Shchekhorskii, A. I.

(1977) "Contour-solid theorems for holomorphic functions," in: *Finite-Difference Smoothnesses in Problems of the Theory of Functions,* Preprint No. 77.10, Institute of Mathematics, Ukrainian Academy of Sciences, Kiev (1977), pp. 23–27 [in Russian].

(1982) *Majorant Properties of Holomorphic Functions of the Hölder Class,* Preprint No. 82.41, Institute of Mathematics, Ukrainian Academy of Sciences, Kiev (1982) [in Russian].

Shevchuk, I. A.

(1973) "On the constructive characteristic of functions of the classes $D^r H^{\omega_2(t)}$ on closed sets with piecewise-smooth boundary," *Ukr. Mat. Zh.,* **25**, No. 1, 81–90 (1973).

(1976) "Some remarks on functions of the modulus-of-continuity type of order $k \geq 2$," in: *Problems of the Theory of Approximation of Functions and Its Applications,* Institute of Mathematics, Ukrainian Academy of Sciences, Kiev (1976), pp. 194–199 [in Russian].

(1976a) "Solid and contour properties of analytic functions in terms of the second modulus of continuity," in: *Problems of the Theory of Approximation of Functions and Its Applications,* Institute of Mathematics, Ukrainian Academy of Sciences, Kiev (1976), pp. 202–204 [in Russian].

(1977) "On the classes of functions $H_{k,m}^{\varphi(t)}$ on sets $\mathfrak{M}$ of the complex plane," in: *Proceedings of the International Conference on the Theory of Approximation of Functions (Kaluga, 1977),* Nauka, Moscow (1977), pp. 409–412 [in Russian].

(1979) "Constructive characteristic of functions continuous on a set $\mathfrak{M} \subset \mathbb{C}$ for the kth modulus of continuity," *Mat. Zametki,* **25**, No. 2, 225–247 (1979).

(1980) "Contour-solid theorems for Hölder classes," in: *Problems of the Theory of Approximation of Functions,* Institute of Mathematics, Ukrainian Academy of Sciences, Kiev (1980), pp. 186–189 [in Russian].

(1984) *Constructive Description of Traces of Differentiable Functions of Real Variables,* Preprint No. 83.19, Institute of Mathematics, Ukrainian Academy of Sciences, Kiev (1984) [in Russian].

(1984a) "Extension of functions which are traces of functions belonging to H_k^{φ} on arbitrary subset of the line," *Analysis Math.*, No. 3, 249–273 (1984).

(1986) "On the uniform approximation of functions on a segment," *Mat. Zametki,* **40**, No. 1, 36–48 (1986).

(1989) "On coapproximation of monotone functions," *Dokl. Akad. Nauk SSSR,* **308**, No. 3, 537–541 (1989).

(1989a) "On the sharpness of an approximation criterion of smoothness for functions on a segment," in: *Approximation and Function Spaces,* Vol. 22, Banach Center Publ., PWN-Polish Scientific Publishers, Warsaw (1989), pp. 401–411.

(1992) "Approximation of monotone functions by monotone polynomials," *Mat. Sb.,* **183**, No. 5, 63–78 (1992).

Shevchuk, L. B.

(1974) *Algorithm and Software for the Numerical Construction of Φ-Extremal Polynomials on a Computer,* Preprint, Kiev (1974), pp. 1–38 [in Russian].

Shirokov, N. A.

(1972) "On the uniform approximation of functions on closed sets having finitely many corner points with nonzero exterior angles," *Dokl. Akad. Nauk SSSR,* **205**, No. 4, 798–800 (1972).

Shokhat, Ya. A.

(1918) *Investigation of One Class of Polynomials Least Deviating from Zero,* Ural Mining Institute, Ekaterinburg (1918) [in Russian].

Shvai, A. I.

(1967) *On the Approximation of Functions of the Classes $W^r H^{\omega}$ Defined on Closed Sets with Arbitrary Angles Different from 2π,* Candidate-Degree Thesis, Kiev (1967) [in Russian].

Shvartsman, P. A.

(1982) "On the traces of functions of two variables that satisfy the Zygmund condition," in: *Investigations in the Theory of Functions of Many Real Variables,* Yaroslavl University, Yaroslavl (1982), pp. 145–168 [in Russian].

(1984) "Lipschitz sections of many-valued mappings and traces of functions of the Zygmund class on an arbitrary compact set," *Dokl. Akad. Nauk SSSR,* **276**, No. 3, 559–562 (1984).

Shvedov, A. S.

(1979) "Jackson theorem in L_p, $0 < p < 1$, for algebraic polynomials and order of comonotone approximation," *Mat. Zametki,* **25**, No. 1, 107–117 (1979).

(1980) "Comonotone approximation of functions by polynomials," *Dokl. Akad. Nauk SSSR,* **250**, No. 1, 39–42 (1980).

(1981) "Order of coapproximation of functions by algebraic polynomials," *Mat. Zametki,* **29**, No. 1, 117–130 (1981).

(1981a) "Coconvex approximation of functions of many variables by polynomials," *Mat. Sb.,* **115**, No. 4, 577–589 (1981).

Singer, I.

(1957) "Sur la meilleure approximation des fonctions abstraites continues à valeurs dans un espace de Banach," *Rev. Math. Pures Appl.,* **2**, 245–262 (1957).

Smirnov, G. S.

(1973) *On a Criterion for an Element of the Best Approximation of an Abstract Function with Values in a Banach Space,* Preprint No. 73-8, Institute of Mathematics, Ukrainian Academy of Sciences, Kiev (1973) [in Russian].

Smirnov, V. I., and Lebedev, N. A.

(1964) *Constructive Theory of Functions of Complex Variables,* Nauka, Moscow (1964) [in Russian].

Stechkin, S. B.

(1948) "Generalization of some Bernstein inequalities," *Dokl. Akad. Nauk SSSR,* **60**, No. 9, 1511–1514 (1948).

(1949) "On the order of the best approximation of continuous functions," *Dokl. Akad. Nauk SSSR,* **65**, No. 2, 135–138 (1949).

(1951) "On de la Vallée Poussin sums," *Dokl. Akad. Nauk SSSR,* **80**, No. 4, 545– 548 (1951).

(1951a) "On the order of the best approximations of continuous functions," *Izv. Akad. Nauk SSSR, Ser. Mat.,* **15**, No. 3, 219–242 (1951).

(1951b) "Bernstein and Rogosinski summation methods," in: G. H. Hardy, *Divergent Series,* Inostrannaya Literatura, Moscow (1951), pp. 479–492 [Russian translation].

(1952) "On the best approximation of periodic functions by trigonometric polynomials," *Dokl. Akad. Nauk SSSR,* **83**, No. 5, 651–654 (1952).

(1953) "Estimate for the remainder of a Taylor series for some classes of analytic functions," *Izv. Akad. Nauk SSSR, Ser. Mat.,* **17**, No. 6, 461–472 (1953).

(1955) "On the absolute convergence of Fourier series. II," *Izv. Akad. Nauk SSSR, Ser. Mat.,* **19**, 221–246 (1955).

(1956) "Some remarks on trigonometric polynomials," *Usp. Mat. Nauk,* **10**, Issue 1, 159–166 (1956).

(1956a) "On the best approximation of some classes of periodic functions by trigonometric polynomials," *Izv. Akad. Nauk SSSR,* **20**, No. 5, 643–648 (1956).

(1961) "On the approximation of periodic functions by Fejér sums," *Tr. Mat. Inst. Akad. Nauk SSSR,* **62**, 48–60 (1961).

(1971) "On the approximation of continuous periodic functions by Favard sums," *Tr. Mat. Inst. Steklov.,* **109**, 26–39 (1971).

Stechkin, S. B., and Subbotin, Yu. N.

(1976) *Splines in Computational Mathematics,* Nauka, Moscow (1976) [in Russian].

Stein, E. M.

(1970) *Singular Integrals and Differential Properties of Functions,* Princeton University, Princeton (1970).

Stepanets, A. I.

(1972) "On one Kolmogorov problem in the case of functions of two variables," *Ukr. Mat. Zh.,* **24**, No. 5, 653–655 (1972).

(1972a) "Approximation of functions satisfying the Lipschitz conditions by Fourier sums," *Ukr. Mat. Zh.,* **24**, No. 6, 781–799 (1972).

(1973) "Approximation of some classes of periodic functions of two variables by Fourier sums," *Ukr. Mat. Zh.,* **25**, No. 5, 599–609 (1973).

(1974) "Approximation of continuous periodic functions of many variables by Riesz spherical means," *Mat. Zametki,* **15**, Issue 5, 831–832 (1974).

(1974a) *Approximation of Periodic Functions by Riesz Sums,* Preprint No. 74-2, Institute of Mathematics, Ukrainian Academy of Sciences, Kiev (1974), pp. 3–47 [in Russian].

(1974b) *Investigation of Extremal Problems in the Theory of Summation of Fourier Series,* Doctoral-Degree Thesis, Kiev (1974) [in Russian].

Stolyarchuk, V. K.

(1974) "On the construction of polynomials that realize a nearly-best approximation of the functions $\operatorname{si} x$ and $\Phi(x) = \dfrac{1}{2\pi}\displaystyle\int_0^x e^{-t^2}\,dt$," *Ukr. Mat. Zh.,* **26**, No. 2, 215–224 (1974).

(1974a) "On the uniform approximation of Bessel functions with entire index by polynomials," *Ukr. Mat. Zh.,* **26**, No. 5, 670–674 (1974).

Stone, M.

(1937) "Applications of the theory of Boolean rings to general topology," *Trans. Amer. Math. Soc.,* **41**, 375–481 (1937).

Storozhenko, É. A.

(1977) "Approximation of functions of the class $L^p(\mathbb{R}^k)$, $0 < p < 1$, by algebraic polynomials," *Izv. Akad. Nauk SSSR, Ser. Mat.,* **41**, 652–662 (1977).

Storozhenko, É. A., and P. Oswald

(1978) "Jackson theorem in the spaces $L^p(\mathbb{R}^k)$, $0 < p < 1$," *Sib. Mat. Zh.*, **19**, No. 4, 888–901 (1978).

Stovbun, A. V.

(1969) "On the uniform approximation in a closed domain with the use of generalized Faber polynomials," *Ukr. Mat. Zh.*, **21**, No. 6, 849–857 (1969).

Subbotin, Yu. N.

(1965) "On the relationship between finite differences and the corresponding derivatives," *Tr. Steklov Mat. Inst.*, **78**, 24–43 (1965).

Sun Yun-shen

(1958) "On the best approximation of functions represented in the form of convolution," *Dokl. Akad. Nauk SSSR,* **118**, No. 2, 247–250 (1958).

(1959) "On the best approximation of periodic differentiable functions by trigonometric polynomials," *Izv. Akad. Nauk SSSR, Ser. Mat.*, **23**, No. 1, 67–92 (1959).

(1961) "On the best approximation of periodic differentiable functions by trigonometric polynomials," *Izv. Akad. Nauk SSSR, Ser. Mat.*, **25**, No. 1, 143–152 (1961).

Szegő, G.

(1921) "Über orthogonale Polynome, die zu einer gegebenen Kurve der komplexen Ebene gehören," *Math. Zeitschr.*, **9**, 218–270 (1921).

(1925) "Über einen Satz von A. Markoff," *Math. Zeitschr.*, **23**, 45–61 (1925).

(1959) *Orthogonal Polynomials,* American Mathematical Society, New York (1959).

Szegő, G., and Zygmund, A.

(1954) "On certain mean values of polynomials," *J. Analyse Math.*, **3**, 225–244 (1954).

Szőkefalvi-Nagy, B.

(1938) "Über gewisse Extremalfragen bei transformierten trigonometrischen Entwicklungen. I. Periodischer Fall," *Ber. Math.-Phys. Kl. Sächs. Akad. Wiss. Lpz.*, **90**, 103–134 (1938).

(1948) "Sur une classe générale de procédés de sommation pour les séries de Fourier," *Hungaria Acta Math.*, **1**, No. 3, 14–52 (1948).

(1949) "Méthodes de sommation des séries de Fourier. II," *Časopis Pěst. Mat. Fys.*, **74**, 210–219 (1949).

(1950) "Méthodes de sommation des séries de Fourier. I," *Acta Sci. Math. Szeged,* **12**, Pars 13, 204–210 (1950).

(1950a) "Méthodes de sommation des séries de Fourier. III," *Acta Sci. Math. Szeged,* **13**, 247–351 (1950).

Taikov, L. V.

(1963) "On best linear methods for approximation of functions of the classes B^r and H^r," *Usp. Mat. Nauk,* **18**, No. 4, 183–189 (1963).

(1967) "On the best approximation in the mean for some classes of analytic functions," *Mat. Zametki,* **1**, No. 2, 155–162 (1967).

(1968) "Inequalities of Kolmogorov type and best formulas of numerical differentiation," *Mat. Zametki,* **4**, No. 2, 233–238 (1968).

Takev, M. D.

(1988) "Theorem of Whitney type in $\mathbb{R}^2$," in: *Proceedings of the International Conference on Constructive Theory of Functions (Varna, May 24–31, 1988),* Bulgarian Academy of Sciences, Sofia (1988).

Tamrazov, P. M.

(1971) "Solid inverse theorems of polynomial approximation for regular compact sets of the complex plane," *Dokl. Akad. Nauk SSSR,* **198**, No. 3, 540–542 (1971).

(1973) "Contour and solid structural properties of holomorphic functions of complex variables," *Usp. Mat. Nauk,* **28**, Issue 1 (169), 131–161 (1973).

(1975) *Smoothnesses and Polynomial Approximations,* Naukova Dumka, Kiev (1975) [in Russian].

(1978) "Finite-difference smoothness and approximation theory," in: *Colloquia Mathematica Societatis János Bolyai "Fourier Analysis and Approximation Theory" (Hungary, 1976),* Budapest (1978), **19**, pp. 827–843.

(1983) "Contour-solid properties of holomorphic functions and mappings with bilogarithmic majorants," *Dokl. Akad. Nauk SSSR,* **270**, No. 4, 791–795 (1983).

Tamrazov, P. M., and Belyi, V. I.

(1980) "Polynomial approximations and moduli of smoothness of functions with quasiconformal boundary," *Sib. Mat. Zh.,* **21**, No. 3, 162–176 (1980).

Telyakovskii, S. A.

(1958) "Approximation of differentiable functions by de la Vallée Poussin sums," *Dokl. Akad. Nauk SSSR,* **121**, No. 3, 426–429 (1958).

(1960) "Approximation of functions differentiable in the Weyl sense by de la Vallée Poussin sums," *Dokl. Akad. Nauk SSSR,* **131**, No. 2, 259–262 (1960).

(1960a) "On the approximation of differentiable functions by linear means of their Fourier series," *Izv. Akad. Nauk SSSR, Ser. Mat.,* **24**, No. 2, 213–242 (1960).

(1961) "On the norms of trigonometric polynomials and approximation of differentiable functions by linear means of their Fourier series. I," *Tr. Mat. Inst. Akad. Nauk SSSR,* **62**, 61–97 (1961).

(1963) "On the norms of trigonometric polynomials and approximation of differentiable functions by linear means of their Fourier series. II," *Izv. Akad. Nauk SSSR, Ser. Mat.,* **27**, No. 2, 253–272 (1963).

(1964) "Integrability conditions for trigonometric series and their application to the study of linear methods of summation of Fourier series," *Izv. Akad. Nauk SSSR, Ser. Mat.,* **28**, No. 6, 1209–1236 (1964).

(1966) "Two theorems on approximation of functions by algebraic polynomials," *Mat. Sb.,* **79**, No. 2, 252–265 (1966).

Tikhomirov, V. M.

(1976) *Some Problems of Approximation Theory,* Moscow University, Moscow (1976) [in Russian].

Timan, A. F.

(1951) "Improvement of the Jackson theorem on the best approximation of continuous functions by polynomials on a finite segment of the real axis," *Dokl. Akad. Nauk SSSR,* **78**, No. 1, 17–20 (1951).

(1953) "Approximation properties of linear methods of summation of Fourier series," *Izv. Akad. Nauk SSSR, Ser. Mat.,* **17**, No. 2, 99–134 (1953).

(1957) "Inverse theorems of the constructive theory of functions defined on a finite segment of the real axis," *Dokl. Akad. Nauk SSSR,* **116**, No. 5, 762–765 (1957).

(1960) *Theory of Approximation of Functions of Real Variables,* Fizmatgiz, Moscow (1960) [in Russian].

Titchmarsh, E. C.

(1937) *Introduction to the Theory of Fourier Integrals,* Clarendon, Oxford (1937).

Tonelli, L.

(1908) "I polinomi d'approssimazione di Tchebychev," *Ann. Mat.,* **15**, No. 3, 47–119 (1908).

Trigub, R. M.

(1960) "Approximation of functions with given modulus of smoothness in the exterior of a segment and a half-axis," *Dokl. Akad. Nauk SSSR,* **132**, No. 2, 303–306 (1960).

(1962) "Approximation of functions by polynomials with integer coefficients," *Izv. Akad. Nauk SSSR, Ser. Mat.,* **26**, No. 2, 261–280 (1962).

(1965) "Constructive characteristic of some classes of functions," *Izv. Akad. Nauk SSSR, Ser. Mat.,* **29**, No. 3, 615–630 (1965).

(1973) "Characteristic of the Lipschitz classes of integer order on a segment on the basis of the rate of polynomial approximation," in: *Theory of Functions, Functional Analysis, and Their Applications,* Issue 18, Kharkov (1973), pp. 63–70 [in Russian].

(1974) "Relationship between the summability and the absolute convergence of series and the Fourier transforms," *Dokl. Akad. Nauk SSSR,* **217**, No. 1, 34–37 (1974).

Trokhimchuk, Yu. Yu.

(1979) "On differential properties of real and complex functions," *Ukr. Mat. Zh.,* **31**, No. 4, 465–469 (1979).

Turan, P.

(1952) "On a trigonometric sum," *Ann. Soc. Polon. Math.,* **25**, 155–161 (1952).

Turetskii, A. Kh.

(1960) "On classes of saturation for some methods of summation of Fourier series of continuous periodic functions," *Usp. Mat. Nauk,* **15**, Issue 6, 149–156 (1960).

Ul'yanov, P. L.

(1967) "On the absolute and uniform convergence of the Fourier series," *Mat. Sb.,* **72**, No. 2, 193–225 (1967).

(1968) "Imbedding of some classes of functions," *Izv. Akad. Nauk SSSR, Ser. Mat.,* **32**, No. 3, 649–686 (1968).

Vallée Poussin, C.-J. de la

(1910) "Sur les polynômes d'approximation et la représentation approchée d'un angle," *Bull. Acad. R. Belg.,* No. 12, 808–844 (1910).

(1911) "Sur les polynômes d'approximation à une variable complexe," *Bull. Acad. R. Belg.,* 199–211 (1911).

(1919) *Leçons sur l'Approximation des Fonctions d'une Variable Réelle,* Gauthier-Villars, Paris (1919).

Verbitskii, L. L.

(1940) "On one method for summation of interpolation polynomials," *Nauch. Zap. Dnepropetrovsk. Univ.,* 113–119 (1940).

Videnskii, V. S.

(1956) "On uniform approximation in the complex plane," *Usp. Mat. Nauk,* **11**, No. 5 (71), 169–175 (1956).

(1960) "Qualitative problems in the theory of the best approximation of functions of complex variables," in: *Investigations of Contemporary Problems in the Theory of Functions of Complex Variables,* Fizmatgiz, Moscow (1960), pp. 258–272 [in Russian].

Vitushkin, A. G.

(1966) "Conditions on a set that are necessary and sufficient for any continuous function analytic at its interior points to admit a uniform approximation by rational fractions," *Dokl. Akad. Nauk SSSR,* **171**, No. 6, 1255–1258 (1966).

Volkov, Yu. I.

(1965) "Some problems of approximation of functions of complex variables in domains with corners," in: *Proceedings of the First Republican Mathematical Conference of Young Researchers,* Issue 2, Institute of Mathematics, Ukrainian Academy of Sciences, Kiev (1965), pp. 130–134 [in Russian].

(1965a) "On the constructive characteristic of functions of complex variables in domains with piecewise-smooth boundary," *Ukr. Mat. Zh.,* **17**, No. 3, 115–119 (1965).

Vorob'ev, N. N.

(1966) "On approximation of analytic functions by partial sums of Faber series," *Dokl. Akad. Nauk Ukr. SSR, Ser. A,* No. 2, 138–141 (1966).

(1967) "A remark on the Dzyadyk theorem on approximation of functions of the class $W^r H^\omega$," *Dokl. Akad. Nauk Ukr. SSR, Ser. A,* No. 1, 6–9 (1967).

(1968) "On simultaneous approximation of functions and their derivatives in the complex plane," *Ukr. Mat. Zh.,* **20**, No. 1, 113–131 (1968).

Vorob'ev, N. N., and Polyakov, R. V.

(1968) "On the constructive characteristic of continuous functions defined on smooth arcs," *Ukr. Mat. Zh.,* **20**, No. 2, 750–758 (1968).

Voronovskaya, E. V.

(1932) "Determination of the asymptotic form of the approximation of functions by Bernstein polynomials," *Dokl. Akad. Nauk SSSR, Ser. A,* No. 4, 74–85 (1932).

(1963) *Method of Functionals and Its Applications,* Leningrad Institute of Electrical Engineering and Communications, Leningrad (1963) [in Russian].

Wallin, H.

(1979) "Extensions of functions preserving moduli of smoothness," *Univ. Umea, Dept. Math.,* No. 2 (1979).

Walsh, J. L.

(1926) "Über die Entwicklung einer analytischen Funktion nach Polynomen," *Math. Ann.,* **96**, 430–436 (1926).

(1926a) "Über den Grad der Approximation einer analytischen Funktion," *Münchener Berichte,* 223–229 (1926).

(1961) *Interpolation and Approximation by Rational Functions in the Complex Domain,* American Mathematical Society, Providence, RI (1961).

Walsh, J. L., and Elliott, H. M.

(1950) "Polynomial approximation to harmonic and analytic functions: generalized continuity conditions," *Trans. Amer. Math. Soc.,* **68**, No. 2, 183–203 (1950).

Walsh, J. L., and Sewell, W.

(1940) "Sufficient condition for various degrees of approximation by polynomials," *Duke Math. J.,* **6**, No. 3, 658–705 (1940).

Wang Xing-hua

(1964) "The exact constant of approximation of continuous functions by the Jackson singular integral," *Acta Math. Sinica,* **14**, No. 2, 231–237 (1964).

Warschawski, S.

(1932) "Über das Randverhalten der Ableitung der Abbildungsfunktion bei konformer Abbildung," *Math. Zeitschr.,* **35**, 321–456 (1932).

(1934) "Bemerkung zu meiner Arbeit: Über das Randverhalten der Ableitung der Abbildungsfunktion bei konformer Abbildung," *Math. Zeitschr.,* **38**, No. 5, 669–683 (1934).

(1942) "On conformal mapping of infinite strips," *Trans. Amer. Math. Soc.,* **51**, 280–335 (1942).

Weierstrass, K.

(1885) "Über die analytische Darstellbarkeit sogenannter willkürlicher Funktionen einer reellen Veränderlichen," *Sitzungsberichte König. Pr. Akad. Wiss. Berlin,* 633–639, 789–805 (1885).

Western, P. W.

(1948) "Inequalities of the Markoff and Bernstein type for integral norms," *Duke Math. J.,* **15**, No. 3, 839–870 (1948).

Weyl, H.

(1917) "Bemerkungen zum Begriff der Differentialquotienten gebrochener Ordnung," *Vier. Natur. Gesellschaft,* **62**, 296–302 (1917).

Whitney, H.

(1934) "Analytic extensions of differential functions defined in closed sets," *Trans. Amer. Math. Soc.,* **36**, No. 1, 63–89 (1934).

(1934a) "Differential functions defined in closed sets," *Trans. Amer. Math. Soc.,* **36**, No. 2, 369–386 (1934).

(1957) "On functions with bounded nth differences," *J. Math. Pures Appl., IX Ser.,* **36**, 67–95 (1957).

(1959) "On bounded functions with bounded nth differences," *Proc. Amer. Math. Soc.,* **10**, No. 3, 480–481 (1959).

Whittaker, E. T., and Watson, G. N.

(1927) *A Course of Modern Analysis. An Introduction to the General Theory of Infinite Processes and of Analytic Functions; with an Account of the Principal Transcendental Functions,* Cambridge University, Cambridge (1927).

Wronicz, Z.

(1984) "Moduli of smoothness associated with Chebyshev systems and approximation by L-splines," *Constr. Theory Funct. 84,* Sofia, 906–916 (1984).

Xie, T.

(1985) "On two problems of Hasson," *Approxim. Theory Appl.,* **1**, No. 2, 137– 144 (1985).

Yudin, V. A.

(1979) "On the modulus of continuity in L_2," *Sib. Mat. Zh.,* **20**, No. 2, 449–450 (1979).

Zamanski, M.

(1949) "Classes de saturation de certains procédés d'approximation des séries de Fourier des fonctions continues périodiques et application à quelques problèmes d'approximation," *Ann. Sci. Ecole Norm Sup. (3),* **66**, No. 1, 19–93 (1949).

(1950) "Classes de saturation des procédés de sommation des séries de Fourier et applications des séries trigonométriques," *Ann. Sci. Ecole Norm Sup. (3),* **67**, No. 1, 161–198 (1950).

Zaritskaya, Z. V.

(1967) "On approximation of functions by linear positive operators in the metric of L_p," *Dokl. Akad. Nauk Ukr. SSR, Ser. A,* No. 1, 14–17 (1967).

Zhuk, V. V., and Natanson, G. I.

(1983) *Trigonometric Fourier Series and Elements of Approximation Theory,* Leningrad State University, Leningrad (1983) [in Russian].

Zolotarev, E. I.

(1877) "Applications of elliptic functions to problems of functions least and most deviating from zero," *Zap. St.Peterburg. Akad. Nauk,* **30**, No. 5 (1877), in: E. I. Zolotarev, *Complete Works,* Issue 2, Academy of Sciences of the USSR, Leningrad (1932), pp. 1–59 [in Russian].

Zukhovitskii, S. I., and Krein, M. G.

(1950) "A remark on one possible generalization of the Haar and Kolmogorov theorems," *Usp. Mat. Nauk,* **5**, No. 1, 217–229 (1950).

Zukhovitskii, S. I., and Stechkin, S. B.

(1956) "On the approximation of abstract functions in a Hilbert space," *Dokl. Akad. Nauk SSSR,* **106**, No. 3, 385–388 (1956).

Zygmund, A.

(1924) "O module ciągłości sumy szeregu sprzężonego z szeregiem Fouriera," *Prace Mat.-Fiz.,* **33**, 25–132 (1924).

(1945) "Smooth functions," *Duke Math. J.,* **12**, No. 1, 46–76 (1945).

(1959) *Trigonometric Series,* Cambridge University, Cambridge (1959).

Subject index

Bei Fragen zur Produktsicherheit wenden Sie sich bitte an:
If you have any questions regarding product safety,
please contact:

Walter de Gruyter GmbH
Genthiner Straße 13
10785 Berlin
productsafety@degruyterbrill.com